Nanomaterials: Ecotoxicity, Safety, and Public Perception

Mahendra Rai · Jayanta Kumar Biswas
Editors

Nanomaterials: Ecotoxicity, Safety, and Public Perception

Springer

Editors
Mahendra Rai
Department of Biotechnology
Sant Gadge Baba Amravati University
Amravati, Maharashtra, India

Jayanta Kumar Biswas
International Centre for Ecological Engineering, Department of Ecological Studies
University of Kalyani
Kalyani, Nadia, West Bengal, India

ISBN 978-3-030-05143-3 ISBN 978-3-030-05144-0 (eBook)
https://doi.org/10.1007/978-3-030-05144-0

Library of Congress Control Number: 2018962789

This Springer imprint is published by the registered company Springer Nature Switzerland AG
The registered company address is: Gewerbestrasse 11, 6330 Cham, Switzerland

Foreword

Nanotechnology is an important and enabling technology today. With the confluence of ideas from diverse disciplines, it is anticipated to positively affect various facets of our lives. Nanomaterials derive their strength by harnessing the property, which with decrease in particle size surface area of materials increases, and some novel properties emerge. This fundamental premise paves way to many possibilities as envisioned by Richard Feynman, "There is Plenty of Room at the Bottom" for "manipulating and controlling things on a small scale". Nanomaterials are expected to enable many innovations in the form of diverse devices, processes and products in the domains of medicine, energy, environment, and others.

The last two decades or so have witnessed a burst of interest and activity in synthesis, characterization, and application of nanomaterials. There is much scope for developing ecotoxicological information. The nanomaterials could interact with different biotic and abiotic factors in the environment and cause adverse biological and ecological effects at different levels. To assess environmental risks, scientific data and knowledge on transport of nanomaterials in the environment and their effects on health are to be properly examined.

Future development of nanotechnology should be accompanied with research on (eco)toxicity associated with possible risks by analyzing behavior and fate of nanomaterials in the ecosystem inclusive of its biotic organisms and human health. There must be a reasonable concern about plausible and potential hazards, based on scientific investigation.

Given the vibrancy of nanoscience and technology, the edited book entitled *Nanomaterials: Ecotoxicity, Safety, and Public Perception* is a topical and useful book providing the state of knowledge on ecotoxicity of nanomaterials and safety issues. The book consists of 17 chapters arranged in three sections dealing with ecotoxicity, risk, and safety issues, regulations and public perceptions. All the chapters have been written by experts from different countries, who have offered authoritative treatments of nanoecotoxicological aspects. It has dealt with nanomaterials in food, soil, medicine, and unintentionally produced nanomaterials, and addresses key issues of ecotoxic effects of nanomaterials on life and environment including humans. It has iterated the need for an increased level of understanding of risks and vulnerabilities that are likely to influence public, regulatory, and non-governmental activities. The book has been designed with adequate coverage for a multisectorial audience of academics, researchers, students, professionals, and policy-makers. I congratulate the editors for their endeavor in producing this useful book.

Bangalore, India

C. N. R. Rao
Linus Pauling Research Professor

Preface

Nanoscience and nanotechnology refer to the control and manipulation of matter at nanometer dimensions for tailoring and hybridizing the physical, biological, and engineering properties of matter. It is a burgeoning, vibrant, and promising field which has offered the possibilities and stunning new achievements as envisioned by Richard Feynman (1959): "There is plenty of room at the bottom". Nanotechnology is not only an emerging field of study, it has become an industry with transforming effects. The use of nanomaterials is expected to offer great promise for diverse devices, procedures, and products in the domains of consumer goods, medicines, energy, environment, and many others. But there is the other side of the coin; the environment is prone to suffer from nanopollution and toxic insult due to release and build up of nanocontaminants during different phases of lifecycle of nanomaterials, viz., production, transportation, disposal, entry into ecosystems, accumulation, bioconcentration, transfer across trophic boundaries, etc. The NMs raise the possibility that they could interact with and cause adverse biological effects at cellular, subcellular, and molecular levels. Assessing potential environmental/ecological risks requires quality information on transport and fate of nanoparticles in the environment, exposures, and vulnerabilities of organisms to the nanomaterials and standard methods for assessing toxicity for aquatic or terrestrial organisms and human health. Nanomaterial toxicity is extremely complex, and our understanding about ecotoxicity is very fussy and incomplete. The systematic risk characterization and evaluation of the safety of nanomaterials require a multidisciplinary approach and convergence of knowledge and efforts from researchers and experts from toxicology, biotechnology, materials science, chemistry, physics, engineering, and other branches of life sciences. Although studies are beginning to appear in the literature addressing the toxicity of various nanomaterials and their potential for exposure, at this stage definitive statements regarding the impacts of nanomaterials on human health and the environment remain sketchy and scanty. Because of our limited understanding of the environmental implications of nanomaterials, there remains a gulf between their promises and pitfalls although the flourishing development of new nanomaterials as well as emerging new applications. No doubt present time demands a clarion call for an increased level of precautions with regard to nanomaterials, as has happened with other emerging contaminants and technologies (e.g., biotechnology). The perceptions of risks and of benefits will vary and are likely to influence public, regulatory, and non-governmental activities regarding risk and benefit evaluations. The Triple Helix, i.e., academia, government, and industry, needs to be extended to a fourth dimension, the civil society, because of the latter's role and relevance in today's world. Since people are receptive to fictional representations as they shape their imaginaries, the "nanomania" has rather alienated the entertainment and leisure industry, and given rise to misconceptions, misunderstandings, and distorted views regarding this particular technology rather than contributed to a rational and genuine discussion. *Quo vadis*? Where will we go? Systematic identification and assessment of the risks posed by any new technology are essential. Systematic research into the flip side of this new technology, especially the risks that it entails, is essential. A prudent, integrated, and holistic approach is required to develop best practices based on the scientific understanding about what we know and what we don't know but need to know so that the engines of creation would not become the engines of destruction.

With suitable breadth of coverage, this edited book has been *designed, written, and organized* for the broad-spectrum multisectorial audience of researchers, students, professionals, and policy-makers. It is expected to be a pointer and reference resource and will cater to the needs of the researchers, professionals as well as the graduate and postgraduate students of nearly every discipline of science and engineering—toxicology, biotechnology, chemistry, physics, biology, medical science, material science & engineering, environmental science, and environmental engineering, who are interested to know the ecotoxicological aspects of the nanomaterial and the potential hazards and risks involved therein. This book would be an inspirational comprehensive resource for researchers, environmental professionals and practitioners as well as industrial portfolio holders, who are interested to earn a working knowledge of this fascinating field of Nanotechnology and Nanotoxicology.

This book includes important chapters written by specialists and experts in the concerned topic, and hopefully it will win wide acceptance among the undergraduate and graduate students of chemistry, biochemistry, pharmacology, microbiology, biotechnology, and allied subjects. It will be very useful for researchers working in nanotechnology, nanobiotechnology, toxicology, microbiology, ecology, and other related fields.

Last but not the least, we hope, this edited book surely will evoke some surprise and shock, hope and despair, and stimulate resolution and action through concerted and collaborative efforts of appropriate bodies to shape the nanoworld and save Mother Nature.

Amravati, India — Mahendra Rai
Kalyani, Nadia, India — Jayanta Kumar Biswas

Contents

Editors and Contributors

About the Editors

Mahendra Rai is a Senior Professor and Basic Science Research Faculty Fellow (UGC) at the Department of Biotechnology, Sant Gadge Baba Amravati University at Amravati, Maharashtra, India. He was a Visiting Scientist at the Department of Bioenergetics, University of Geneva, Switzerland and at the Department of Plant Protection of Debrecen University, Debrecen, Hungary. He visited the Department of Chemical Biology, University of Campinas, Brazil under Indo-Brazil Research Programme (DST-CNPq collaboration). He was Visiting Professor in Nicolaus Copernicus University, Torun, Poland and was invited by State University of Campinas, Brazil. He was also Visiting Professor in Ostrava, Czech Republic, and Department of Pharmacognosy, University of Rosario, Argentina.

His area of expertise includes microbial biotechnology and nanobiotechnology. His present research interests are nanobiotechnology in medicine and agriculture. He has published more than 380 research papers in national and international journals. In addition, he has edited/authored more than 47 books and 6 patents.

Jayanta Kumar Biswas is an Associate Professor at Department of Ecological Studies and International Centre for Ecological Engineering, University of Kalyani. He obtained M.Sc. in Zoology, M.Phil. in Ecology, and Ph.D. on ecotechnological management of aquatic systems. His interest and expertise span the following focal areas: ecotoxicology and bioremediation of toxic metal(loid)s; ecological engineering and ecotechnological remediation of water and soil contaminants; environmental microbiology; nanobiotechnology. He received many awards and fellowships including Fellow, National Environmentalists Association, Zoological Society of India, Zoological Society, Kolkata, Senior Scientist Award & Gold Medal (ZSI), Best Research Paper Award (AESA), Outstanding Reviewer Award (Chemosphere (Elsevier) & Environmental Geochemistry and Health (Springer); National Merit Scholarship (MHRD, GoI), etc. He is credited with publishing 125 research papers in reputed journals. He is serving as editorial board member of several international journals of repute, namely Science of the Total Environment (Elsevier); Environmental Geochemistry and Health (Springer); Current Pollution Reports (Springer); Energy & Environment (SAGE).

Contributors

Nadeesh M. Adassooriya Faculty of Applied Science, University of Sri Jayewardenepura, Nugegoda, Sri Lanka

M. Bhuvaneshwari Centre for Nanobiotechnology, Vellore Institute of Technology, Vellore, India

Jayanta Kumar Biswas Enviromicrobiology, Ecotoxicology and Ecotechnology Research Unit, Department of Ecological Studies, University of Kalyani, Kalyani, Nadia, West Bengal, India; International Centre for Ecological Engineering, University of Kalyani, Kalyani, West Bengal, India

Soumyajit Biswas Department of Biochemistry & Biophysics, University of Kalyani, Kalyani, Nadia, West Bengal, India

Adriano Brandelli Centro de Nanociência e Nanotecnologia, Universidade Federal do Rio Grande do Sul, Porto Alegre, Brazil

N. Chandrasekaran Centre for Nanobiotechnology, Vellore Institute of Technology, Vellore, India

Ana F. M. Costa Department of Bioprocess and Biotechnology, School of Pharmaceutical Sciences of São Paulo State University-UNESP, Araraquara, São Paulo, Brazil

Maria Angélica M. Costa Department of Biochemistry and Chemical Technology, Institute of Chemistry, São Paulo State University-UNESP, Araraquara, São Paulo, Brazil

Kelly J. Dussán Department of Biochemistry and Chemical Technology, Institute of Chemistry, São Paulo State University-UNESP, Araraquara, São Paulo, Brazil

Wilson Engelmann Universidade do Vale do Rio dos Sinos – UNISINOS, Dois Irmãos, Brazil

Henrique M. Fogarin Department of Biochemistry and Chemical Technology, Institute of Chemistry, São Paulo State University-UNESP, Araraquara, São Paulo, Brazil

Patrycja Golinska Department of Microbiology, Nicolaus Copernicus University, Toruń, Poland

Indarchand Gupta Department of Biotechnology, Government Institute of Science, Aurangabad, Maharashtra, India

Raquel Von Hohendorff Universidade do Vale do Rio dos Sinos – UNISINOS, São Leopoldo, Brazil

Avinash P. Ingle Department of Biotechnology, Engineering School of Lorena, University of Sao Paulo, Lorena, SP, Brazil; Nanobiotechnology Laboratory, Department of Biotechnology, SGB Amravati University, Amravati, Maharashtra, India

V. Iswarya Centre for Nanobiotechnology, Vellore Institute of Technology, Vellore, India

Bushra Jamil Department of Bio-Sciences, National University of Medical Sciences, The Mall, Rawalpindi, Pakistan

Rabia Javed Department of Bio-Sciences, National University of Medical Sciences, The Mall, Rawalpindi, Pakistan

Talita Martins Lacerda Department of Biotechnology, Engineering School of Lorena (EEL), São Paulo University (USP), CEP, Lorena, Brazil

Daniele Weber S. Leal Faculdades Integradas de Taquara – FACCAT and JUSNANO Reseacher (UNISINOS), Porto Alegre, Brazil

Michele Lima-Souza Department of Analytical Chemistry, Institute of Chemistry, São Paulo State University-UNESP, Araraquara, São Paulo, Brazil

Sanchita Mandal Future Industries Institute, University of South Australia, Mawson Lakes, SA, Australia

K. M. Manjaiah Division of Soil Science and Agricultural Chemistry, ICAR-Indian Agricultural Research Institute, New Delhi, India

Paulo Ricardo Franco Marcelino Department of Biotechnology, Engineering School of Lorena (EEL), São Paulo University (USP), CEP, Lorena, Brazil

G. B. Meshkov Faculty of Physics, Lomonosov Moscow State University, Moscow, Russia

Monojit Mondal Enviromicrobiology, Ecotoxicology and Ecotechnology Research Unit, Department of Ecological Studies, University of Kalyani, Kalyani, Nadia, West Bengal, India

Mariete Barbosa Moreira Chemistry Institute, Universidade Estadual Paulista Júlio de Mesquita Filho (UNESP), CEP, Araraquara, Brazil

Maria Jose Morilla Department of Science and Technology, Nanomedicines Research and Development Center, Quilmes National University, Buenos Aires, Argentina

Amitava Mukherjee Centre for Nanobiotechnology, Vellore Institute of Technology, Vellore, India

Raj Mukhopadhyay Division of Irrigation and Drainage Engineering, ICAR-Central Soil Salinity Research Institute, Karnal, India

Mikhail A. Novikov East-Siberian Institution of Medical and Ecological Research, Angarsk, Irkutsk region, Russia

Semih Otles Engineering Faculty, Food Engineering Department, Ege University, Bornova, Izmir, Turkey

Lorena O. Pires Department of Biochemistry and Chemical Technology, Institute of Chemistry, São Paulo State University-UNESP, Araraquara, São Paulo, Brazil

Yu V. Plekhanova FSBIS G.K. Skryabin Institute of Biochemistry and Physiology of Microorganisms, Russian Academy of Sciences, Pushchino, Russia

Asma Saleem Qazi Department of Bio-Sciences, National University of Medical Sciences, The Mall, Rawalpindi, Pakistan

Mahendra Rai Nanobiotechnology Laboratory, Department of Biotechnology, SGB Amravati University, Amravati, Maharashtra, India

S. K. Rajkishore Department of Nano Science & Technology, TNAU, Coimbatore, India

A. N. Reshetilov FSBIS G.K. Skryabin Institute of Biochemistry and Physiology of Microorganisms, Russian Academy of Sciences, Pushchino, Russia

Ryan Rienzie Faculty of Agriculture, University of Peradeniya, Peradeniya, Sri Lanka

Eder Lilia Romero Department of Science and Technology, Nanomedicines Research and Development Center, Quilmes National University, Buenos Aires, Argentina

Buket Yalcin Sahyar Whirlpool Corporation, Manisa, Turkey

Binoy Sarkar Future Industries Institute, University of South Australia, Mawson Lakes, SA, Australia; Department of Animal and Plant Sciences, The University of Sheffield, Sheffield, UK

Michael G. Shurygin Irkutsk Scientific Center of Surgery and Traumatology, Irkutsk, Russia

Irina A. Shurygina Irkutsk Scientific Center of Surgery and Traumatology, Irkutsk, Russia

Débora D. V. Silva Department of Biochemistry and Chemical Technology, Institute of Chemistry, São Paulo State University-UNESP, Araraquara, São Paulo, Brazil

Silvio Silvério da Silva Department of Biotechnology, Engineering School of Lorena (EEL), São Paulo University (USP), CEP, Lorena, Brazil

Olga V. Sinitsyna Laboratory for Physical Chemistry of Polymers, A. N. Nesmeyanov Institute of Organoelement Compounds of Russian Academy of Sciences, Moscow, Russia

Larisa M. Sosedova East-Siberian Institution of Medical and Ecological Research, Angarsk, Irkutsk region, Russia

Kizhaeral S. Subramanian Department of Nano Science & Technology, TNAU, Coimbatore, India

Magdalena Świecimska Department of Microbiology, Nicolaus Copernicus University, Toruń, Poland

Muhammad Ali Syed Department of Microbiology, University of Haripur, Haripur, Pakistan

S. E. Tarasov FSBIS G.K. Skryabin Institute of Biochemistry and Physiology of Microorganisms, Russian Academy of Sciences, Pushchino, Russia

Eugeniy A. Titov East-Siberian Institution of Medical and Ecological Research, Angarsk, Irkutsk region, Russia

Magdalena Wypij Department of Microbiology, Nicolaus Copernicus University, Toruń, Poland

I. V. Yaminsky Laboratory for Physical Chemistry of Polymers, A. N. Nesmeyanov Institute of Organoelement Compounds of Russian Academy of Sciences, Moscow, Russia; Faculty of Physics, Lomonosov Moscow State University, Moscow, Russia

Part I
Ecotoxicity

Chapter 1
Nanomaterials: What Are They, Why They Cause Ecotoxicity, and How This Can Be Dealt With?

Mahendra Rai, Indarchand Gupta, Avinash P. Ingle, Jayanta Kumar Biswas and Olga V. Sinitsyna

Abstract Nanomaterials have been benefiting human by their wide applications in different fields. Till date, many types of natural and engineered nanomaterials have been reported. Each of them has specific characteristics, which are helpful in deciding their use for particular application. Although they are beneficial to the human beings, there is probability of their harmful effects on the ecosystem. After the desired use of the nanomaterials, they are routinely disposed of into the environment either intentionally or unintentionally. This scenario can create the harmful environment for the whole ecosystem. The ecotoxicity of nanomaterials is an imperative point to be considered for the safety of flora and fauna. Hence, with argument on their characteristics and applications, their safety for human and environment should also be considered. Therefore, the present chapter introduces the nanomaterials, encompasses the discussion on major types of nanomaterials, which are being available naturally and others that are synthesized artificially. The parameters which make

---

M. Rai (✉)
Nanobiotechnology Laboratory, Department of Biotechnology, SGB Amravati University, Amravati 444602, Maharashtra, India
e-mail: pmkrai@hotmail.com; mahendrarai@sgbau.ac.in

I. Gupta
Department of Biotechnology, Government Institute of Science, Nipatniranjan Nagar, Caves Road, Aurangabad 431004, Maharashtra, India
e-mail: indarchandgupta@gmail.com

A. P. Ingle
Department of Biotechnology, Engineering School of Lorena, University of Sao Paulo, Estrada municipal do Campinho, sn, Lorena SP-12602-810, Brazil
e-mail: ingleavinash14@gmail.com

J. K. Biswas
Enviromicrobiology, Ecotoxicology and Ecotechnology Research Unit, Department of Ecological Studies, University of Kalyani, Kalyani, Nadia 741235, West Bengal, India
e-mail: biswajoy2008@gmail.com

O. V. Sinitsyna
Laboratory for Physical Chemistry of Polymers, A. N. Nesmeyanov Institute of Organoelement Compounds of Russian Academy of Sciences, Moscow, Russia
e-mail: sinitsyna@gmail.com

M. Rai and J. K. Biswas (eds.), *Nanomaterials: Ecotoxicity, Safety, and Public Perception*, https://doi.org/10.1007/978-3-030-05144-0_1

nanoparticles harmful to ecosystem have also been discussed. Moreover, the special emphasis is given on how the scientific community can deal the situation to avoid the harmful effects of nanoparticles so that it can be beneficial to mankind without causing any damage to ecosystem.

Keywords Nanoparticles · Engineered nanomaterials · Toxic effects · Ecotoxicity

1.1 Introduction

Materials, which consist of particles with at least one dimension smaller than 100 nm, are called nanomaterials. The era of the increasing popularity of the nanomaterials began with the discovery of fullerenes in 1985 by Harold Kroto, James Heath, Sean O'Brien, Robert Curl, and Richard Smalley (Kroto et al. 1985). According to The Nanodatabase (2018), commercial products containing fullerenes are already on the market. These are rackets for tennis and badminton, and skin care products. Another database, The Nanotechnology Consumer Product Inventory, was created in 2005 to document the emergence of nanotechnology products in the consumer market (Vance et al. 2015). The number of products in the Inventory is constantly growing. The most advanced category for the introduction of nanomaterials is the Health and Fitness. Oxides of titanium, silicon and zinc are the most produced nanomaterials worldwide, and silver nanoparticles (AgNPs) are the most popular additive in products (Vance et al. 2015). The nanoparticles act as antimicrobial agents, pigments, UV filters, protective coatings, reinforcing additives, etc. The global market of the nanomaterials is growing rapidly. According to a pessimistic forecast, its value will reach $11.3 billion in 2020, and an optimistic forecast predicts the market value of $55.0 billion in 2022 (Inshakova and Inshakov 2017).

The active use of the nanomaterials raises the question of their safety for the environment. The first publication considering the potential ecological impacts of the engineered nanomaterials appeared after more than 15 years of the widespread development of nanotechnology (Colvin 2003; Skjolding et al. 2016). A new discipline, nanoecotoxicology, began to study the environmental effects of nanoscale materials. The evaluation of the behavior and fate of the nanomaterials in complex environmental matrices is an extremely challenging problem (Boyes et al. 2017). The possible release of the nanoparticles into the environment should be considered across the entire life-cycle of a product, including manufacture, use, and disposal. The nanoparticles can possess not only advantageous but also unique toxic properties different from the bulk materials. Potential adverse effects should be taken into account from the cellular level to the whole ecosystem. The environmental hazard evaluation, based on dose-by-mass, may not be valid for the nanomaterials, and the question on the relevant dose-metrics in nanoecotoxicology is still open.

How to use the gifts of nanotechnology and protect the environment from the possible harmful effects is one of the most pressing issues of today. In the present chapter, we try to answer these major questions.

1.2 What Are Nanomaterials?

Nanomaterials are considered as building blocks of nanotechnology. According to European Commission, a nanomaterial can be defined as any natural, incidental or manufactured material containing particles, in an unbound state or as an aggregate or as an agglomerate, and where, about 50% or more of the particles has one or more external dimensions is in the size range 1–100 nm (European Commission 2011). Nanomaterials can be present in the form of particles, tubes, rods or fibers. The nanomaterials having same composition to its known materials in bulk form may possess different physico-chemical properties that the same materials have in bulk form, and hence, such nanomaterials may behave differently, if enter into the body, and may thus pose different potential hazards (Chattopadhyay 2018).

1.2.1 Types of Nanomaterials

Nanomaterials are mainly classified into three major categories which are as follows:

1.2.1.1 Natural Nanomaterials

The natural nanomaterials are already present in nature or produced by natural systems. Biological systems like viruses particularly the capsid of viruses is the classical example of natural nanomaterials (Petrescu and Blum 2018). In addition, the wax crystals (tubules) covering a nasturtium or lotus leaf (Genzer and Efimenko 2006; Koch et al. 2006; Ensikat et al. 2011), the "tiny hairs" on the bottom of gecko feet (Autumn 2006; Ge et al. 2007), wing scales of butterfly (Kertézs et al. 2006; Wu et al. 2010), the blue hue of tarantulas (Hsiung et al. 2015), spider and spider-mite silk fibers (Silva and Rech 2013; Goto et al. 2016), the fine structure of foraminifera (mainly chalk) (Giere 2009), etc. are all the examples of naturally occurring nanomaterials (nanotechnology). In addition, products generated after forest fires, volcanic ash, ocean spray, and the radioactive decay of radon gas are some other natural sources, which can generate nanoparticles (Chattopadhyay 2018). Figure 1.1 shows the structure of some important naturally occurring nanomaterials.

1.2.1.2 Incidental Nanomaterials

Nanomaterials which are incidentally produced as a byproduct of various mechanical or industrial processes are usually called as incidental nanomaterials. The important sources responsible for the generation of these nanomaterials mainly include vehicle engine exhausts, welding fumes, combustion processes from domestic solid fuel heating and cooking. Some of the incidental nanomaterials also cause environmental pollution (Chattopadhyay 2018).

Fig. 1.1 Natural nanomaterials (**I**) (A) Nasturtium (*Nasturtium officinale*) leaves (B) SEM image of the upper leaf side of nasturtium showing wax tubules [Reprinted from Koch et al. 2006 with permission from ACS] (C) Lotus (*Nelumbo nucifera*) leaves (D) SEM image of the upper leaf side of lotus showing papillae, wax clusters and wax tubules [adapted from Ensikat et al. (2011) as open access article] (**II**) (E) Image of gecko (F) Scanning electron micrograph of tiny nano hairs on the bottom of gecko feet [Reprinted from Autumn (2006) as open access article] (**III**) (G) Butter fly (H) Optical microscope image of wings scales (I) TEM image of transverse section of one scale [K & L are reprinted from Wu et al. (2010) as open access article] (**IV**) (J) Electron micrograph of *Poecilotheria metallica* (blue hue tarantulas) hairs showing symmetric array of rod-like, tubular folding projecting longitudinally on its periphery (K) Organized multilayered nanostructures in hairs of *P. metallica* [both J and K are adapted from online report available at https://phys.org/news/2015-11-blue-hue-tarantulas.html] (**V**) (L) Spider (*Nephila clavata*) (M) and (N) SEM images of spider silk showing size in nanoscale [Reprinted from Goto et al. (2016) an open access article]

1.2.1.3 Engineered Nanomaterials

Engineered nanomaterials are those materials, which are manufactured by humans and intentionally engineered to have certain required properties. Till date, many nanomaterials have been developed for a variety of applications which mainly include nanoparticles, liposomes, solid lipid nanoparticles, dendrimers, polymeric nanoparticles (nanospheres, nanocapsules), nanomicelles, fullerenes, carbon nanotubes, quantum dots, etc. Each of the above mentioned engineered nanomaterials are briefly discussed below.

Nanoparticles

The particles of metals or non-metals having the size range between 1 and 100 nm are known as nanoparticles. Some of the nanoparticles are surrounded with the interfacial layer which is considered as an integral part of nanoparticles, fundamentally affecting all of their properties. This layer generally consists of ions or inorganic and organic molecules. Organic molecules coating nanoparticles are known as stabilizers, capping and surface ligands (Fig. 1.2A). The nanoparticles can be used for various applications in different sectors including agriculture, medicine and electronics, etc.

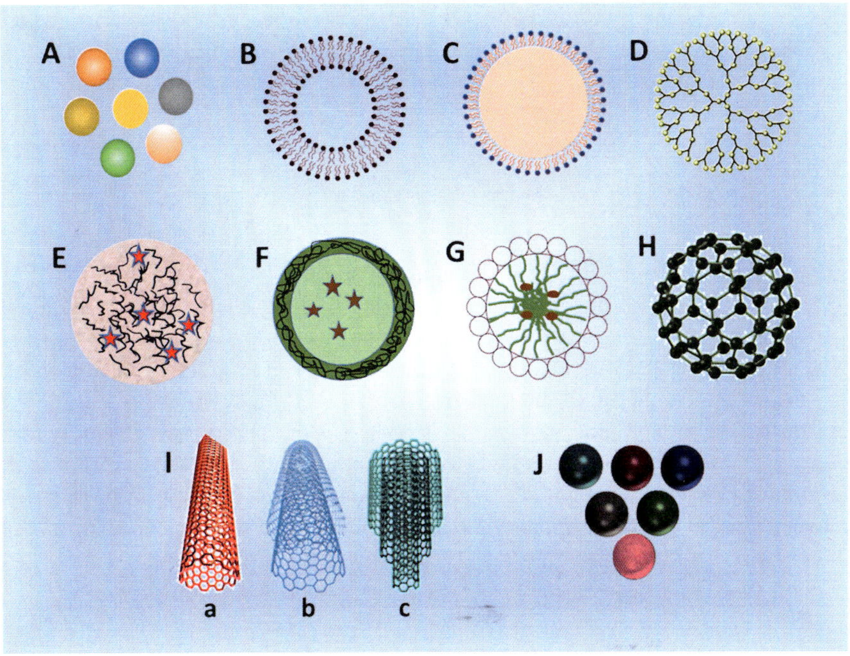

Fig. 1.2 Schematic representation of various engineered nanomaterials (A) Nanoparticles (self-prepared) (B) Liposomes (self-prepared) (C) Solid lipid nanoparticles (self-prepared) (D) Dendrimers (adapted from Singh et al. (2014); an open access article) (E) Polymer nanoparticles (F) Nanomicelles (self-prepared) (G) Fullerene (adapted from general article available at http://www.dancingwithwater.com/products/shungite-fullerenes-for-water/) (H) Carbon nanotubes (a) single-walled carbon nanotube (adapted and modified from general article available at https://dooko.es/nanotubos-de-carbono-dooko/) (b) double-walled carbon nanotube (adapted and modified from general article available at https://nanografi.com/blog/doublewalled-carbon-nanotubes-overview/) (c) multi-walled carbon nanotube (adapted and modified from https://www.indiamart.com/proddetail/multi-walled-carbon-nanotube-16611189291.html; open access general article) (I) Quantum dots (self-prepared)

Liposomes

Liposomes are spherical vesicles and composite structures mainly composed of one or more phospholipid bilayers (Fig. 1.2B). Liposomes were for the first time discovered in 1965 by Bangham (Muller et al. 2000). These are generally used as a carrier system for administration of nutrients and pharmaceutical drugs due to their small size and hydrophobic and hydrophilic character (Akbarzadeh et al. 2013). It was proposed that phospholipids of liposomes spontaneously form closed structures when they are hydrated in aqueous solutions. Hence, liposomes having one or more phospholipid bilayer membranes can easily transport aqueous or lipid drugs, depending on the nature of those drugs (Chrai et al. 2001). Recently, liposomes are extensively used in cosmetic and pharmaceutical industries. In addition, liposomes are widely studied as a delivery system for encapsulation of various unstable molecules like antimicrobials, antioxidants, flavors and bioactive elements to protect their functionality (Benech et al. 2002; Shehata et al. 2008).

Solid-Lipid Nanoparticles

Solid lipid nanoparticles (SLN) are sub-micron colloidal carriers ranging from 50 to 1000 nm (Fig. 1.2C). They are composed of solid lipid core matrix (physiological lipid) that can solubilize lipophilic molecules. The physiological lipids are dispersed in water or in aqueous surfactant solution (Ekambaram et al. 2012). It was proposed that SLN were for the first time introduced in 1991 as the alternative delivery system to traditional colloidal carriers, such as emulsions, liposomes and polymeric micro- and nanoparticles. Nowadays, these nanomaterials are at the forefront of the developing field of nanotechnology having a number of potential applications in drug delivery, clinical medicine and research due to their unique size-dependent properties (Mukherjee et al. 2009).

Dendrimers

Dendrimers are nano-sized, radially symmetric molecules having spherical three-dimensional morphology with well-defined, homogeneous and monodisperse structure that has a typically symmetric core, an inner shell, and an outer shell (Fig. 1.2D) (Srinivasa-Gopalan et al. 2007). The name comes from the Greek word dendron which means "tree". Dendrimers were discovered by Fritz Vogtle for the first time in 1978. Further, by Donald Tomalia and co-workers in the early 1980s, and at the same time, but independently by George R. Newkome (Abbasi et al. 2014). Dendrimers have attracted a great deal of attention due to their broad range of applications in different fields. The special features make the dendrimers promising candidates for many applications. Dendrimers are engineered nanomaterials having combination of many functional groups and a compact molecular structure (Tomalia and Frechet

2002). Nowadays, dendrimers are promisingly used for anticancer therapies and diagnostic imaging (Abbasi et al. 2014).

Polymer Nanoparticles

Polymer nanoparticles are generally defined as colloidal systems usually having size range around 5–10 nm to an upper size limit of 1000 nm. It is a common term used for all polymer-based nanomaterials but it is specially applied for nanospheres and nanocapsules (Fig. 1.2E, F) (Lu et al. 2011). It was proposed that Paul Ehrlich for the first time developed polymeric nanoparticles with experimental efforts by Ursula Scheffel. Further, extensive work had been carried out by the group of Peter Speiser at ETH Zürich (i.e. Science, Technology, Engineering and Mathematics University, Zürich) Zürich, Switzerland in the late 1960s and early 1970s (Mallakpour and Behranvand 2016). In the last few years polymeric nanoparticles have attracted considerable interest all over the world due to their unique properties such as small size and large surface area (Farokhzad and Langer 2009). It has extensive applications in various fields such as drug delivery systems, biosensors, catalysts, nanocomposites, agriculture, environment, etc. (Rao and Geckeler 2011; Crucho and Barros 2017).

Nanomicelles

Structurally, nanomicelles are the colloidal materials composed of amphiphilic monomers, and have two regions; a small hydrophobic head and a long hydrophilic tail having size in the range of 5–100 nm (Fig. 1.2G). The hydrophobic head generally interacts with hydrophobic drugs/agents, whereas the hydrophilic tail helps surrounding with water and enhances solubility (Trinh et al. 2017). As far the applications of nanomicelles in healthcare are concerned, there is possibility of encapsulation of both hydrophobic drugs and imaging agents in nanomicelles to improve the sensitivity of imaging and diagnosis. Nanomicelles possess various advantageous characteristics, which mainly include structural stability, less toxicity, ability to entrap large amounts of hydrophobic drugs/contrast agents, solubility in water, conjugation with target ligand, and stimuli-sensitive regions, etc. (Amirmahani et al. 2017; Trinh et al. 2017).

Fullerenes

A fullerene is a hollow sphere made up of allotropes of carbon molecule (Fig. 1.2H). Spherical fullerenes also referred to as Buckminsterfullerenes or buckyballs, resemble to the football. Moreover, cylindrical fullerenes are also called carbon nanotubes. Fullerenes are similar in structure to graphite, which is composed of stacked graphene sheets of linked hexagonal rings (Schwerdtfeger et al. 2015). For the first-time fullerene molecule was manufactured in 1985 by Richard Smalley, Robert Curl,

James Heath, Sean O'Brien, and Harold Kroto at Rice University, Texas, USA. It is also called as buckminsterfullerene (C60), which was named after Richard Buckminster Fuller, a noted architectural modeler, who popularized the geodesic dome. Various unexceptional properties of fullerene like size, hydrophobicity, three-dimensionality and electronic configurations coupled them as appealing nanomaterial for various applications. The unique carbon cage structure of fullerene makes them a potential therapeutic agent (Bakry et al. 2007).

Carbon Nanotubes (CNTs)

CNTs are also allotropes of carbon like fullerene with a cylindrical nanostructure. It was proposed that carbon nanotubes were investigated in 1991, and this discovery opened up a new era in materials science. Carbon nanotubes are categorized into different types depending upon the number of tubes present in it. The CNT having single tube is called as single-walled carbon nanotube (SWCNT) (Fig. 1.2Ia). It is like a regular straw. The CNT having one outer and one inner tube is called as double-walled carbon nanotubes (DWCNTs) (Fig. 1.2Ib), and a third type is multi-walled nanotubes (MWNT) (Fig. 1.2Ic) which is collection of many (100 or more) nested tubes of continuously increasing diameters (Eatemadi et al. 2014). CNTs have unusual and extraordinary thermal conductivity, mechanical, and electrical properties. In addition, the exceptional strength and stiffness make CNTs as nanomaterials of choice for various applications as additives to different structural materials. For instance, CNTs used in the preparation of baseball bats, golf clubs, car parts or damascus steel (Gullapalli and Wong 2011).

Quantum Dots

Quantum dots (QDs) are referred to as very small semiconductor particles having size in the range of few nanometers only (Fig. 1.2J). QDs are considered as central theme in nanotechnology. Many types of QDS will emit light of specific frequencies if electricity or light is applied to them (Khaledi-Nasab et al. 2014). It possesses size and shape dependent optical and electronic properties, which are different from those of larger particles (Sabaeian and Khaledi-Nasab 2012). These properties are intermediate between those of bulk semiconductors and those of discrete molecules. Due to highly tunable properties, QDs are reported to have potential applications in development of transistors, solar cells, LEDs, diode lasers and second-harmonic generation, quantum computing and medical imaging (Ramírez et al. 2015). In addition, QDs can also be used in the inkjet printing and spin-coating because the extreme small size of QDs helps it to be suspended in solution (Coe-Sullivan et al. 2005).

Among the above-mentioned nanomaterials, engineered nanomaterials which are prepared with certain required properties are being recently extensively used for enormous applications in different sectors including healthcare, electronics, cosmetics, textiles, environmental protection, etc. The increasing number of products produced using nanomaterials is entering into the market. Therefore, the extensive use of nanomaterials increases the chances of their release in the environment and cause ecotoxicity. Hence, in this chapter, we have mainly focused on the ecotoxicity of engineered nanomaterials.

1.3 Why Nanomaterial Causes Ecotoxicity?

Research reports on nanomaterials have suggested that they have some characteristics which play deciding role in their toxicity. It involves their physico-chemical properties such as size, surface area, dissolution and increased agglomeration. Many reports have indicated the ecotoxicity of nanomaterials. However, the extent of toxicity to each component of the ecosystem varies. Even some nanoparticles are toxic to organism at lower concentration. Therefore, such variation demands the detailed investigation of the factors, which play important role in determining nanomaterials to be toxic. But relatively a few reports are available regarding the ecotoxicity of every nanomaterial. However, based on the available reports, it is known that mobility, transfer and uptake mechanism used to play major role in their ecotoxicity.

As soon as nanomaterials are released in the environment, they possibly interact with its component in various ways. As far as exposure to humans and animals is concerned, these nanomaterials are taken up by alimentary canal, and the exposure occurs through ingestion. Similarly, respiratory and dermal exposure take place via inhalation and injection, respectively. Nanomaterials spread in air, due to gravity, can get condensed or aggregated. The aggregation will lead to increase in the size of particle. Therefore, nanomaterial in suspended or in aggregated form, can enter into the respiratory system through air. However, the aggregated nanoparticles in air, may get settled down to soil and hence they can be spread or transported through the soil and water bodies.

1.3.1 Size

The size of nanoparticles play an important role in determining their toxicity. The nanoparticles' size in the range of 1–100 nm and shape is of interest for biological interface. Many studies have shown that materials under 100 nm cause harmful effects to the biological system. They cause more damage as compared to their bulk counterpart. For example, inhalation toxicity studies of titanium dioxide nanoparticles (TiO_2NPs) of 20 and 250 nm size have shown that smaller nanoparticle exerts high inflammatory reaction as compared to the bigger material. Further investigation has demonstrated that the exposure prolonged their retention in tissue, augmented translocation and greater tissue damage. Moreover, the nanomaterials of diameter under 12 nm can easily pass through the blood-brain barrier (Oberdorster et al. 2004; Sarin et al. 2008; Sonavane et al. 2008). Similarly, the materials of the 30 nm or less in diameter can be endocytosed by cells (Conner and Schmid 2003). It is also well known that smaller size nanoparticles have higher surface area. As compared to large particles with equal dose, smaller nanoparticles cause the dose-dependent increase in oxidation and DNA damage (Donaldson and Stone 2003). Therefore, any biological application or release of such materials in the environment can lead to hazardous effects.

1.3.2 Agglomeration

Nanoparticles can behave differently in the different medium. Therefore, it has high chances to undergo agglomeration in the test medium. However, it is almost difficult to disperse nanoparticles in water. Hence, due to high surface activity, agglomeration property is generally found in almost all nanoparticles. To prevent agglomeration, an anti-dispersant material has to be added. Therefore, it implies that nanoparticles, when released into natural water bodies, can easily get agglomerated. In such cases, they will appear as a pollutant and can show the hazardous effects on the biological components of the water ecosystem. The presence of inorganic salts, ions and pH in the dispersing medium can also affect the extent of agglomeration and rate of sedimentation. The agglomeration of nanoparticles depends on their surface charge and pH of the surrounding medium. The pH of the natural water bodies vary from alkaline to acidic. Whereas, the inner body pH of organism changes from the surrounding environment. Consequently, change in pH and ionic strength of the stomach may cause the nanoparticles agglomeration, affecting their uptake. In laboratory conditions, agglomeration can affect their bioavailability. However, if released in the environment it can lead to hazardous effects. Herein, they can show the homogeneous and heterogeneous agglomeration of nanoparticles with natural colloidal components of ecosystem (Rana and Kalaichelvan 2013). It is noteworthy that intracellular agglomeration of nanoparticles cannot be avoided. It leads to a decrease in total surface area. They were reported in lysosome, endosomes and vesicles of cell (Kim et al. 2009).

1.3.3 Dissolution

It is generally accepted that the dissolution leading to ion release in the medium plays a significant role in their effect on the biological system. When nanomaterials are released into the natural water bodies, they are bound to release respective ions. In terms of absolute silver content, a lethal concentration of AgNPs was reported to be about three times higher as compared to the silver ions. The rate of dissolution in the surrounding medium governs the concentration of ion release. Hence, it will be the key component in exerting the toxic effect to the target. The release of ions also mainly depends upon chemical nature (metallic and non-metallic), particle size, surface functionalization, crystal structure, temperature and presence of biomolecules and salt in the dispersion medium (Li et al. 2011). PVP-stabilized nanoparticles dissolve better as compared to citrate-stabilized nanoparticles (Kittler et al. 2010). It is claimed that the citrate coating reduces the outgoing silver ions, hence avoiding its release (Kittler et al. 2010). Similarly, copper oxide nanoparticles (CuONPs) treatment to cowpea seeds show toxic effect majorly due to increased release of Cu^{+} ions in the plant, causing the symplastic connections among cells via blocking plasmodesmata (Yuan et al. 2008).

Prolonged storage of nanoparticles lead to decrease in their size, suggesting the release of ions in the medium. It may reduce the nanoparticle diameter by 33% (Kittler et al. 2010). AgNPs stored for 6 months in dispersion medium has been shown to have lethal concentration 20 times smaller as compared to the freshly prepared nanoparticles. This study clearly suggests the role of release of silver ions in its toxicity when stored for longer period. However, many studies have reported the toxicity of nanomaterials to almost all components of the ecosystem, but none of them has emerged with a simple conclusion. The reason of this may include the diversity in particle properties, selection of toxicological model, cell type, dosing parameters and the type of biochemical methods used to study the toxicity. Additionally, the majority of studies have been performed in vitro which may differ actually in vivo. Some of the studies reporting the ecotoxic effect of various nanoparticles have been overviewed in Table 1.1.

Table 1.1 Overview of ecotoxic effects of various nanomaterials

Type of nanomaterial	Test organism	Size (nm)	Observation	Reference
TiO_2	*Daphnia magna*	25	Smaller nanoparticles exerts larger effect	Hund-Rinke and Simon (2006)
PVP coated AgNPs	Human mesenchymal cells	50–20	Dissolution of AgNPs lead to toxicity	Kittler et al. (2010)
CeO_2	*Daphnia magna*	5–90	Amount of Ce remaining suspended correlates with toxicity	Alam et al. (2016)
CuNP	*Danio rerio*	80	Partially by dissolution and majorly by agglomeration	Griffitt et al. (2007)
CuONP	*Tigriopus fulvus*, *Artemia franciscana*, *Paracentrotus lividus*, *Brachionus plicatilis*	25–55	Agglomeration and ion release	Rotini et al. (2018)
ZnONP	*Dunaliella tertiolecta*, *Vibrio fischeri*, *Artemia salina*	<100	Zinc ions and interactions of particle/aggregates with target organisms	Schiavo et al. (2018)
TiO_2NP ZnONP	*Daphnia magna*	20–50	Release of respective ions	Gokçe et al. (2018)
ZnONP	*Plectranthus barbatus*, *Ceriodaphnia cornuta*	30–60	Particle dissolution, agglomeration	Vijayakumar et al. (2017)
TiO_2 NPs	*Mytilus coruscus*	20–30	Agglomeration	Hu et al. (2017)

1.4 How to Deal with Nanoparticles' Toxicity

In order to provide safe nano-based technology to end user with minimum or no harmful effects, it is very important to search for the remedies to suppress their toxic effects. Hence, there is always discussion in the scientific community to minimize the toxicity of nanoparticles. For this purpose, the properties which make nanoparticle toxic should be targeted. As discussed above, dissolution of nanoparticles in the surrounding medium plays a major role in their toxicity. When nanoparticles are released in environment such as natural water bodies, there is a high probability of their interaction with the compounds and biomolecules present in the medium. For instance, the presence of Nitrogen (N) and Sulfur (S) in the surrounding medium have been reported to reduce the ecotoxicity of AgNPs by restricting the rate of ion release from the nanoparticles (Gunsolus et al. 2015). It will definitely help nanoparticles to retain all of their physico-chemical properties. Likewise, N-acetylcysteine (NAC), a glutathione precursor, has been well known potent antioxidant. It is required to protect a biological system against oxidative stress. NAC has been reported to protect against the cytotoxic and genotoxic effects of AgNPs. Pretreatment of HepG2 cell lines by NAC, followed by AgNPs exposure, has demonstrated the results equivalent to the unexposed control cells (Kim et al. 2009). Hence, use of such antioxidant compounds can help in reducing the ecotoxic effects of nanoparticles.

Another important feature which should be taken into consideration is its agglomeration property. To avoid agglomeration a compound stabilizing nanoparticle surface can be very useful. Such molecules should be either present in the surrounding medium or it should be added as a capping agent on the nanoparticles. AuNPs with negatively charged ligands such as citrate has been shown to be less toxic as compared to those AuNPs carrying positively charged ligands. Herein citrate at neutral pH, shields AuNPs, thereby preventing their aggregation. Moreover, the conjugation of AuNPs with bovine serum albumin (BSA) in 1:1 ratio, also reported to make them non-toxic to *Daphnia*. BSA prevented the aggregation of AuNPs, which makes it more stable, and thus indirectly avoids the toxic effect of AuNPs (Mattsson et al. 2018).

1.5 Conclusions

The toxicity of nanoparticles has been a hot topic of research because nanoparticles have been gaining a great deal of attention for many type of applications. Even some of the nano-based products have been already available in the market. They are the point of attraction due to their important properties, which make them more attractive in use as compared to particles with bigger dimensions. However, after the use of these nanoparticles by the consumers, there are huge chances of their transfer to the ecosystem surrounding them by release into water bodies and soil environment. Due to their unique properties, they may interact actively with the components of the ecosystem. Hence, it is of utmost importance to study the behavior of nanoparticles released into the ecosystem. There is also a pressing need to understand the agglomeration kinetics of nanoparticles to develop insight into direct and indirect biological effects in the environment, especially in the aquatic environment.

Nanoparticles possess unique characteristics. Size, surface reactivity and stability are the major factors in deciding their activity. Smaller size nanoparticles with high surface activity and high possibility of agglomeration exhibit major toxicity. Most of the research on ecotoxicity was performed using few higher plants and animals such as mice, fish, etc. as model systems. However, these systems do not represent complete ecosystem, and therefore there is a need of using each and every component of ecosystem as target for ecotoxicity studies.

Finally, the development of nanotechnology-based products also requires to be safe for human and environment. Nanoparticles when interact with components of the ecosystem have the higher chances to be toxic, majorly governed by their dissolution in the surrounding medium. To avoid toxic effects, use of materials preventing the ion release from nanoparticles has promising results. More focus on such issues can definitely assure the maximum intended use of nanoparticles. Hence, it can lead to the development of nano-based product with minimum or no ecotoxic effect and with wide use for the consumer.

References

Abbasi E, Aval SF, Akbarzadeh A, Milani M, Nasrabadi HT, Joo SW, Hanifehpour Y, Nejati-Koshki K, Pashaei-Asl R (2014) Dendrimers: synthesis, applications, and properties. Nanoscale Res Lett 9(1):247

Akbarzadeh A, Rezaei-Sadabady R, Davaran S, Joo SW, Zarghami N, Hanifehpour Y, Samiei M, Kouhi M, Nejati-Koshki K (2013) Liposome: classification, preparation, and applications. Nanoscale Res Lett 8(1):102. https://doi.org/10.1186/1556-276x-8-102

Alam B, Philippe A, Rosenfeldt RR, Seitz F, Dey S, Bundschuh M, Schaumann GE, Brenner SA (2016) Synthesis, characterization, and ecotoxicity of CeO_2 nanoparticles with differing properties. J Nanopart Res. https://doi.org/10.1007/s11051-016-3613-9

Amirmahani N, Mahmoodi NO, Galangash MM, Ghavidast A (2017) Advances in nanomicelles for sustained drug delivery. J Ind Eng Chem 55:21–34

Autumn K (2006) How gecko toes stick. Am Sci 94:124–132

Bakry R, Vallant RM, Najam-ul-Haq M, Rainer M, Szabo Z, Huck CW, Bonn GK (2007) Medicinal applications of fullerenes. Int J Nanomed 2(4):639–649

Benech RO, Kheadr EE, Laridi R, Lacroix C, Fliss I (2002) Inhibition of listeria innocua in cheddar cheese by addition of nisin Z in liposomes or by in situ production in mixed culture. Appl Environ Microbiol 68:3683–3690

Boyes WK, Thornton BLM, Al-Abed SR, Andersen CP, Bouchard DC, Burgess RM, Hubal EAC, Ho KT, Hughes MF, Kitchin K, Reichman JR, Rogers KR, Ross JA, Rygiewicz PT, Scheckel KG, Thai SF, Zepp RG, Zucker RM (2017) A comprehensive framework for evaluating the environmental health and safety implications of engineered nanomaterials. Crit Rev Toxicol 47(9):767–810

Chattopadhyay GP (2018) Technologies in the era of singularity. Notion Press, Chennai, India, p 206

Chrai SS, Murari R, Imran A (2001) Liposomes: a review. Bio Pharm 14(11):10–14

Coe-Sullivan S, Steckel JS, Woo WK, Bawendi MG, Bulović V (2005) Large-area ordered quantum-dot monolayers via phase separation during spin-casting. Adv Funct Mater 15(7):1117–1124
Colvin VL (2003) The potential environmental impact of engineered nanomaterials. Nat Biotechnol 21(10):1166–1170
Conner SD, Schmid SL (2003) Regulated portals of entry into the cell. Nature 422(6927):37–44
Cruchoa CIC, Barros MT (2017) Polymeric nanoparticles: a study on the preparation variables and characterization methods. Mater Sci Eng C 80:771–784
Donaldson K, Stone V (2003) Current hypotheses on the mechanism of toxicity of ultrafine particles. Ann Ist Super Sanità 39:405–410
Eatemadi A, Daraee H, Karimkhanloo H, Kouhi M, Zarghami N, Akbarzadeh A, Abasi M, Hanifehpour Y, Joo SW (2014) Carbon nanotubes: properties, synthesis, purification, and medical applications. Nanoscale Res Lett 9(1):393
Ekambaram P, Sathali AAH, Priyanka K (2012) Solid lipid nanoparticles: a review. Sci Revs Chem Commun 2(1):80–102
Ensikat HJ, Ditsche-Kuru P, Neinhuis C, Barthlott W (2011) Superhydrophobicity in perfection: the outstanding properties of the lotus leaf. Beilstein J Nanotechnol 2:152–161
European Commission (2011) Nanomaterials European Commission. Last updated 18 October 2011
Farokhzad OC, Langer R (2009) Impact of nanotechnology on drug delivery. ACS Nano 3:16–20
Ge L, Sethi S, Ci L, Ajayan PM, Dhinojwala A (2007) Carbon nanotube-based synthetic gecko tapes. Proc Natl Acad Sci USA 104:10792–10795
Genzer J, Efimenko K (2006) Recent developments in superhydrophobic surfaces and their relevance to marine fouling: a review. Biofouling 22(5):339–360
Giere O (2009) Meiobenthology: the microscopic motile fauna of aquatic sediments, 2nd ed. Springer, Berlin, Germany, pp 527. ISBN: 978-3540686576
Gokçe D, Köytepe S, Özcan I (2018) Effects of nanoparticles on *Daphnia magna* population dynamics. Chem Ecol 34(4):301–323
Goto H, Kikuchi R, Wang A (2016) Spider silk/polyaniline composite wire. Fibers 4(2):12. https://doi.org/10.3390/fib4020012
Griffitt RJ, Weil R, Hyndman KA, Denslow ND, Powers K, Taylor D, Barber DS (2007) Exposure to copper nanoparticles causes gill injury and acute lethality in zebrafish (*Danio rerio*). Environ Sci Technol 41(23):8178–8186
Gullapalli S, Wong MS (2011) Nanotechnology: a guide to nano-objects. Chem Eng Prog 107(5):28–32
Gunsolus IL, Mousavi MPS, Hussein K, Bühlmann P, Haynes CL (2015) Effects of humic and fulvic acids on silver nanoparticle stability, dissolution, and toxicity. Environ Sci Technol 49:8078–8086
Hsiung BK, Deheyn DD, Shawkey MD, Blackledge TA (2015) Blue reflectance in tarantulas is evolutionarily conserved despite nanostructural diversity. Sci Adv 1(10):e1500709. https://doi.org/10.1126/sciadv.1500709
Hu M, Lin D, Shang Y, Hu Y, Lu W, Huang X, Ning K, Chen Y, Wang Y (2017) CO_2-induced pH reduction increases physiological toxicity of nano-TiO_2 in the mussel Mytilus coruscus. Sci Rep 7:40015. https://doi.org/10.1038/srep40015
Hund-Rinke K, Simon M (2006) Ecotoxic effect of photocatalytic active nanoparticles TiO_2 on algae and daphnids. Environ Sci Poll Res 13(4):1–8
Inshakova E, Inshakov O (2017) World market for nanomaterials: structure and trends. In: MATEC web of conferences, vol 129, p 02013
Kertézs K, Zs Bálint, Vértesy Z, Márk GI, Louse V, Vigneron JP, Biró LP (2006) Photonic crystal type structures of biological origin: structural and spectral characterization. Curr Appl Phys 6:252–258
Khaledi-Nasab A, Sabaeian M, Sahrai M, Fallahi V (2014) Kerr nonlinearity due to intersubband transitions in a three-level InAs/GaAs quantum dot: the impact of a wetting layer on dispersion curves. J Optics 16(5):055004. https://doi.org/10.1088/2040-8978/16/5/055004
Kim S, Choi JE, Choi J, Chung KH, Park K, Yi J, Ryu DY (2009) Oxidative stress-dependent toxicity of silver nanoparticles in human hepatoma cells. Toxicol In Vitro 23(6):1076–1084

Kittler S, Greulich C, Diendorf J, Köller M, Epple M (2010) Toxicity of silver nanoparticles increases during storage because of slow dissolution under release of silver ions. Chem Mater 22(16):4548–4554

Koch K, Dommisse A, Barthlott W (2006) Chemistry and crystal growth of plant wax tubules of lotus (*Nelumbo nucifera*) and nasturtium (*Tropaeolum majus*) leaves on technical substrates. Cryst Growth Des 6(11):2571–2578

Kroto HW, Heath JR, O'Brien SC, Curl RF, Smalley RE (1985) C60: Buckminsterfullerene. Nature 318:162–163

Li M, Zhu L, Lin D (2011) Toxicity of ZnO nanoparticles to *Escherichia coli*: mechanism and the influence of medium components. Environ Sci Technol 45(5):1977–1983

Lu XY, Wu DC, Li ZJ, Chen GQ (2011) Polymer nanoparticles. Prog Mol Biol Transl Sci 104:299–323

Mallakpour S, Behranvand V (2016) Polymeric nanoparticles: Recent development in synthesis and application. Express Polym Lett 10(11):895–913

Mattsson K, Aguilar R, Torstensson O, Perry D, Bernfur K, Linse S, Hansson L-A, Åkerfeldt KS, Cedervall T (2018) Disaggregation of gold nanoparticles by *Daphnia magna*. Nanotoxicology. https://doi.org/10.1080/17435390.2018.1485982

Mukherjee S, Ray S, Thakur RS (2009) Solid lipid nanoparticles: a modern formulation approach in drug delivery system. Ind J Pharm Sci 71(4):349–358

Muller RH, Mader K, Gohla S (2000) Solid lipid nanoparticles (SLN) for controlled drug delivery: a review of the state of the art. Eur J Pharm Biopharm 50:161–177

Oberdörster G, Sharp Z, Atudorei V, Elder A, Gelein R, Kreyling W, Cox C (2004) Translocation of inhaled ultrafine particles to the brain. Inhal Toxicol 16(6–7):437–445

Petrescu DS, Blum AS (2018) Viral-based nanomaterials for plasmonic and photonic materials and devices. WIREs Nanomed Nanobiotechnol 10: e1508. https://doi.org/10.1002/wnan.1508

Ramírez HY, Flórez J, Camacho AS (2015) Efficient control of coulomb enhanced second harmonic generation from excitonic transitions in quantum dot ensembles. Phys Chem Chem Phys 17(37):23938. https://doi.org/10.1039/C5CP03349G

Rana S, Kalaichelvan PT (2013) Ecotoxicity of nanoparticles. ISRN Toxicol 2013, Article ID 574648, 11 pp. http://dx.doi.org/10.1155/2013/574648

Rao JP, Geckeler KE (2011) Polymer nanoparticles: Preparation techniques and size-control parameters. Prog Polym Sci 36:887–913

Rotini A, Gallo A, Parlapiano I, Berducci MT, Boni R, Tosti E, Prato E, Maggi C, Cicero AM, Migliore L, Manfra L (2018) Insights into the CuO nanoparticle ecotoxicity with suitable marine model species. Ecotoxicol Environ Saf 147:852–860

Sabaeian M, Khaledi-Nasab A (2012) Size-dependent intersubband optical properties of dome-shaped InAs/GaAs quantum dots with wetting layer. Appl Optics 51(18):4176–4185

Sarin H, Kanevsky AS, Wu H, Brimacombe KR, Fung SH, Sousa AA, Auh S, Wilson CM, Sharma K, Aronova MA, Leapman RD, Griffiths GL, Hall MD (2008) Effective transvascular delivery of nanoparticles across the blood-brain tumor barrier into malignant glioma cells. J Transl Med 6:80. https://doi.org/10.1186/1479-5876-6-80

Schiavo S, Oliviero M, Li J, Manzo S (2018) Testing ZnO nanoparticle ecotoxicity: linking time variable exposure to effects on different marine model organisms. Environ Sci Poll Res Int 25(5):4871–4880

Schwerdtfeger P, Wirz LN, Avery J (2015) The topology of fullerenes. WIREs Comput Mol Sci 5:96–145

Shehata T, Ogawara K, Higaki K, Kimura T (2008) Prolongation of residence time of liposome by surface-modification with mixture of hydrophilic polymers. Int J Pharm 359:272–279. https://doi.org/10.1016/j.ijpharm.2008.04.004

Silva LP, Rech EL (2013) Unravelling the biodiversity of nanoscale signatures of spider silk fibres. Nat Commun 4:1–9. https://doi.org/10.1038/ncomms4014

Singh U, Dar MM, Hashmi AA (2014) Dendrimers: synthetic strategies, properties and applications. Orient J Chem 30(3):911–922

Skjolding LM, Sørensen SN, Hartmann NB, Hjorth R, Hansen SF, Baun A (2016) Aquatic ecotoxicity testing of nanoparticles—The quest to disclose nanoparticle effects. Angew Chem Int Ed 55:15224–15239

Sonavane G, Tomoda K, Makino K (2008) Biodistribution of colloidal gold nanoparticles after intravenous administration: effect of particle size. Colloids Surf B 66(2):274–280

Srinivasa-Gopalan S, Yarema KJ (2007) Nanotechnologies for the life sciences: dendrimers in cancer treatment and diagnosis, vol 7. Wiley, New York

The Nanodatabase (2018). http://nanodb.dk

Tomalia DA, Frechet JMJ (2002) Discovery of dendrimers and dendritic polymers: a brief historical perspective. J Polym Sci Part A 40:2719–2728

Trinh HM, Joseph M, Cholkar K, Mitra R, Mitra AK (2017) Nanomicelles in diagnosis and drug delivery. In: Mitra A, Cholkar K, Mandal A (eds) Emerging nanotechnologies for diagnostics, drug delivery and medical devices. Elsevier, Amsterdam, Netherlands, pp 45–58

Vance ME, Kuiken T, Vejerano EP, McGinnis SP, Hochella MF Jr, Rejeski D, Hull MS (2015) Nanotechnology in the real world: redeveloping the nanomaterial consumer products inventory. Beilstein J Nanotechnol 6:1769–1780

Vijayakumar S, Malaikozhundan B, Shanthi S, Vaseeharan B, Thajuddin N (2017) Control of biofilm forming clinically important bacteria by green synthesized ZnO nanoparticles and its ecotoxicity on *Ceriodaphnia cornuta*. Microbial Pathogen 107:88–97

Wu W, Shi T, Liao G, Zuo H (2010) Research on spectral reflection characteristics of nanostructures in morpho butterfly wing scale. J Phys Conf Ser 276:012049. https://doi.org/10.1088/1742-6596/276/1/012049

Yuan JS, Galbraith DW, Dai SY, Griffin P, Stewart CN Jr (2008) Plant systems biology comes of age. Trends Plant Sci 13:165–171

Chapter 2
Nano-bio Interactions and Ecotoxicity in Aquatic Environment: Plenty of Room at the Bottom but Tyranny at the Top!

Jayanta Kumar Biswas, Mahendra Rai, Avinash P. Ingle, Monojit Mondal and Soumyajit Biswas

Abstract Nanotechnology has evolved as an outstanding versatile technology expected to become ubiquitous and to revolutionize the functionality of products in various sectors. The hallmarks of nanomaterials are higher surface area to volume ratio and greater interfacial nature that pave way to their unique fascinating features and functionalities. The environmental fate and behaviour of nanoparticles entail bioavailability, uptake, internalization, and toxicity, which are conditioned by and interactions of the chemistry of both the nanomaterial and the ambient aquatic environment, and the biology of the organisms. Oxidative stress is a predictive paradigm for ambient nanomaterials' toxicity. Nanoparticles can cross trophic boundaries via bioconcentration and biomagnifications through food chain. Majority of the nanotoxicological studies undertaken so far are descriptive or "proof-of-principle" experiments, which have tried to document toxic effects on individual organisms, bearing hardly any concrete ecological implications. To bridge the knowledge gap an urgent

J. K. Biswas (✉) · M. Mondal
Enviromicrobiology, Ecotoxicology and Ecotechnology Research Unit, Department of Ecological Studies, University of Kalyani, Kalyani, Nadia 741235, West Bengal, India
e-mail: biswajoy2008@gmail.com

M. Mondal
e-mail: monojit.mb@gmail.com

J. K. Biswas
International Centre for Ecological Engineering, University of Kalyani, Kalyani 741235, West Bengal, India

M. Rai · A. P. Ingle
Nanobiotechnology Laboratory, Department of Biotechnology, SGB Amravati University, Amravati 444602, Maharashtra, India
e-mail: mahendrarai7@gmail.com; pmkrai@hotmail.com

A. P. Ingle
e-mail: ingleavinash14@gmail.com

S. Biswas
Department of Biochemistry & Biophysics, University of Kalyani, Kalyani, Nadia 741235, West Bengal, India
e-mail: sb.biswas4@gmail.com

M. Rai and J. K. Biswas (eds.), *Nanomaterials: Ecotoxicity, Safety, and Public Perception*, https://doi.org/10.1007/978-3-030-05144-0_2

need is to undertake comprehensive studies for unveiling interactions and effects of NPs on different species belonging to different trophic levels of the aquatic ecosystem, and their toxicological responses from genetic to systemic levels. The life cycle assessment and ecological risk assessment of engineered nanomaterials are imperatives for the establishment and implementation of effective and protective regulatory policy. This chapter provides an illustrated account of promises and pitfalls of nanomaterials on an ecotoxicological canvas, with a focus on their toxic effects on life at hierarchical levels of both biological and ecological organization.

Keywords Nanomaterials · Nanoparticles · Nanopollution
Nano-bio-eco interactions · Ecotoxicity · Oxidative stress · Ecological risk

2.1 Introduction

Nanotechnology as "out of the box" technology: Small is big!

Nanotechnology is the scientific art of manipulating materials at atomic, molecular and macromolecular scales, with at least one dimension in the size range of 1–100 nm for fabrication, control, characterization and application of new generation of materials, structures, devices and products, which are smaller, lighter, stronger, quicker, and smarter (SCENHIR 2007; Biswas and Sarkar 2018). Nanotechnology is regarded as one of the greatest human innovations, having huge potential for transforming life and solve problems. Nanotechnology has evoved as a revolutionary technology triggering the neo-technological revolution in our daily lives and the materials we use. It's an enabling technology with an immense potential of offering an all-embracing innovation which would have impacts on almost every sphere of research and human's search for smarter stuffs, materials and machines. As a path-breaking platform technology nanotechnology is predicted to become ubiquitous and expected to revolutionize the functionality of products. It has already grown globally as a multibillion dollar industry (Oberdorster et al. 2005), and rapidly expanding and advancing frontier field of cutting edge research. This has already yielded a vast array of commercially available nanomaterials (NMs)/nanoproducts (NPs) in the forms of electronics, optics, textiles, medical devices, healthcare, cosmetics, nutrition, drug designing, food packaging, fuel cells, next-gen batteries, catalysts, biosensors, stain-resistant clothing, paints, self-cleaning windows, etc., which find wide applications in different fields of information technology, energy generation, environmental management, biomedical applications, food, agriculture, biotechnology, water treatment, etc. Such materials may be carbon-based such as carbon spheres (e.g., fullerenes), carbon nanotubes, metal-based NPs, composite NMs or multilayer NPs (e.g., platinum core–silica shell), or NPs with capping or coating (e.g., functionalized zinc oxide NPs). The most common NMs in use are carbon (e.g. fullerenes, nanotubes), silver, silica, titanium dioxide, zinc oxide, and cerium oxide (Contado 2015).

The present chapter provides an illustrated account of promises and pitfalls of nanomaterials drawn on an ecotoxicological canvas, with a focus on their toxic effects on life at its hierachical levels from molecular to systemic organization of organisms living in an aquatic environment as well as on life forms hierarchically linked with each other to form food chain in an ecological organization. It also attempts to draw attention to the fact that the present state of the knowledge of toxicity of nanomaterials in general and ecotoxicity in particular, is lagging far behind the state-of-the-art of nanotechnology, leaving serious dearth of ecotoxicological data on existing and emerging nanomaterials.

2.2 Plenty of Room for Nano-versatility: Materials and Products Aplenty

Scientists have drawn inspiration and emulation from Mother Nature who offers living examples of an extraordinary array of biomolecular nanodevices, such as molecular motors and protein membranes. The cell itself is a nano masterpiece, a nano factory par excellence! The underlying truth behind genesis of wonderful "novel and unique" features and fascinating functionalities among NMs and NPs is the simple science: Size matters the most for materials to furnish fascinating features; as objects get smaller they have higher surface area to volume ratio and greater interfacial nature of the chemical material of which they are built, which paves ways to new and newer physical features and functionalities springing up surprises, wonders unlimited. The unique properties of NPs and NMs are categorized into the following: (i) better catalytic efficiency due to greater surface-to-volume ratio, (ii) higher electrical conductivity in ceramics and magnetic nanocomposites, (iii) higher magnetic coercivity up to a critical grain size, (iv) enhanced mechanical hardness of metals and superplasticity of ceramics, (v) spectral shift of optical absorption and fluorescence properties and higher quantum efficiency of semiconductor crystals, (vi) increased sterical selectivity enabling hollow spheres for targeted drug transportation and controlled delivery, and (vii) biological properties such as increased permeability through biological barriers (plasma membrane, blood-brain barrier, etc.) and improved biocompatibility. Figure 2.1 beautifully reflects how Richard Feynman's dream of "plenty of room at the bottom" is coming true through nanotechnological engines of creation in realizing myriad possibilities of direct manipulation of individual atoms and molecules at the nanoscale in interesting ramifications and amazing ways and means. Sectorial applications of nanotechnology encompass the following: (i) Automobile industry: lightweight construction, paintings, catalysts, sensors, tires (fillers), coating for wind screen and car bodies; (ii) Chemical industry: fillers for paint systems, coating systems, paper impregnation, switchable adhesives, magnetic fluids; (iii) Engineering: anti-blocking coatings, scratch-resistant coatings for machines and tools, and lubricant-free bearings; (iv) Electronic industry: data-memory (M-RAM, GMR-HD), displays (OLED, FED), laser diodes, glass fibres, optical switches, IR-blocking

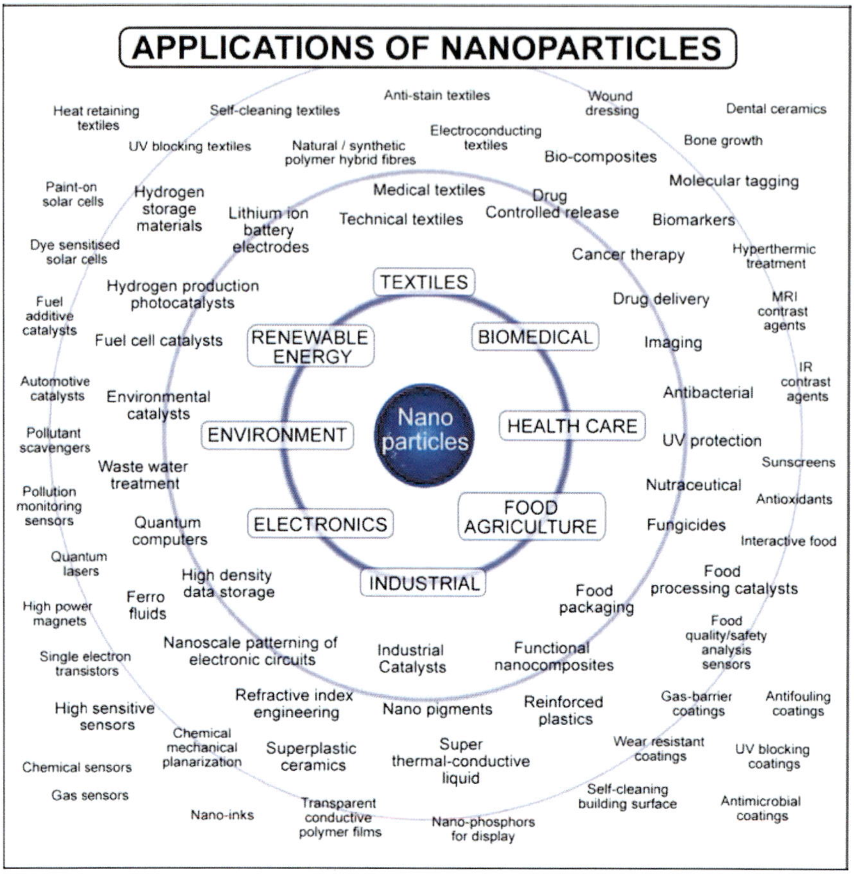

Fig. 2.1 Applications of nanoparticles in myriad of fields and prolific purposes (adapted from Tsuzuki 2009)

filters, conductive (and antistatic) coatings; (v) Construction: construction materials, insulating material, flame retardants, surface-functionalized building materials, facade coatings, groove mortar; (vi) Medicine: drug delivery systems, medical tests and probes, prostheses and implants, antimicrobial agents and coatings, anticancer agents; (vii) Textiles: enhanced textiles, such as breathable, waterproof and stain resistant smart fabrics, surface processing, cloths; (viii) Energy: fuel cells, solar cells, batteries, capacitors; (ix) Cosmetics: sunscreens, lipsticks, skin creams, tooth paste; (x) Food and drinks: packaging materials, storage life sensors, adhesives, clarifiers; (xi) Household: ceramic coatings for irons, odour catalyse, cleaning agents; (xii) Sports/outdoor: anti-gigging agents for glasses/goggles, antifouling coatings for boats/ships, reinforced tennis rackets and balls.

Examples galore, to mention a few, carbon fibres and bundles of multi-walled CNTs are used in polymers to control or enhance conductivity, with applications such as antistatic packaging. Carbon black is used as a filler to reinforce car tyres. While clays containing naturally occurring nanoparticles are used as construction material clay particle based composites—containing plastics and nano-sized flakes of clay find usage in car bumpers. Nanocoatings and nanostructured surfaces are represented in molecular beam epitaxy or metal oxide chemical vapor deposition for optoelectronic devices, or in catalytically active and chemically functionalized surfaces. The self-cleaning window has been developed with coating of highly activated titanium dioxide, engineered to be highly hydrophobic (water repellent) and antibacterial coatings based on nanoparticulate oxides that catalytically destroy chemical agents. Tougher and harder cutting tools made of nanocrystalline materials, such as tungsten carbide, tantalum carbide and titanium carbide, are more wear and erosion-resistant, and long lasting attracting applications in the drills used to bore holes in circuit boards. Manufactured metal oxide nanomaterials are among the most widely used engineered NMs in the fields of biotechnology, biosensing, clinical diagnosis and therapy, food safety, water and sewage treatment. As they improve therapeutic index of drugs, recently, more than 20 types of NPs are being used in clinical applications for drug delivery and diagnostic applications inside different organs of the human body, and various other types are under their development stages. NPs-based temperature sensors composed of NPs like Cd, Te, and Se are widely used for measurement of temperature changes in biochemical reactions, chemical reactions, and thermal fluid-based materials and so on. Compared to CNTs, SWCNTs are attracting more attention because of their outstanding electrical, mechanical, thermal, sensing, and optical properties that enable them to be utilized in many applications, such as chemistry, energy, electronics, optoelectronics, biomaterial, etc. Fluorescent semiconductor NMs, also known as QDs, often made up of groups II and VI elements (e.g., CdSe, CdTe, or ZnO) or groups III and V elements (e.g., InP) are powerful nanoscale light source. Many nanoemulsions have also been formulated to increase the bioavailability of herbicide and pesticide. Engineered polymeric nanoparticles have been used in bioremediation of hydrophobic contaminants and soil remediation. Biogenic uranite nanoparticles have been used for uranium bioremediation. Visible light photocatalysis of transition metal oxides, another nanoscale technological development, produces nanoparticles, nanoporous fibres and nanoporous foams that can be used for microbial disinfection and for the removal of organic contaminants like PPCPs and EDCs. Natural protein, carbohydrate and fat molecules have been modified with nanotechnology, and used in food packaging and food ingredients including food additives, nutraceuticals etc.

2.3 Nanopollution and Ecotoxicity: Potential Pitfalls in the Horizon

Like Nature nanotech applications have both top down and bottom up implications. Nothing is perfect in this world, and every good thing has its downsides as well. The same is true for nanotechnology having two sides of a coin. As of today, nanotechnology is mainly at the top-down stage: the nanostructures, nanomaterials and nanoparticles are made by machining and etching techniques. NPs and NM products may introduce enhanced biological effects at cellular and molecular levels of the affected organism. Since in an ecosystem of hierarchical organizations no individual population is living in isolation but connected with other organisms in trophic bonds or heterospecific interactive relationships, as well as interacting with the abiotic factors of the ambience to form the community structure, the potential pitfalls or undesired effects of NMs in biological systems would have greater ecological implications for trophic transmission and cascade effects. They may get transferred thorough the food chain in bottom-up fashion often concentrating and amplifying on some occasions (Drobne 2007; Fard et al. 2015).

Nanoparticles are actually not new and have a history as long as that of the Earth itself. NPs can be produced either intentionally as engineered nanoparticles (ENPs), or from natural sources such as aerosols from volcanic eruptions, forest fires, pollen fragments, and viruses as well as anthropogenic sources such as power plants, vehicles, coal combustion, frying, and welding, and include nanoparticles such as soot, black or elemental carbon. Since anthropogenic NPs do not appear in nature, living organisms do not have proper means to deal with nanocontaminants. Stemmed and fuelled from their clear commercial interest for a vast array of applications primarily due to their nano-specific and tuneable physicochemical properties current time has witnessed an unprecedented exponential growth of ENPs in the environment. As human society becomes attracted more towards harnessing potential benefits of nanotechnology, production of NMs, and generation and use of new generation NP-enabled products increase. Consequently, scores of synthetic/engineered NPs will emerge as new environmental contaminants because at the end-of-life of the NMs they will end up in waste streams (Keller and Lazareva 2014; Part et al. 2015; Giese et al. 2018). Nanopollution is a generic term for environmental contamination due to NMs during different phases of their life cycles from genesis to disposal. Disposed via domestic wastes or down the drain NPs start their journey and follow diverse routes to reach an environmental media, water bodies being the most likely ones. Nanopollution is an 'invisible' pollution and considered as the most difficult pollution to be managed and controlled (Gao et al. 2015). Once being released into the aquatic environment, NPs in fact are subject to environmentally driven modifications (dissolution, aggregation and sedimentation) which may affect the bioavailability of metallic NPs and then their toxicity to specific organisms.

2.4 Nano-eco Interactions: Aggregation, Deposition and Mobility of NPs in Aquatic Systems

Nano-eco interactions (Fig. 2.2) generally involve physicochemical interactions among NMs and abiotic environments (Klaine et al. 2008; Keller et al. 2010; Selck et al. 2016). The environmental fate and behaviour of NPs (Fig. 2.2) entail bioavail-

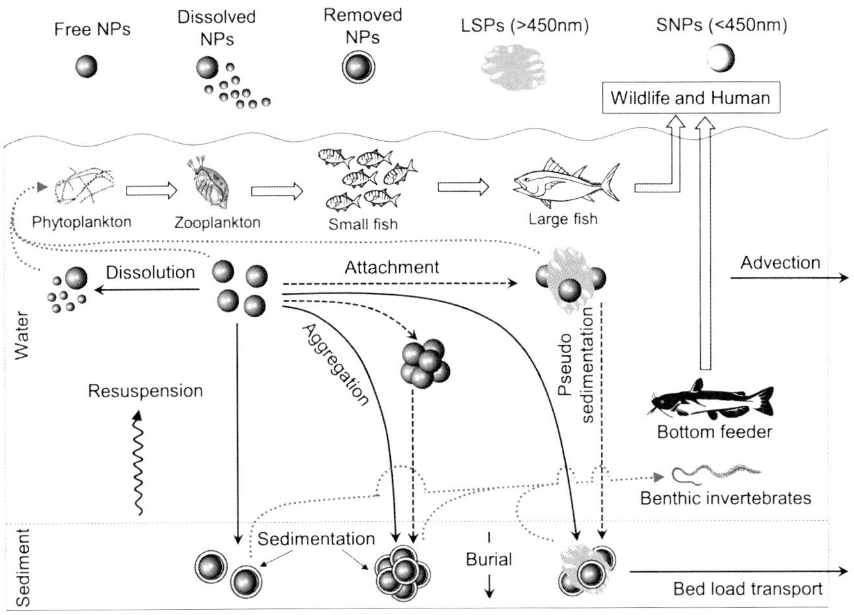

Fig. 2.2 Nano-eco interactions and fate of nanoparticles in an aquatic environment

ability, uptake, internalization, and toxicity, which are controlled by the prevailing physicochemical conditions such as particle size, surface area, composition and oxidation state, coating, pH, EC, ionic strength, hardness, dissolved organic carbon, humic and fulvic acids, etc. (Dunphy Guzman et al. 2006; Klaine et al. 2008; Hou et al. 2016; Pu et al. 2016). The most important processes affecting NPs fate in aquatic systems are agglomeration and aggregation, dissolution, redox reactions and transformation into new solid phases. Particle aggregation and deposition are intimately linked phenomena (Wiesner et al. 2006), and determine mobility and bioavailability of nanomaterials. Aggregation results from interaction between two mobile objects, whereas deposition is the attachment of a mobile particle to an immobile phase (Elimelech and Omelia 1990). Nanoparticle aggregation and deposition occur in two steps. First, NPs are transported to the a surface, and aggregate under a suite of physicochemical conditions. particle surface characteristics (surface charge, surface functional groups, Hamaker constant, and wettability) interact with the ambient aquatic conditions such as pH, ionic strength, temperature, etc. In deposition, the surface behaves an immobile "collector" where particles accumulate. The surface properties of NPs play pivotal roles in influencing their stability and mobility as colloidal suspensions or their aggregation into larger particles and removed from the water column, and transported, adsorbed and/or transformed in the sediments (Meesters et al. 2013). However, interplay of steric repulsion/attraction, hydration effects, hydrophobic interactions, magnetic attraction, or particle surface charge involved determines the complexity of the deposition/aggregation process (Metcalfe et al. 2006). Considerable sedimentation of NP aggregates can be expected, and sediments serve as significant sink for NPs while benthic organisms act as their key receptors (Baun et al. 2008).

The environmental fate and bioavailability of NPs depend on their interactions with aquatic colloids and subsequent modulation (Klaine et al. 2008; Peijnenburg et al. 2015). As a ubiquitous component of aquatic systems, natural organic matter (NOM) may influence the surface speciation and charge of NPs, and thus affect their aggregation/deposition properties (Buffle et al. 1998). They may be sorbed to the surfaces of ENPs as a result of electrostatic, hydrogen bonding, and hydrophobic interactions (Ojamae et al. 2006). The formation of larger aggregates by high molecular-weight NOM compounds (humic and fulvic acids) facilitates the removal of ENPs into sediments decreasing their bioavailability and nanotoxicity (Lin et al. 2006; Wang et al. 2011). In contrast, solubilization by natural surfactants such as lower-molecular-weight NOM compounds will increase their mobility and further the bioavailability of ENPs (Fig. 2.2). Metallic ENPs often appear with inorganic or organic coatings of citrate, cysteine, carbonate or surfactants such as sodium dodecyl sulfate to maintain the stability of the colloidal suspension (Mafune et al. 2000). Nonmetallic ENPs such as carbon nanotubes and fullerenes have hydrophobic surfaces, hence are not readily dissolved in water. These ENPs may be solubilized by functionalization with polar groups at their surfaces. Otherwise, the surfaces of hydrophobic carbon nanotubes are likely to interact preferentially with hydrophobic or amphiphilic compounds.

Due to their high surface area to volume ratio, NPs play an important role in the solid/water partitioning of contaminants. They are analogous to humic substances in sorbing pollutants. NPs play dual additional role as transfer vectors and facilitator of entry of NP-sorbed contaminants into cells of organisms potentiating toxic effects. At nano range, the properties of NMs differ substantially from the bulk materials due to the increased specific surface area and reactivity, which may lead to amplified bioavailability and toxicity (Selck et al. 2016; Skjolding et al. 2016). The solubilisation of metal-containing NPs can be a key factor in their toxicity (Wang et al. 2009; Kahru and Dubourguier 2010). Dissolution reactions lead to the release of toxic ions (Wang et al. 2016). The bactericidal effect of silver NPs was also due to the release of silver ions (Morones et al. 2005).

2.5 Nano-bio Interactions

Since man-made NPs do not naturally appear in nature, organisms may not have appropriate defensive means and mechanisms to deal with these invisible nano-invaders. The surface properties of NPs and their biocompatibility depend on the charges carried by the particle and its chemical reactivity. The routes of exposure and uptake of NPs by aquatic organisms include dermal, ingestion, inhalation through skin, diet and gills. Filter feeding animals (e.g. forage fishes, crustaceans and molluscs) feeding on suspended matter and food particles from water are particularly vulnerable to exposure and uptake of NPs dispersed in the water column. Bottom dwelling and benthivorous animals (*Chironomus* sp., *Arenicola* sp.) and biofilm feeders may be more prone to be affected by sediment NP exposure.

2.5.1 Nanotoxicity at Cellular and Molecular Level

The unique physicochemical properties of engineered NM are attributable to their small size, large surface area, durability, chemical composition, crystallinity, electronic properties, surface reactivity, surface groups, surface coatings, solubility, shape, and aggregation. Although impressive from a physicochemical viewpoint, the novel properties of NM raise the possibility of interacting with and causing of damage to biological components at molecular, subcellular, cellular, physiological and systemic levels (Nel et al. 2006). These potentially harmful effects could be enhanced by the ability of nanoparticles to be taken up and travel through the body, deposit in target organs, penetrate cell membranes, lodge in mitochondria, and trigger harmful responses.

NP-cell interactions (Fig. 2.3) are mainly dictated by the different surface properties of NPs. Unique intrinsic properties of NPs including high tensile strength, high conductivity, physiochemical, electrical and thermal properties make them possible to interact with the cell. Different chemical moieties present on the NP surface also decide the NP interaction with cell and lipid bilayer (Verma and Stellacci 2010). In prokaryotic cells, silver and zinc oxide NPs electrostatically interact with the bacterial cell surface and cause toxicity. Such interactions also result in morphological and mitochondrial perturbations as well as cytoplasmic accumulation of NPs within the cells (Sinha et al. 2011). The semi-permeable plasma membrane of eukaryotes selectively allows a few nano-sized molecules across the lipid membrane either through specific membrane transport protein channels or by endocytosis (Fig. 2.3) (Conner and Schmid 2003). For engineered NPs, crossing the lipid bilayer is difficult, and only cationic nanoparticles can penetrate by creating pores in the cell membranes. The internalization of anionic NPs occurs through their nonspecific binding and clustering on cationic sites present on the cell membrane followed by endocytosis (Shi et al. 2007, 2011). The cationic NPs bind actively to the negatively charged groups on the cell surface and move across the plasma membrane at a higher rate as compared to the cationic or neutral NPs (Martin et al. 2008). They can also be internalized through the clathrin-mediated pathway and upon inhibition of these endocytic pathways, compensatory endocytosis takes place for faster internalization (Lee et al. 2012). The uptake pathways include the clathrin-mediated, caveloe-mediated, and lipid raft-mediated endocytosis and phagocytosis, as well as pinocytosis and macropinocytosis (Conner and Schmid 2003; Panzarini et al. 2018).

In fact, one of the first biological interactions that take place when a nanoparticle penetrates or enters a tissue is contact with the surface membrane of the target cell. This interaction can lead to membrane damage based on particle properties such as hydrophobicity, cationic charge, or detergent activity that allow the particle contact,

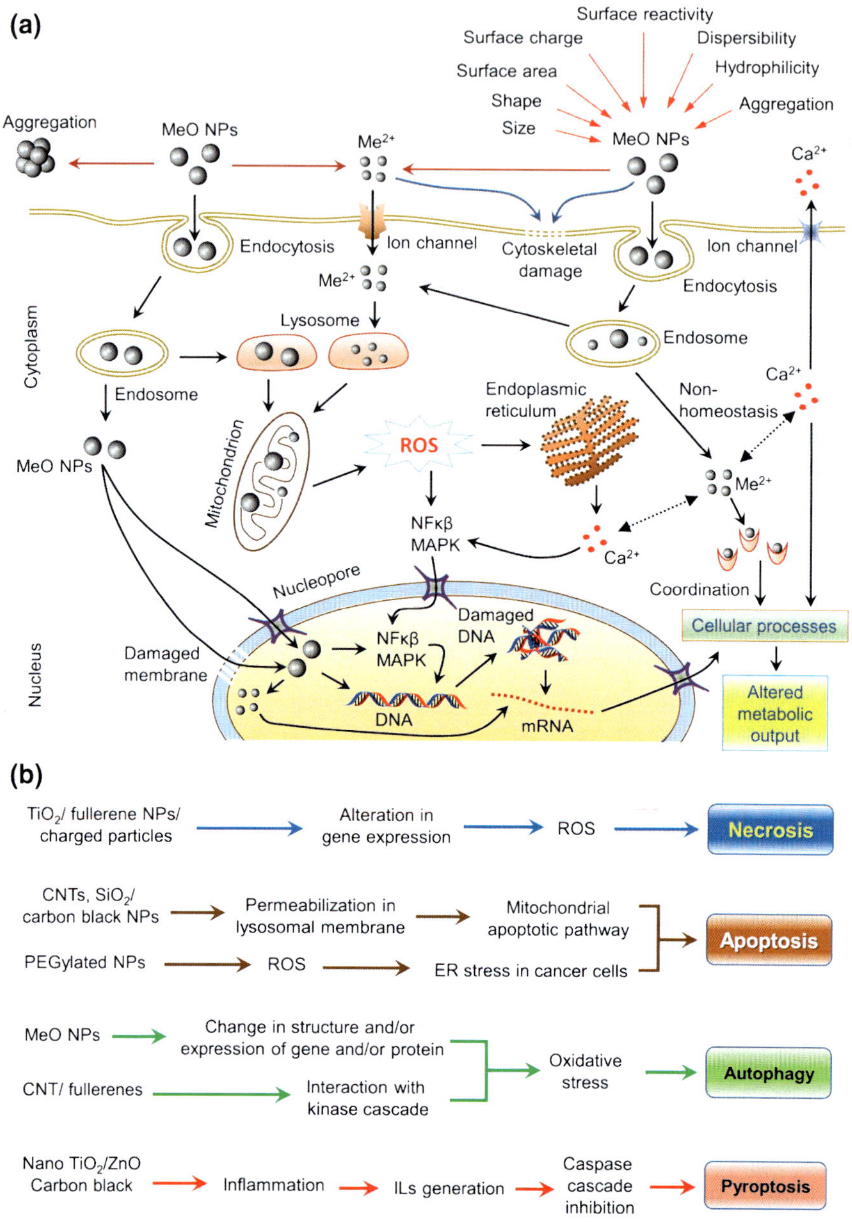

Fig. 2.3 Cascade of nano-biointeractions: **a** different levels of cellular and molecular toxicity, **b** nanoparticle based variations of toxicity pathways

penetration, or disruption of membrane integrity. The cell may respond by leakage of intracellular content, release of cellular enzymes, intracellular Ca^{2+} release, and the induction of apoptosis. Protein denaturation or degradation at the organic/inorganic interphase can lead to splitting of covalent bonds responsible for protein structure, such as disulfide bonds, and cause structural and functional changes in proteins at the primary, secondary, or tertiary level. This could lead to loss or interference of vital enzyme activities (Vertegel et al. 2004). Cellular uptake of certain NPs may have access to the nucleus, where DNA damage could result in mutagenesis and carcinogenesis in some cases (Bhattacharya et al. 2009).

Oxidative stress is a predictive paradigm for ambient NMs toxicity. The interaction of NMs with cellular chemistry can induce generation of reactive oxygen species (ROS) and toxic free radicals (Drobne 2007; Fard et al. 2015), which are extremely reactive and highly unstable, and attack and snatch energy from available cell molecules (Nel et al. 2009; Hou et al. 2016). ROS are the key signaling molecules for several pathways and they are generally produced as the by-products of mitochondrial electron transport chain. Free radicals are able to cause oxidative stress, and damage to cell membrane, proteins, and DNA ((Fu et al. 2014; Sabella et al. 2014; He et al. 2015; Khanna et al. 2015; Hou et al. 2016). Oxidative stress refers to a state in which cellular GSH is depleted while oxidized glutathione (GSSG) accumulates. Under normal coupling conditions, ROS are generated at low frequency, mostly in mitochondria, and are easily neutralized by antioxidant defense mechanisms such as the glutathione (GSH)/glutathione disulfide (GSSG) redox couple. Cells respond to oxidative stress by mounting a number of protective and injurious responses. Depending on the stress level, there can be a protective or an injurious outcome (Nel et al. 2006, 2009). The protective (Tier 1) responses are elicited by even minor changes in the cellular redox equilibrium. Phase II enzymes (e.g. glutathione-S-transferase, NADPH quinone oxidoreductase, catalase, superoxide dismutase, glutathione peroxidase, UDP-glucoronosyltransferase exert antioxidant, detoxification and anti-inflammatory effects that are responsible for preventing or slowing down the effects of oxidative injury (Park et al. 2008; Dayem et al. 2017). While small amounts of ROS could be buffered by the antioxidant defense pathways in the cell, excess amounts of ROS could lead to a drop in the GSH/GSSG ratio. A reduced or compromised phase II response promotes susceptibility to oxidant injury. Ambient nanoparticles can elicit further ROS production in mitochondria, in addition to ROS generation by catalytic conversion pathways and NADPH oxidase activation. A higher level of oxidative stress can lead to a Tier 2 response, where active free radical species lead to an imbalance in downstream biochemical processes by triggering DNA damage, altered cellular signaling and programmed cell death. This is characterized by pro-inflammatory effects that follow the activation of intracellular signaling cascades, including the MAP kinase and NF-β signaling cascades (Wang et al. 2016). These signaling cascades are responsible for the activation of a number of cytokines, chemokines, and adhesion molecules that play a role in local and systemic inflammatory responses (Wang et al. 2011). Tier 3 responses involve mitochondrial perturbation manifested in changes in mitochondrial membrane potential, calcium uptake, O_2^- production, enhanced ROS generation, depletion of cellular GSH content, cardiolipin integrity, interference in mitochondrial electron transfer, failure in ATP generation, and activation of apoptosis, aponecrosis and cytotoxicity pathways (Wang et al. 2011).

2.5.2 Nanotoxicity at Organismal System Level

Silver ions released from AgNPs can reduce cell growth, photosynthesis and chlorophyll production of the algae *Thalassiosira weissflogii* (Miao et al. 2007). Significant inhibitions of growth and reproduction of zooplankton *Daphnia* sp. were observed upon exposure to silver nanoparticles (Zhao and Wang 2010). AgNPs affected early development in fish including spinal cord deformities, cardiac arrhythmia and survival (Fabrega et al. 2011). Upon exposure to waterborne CuO NPs zebra fish (*D. rerio*) juveniles exhibited histological damages, Cu accumulation in gill (Griffitt et al. 2008). Lethal and sub-lethal effects were noticed for *O. latipes* and medaka fish (*Oryziaslatipes*) against FeONPs (Chen et al. 2011, 2013) whereas Fe_2O_3NPs were found to affect embryonic development of zebrafish resulting in embryos mortality, hatching delay, and malformation (Zhu et al. 2011). There appeared significant hematological changes in *Oreochromis mossambicus* subjected tobiosynthetized Fe_2O_3 NPs (96 h) (Karthikeyeni et al. 2013). *Cyprinus carpio* showed reduced growth and Cu accumulation in different tissues and potential neurotoxicity for juveniles to sub-lethal concentrations of CuO NPs (Zhao et al. 2011). Hao et al. (2013) highlighted severe histopathological alterations in case of juvenile of *C. carpio* exposed to ZnO NPs. Du et al. (2017) noted significant decrease in survival, hatching, and larval growth rate in zebrafish embryos exposed to different concentrations of ZnO NPs.

2.5.3 Nanotoxicity at Ecosystem Level

While unicellular organisms do absorb contaminants directly through cell walls and cell membranes, multicellular organisms such as invertebrates and fish can absorb contaminants directly from the water and via food ingestion. Bioaccumulation of NPs in individual organism occurs from its direct exposure to contaminated aquatic medium and contaminated food, and it is the ratio of a contaminant concentration in an organism to the contaminant concentration in the environment. On the other hand bioconcentration of NPs in individual organism refers to the accumulation of a contaminant in an organism solely from its environment (i.e. water). But biomagnification occurs through the food chain, where NPs gets transferred from one organisms of a specific trophic level to another of higher trophic level, and then to successive higher trophic levels. The biomagnification factor (BMF) refers to the ratio of a contaminant concentration in a predator to the concentration in its prey. BMFs exceeding one generally indicate the occurrence of biomagnification in a trophic chain. It does not occur for all contaminants, but for persistent and recalcitrant contaminants. The trophic transfer factor (TTF) is the ratio of the concentration of a substance in an organism's tissue to the concentration of that substance in organism's food.

Some nanoparticles like AgNPs have potential impacts on ecologically important aquatic bacteria (Sheng and Liu 2017). Beddow et al. (2014) showed that AgNPs adversely affected activities of ammonia oxidizing bacteria and thus have the potential to severely impact key microbially driven processes such as nitrification in the environment. The detrimental effects of CuO NPs on soil microbial community were expressed in terms of reduced hydrolytic activities, dehydrogenase oxidative potential and community composition (Frenk et al. 2013). Doiron et al. (2012) studied the adverse effects of polymer-coated AgNPs on marine microbial communities, resulting in prolonged lag phase for adaptation and erosion of community richness. Colman et al. (2012) observed that NP exposure leads to changes in community compositions and diversities which have significant bearing on community activity and functioning. The accumulation of NPs on the surface of photosynthetic organisms may inhibit photosynthetic activity due to shading effects, i.e., reduced light availability, rather than because of toxicity of the NPs. AgNP was found to reduce cell-specific growth rate, photosystem II quantum yield, and chlorophyll a content of a marine (*Thalassiosira weissflogii*) and freshwater phytoplankton (*Chlamydomonas reinhardtii*) accounted for released Ag^+ (Miaoet al. 2007; Navarro et al. 2007). The nanoparticle aggregate adhesion may cause physical effects and loss of mobility in zooplankton *Daphnia magna* (Lovern et al. 2007). The toxicity, bioaccumulation and biomagnification of material containing silver nanoparticles were studied in a model food chain comprising green algae (*Chlorella* sp.), water flea (*Moina macrocopa*), blood worm (*Chironomus* spp.) and silver barb (*Barbonymus gonionotus*) (Yoo-iam et al. 2014). The food chain transfer of AgNPs occurred only from *Chlorella* sp. to *M. macrocopa*, and there was no evidence of biomagnification from food sources to consumers in a simple tropical food chain. The fish *D. rerio* showed TiO_2 NP uptake via ingestion of contaminated daphnids (Zhu et al. 2010). Holbrook et al. (2008) studied the transfer of QDs from the bacteria *Escherichia coli* to the protozoan *Tetrahymena pyriformis*.

2.6 Conclusion and Future Scope of Research

Developments in nanotechnology appear to be revolutionary rather than evolutionary, resulting in rapid proliferation of extremely broad spectrum of application of new materials and contaminants in the environment, where their possible ecotoxicological impacts remain unknown. Based on the present state of knowledge it is very difficult to predict how nanoparticles would behave in the environment (dispersed/aggregated/agglomerated/ionized state), which gets more complicated when their persistence, residence time, longevity, partitioning, transformation and fate in the environment and organisms' body are unknown. Existing knowledge gap, insufficient research and unanticipated side effects have presented major uncertainties. The diversity of nanoproducts and current lack of experimental data on many species indicate that the scientific community is currently some way off from being able to show protection of most of the organisms most of the time in the natural environment against the new contaminants coming from the evolving nanotechnology. The current ecotoxicological bioassays involving only a few species as representatives for a vast biodiversity e.g., crustaceans, is of course a gross simplification of an ecosystem as well as a shortcoming in such risk assessment (OECD 2012). Indeed, as no single test or species of living organism shows uniform sensitivity to all chemical compounds, a battery of bioassays with different sensitivity profiles should be adopted to assure adequate evaluation of the ecotoxicological situation. Especially, issues gaining amplified attention are the establishment of toxicological profiles of engineered nanomaterials with regard to the nano-bio--eco interface, which entails the cataloging of interactions among nanomaterials, biotic, and abiotic environmental factors of the ecosystem. Toxicological risk assessment of NPs involves hazard identification, dose-response assessment, exposure assessment, and risk characterisation. Both in vivo and in vitro toxicity methods are required for a comprehensive toxicity assessment of NPs. Given the growing variety of NP interacting with the diverse aquatic species under complex aquatic regime, an urgent need for sound nanoecotoxicological risk assessment is to understand the mechanisms that govern the fate and toxicity of NPs in aquatic environments and their interactions at the nano-bio-eco interface with an emphasis on the quantitative structure–activity relationship (QSAR). Since the chemistry of different types of water bodies is heterogeneous in nature with variations in pH, Eh, dissolved oxygen, salinity, hardness, ionic strength, or organic ligands, and other physicochemical parameters, research on the fate of NPs in the aquatic environment should be dealt in the light of limnology and biogeochemistry. Till date there have been little systematic ecotoxicological studies investigating influence of alterations in the profile of water chemistry on ecotoxicity of NPs. The problem becomes more complex, because nanoparticles can significantly change their structure, shape, and size as a result of aggregation, solubilisation or adsorption phenomena. Despite a little preliminary studies, synergistic effects due to simultaneous exposure to multiple nanoparticles as well as toxic effects from genetic to systemic levels are not yet well known, and further research should be carried out to fill this knowledge gap. There is little information on role of NPs as vectors of chemicals, microorganisms, and their interactions with other stressors. Examination of the distinct contribution of the particulate and dissolved forms of NPs in both the abiotic environment and biota is of great significance for framing regulatory norms and risk assessment. Majority of the nanotoxicological studies undertaken so far are descriptive or "proof-of-principle" experiments which have tried to document toxic effects on individual organisms, bearing hardly any concrete ecological implications. Indeed nanoparticles accumulated by primary producers could be transferred to the successive trophic levels via bioaccumation, bioconcentration and biomagnification, leading to cascade and ripple effects. To bridge the knowledge gap an urgent need is to undertake comprehensive studies for unveiling interactions and effects of NPs on different species belonging to different trophic levels of the aquatic ecosystem, including bacteria, phytoplankton, zooplankton, and fish, their toxicological responses from genetic to systemic levels. It finds particular interest in aquaculture system where human beings may face more ecotoxicological risk as end user of fish. The environmental risk assessment of NMs could be performed using the existing tiered approach and regulatory framework, but with modifications to methodology including chemical characterisation of the materials being used. There are many challenges ahead, and controversies (e.g., reference substances for ecotoxicology), but knowledge transfer from mammalian toxicology, colloid chemistry, as well as material and geological sciences, will enable ecotoxicology studies to move forward in the new multi-disciplinary field. Obviously, the scientific debate on the environmental safety of NMs needs to adopt a multi-disciplinary approach involving physicists, chemists, material scientists, biologists, toxicologists, risk assessors, regulators and policy makers. A lifecycle assessment of ENPs released into the environment is imperative for the establishment and implementation of effective and protective regulatory policy. The development of a strong collaboration among industry, government and scientific community would facilitate the development of an appropriate ecotoxicological protocol, hazard identification, risk assessment and protective measures.

References

Baun A, Hartmann NB, Grieger K, Kusk KO (2008) Ecotoxicity of engineered nanoparticles to aquatic invertebrates: a brief review and recommendations for future toxicity testing. Ecotoxicology 17:387–395

Beddow J, Stolpe B, Cole P, Lead JR, Sapp M, Lyons BP, Colbeck I, Whitby C (2014) Effects of engineered silver nanoparticles on the growth and activity of ecologically important microbes. Environ Microbiol Rep 6(5):448–458

Bhattacharya K, Davoren M, Boertz J, Schins R, Hoffmann E, Dopp E (2009) Titanium dioxide nanoparticles induce oxidative stress and DNA-adduct formation but not DNA-breakage in human lung cells. Part Fiber Toxicol 6:17

Biswas JK, Sarkar D (2018) Nanopollution in the aquatic environment and ecotoxicity: no nano issue! Curr Poll Rep (in press)

Buffle J, Wilkinson KJ, Stoll S, Filella M, Zhang JW (1998) A generalized description of aquatic colloidal interactions: the three-colloidal component approach. Environ Sci Technol 32:2887–2899

Conner SD, Schmid SL (2003) Regulated portals of entry into the cell. Nature 422:37–44. https://doi.org/10.1038/nature01451

Contado C (2015) Nanomaterials in consumer products: a challenging analytical problem. Front Chem 3:48. https://doi.org/10.3389/fchem.2015.00048

Chen H, Yada R (2011) Nanotechnologies in agriculture: new tools for sustainable development. Trends Food Sci Technol 22:585–594. https://doi.org/10.1016/j.tifs.2011.09.004

Cheng LC, Jiang X, Wang J, Chen C, Liu RS (2013) Nano-bio effects: interaction of nanomaterials with cells. Nanoscale 7:3547–3569

Colman BP, Wang S-Y, Auffan M, Wiesner MR, Bernhardt ES (2012) Antimicrobial effects of commercial silver nanoparticles are attenuated in natural streamwater and sediment. Ecotoxicol Lond Engl 21:1867–1877

Dayem AA, Hossain MK, Lee SB, Kim K, Saha SK, Yang G-M, Choi HY, Cho S-G (2017) The role of reactive oxygen species (ROS) in the biological activities of metallic nanoparticles. Int J Mol Sci 18:120. https://doi.org/10.3390/ijms18010120

Doiron K, Pelletier E, Lemarchand K (2012) Impact of polymer-coated silver nanoparticles on marine microbial communities: a microcosm study. Aquat Toxicol Amst Neth 124–125:22–27

Drobne D (2007) Nanotoxicology for safe and sustainable nanotechnology. Arh Hig Rada Toksikol 58:471–478

Du J, Cai J, Wang S, You H (2017) Oxidative stress and apotosis to zebrafish (Danio rerio) embryos exposed to perfluorooctane sulfonate (PFOS) and ZnO nanoparticles. Int J Occup Med Environ Health 30(2):213–229

Dunphy Guzman KA, Finnegan MP, Banfield JF (2006) Influence of surface potential on aggregation and transport of titania nanoparticles. Environ Sci Technol 40:7688–7693
Elimelech M, Omelia CR (1990) Effect of particle-size on collision efficiency in the deposition of Brownian particles with electrostatic energy barriers. Langmuir 6:1153–1163
Fard JK, Jafari S, Eghbal MA (2015) A review of molecular mechanisms involved in toxicity of nanoparticles. Adv Pharm Bull 5:447–454
Fabrega J, Samuel Luoma R, Tyler Charles, Tamava R, Galloway J (2011) Silver nanoparticles behaviour and effects in the aquatic environment. Environ Int 37:517–531
Frenk S, Ben-Moshe T, Dror I, Berkowitz B, Minz D (2013) Effect of metal oxide nanoparticles on microbial community structure and function in two different soil types. PLoS ONE 8:e84441. https://doi.org/10.1371/journal.pone.0084441
Fu PP, Xia Q, Hwang H-M, Ray PC, Yu H (2014) Mechanisms of nanotoxicity: generation of reactive oxygen species. J Food Drug Anal 22:64–75
Gao Y, Yang T, Jin J (2015) Nanoparticle pollution and associated increasing potential risks on environment and human health: a case study of China. Environ Sci Pollut Res 22:19297–19306
Giese B, Klaessig F, Park B, Kaegi R, Steinfeldt M, Wigger H, von Gleich A, Gottschalk F (2018) Risks, release and concentrations of engineered nanomaterial in the environment. Sci Rep 8:1565. https://doi.org/10.1038/s41598-018-19275-4
Griffitt RJ, Luo J, Gao J et al (2008) Effects of particle composition and species on toxicity of metallic nanomaterials in aquatic organisms. Environ Toxicol Chem 27:1972–1978. https://doi.org/10.1897/08-002.1
Hao L Chen L, Hao J, Zhong N (2013) Bioaccumulation and sub-acute toxicity of zinc oxide nanoparticles in juvenile carp (Cyprinus carpio): a comparative study with its bulk. Ecotoxicol Environ Saf 91:52–60
He X, Aker WG, Fub PP, Hwang H-M (2015) Toxicity of engineered metal oxide nanomaterials mediated by nano–bio–eco–interactions: a review and perspective. Environ Sci Nano. https://doi.org/10.1039/c5en00094g
Hou J, Wang X, Hayat T, Wang X (2016) Ecotoxicological effects and mechanism of CuO nanoparticles to individual organism. Environ Pollut. http://dx.doi.org/10.1016/j.envpol.2016.11.066
Holbrook RD, Murphy KE, Morrow JB, Cole KD (2008) Trophic transfer of nanoparticles in a simplified invertebrate food web. Nat Nanotechnol 3:352–355
Kahru A, Dubourguier H-C (2010) From ecotoxicology to nanoecotoxicology. Toxicology 269:105–119
Karthikeyeni S, Siva Vijayakumar T, Vasanth S, Ganesh A, Manimegalai M, Subramanian P (2013) Biosynthesis of iron oxide nanoparticles and its haematological effects on fresh water fish Oreochromis mossambicus. J Acad Indus Res 1(10):645–649
Keller AA, Lazareva A (2014) Predicted releases of engineered nanomaterials: from global to regional to local. Environ Sci Technol Lett 1:65–70
Keller AA, Wang H, Zhou D, Lenihan HS, Cherr G, Cardinale BJ, Miller R, Ji Z (2010) Stability and aggregation of metal oxide nanoparticles in natural aqueous matrices. Environ Sci Technol 44:1962–1967
Khanna P, Ong C, Bay BH, Baeg GH (2015) Nanotoxicity: an interplay of oxidative stress, inflammation and cell death. Nanomaterials 5:1163–1180
Klaine SJ, Alvarez PJJ, Batley GE, Fernandes TE, Handy RD, Lyon DY, Mahendra S, McLaughlinand MJ, Lead JR (2008) Nanomaterials in the environment: behavior, fate, bioavailability, and effects. Environ Toxicol Chem 27:1825–1851
Lee JC, Son YO, Pratheeshkumar P, Shi X (2012) Oxidative stress and metal carcinogenesis. Free Radic Biol Med. https://doi.org/10.1016/j.freeradbiomed.2012.06.002
Lin W, Huang YW, Zhou XD, Ma Y (2006) In vitro toxicity of silica nanoparticles in human lung cancer cells. Toxicol Appl Pharmacol 217(3):252–259
Lovern SB, Strickler JR, Klaper R (2007) Behavioral and physiological changes in Daphnia magna when exposed to nanoparticle suspensions (titanium dioxide, nano-C60, and C60HxC70Hx). Environ Sci Technol 41:4465–4470

Mafune F, Kohno J, Takeda Y, Kondow T, Sawabe H (2000) Structure and stability of silver nanoparticles in aqueous solution produced by laser ablation. J Phys Chem B 104:8333–8337
Martin AL, Bernas LM, Rutt BK, Foster PJ, Gillies ER (2008) Enhanced cell uptake of superparamagnetic iron oxide nanoparticles functionalized with dendritic guanidines. Bioconjug Chem 19:2375–2384. https://doi.org/10.1021/bc800209u
Meesters JAJ, Veltman K, Hendriks AJ, van de Meent D (2013) Environmental exposure assessment of engineered nanoparticles: why REACH needs adjustment. Integr Environ Assess Manag 9(3):15–26
Metcalfe AM, Stoll S, Burd A (2006) The effect of inhomogeneous stickiness on polymer aggregation. J Colloid Interface Sci 298:629–638
Miao AJ, Quigg A, Schwehr K, Xu C, Santschi P (2007) Engineered silver nanoparticles (ESNs) in coastal marine environments: bioavailability and toxic effects to the phytoplankton *Thalassiosira weissflogii*. In: 2nd international conference on the environmental effects of nanoparticles and nanomaterials, 24th–25th September, London, UK
Morones JR, Elechiguerra JL, Camacho A, Holt K, Kouri JB, Ramirez JT, Yacaman MJ (2005) The bactericidal effect of silver nanoparticles. Nanotechnology 16:2346–2353
Navarro E, Piccapietra F, Wagner B, Kaegi R, Odzak N, Sigg L, BehraR (2007) Toxicity mechanisms of silver nanoparticles to *Chlamydomonas reinhardtii*. In: 2nd international conference on the environmental effects of nanoparticles and nanomaterials (oral presentation), London, UK
Nel A, Xia T, Mädler L, Li N (2006) Toxic potential of materials at the nanolevel. Science 311:622–627
Nel AE, Mädler L, Velegol D, Xia T, Hoek EMV, Somasundaran P, Klaessig F, Castranova V, Thompson M (2009) Understanding biophysicochemical interactions at the nano-bio interface. Nat Mater 8:543–557
Oberdorster G, Oberdorster E, Oberdorster J (2005) Nanotoxicology: an emerging discipline evolving from studies of ultrafine particles. Environ Health Perspect 113:823–839
OECD (2012) Important issues on risk assessment of manufactured nanomaterials, Series on the safety of manufactured nanomaterials No. 33 E. OECD Environment Directorate, Health and Safety Division, Ed. Organisation for Economic Co-operation and Development, Paris
Ojamae L, Aulin C, Pedersen H, Kall PO (2006) IR and quantum-chemical studies of carboxylic acid and glycine adsorption on utile TiO_2 nanoparticles. J Colloid Interface Sci 296:71–78
Panzarini E, Mariano S, Carata E, Mura F, Rossi M, Dini L (2018) Intracellular transport of silver and gold nanoparticles and biological responses: an update. Int J Mol Sci 19(5):1305. https://doi.org/10.3390/ijms19051305
Park EJ, Yi J, Chung KH, Ryu DY, Choi J, Park K (2008) Oxidative stress and apoptosis induced by titanium dioxide nanoparticles in cultured BEAS-2B cells. Toxicol Lett 180(3):222–229
Part F, Zecha G, Causon T, Sinner E-K, Huber-Humer M (2015) Current limitations and challenges in nanowaste detection, characterisation and monitoring. Waste Manage 43:407–420
Peijnenburg WJGM, Baalousha M, Chen J, Chaudry Q, Von der kammer F, Kuhlbusch TAJ, Lead J, Nickel C, Quik JTK, Renker M, Wang Z, Koelmans AA (2015) A review of the properties and processes determining the fate of engineered nanomaterials in the aquatic environment. Crit Rev Environ Sci Technol 45(19):2084–2134
Pu Y, Tang F, Adam P-M, Iaratte B, Ionescu RE (2016) Fate and characterization factors of nanoparticles in seventeen sub-continental freshwaters: a case study on copper nanoparticles. Environ Sci Technol. https://doi.org/10.1021/acs.est.5b06300
Sabella S, Carney RP, Brunetti V, Malvindi MA, Al-Juffali N, Vecchio G, Janes SM, Bakr OM, Cingolani R, Stellacci F, Pompa PP (2014) A general mechanism for intracellular toxicity of metal-containing nanoparticles. Nanoscale 6:7052–7061
SCENHIR (2007) Opinion on: the scientific aspects of the existing and proposed definition relating to products of nanoscience and nanotechnologies. In: Brussels: European commission health and consumer protection directorate general

Selck H, Handy RD, Fernandes TF, Klaine SJ, Petersen EJ (2016) Nanomaterials in the aquatic environment: a European Union-United States perspective on the status of ecotoxicity testing, research priorities, and challenges ahead. Environ Toxicol Chem 35(5):1055–1067

Sheng Z, Liu Y (2017) Potential impacts of silver nanoparticles on bacteria in the aquatic environment. J Environ Manage 191:290–296

Shi X, Thomas TP, Myc L, Kotlyar A, Baker JR (2007) Synthesis, characterization, and intracellular uptake of carboxyl-terminated poly (amidoamine) dendrimer-stabilized iron oxide nanoparticles. Phys Chem Chem Phys 9:5712–5720. https://doi.org/10.1039/b709147h

Shi X, von dem Bussche A, Hurt RH, Kane AB, Gao H (2011) Cell entry of one-dimensional nanomaterials occurs by tip recognition and rotation. Nat Nanotechnol 6:714–719

Sinha R, Karan R, Sinha A, Khare SK (2011) Interaction and nanotoxic effect of ZnO and Ag nanoparticles on mesophilic and halophilic bacterial cells. Bioresour Technol 102(2):1516–1520

Skjolding LM, Sørensen SN, Hartmann NB, Hjorth R, Hansen SF, Baun A (2016) A critical review of aquatic ecotoxicity testing of nanoparticles—The quest for disclosing nanoparticle effects. Angew Chem Int Ed. https://doi.org/10.1002/anie.201604964

Tsuzuki T (2009) Commercial production of inorganic nanoparticles. Int J Nanotechnol 6:567–578

Verma A, Stellacci F (2010) Effect of surface properties on nanoparticle-cell interactions. Small 6:12–21. https://doi.org/10.1002/smll.200901158

Vertegel AA, Siegel RW, Dordick JS (2004) Silica nanoparticle size influences the structure and enzymatic activity of adsorbed lysozyme. Langmuir 20:6800–6807. https://doi.org/10.1021/la0497200

Wang D, Sun L, Liu W, Chang W, Gao X, Wang Z (2009) Photoinduced DNA cleavage by alpha-, beta-, and gamma-cyclodextrin-bicapped C60 supramolecular complexes. Environ Sci Technol 43:5825–5829

Wang Z, Li J, Zhao J, Xing B (2011) Toxicity and internalization of CuO nanoparticles to prokaryotic alga *Microcystis aeruginosa* as affected by dissolved organic matter. Environ Sci Technol 45:6032–6040

Wang D, Lin Z, Wang T, Yao Z, Qin M, Zheng S, Lu W (2016) Where does the toxicity of metal oxide nanoparticles come from: the nanoparticles, the ions, or a combination of both? J Hazard Mater. http://dx.doi.org/10.1016/j.jhazmat.2016.01.066

Wiesner MR, Lowry GV, Alvarez P, Dionysiou D, Biswas P (2006) Assessing the risks of manufactured nanomaterials. Environ SciTechnol 40:4336–4345

Yoo-iam M, Chaichana R, Satapanajaru T (2014) Toxicity, bioaccumulation and biomagnification of silver nanoparticles in green algae (*Chlorella* sp.), water flea (*Moina macrocopa*), blood worm (*Chironomus* spp.) and silver barb (*Barbonymus gonionotus*). Chem Spec Bioavail 26(4):257–265

Zhao C-M, Wang W-X (2010) Biokinetic uptake and efflux of silver nanoparticles in *Daphnia magna*. Environ Sci Technol 44:7699–7704

Zhao J, Wang Z, Liu X, Xie X, Zhang K, Xing B (2011) Distribution of CuO nanoparticles in juvenile carp (Cyprinus carpio) and their potential toxicity. J Haz Mater 197:304–310

Zhu M-T, Wang B, Wang Y, Yuan L, Wang H-J, Wang M, Ouyang H, Chai Z-F, Feng W-Y, Zhao Y-L (2011) Endothelial dysfunction and inflammation induced by iron oxide nanoparticle exposure: risk factors for early atherosclerosis. Toxicol Lett 203:162–171. https://doi.org/10.1016/j.toxlet.2011.03.021

Zhu X, Chang Y, Chen Y (2010) Toxicity and bioaccumulation of TiO_2 nanoparticle aggregates in *Daphnia magna*. Chemosphere 78:209–215

Chapter 3
Chemical Structure and Toxicity of Nanomaterials Used in Food and Food Products

Semih Otles and Buket Yalcin Sahyar

Abstract There are many areas for potential applications of nanomaterials in food industry, such as food processing, food packaging and nutrient supplementation. Food nanotechnology is still a growing area, which still needs clarification, definition and standardization from larger nanotechnology spectrum. Due to the lack of knowledge on nanomaterials and their usage inside complex matrix of food, health risk and toxicity of nanomaterials is becoming main parameters, which need to be defined, identified, and controlled for every single and novel application. Unfortunately, food matrix is complex and application of nanomaterials in the food matrix is also novel and unique for every specific application that is why every single specific application need to have its own investigation, classification, determination and toxicological studies. This chapter summarizes the application of nanomaterials in food industry and potential toxicological effect of nanomaterials especially regarding food industrial applications.

Keywords Food · Nanomaterials · Nanotechnology · Structure and toxicology

3.1 Introduction

Recently, nanotechnology has been considered as to be interesting technology that has fundamentally changed the food industry. It is a technology, which deals with the atoms, molecules on the nanoscale and/or range with the size of 1–100 nm to create and use materials that have novel properties. Nanomaterials have unique properties unlike their macroscale counterparts due to the high surface to volume ratio. And also

S. Otles (✉)
Engineering Faculty, Food Engineering Department, Ege University, 35100 Bornova, Izmir, Turkey
e-mail: semih.otles@gmail.com

B. Y. Sahyar
Whirlpool Corporation, OSB 2. Kisim, 45030 Manisa, Turkey
e-mail: buket.yalcin.35@gmail.com

M. Rai and J. K. Biswas (eds.), *Nanomaterials: Ecotoxicity, Safety, and Public Perception*, https://doi.org/10.1007/978-3-030-05144-0_3

their physicochemical properties like color, solubility, strength, diffusivity, toxicity, magnetic, optical and thermodynamic are novel and different from their macroscale counterparts. Nanotechnology presents a wide range of opportunities for the development and application of structures, materials, or system with new properties for food industry (Singh et al. 2017).

Consumer demands and concerns about food quality and health benefits are compelling the researchers to find a way to enhance food quality while maintaining the nutritional value of the product. The demand of nanomaterials, which contain essential elements and are non-toxic, has increased in the food industry. These nanomaterials have been also found to be stable at high temperature and pressure. Nanotechnology represents a diversity of solutions from food manufacturing, processing to packaging. Nanomaterials bring about a great difference not only in the food safety and quality but also in health benefits that food delivers. Many studies are coming up with novel techniques, methods, and products that have direct applications of nanotechnology in food industry (Singh et al. 2017).

There are successful examples of nanomaterials' application in food industry, specifically food packaging industry. These materials have permission from authorities such as US FDA (Food and Drug Administration). But there are reports and details still missing, which are deeply focused on toxicity and long-term risk assessment of these nanomaterials. Another promising area for application of nanomaterials is biosensors, which can be directly or indirectly related with food and food packaging and applied to monitor food allergens, proteins, food-borne pathogens, etc. These nanomaterial-based biosensors are safer than other options due to their indirect contact with food media and gas detection mechanism.

In this chapter, nanomaterials definition, classification and characterization approaches have been discussed with special reference to food nanomaterials and characterization. Some examples are given for nanomaterials which are generally used in food industrial applications. Toxicological effects of nanomaterials used in food industrial purposes are dealt. Gaps and open points for nanomaterials structure are also referred.

3.2 Nanomaterials

Nanomaterials are attractive for science and technology due to their unique structure and behavior on the human body, which is not well-known. Specially, chemical and biochemical structure of nanomaterials are taken into consideration to inhibit toxicity and hazard capabilities. New characterization and validation methods need to be developed and/or improved to get well-known nanomaterials before they are going to be the main agents of manufacturing and medicine (Nel et al. 2006; Jones and Grainger 2009). The production and use of nanomaterials still are developing areas, which have raised many concerns among scientific, public and regulatory authorities (Dhawan and Sharma 2010).

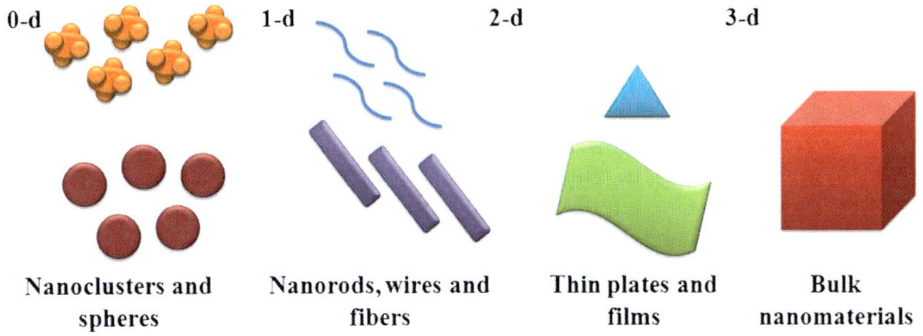

Fig. 3.1 Classification of nanomaterials

Classification of nanomaterials can be done by several methods (Fig. 3.1). Traditional classification can be done by using traditional pharmacological and toxicological approaches; where physicochemical characterization involves dose response, acute and chronic exposure (Warheit 2008) whereas physiological reactivity entails molecular, cellular, organ and tissue toxicity (Nel et al. 2006; Takagi et al. 2008; Poland et al. 2008; Kostarelos et al. 2008; Jones and Grainger 2009). Based on genesis nanoparticles are of two main categories: natural and anthropogenic nanoparticles. Natural nanoparticles exist in the environment as soil colloids, airborne nanocrystals of sea salts, fullerenes, carbon nanotubes and biogenic magnetite (Dhawan and Sharma 2010). Characterization of nanoparticles is based on their dimension, morphology, composition, uniformity, and agglomeration (Buzea et al. 2007).

Hence, nanomaterials applicable in the food industry are generally focused on food packaging. The use of polymers as food packaging material is an important example for industrial usage of nanomaterials. Nanomaterials can provide novel and efficient properties into food packaging systems. The benefits of nanopolymer materials for food packaging system can be summarized as:

- Efficient performances,
- Economic energy inputs for production, transportation and storage,
- Increased biodegradability and barrier protection to gases and light,
- Reduction of volume of waste material to be disposed of in landfills contributing to decreased CO_2 emissions.

While main applications of polymer nanomaterials for food packaging systems have been summarized by Silvestre et al. (2011) as stated below.

- The presence of nanoparticles in the polymer matrix materials improves the packaging properties such as polymer-flexibility, gas barrier properties, temperature/moisture stability,
- The presence of nanoparticles allowes packages to interact with food and the environment, and play a dynamic role in food preservation.

- The presence of nanodevices in the polymer matrix can monitor the condition of packaged food or the environment surrounding the food, and can also act as a guard against fraudulent imitation.

3.3 Structure of Nanomaterials

Identifying the structure of nanomaterials is not easy, and direct way. Structure of different types of nanomaterials varies according to usage and influence area. The following approaches are used to identify structures of nanomaterials and characterization of nanomaterials.

Physical characterization: Material properties, cleaning procedures, scientific validity, and variability from batch to batch and solution properties such as intrinsic aqueous stability, flocculation and aggregation need to be clarified and identified before usage of nanomaterials in many kind of applications. Quality control of nanomaterials needs to be done and physical characterization is really important, especially before critical biological usage of nanomaterials and before starting in vitro or in vivo testing. Characterization analysis of nanomaterials should involve purity certification, aqueous leachable, and electron microscopy of bulk nanophase morphology, polydispersity, intrinsic aggregation, soluble, bulk-phase thermal analysis, redox or conductivity behaviour, and spectroscopy to provide some quality control of nanomaterials physical and chemical states, and stability (Jones and Grainger 2009).

Surface contamination: Surface contamination is a parameter between physical, chemical and physicochemical characterization. Surface contamination of nanomaterials is a thermodynamic process driven by surface energy reduction through the adsorption of external air and water borne contaminants on nanomaterials' surface through physico-chemical processes. Surface contamination of nanomaterials results from surface corrosion, oxidation, charging or electron transfer reaction between nanomaterials. Such is expected to reduce surface free energy of the solid interface, due to highly reactive nature of nanophase surfaces. Alumina, silica, and activated charcoals are some example of finely dispersed materials, which are already used as adsorbent beds to remove dissolved species. Nanomaterials can also be used as adsorbent beds with substantially increased surface areas. As a consequence, contamination of nanomaterials in both aqueous media and air is likely by spontaneous adsorption from ambient phases at a degree of 200–1000 m^2/g surface areas. Nanomaterial surface can unexpectedly introduce high levels of contaminants by reversible adsorption of aqueous or airborne species, hence nanomaterials represent surface leachate to the solution more than their purified situation. This could be resulted in transportation of high levels of toxic material or species into human beings via adsorption. Colloidal stability of the nanomaterial may alter by surfactant or solute adsorption to start aggregation. Adsorption of toxic endogenous adsorbates from intrinsically toxic nanomaterials also needs to be controlled by surface analysis. Surface modification agents can be used to stabilize nanomaterials, need to be deeply identified for their possible toxicity. Nanomaterial uptake into biological systems is influenced by surface chemistry and properties, surface aggregation and contamination. Thus, nanomaterials interactions with living systems can be identified and understood by in terms of surface composition, oxidation, corrosion, charging or electron transfer reactions of the nanomaterial (Jones and Grainger 2009). Characterization of nanoparticles used in the food industry can give another perspective to understand nanomaterial structure.

Physicochemical characterization of nanoparticles in food: There are two main approaches for the production of nanoparticles. One of them is bottom up, which can be defined as assembling materials from a molecular or atomic level via chemical reactions into nanoparticles by allowing for the precursor particles to grow in size. Second one is top down, which can be defined as a bulk material breaking into small pieces by using chemical, mechanical or other form of energy (Oberdorster et al. 2005; Powers et al. 2006). These approaches are important for better understanding of possible benefits and potential toxicity of nanoparticles for biological systems (The Royal Society and the Royal Academy of Engineering 2004). Production steps and parameters need to be clearly identified; the composition, purity of nanoparticles and physical, chemical and physicochemical changes of nanoparticles need to be monitored to see potential risks during industrial usage (Powers et al. 2007). Based on scientific studies (Oberdorster et al. 2005; Bouwmeester et al. 2009), the following three criteria should be considered as prioritization of characterization of nanoparticles: (i) the substance within which a material is being evaluated, (ii) the importance of measuring a specific parameter within that substance, and (iii) the feasibility of measuring the parameter within a specific substance.

Complexity of food matrix is an important parameter that directly affects the characterization of nanoparticles. Nanoparticles can easily change their functionalities (particle size, size distribution, potential agglomeration and surface charge) by interacting with different biological matrices (Powers et al. 2006). Nanoparticles' interaction with complex matrices can also easily affect nanoparticles' behavior by dilution, heating, freezing-thawing cycle (Oberdorster et al. 2005). There are still lacking of analytical methods to characterize nanoparticles and define nanomaterials (Bouwmeester et al. 2009).

Here are main elements, which can be easily used and applied at nanoscale, and have proven benefits in food industrial usage as nanomaterials.

Silver (Ag): Silver has strong antimicrobial effect and the antimicrobial mechanism of silver is a promising area, specifically at nanoscale as nanosilver. In Fig. 3.2 an example for chemical production of nanosilver is given.

Antimicrobial effect mechanism of silver can be classified as: (a) interference with vital cellular processes by binding to sulfhydryl or disulfide functional groups on the surfaces of membrane proteins and other enzymes, (b) disruption of DNA replication, (c) oxidative stress through the catalysis of reactive oxygen species (ROS) formation.

Silver antimicrobial effect can easily be seen on Gram-negative bacteria such as *E. coli* and afterwards on Gram-positive bacteria such as *S. aureus*. Another example for antibacterial activity of silver is silver zeolites, which derives from ability of silver to catalyze the production of reactive oxygen species that causes cell death by creating oxidative stress (Inoue et al. 2002). Antioxidant-rich Bacillus spores are also highly

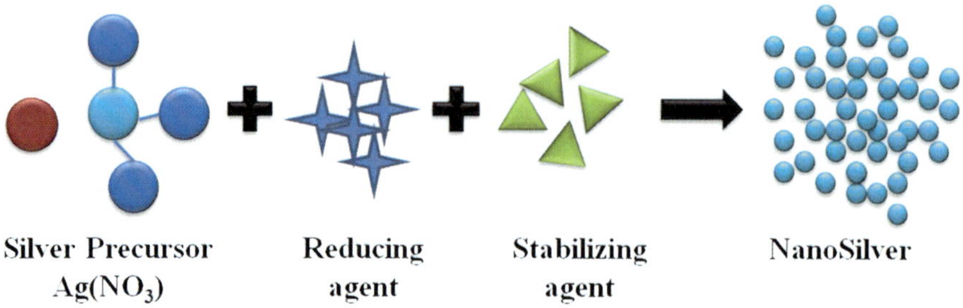

Fig. 3.2 Chemical production of nanosilver

resistant to silver antimicrobials, whereas vegetative and relatively antioxidant-poor Bacillus cells are quite vulnerable (Galeano et al. 2003). It is possible that all of these mechanisms contribute to the antimicrobial activity of silver, which would enlighten its wide effectiveness as well as the rare reports of silver-resistant bacterial strains (Duncan 2011). On the other hand, silver nanoparticles can be used as antibacterial and antifungal agents in a diverse range of applications: air sanitizer sprays, socks, pillows, slippers, face masks, wet wipes, detergent, soap, shampoo, toothpaste, air filters, coatings of refrigerators, vacuum cleaners, washing machines, food storage containers, cellular phones, and even in liquid condoms (Buzea et al. 2007).

Moreover, packaging material, which includes silver based nanoparticles can be classified into 2 categories L non-degradable polymeric, and biodegradable edible coating film made by either a polymer or a stabilizing agent (Carbone et al. 2016). Some silver based nanoparticle including packaging samples are applicable to food industry. Silver and ZnO nanoparticles used on LDPE (low-density polyethylene) polymer matrix is an example, which preserve and extend the shelf life of orange juice (Emamifar et al. 2010). This active packaging system represents very effective antimicrobial in combination with heat treatment at the pasteurization temperature (55–65 °C). The antimicrobial activity of LDPE with silver nanoparticles is found more effective against yeast and molds compared to the other active polymer matrices containing ZnO nanoparticles. The antimicrobial activity of silver based nanoparticles allowed decrease in pasteurization temperature range for orange juice. Researches showed that silver based nanoparticle used on LDPE was able to maintain the sensory, physicochemical and physiological qualities of barberry and strawberry fruits at a higher level compared with normal packaging realized with polyethylene bags (Yang et al. 2010; Motlagh et al. 2012). Silver based nanoparticles incorporated into pullulan films (also includes essential oils) can be used to control pathogens such as *S. aureus* and *L. monocytogenes* on meat and poultry products (Morsy et al. 2014; Carbone et al. 2016).

Titanium dioxide (TiO_2): Titanium dioxide (TiO_2) are promising nanoparticles owing to their antimicrobial properties (Kim et al. 2003; Robertson et al. 2005; Fu et al. 2005; Kangwansupamonkon et al. 2009; Kubacka et al. 2009; Hamal et al. 2010; Kong et al. 2010; Duncan 2011), and are considered biologically inert in both humans

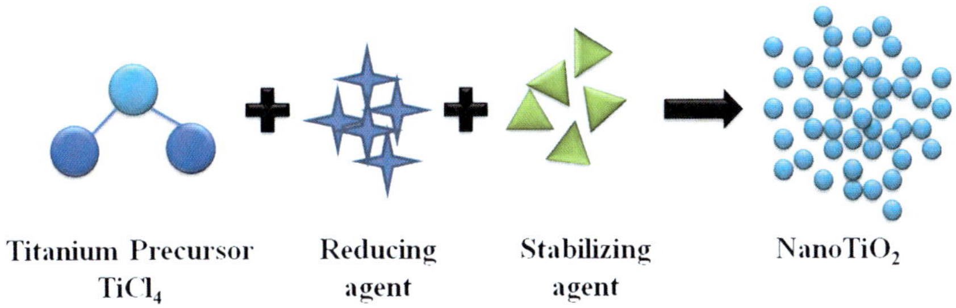

Fig. 3.3 Chemical production of TiO_2 nanoparticles

and animals (Gurr et al. 2005; Buzea et al. 2007). Titanium dioxide is widely used as a white pigment and food colorant. The daily intake of titanium dioxide from food is estimated to be 5.4 mg/person/day (about 90 μg/kg bw/day) in the United Kingdom. In Europe, food-grade titanium dioxide is specified as E171, and approximately 36% of the particles in E171 have diameters of <100 nm. Candies, sweets, and chewing gums have the highest content of titanium (0.01–1 mg titanium per serving. More than the expected number of foods contained titanium dioxide nanoparticles with diameters of <200 nm (Carbone et al. 2016). In Fig. 3.3 an example for chemical production of TiO_2 nanoparticle is given.

The differentiation of TiO_2 nanoparticles from AgNPs (silver nanoparticles) is photocatalytic properties of TiO_2 nanoparticles, and TiO_2-based antimicrobials are only active in the presence of UV light. TiO_2 nanoparticles have been found to be effective against common food borne pathogens (*Salmonella choleraesuis* subsp., *Vibrio parahaemolyticus*, and *Listeria monocytogenes*) under UV illumination but not in the dark (Kim et al. 2003; Duncan 2011). Also, potential usage area for nanostructured (TiO_2) photocatalyst materials is sterilizing equipment from environmental microorganisms especially in health care facilities (Shintani et al. 2006; Buzea et al. 2007).

Apart from Ag and TiO_2 nanoparticles, magnesium oxide (Stoimenov et al. 2002; Huang et al. 2005a, b; Lin et al. 2005), copper and copper oxide (Cioffi et al. 2005; Yoon et al. 2007; Anyaogu et al. 2008; Mary et al. 2009; Cárdenas et al. 2009; Esteban-Tejeda et al. 2009; Ren et al. 2009; Perelshtein et al. 2009; Palza et al. 2010; Rispoli et al. 2010), zinc oxide (Tam et al. 2008; Droval et al. 2008; Jin et al. 2009; Li et al. 2009; Liu et al. 2009; Sevinç and Hanley 2010; Lee, 2009; Bajpai et al. 2010; Chaurasia et al. 2010; Yuvaraj et al. 2010; Xie et al. 2011; Emamifar et al. 2011), cadmium selenide/telluride (Lu et al. 2008, 2009; Priester et al. 2009) and chitosan (Qi et al. 2004; Xing et al. 2009; Lu et al. 2010), and carbon nanotubes (Kang et al. 2007, 2009) are also nanoscale materials which have antimicrobial properties (Duncan 2011; Sharma et al. 2017).

3.4 Nanomaterials' Applications in Food Industry

Recently, nanotechnology can be applied and adapted into agricultural production, food processing, preservation and packaging, and functioned foods (as a food supplement and additive) (Bouwmeester et al. 2009). While food sector applications of nanotechnologies are food ingredient derived to the nanoscale (emulsion stability, nanoscale nutraceuticals and nanoencapsulation), food packaging (nanocomposites, active and intelligent food contact materials, antimicrobial nanopackaging, time temperature indicators etc.) (Cushen et al. 2012).

There are numerous opportunities for application of nanomaterials in food industry and agriculture (such as, pesticide, fertilizer or vaccine delivery; plant and animal pathogen detection; targeted genetic engineering) to food processing (such as, encapsulation of odor or flavor enhancers; food quality or textural improvement; new viscosifying or chelation agents) to food packaging (such as, pathogen and gas sensors; anti-counterfeiting devices, UV-protection, and stronger, more impermeable polymer films) to nutrient supplements (such as, nutraceuticals with higher bioavailability and stability). Food nanotechnology is still a growing area which is not known very well from a greater nanotechnology spectrum (Duncan 2011). There are numerous applications and samples of nanomaterial applications in food industry. Here are some examples for nanomaterials usage in the food technology field.

There are nanomaterials and/or nanoscale particle samples for food industrial usage. Gold nanoparticles (AuNPs) can be used to detect some small organic molecules such as food adulterants and contaminants by using electrochemical detection methods (Duncan 2011). Nanoscale clay particles can be used in multilayer packaging for beer bottles and carbonated drinks. These nanoscale clay particles embedded in plastic bottles and nylon food film stiffen packaging and reduce gas permeability keeping oxygen-sensitive foods fresher, and extend shelf life. Bayer polymer used a low cost nanoclay composite interior coating for paperboard cartons to keep the juice fresh. Nanoclay barrier usage inside PET beer bottles increases shelf-life of beer bottles. Silicon oxide nanocoating can be used to improve shelf-life of PET bottles (Silvestre et al. 2011).

One of the main areas for application of nanomaterials is a food packaging field. The first example is silver nanoparticles and nanocomposites, which are antimicrobial food packaging materials (especially for beverage application). The second example is barrier applications of polymer nanocomposites, which enhance and modify food packaging material's permeability to get the better mechanical structure. Using these nanocomposites cost savings is increased and waste reduction is possible (Duncan 2011). Intelligent and active food packaging is the main focusing areas where nanotechnology is expected to have a large impact. Silver nanoparticles are always first example for food application examples of nanotechnology with its antimicrobial properties. It has already used from FresherLongerTM inside storage containers in a polypropylene base material for inhibition of growth of microorganisms. Another commercial application of silver nanoparticles is Sharper Image® and BlueMoonGoods in the US, silver nanoparticles incorporated into plastic food containers with providing antibacterial and antimicrobial properties that keep food safer, fresher, healthier and tastier claims (Zeng et al. 2006). Silver zeolites (has FDA and EFSA approval for food contact use) are commercial nanoparticles used in the food packaging film.

Biosensors are other interesting area for nanotechnology application by benefits of simplicity and cost effectiveness, reduced power requirements and easier recycling; and finally not necessity of exogenous molecules or labels. Biosensors can work rapidly and give response in minutes to detect allergen proteins, pathogens, spoilage, and chemical contaminants. Biosensors can be used for detection, tracing, tracking and brand collection. Some smart nanopackaging applications are already present and tagging sensor is used on food packaging to give know-how about the product directly to the costumer to warn them about usability of food product, while another tagging system can be used for transportation of food product to control the quality and stability of environmental conditions of storage area during food product chain as barcode systems (Silvestre et al. 2011).

3.5 Toxicological Effects

Toxicology is complex and hard to understand topic for customers specially related to what they eat or inhale to their body via their selection. Selection is an important parameter for customer and food correlation. People cannot select the air intake of their body, but they can select what they put inside their table. Customer knowledge about what they eat and what they put inside their table is increased. That makes toxicological studies need to be focused deeply on food industrial application of nanomaterials. Further, toxicological effects of nanoparticle, nanomaterial and nanoscale systems on human health, environment and food are defined by giving some specific examples. Generally, the toxic effect of materials can be exhibited by tissue inflammation, and altered cellular redox balance toward oxidation, causing abnormal function or cell death. The manipulation of atomic level of nanoscale materials is not easily identifiable and controlled based on the current knowledge. Amidst the limitless diversity of nanomaterials, some represent toxic effects to biological systems, while others represent health benefits (Buzea et al. 2007). Apart from this, epidemiological studies have presented a strong correlation between air pollution levels with nanoparticles and various cancer types, cardiovascular diseases and mortality. Adverse effect of nanoscale systems on human health depends on some individual factors as genetics, existing disease, and also exposure time and amount of nanoparticles. Chemistry, size, shape, agglomeration state and electromagnetic properties of nanoparticle are also important factors, which manipulate adverse effect of nanoscale systems. The toxicity of chemically similar materials could be changed significantly by the manipulation of several physicochemical properties. In order to reduce the knowledge gap between the development and in vivo toxicity of nanoparticles, an effort is needed by the scientific community to study the physiological effects of acute and chronic exposure to nanoparticles. A fundamental understanding of the biological interactions of nanoparticles with cells, proteins, and tissues, are vital to the future design of safe nanotechnologies. Prior to their wider adoption in everyday products and clinical use, nano-products must be represented to have a high degree of biocompatibility, with minimum negative effects on blood components, genetic material, and cell viability (Sharifi et al. 2012). A study on animal and human showed that inhaled nanoparticles are less efficiently removed than larger particles by the macrophage clearance mechanisms in the lungs, causing lung damage, and that nanoparticles can travel through the circulatory, lymphatic, and nervous systems to many tissues and organs, including the brain (Buzea et al. 2007). Recently, application of nanoparticles into consumer products is growing but the knowledge about potential toxicity of nanoparticles is limited. Several studies revealed that toxicity profiles of nanomaterials are not stable, and have deviation, when compared to other chemical analogs (Donaldson et al. 2001; Oberdorster et al. 2005; Nel et al. 2006). The main points for risk assessment are validity and sensitivity of existing test systems (Balbus et al. 2007). The exposure routes are also important parameters needed to be focused. Test results are generally given by exposing a test organism to one or a few types and size of nanoparticles, but in reality animals are generally exposed to many nanoparticles at varied and even higher concentrations. These limit the utility of obtaining data for risk assessment. Extrapolation from one type of nanoparticles to another or from one size to another is not scientifically recommendable (Bouwmeester et al. 2009).

Due to the current differences in world legislation on nanotechnology especially regarding the safety tests, some products are available on the global market, before their safety claims are widely tested and/or approved. This situation makes it difficult to manage nanomaterials with regard to standardizing safety testing for an ever increasing number of nano-based products. There are specially no practical toxicology research and risk assessments for nanomaterials used in the food sector. The safety evaluation of new substances can start with an overview of physical and chemical properties to represent safe handling and storage of the material during industrial use and responsible, ethical disposal (Handy and Shaw 2007). Second step starts with toxicity tests (acute and chronic tests, oral toxicity, dermal toxicity, skin irritation as well as mutagenicity tests). Some materials present toxicity at nanoscale and not at the macroscale. Single-walled carbon nanotubes are examples, which inhibit human embryonic kidney cell proliferation and negatively impacted on cell growth and cell turnover (Cui et al. 2005; Cushen et al. 2012).

Toxicity effects of food ingredient derived to nanoscale (emulsion stability, nanoscale nutraceuticals and nanoencapsulation): Environmental effect and fates of engineered nanomaterials are not well known, especially their stability at nanoscale stage. To indicate the potential environmental effects of nanoscale materials, it is really important to define their species, sensitivity, stability to adverse effects of various other nanomaterials. Thus, characterization of nanomaterials, clarification of nanomaterials production steps (such as emulsion stability), interaction mechanisms with other nano and/or macro materials and stability of the environment are fundamental parameters, which need to be determined for every specific nanomaterial in a particular time period. In a research, zebrafish and algal species were used as models of various tropic levels and feeding strategies. In this research, organisms were exposed to silver, copper, aluminum, nickel, and cobalt and titanium dioxide nanoparticles. The results showed that nanosilver and nanocopper caused toxicity in all tested organisms, with 48 h median lethal concentrations as low as 40 and 60 g/L, respectively, in *Daphnia pulex* adults, whereas titanium dioxide did not cause toxicity. It can be concluded that toxicity of nanomaterial differs among species. Dissolution is also another important parameter in this research, which indicated minor toxicity of silver and copper, while the major toxicity of nickel. Nanoscale forms of these materials were less toxic than soluble forms based on mass added, but other dose metrics should be developed to accurately assess concentration–response relationships for nanoparticle exposures (Griffitt et al. 2008).

Toxicity effects on food packaging (nanocomposites, active and intelligent food contact materials, antimicrobial nanopackaging, time-temperature indicators, etc.): Main toxicity risks for application of nanomaterials in food packaging is migrating into the enclosed food. However, tests on chemical migration from food and beverage packaging can be used as a reference because they are reproducible, reliable and overall well established. Risk assessment studies for food packaging can be focused on migration tests at a basic level (Cushen et al. 2012).

The FDA modified the food additive regulations to allow the direct addition of silver nitrate as a disinfectant to commercially bottled water at concentrations not exceeding 17 lg/kg (FDA, US 2009). Conventional reason regards silver as safe to humans and other higher order organisms when used responsibly, and silver-based pharmaceuticals have few, if any acute or chronic known side-effects at FDA-permitted doses. Silver is considerably effective at penetrating biofilms, which has been a drawback for many molecular antimicrobials (Monteiro et al. 2009). So far, the FDA has approved over a dozen silver-containing zeolites or other substances for use as food contact materials for the purpose of disinfection, as well as numerous silver-coated medical devices (Duncan 2011). Recently it was demonstrated that, silver nanoparticles can cross the gastrointestinal barrier, be absorbed into the blood and cause adverse health effects. In researches, it is numerically detailed that, 56 nm silver nanoparticles can be found in rat blood, the liver, and other organs after 90 days of oral exposure at a dose of 30, 125, or 500 mg/kg bw/day. It was also reported a no-observed-adverse-effect level (NOAEL) at 30 mg/kg bw/day and a lowest-observed-adverse-effect level (LOAEL) at 125 mg/kg bw/day for silver nanoparticles (Higashisaka et al. 2015).

Many researchers focused on the in vivo and in vitro toxicity of titanium dioxide nanoparticles. It was found that oral administration of titanium dioxide nanoparticles (140 nm diameter) in water to rats at 175, 550, 1750 or 5000 mg/kg bw/day resulted in no mortality or no clear body weight losses. In addition, it was investigated that even after titanium dioxide nanoparticles were orally administered to rats for 13 weeks (7 days/week), absorption of the nanoparticles was extremely low, and even at the highest dose (1041.5 mg/kg body weight per day), the amounts of nanoparticles in the sampled organs were not significantly higher than the amounts in the control group (Higashisaka et al. 2015).

3.6 Gaps and Open Points for Nanomaterials' Structure

Scientific knowledge gaps of nanomaterials hinder the current risk assessment of nanoparticles' application in general, and in the food industry in particular. Gaps and open points on safety of nanoparticles mainly focused on the question 'what information is additionally required compared to dossier requirements for conventional chemicals'? Hence another question coming respectively 'are in vitro tests equally applicable for nanoparticles as for conventional chemicals', as the role of in vitro test results for chemical in risk assessment is also still subject to many uncertainties. However, a question coming from a regulatory point of view is 'what information is additionally required and if the current regulatory system within the EU is suited to deal with nanotechnology'. The nanotechnology approach of the EU is 'safe, integrated and responsible' (Communication from the Commission—Towards a European strategy for nanotechnology 2004). The EU has its scientific committees and commission services to perform a scientific and legislative review on the suitability of the existing regulation for nanotechnologies. It becomes clear that there is currently no nano-specific regulation in the EU (Chaudhry et al. 2007), or other countries (Hodge 2007), however there are numbers of regulatory reports such as; US Food and Drug Administration (FDA), Scientific Committee on Emerging and Newly-Identified Health Risks, the National Nanotoxiciology Research Centre, UK and the Organization for Economic Co-operation and Development (OECD).

The US Food and Drug Administration (FDA) was among the first government agencies around the world, who defined nanotechnologies, but the FDA has no nanotechnology specific regulations, as it regulates "products, not technologies". The same matter proceeds true for other important nanotechnology innovative worldwide regions like Japan and China (Chau et al. 2007). Despite the lack of a specific regulation of nanotechnology any new development in nanotechnology will at the moment not be taking place in a regulatory vacuum (Gergely 2010).

On the other hand, principally, the EU regulatory framework also comprises nanotechnologies as accomplished by the Scientific Committee on Emerging and Newly-Identified Health Risks (SCENIHR 2007a). Compatibly with the Health Council of the Netherlands considered that: "the best course of action would be to modify existing laws and rules as and when developments within the fields of Nanoscience and nanotechnologies render such measures necessary" (Health Council Netherlands 2006). Necessity of adjustments of legislation, guidelines and guidance documents is concerned for testing of nanoparticles (SCENIHR 2007a). However, there is still lack of knowledge on this topic, while there is a special need for a restrict definition for nanoparticles specifically from a regulatory perspective. The definition of nanoparticles as 'the order of 100 nm or less' is also arbitrary due to lack of knowledge on the relationship between particle size and kinetics or toxicological effects (SCENIHR 2007b). The size limits and interacting mechanisms of nanoparticles need to be known and identified clearly (Bouwmeester et al. 2009).

However, there are investigations on the health effects of inhaled nanotubes and the surface reactivity and free radical generating potential of nanomaterials are being carried out at the National Nanotoxiciology Research Centre, UK. An international approach to regulate the risks from nanomaterials through the Organization for Economic Co-operation and Development (OECD) has developed. This has drawn research work on occupational exposure undertaken by the Health and Safety laboratories as part of multinational programs in the G8 countries (Canada, France, Germany, Italy, Japan, Russia, UK, and USA). Nanotechnology related terminologies and validated measurement and characterization protocols in addressing the nanosafety issues to consider social and ethical concerns and demands are being undertaken by the British Standards Institution, International and European Committee for Standardization, and OECD (Neethirajan and Jayas 2011).

Potential risk for nonfood related nanomaterials: Nanomaterials can be toxic and cause harmful effects. Research has also revealed that the way of making nanomaterials and processing might produce products without harmful effects (Zhang et al. 2003): such as, toxicity of some substance (e.g. selenium) can be significantly reduced while the particle size was decreased to the nanoscale. It was also known that pure carbon nanotubes implemented to the trachea of mice can cause death; where, nitrogen doped carbon nanotubes reduced their toxicity and risk of death (Roach 2006; Chau et al. 2007).

Potential risk for food related nanomaterials: More investigations on the toxicity of nanoparticles contacting with food products should be performed and present to close the gaps and open points of nanoparticles usage on food and food products (Chau et al. 2007). In the long term, the increased and emerging use of nanomaterials, applications and products of nanotechnology can pose new indirect sources of food contamination with nanoparticles. This risk of exposure can arise from the use of nanoscale veterinary medicines and pesticides, contact of food with nanobased coatings during preparation or processing, or potential migration of nanoparticles from food packaging. Microemulsion based (e.g. Syngenta's herbicide, Flex, and Isagro's fungicide) and microencapsulation based (e.g. Syngenta's insecticide, Karate Zeon, and Dow Agroscience's insecticide, Empire) technologies are already known examples of pesticide formulations. Some agrochemical companies might be working on the development of nanoscale pesticides and veterinary medicines. Nanobased pesticide formulations have been clarified and exposed for safety in handling, controlled delivery, better dispersions in water and potential reduction in the use of active ingredients due to better administer into pests, plants and animals. However, it is still difficult to assess whether nanoparticles in the environment will bioaccumulate/bioconcentrate in the food chain (Evans 2002; Chaudhry et al. 2008).

Recently, an article published on March 27, 2017 the European Commission's (EC) Joint Research Centre (JRC) informs that its researchers have published a review of the legislation on nanomaterials in the EU. The study summarizes how nanomaterials are covered by the EU regulatory framework, listing the different EU regulations that include specific provisions for nanomaterials. Some regulations provide a definition of nanomaterials (e.g. REACH, biocidal products regulation, cosmetic products regulation), while others focus on information requirements and safety assessment (e.g. plastic food contact materials (FCM) regulation, active and intelligent FCM regulation) (Anonymous A 2017; Anonymous B 2017).

3.7 Benefits and Bottlenecks of Nanomaterials Usage

Nanomaterials often show beneficial properties, from UV absorbance in sunscreen to oil-less lubrication of motors. While there are some health risks, human exposure to toxic nanoparticles can be reduced through identifying creation-exposure pathways of toxicants (Buzea et al. 2007). Nanomaterials are used in cosmetics due to their transparent appearance and enhanced efficacy. They are also used in tennis rackets and baseball bats to improve their strength and to make them lighter. Nanomaterials are also used in textile industry to produce stain-, wrinkle- and water-resistant clothing (Thomas et al. 2006). Nanomaterials (such as silver and ZnO) are used in food packaging as well as daily life appliances like washing machines and water purifiers (Bouwmeester et al. 2009). The toxicity and safety aspects of nanomaterials are lagging far behind the rate at which they are being produced. However, nanomaterials are incompletely defined and cannot be controlled and identified entirely. All the questions and gaps are starting from that definition. For now, apart from our knowledge of nanomaterials and their toxic effects, we cannot directly tell, they are not safe to use or they are totally safe and trustable to use. Due to new technological developments and industrial applications, before using nanomaterials in any kind of field, all possibilities for the specific purpose need to be studied deeply, and alternatives need to be checked to eliminate possibility of negative environmental and health effects. Simultaneously, toxicological and ecotoxicological studies are required for every defined nanomaterials, and also regulations and standardizations need to be framed and implemented strictly. All these factors may rise to conflicting results and slow down the growth of this field (Dhawan and Sharma 2010).

3.8 Conclusion

In recent years, usages of nanomaterials in consumer product are becoming phenomenal but the knowledge of potential toxicity of nanomaterials is limited. Numerous studies have indicated that the toxicity profiles of nanomaterials are not stable, and have many deviations, while comparing with other chemical stages. The fundamental point for risk assessment is the sensitivity and validity of existing analysis and test methods, which means specific analysis and test methods are needed. Some of materials present toxicity at nanoscale, while others present toxicity at macro scales. That is why, food industrial applications of nanomaterials are becoming risky due to their adverse human health effects from direct contact and oral exposure, which warrants detailed study.

References

Anonymous A (2017). https://www.foodpackagingforum.org/news/regulation-of-nanomaterials-in-the-eu, 19 Sept 2018

Anonymous B (2017). https://ec.europa.eu/jrc/en/science-update/how-are-nanomaterials-regulated-eu, 19 Sept 2018

Anyaogu KC, Fedorov AV, Neckers DC (2008) Synthesis, characterization, and antifouling potential of functionalized copper nanoparticles. Langmuir 24(8):4340–4346

Bajpai SK, Chand N, Chaurasia V (2010) Investigation of water vapor permeability and antimicrobial property of zinc oxide nanoparticles-loaded chitosan-based edible film. J Appl Polym Sci 115:674–683

Balbus JM, Maynard AD, Colvin VL, Castranova V, Daston GP, Denison RA, Dreher KL, Goering PL, Goldberg AM, Kulinowski KM et al (2007) Meeting report: hazard assessment for nanoparticles—report from an interdisciplinary workshop. Environ Health Perspect 115(11):1654–1659

Bouwmeester H, Dekkers S, Noordam MY, Hagens WI, Bulder AS, de Heer C, ten Voorde SECG, Wijnhoven SWP, Marvin HJP, Sips AJAM (2009) Review of health safety aspects of nanotechnologies in food production. Regul Toxicol Pharmacol 53:52–62

Buzea C, Pacheco II, Robbie K (2007) Nanomaterials and nanoparticles: sources and toxicity. Biointerphases 2(4):17–71

Carbone M, Donia DT, Sabbatella G, Antiochia R (2016) Silver nanoparticles in polymeric matrices for fresh food packaging. J King Saud Univ Sci 28:273–279

Cárdenas G, Díaz J, Meléndrez M, Cruzat C, Cancino AG (2009) Colloidal Cu nanoparticles/chitosan composite film obtained by microwave heating for food package applications. Polym Bull 62:511

Chau C-F, Wu S-H, Yen G-C (2007) The development of regulations for food nanotechnology. Trends Food Sci Technol 18:269–280

Chaudhry Q, George C, Watkins R (2007) Nanotechnology regulation—developments in the United Kingdom. In: Hodge GA, Bowman DM, Ludlow K (eds) New global frontiers in regulation: the age of nanotechnology. Edward Elgar, Cheltenham, pp 212–238

Chaudhry Q, Scotter M, Blackburn J, Ross B, Boxall A, Castle L, Aitken R, Watkins R (2008) Applications and implications of nanotechnologies for the food sector. Food Addit Contam 25(3):241–258

Chaurasia V, Chand N, Bajpai SK (2010) Water sorption properties and antimicrobial action of zinc oxide nanoparticles-loaded cellulose acetate films. J Macromol Sci Part A Pure Appl Chem 47:309–317

Cioffi N, Torsi L, Ditaranto N, Tantillo G, Ghibelli L, Sabbatini L, Bleve-Zacheo T, D'Alessio M, Zambonin PG, Traversa E (2005) Copper nanoparticle/polymer composites with antifungal and bacteriostatic properties. Chem Mater 17(21):5255–5262

Communication from the Commission—Towards a European strategy for nanotechnology (2004) Luxembourg: Office for Official Publications of the European Communities 2004, p 24

Cui D, Tian F, Ozkan CS, Wang M, Gao H (2005) Effect of single wall carbon nanotubes on human HEK293 cells. Toxicol Lett 155(1):73–85

Cushen M, Kerry J, Morris M, Cruz-Romero M, Cummins E (2012) Nanotechnologies in the food industry recent developments, risks and regulation. Trends Food Sci Technol 24:30–46

Dhawan A, Sharma V (2010) Toxicity assessment of nanomaterials: methods and challenges. Anal Bioanal Chem 398:589–605

Donaldson K, Stone V, Clouter A, Renwick L, MacNee W (2001) Ultrafine particles. Occup Environ Med 58(3):211–216. 199

Droval G, Aranberri I, Bilbao A, German L, Verelst M, Dexpert-Ghys J (2008) Antimicrobial activity of nanocomposites: poly(amide) 6 and low density poly(ethylene) filled with zinc oxide e-Polymer:128

Duncan TV (2011) Applications of nanotechnology in food packaging and food safety: barrier materials, antimicrobials and sensors. J Colloid Interface Sci 363:1–24

Emamifar A, Kadivar M, Shahedi M, Soleimanian-Zad S (2010) Evaluation of nanocomposite packaging containing Ag and ZnO on shelf life of fresh orange juice. Innovative Food Sci Emerg Technol 11:742–748
Emamifar A, Kadivar M, Shahedi M, Soleimanian-Zad S (2011) Effect of nanocomposite packaging containing Ag and ZnO on inactivation of Lactobacillus plantarum in orange juice. Food Control 22(3–4):408–413
Esteban-Tejeda L, Malpartida F, Esteban-Cubillo A, Pecharromán C, Moya JS (2009) Antibacterial and antifungal activity of a soda-lime glass containing copper nanoparticles. Nanotechnology 20:50
Evans SM, Ashwood P, Warley A, Berisha F, Thompson RP, Powell JJ (2002) The role of dietary microparticles calcium in apoptosis interleukin-1beta release of intestinal macrophages. Gastroenterology 1235:1543–1553
Fu G, Vary PS, Lin C-T (2005) Anatase TiO_2 nanocomposites for antimicrobial coatings. J Phys Chem B 109:8889–8898
Galeano B, Korff E, Nicholson WL (2003) Inactivation of vegetative cells, but not spores, of Bacillus anthracis, B. cereus, and B. subtilis on stainless steel surfaces coated with an antimicrobial silver- and zinc-containing zeolite formulation. Appl Environ Microbiol 69:4329–4331
Gergely A, Chaudhry Q, Bowman DM (2010) Regulatory perspectives on nanotechnologies in food and food contact materials. In: Hodge GA, Bowman BM, Maynard AD (eds) International handbook on regulating technologies, Edward Elgar, Cheltenham, 321–341
Griffitt RJ, Luo J, Gao J, Bonzongo J-C, Barber DS (2008) Nanomaterials in the environment, effects of particle composition and species on toxicity of metallic nanomaterials in aquatic organisms. Environ Toxicol Chem 27(9):1972–1978
Gurr JR, Wang ASS, Chen CH, Jan KY (2005) Ultrafine titanium dioxide particles in the absence of photoactivation can induce oxidative damage to human bronchial epithelial cells. Toxicology 213:66–73
Hamal DB, Haggstrom JA, Marchin GL, Ikenberry MA, Hohn K, Klabunde KJ (2010) A multifunctional biocide/sporocide and photocatalyst based on titanium dioxide (TiO2) codoped with silver, carbon, and sulfur. Langmuir 26(4):2805–2810
Handy RD, Shaw BJ (2007) Toxic effects of nanoparticles and nanomaterials: implications for public health, risk assessment and the public perception of nanotechnology. Health Risk Soc 9:125–144
Health Council Netherlands (2006) Health significance of nanotechnologies. The Hague: Health Council of the Netherlands, Publication No. 2006/06
Higashisaka K, Yoshioka Y, Tsutsumi Y (2015) Applications and safety of nanomaterials used in the food industry, ©2015 Food Safety Commission, Cabinet Office, Government of Japan. https://doi.org/10.14252/foodsafetyfscj.2015005
Hodge G, Bowman D, Ludlow K (2007) New global frontiers in regulation: In: The age of nanotechnology. Edward Elgar, Cheltenham
Huang L, Li D-Q, Evans DG, Duan X (2005a) Preparation of highly dispersed MgO and its bactericidal properties. Eur Phys J D 34:321–323
Huang L, Li D-Q, Lin Y-J, Wie M, Evans DG, Duan X (2005b) Controllable preparation of Nano-MgO and investigation of its bactericidal properties. J Inorg Biochem 99(5):986–993
Inoue Y, Hoshino M, Takahashi H, Noguchi T, Murata T, Kanzaki Y, Hamashima H, Sasatsu M (2002) Bactericidal activity of Ag-zeolite mediated by reactive oxygen species under aerated conditions. J Inorg Biochem 92:37–42
Jin T, Sun D, Su JY, Zhang H, Sue H-J (2009) Antimicrobial efficacy of zinc oxide quantum dots against Listeria monocytogenes, Salmonella Enteritidis, and Escherichia coli O157:H7. J Food Sci 74:M46–M52
Jones CF, Grainger DW (2009) In vitro assessments of nanomaterial toxicity. Adv Drug Deliv Rev 61:438–456
Kang S, Pinault M, Pfefferle LD, Elimelech M (2007) Single-walled carbon nanotubes exhibit strong antimicrobial activity. Langmuir 23(17):8670–8673

Kang S, Mauter MS, Elimelech M (2009) Microbial cytotoxicity of carbon-based nanomaterials: implications for river water and wastewater effluent. Environ Sci Technol 43(7):2648–2653

Kangwansupamonkon W, Lauruengtana V, Surassmo S, Ruktanonchai U (2009) Antibacterial effect of apatite-coated titanium dioxide for textiles applications. Nanomed Nanotechnol Biol Med 5:240–249

Kim B, Kim D, Cho D, Cho S (2003) Bactericidal effect of TiO2 photocatalyst on selected food-borne pathogenic bacteria. Chemosphere 52(1):277–281

Kong H, Song J, Jang J (2010) Photocatalytic antibacterial capabilities of TiO_2—biocidal polymer nanocomposites synthesized by a surface-initiated photopolymerization. Environ Sci Technol 44(14):5672–5676

Kostarelos K, Bianco A, Prato M (2008) Hype around nanotubes creates unrealistic hopes. Nature 453(7193):280

Kubacka A, Cerrada ML, Serrano C, Fernández-García M, Ferrer M, Fernández-García M (2009) Plasmonic nanoparticle/polymer nanocomposites with enhanced photocatalytic antimicrobial properties. J Phys Chem C 113:9182–9190

Lee S (2009) Multifunctionality of layered fabric systems based on electrospun polyurethane/zinc oxide nanocomposite fibers. J Appl Polym Sci 114:3652–3658

Li X, Xing Y, Jiang Y, Ding Y, Li W (2009) Antimicrobial activities of ZnO powder-coated PVC film to inactivate food pathogens. Int J Food Sci Technol 44(11):2161–2168

Lin Y-J, Li D-Q, Wang G, Huang L, Duan X (2005) Preparation and bactericidal property of MgO nanoparticles on γ-Al_2O_3. J Mater Sci Mater Med 16(1):53–56

Liu Y, He L, Mustapha A, Li H, Hu ZQ, Lin M (2009) Antibacterial activities of zinc oxide nanoparticles against Escherichia coli O157:H7. J Appl Microbiol 107(4):1193–1201

Lu Z, Li CM, Bao H, Qiao Y, Toh Y, Yang X (2008) Mechanism of antimicrobial activity of CdTe quantum dots. Langmuir 24(10):5445–5452

Lu Z, Li CM, Baoi H, Qiao Y, Bao Q (2009) Photophysical mechanism for quantum dots-induced bacterial growth inhibition. J Nanosci Nanotechnol 9(5):3252–3255

Lu Y, Chen Y, Lin H, Wang C, Yang Z (2010) Preperation of chitosan nanoparticles and their application to Antheraea pernyi Silk. J Appl Polym Sci 117:3362–3369

Mary G, Bajpai SK, Chand N (2009) Copper (II) ions and copper nanoparticles-loaded chemically modified cotton cellulose fibers with fair antibacterial properties. J Appl Polym Sci 113:757–766

Monteiro DR, Gorup LF, Takamiya AS, Ruvollo-Filho AC, de Camargo ER, Barbosa DB (2009) The growing importance of materials that prevent microbial adhesion: antimicrobial effect of medical devices containing silver. Int J Antimicrob Agents 34(2):103–110

Morsy MK, Khalaf HH, Sharoba AM, El-Tanahi HH, Cutter CN (2014) Incorporation of essential oils and nanoparticles in pullulan films to control foodborne pathogens on meat and poultry products. J Food Sci 79:M675–M682

Motlagh NV, Mosavian MTH, Mortazavi SA (2012) Effect of polyethylene packaging modified with silver particles on the microbial, sensory and appearance of dried barberry. Packag Technol Sci 26:39–49

Neethirajan S, Jayas DS (2011) Nanotechnology for the food and bioprocessing industries. Food Bioprocess Technol 4:39–47

Nel A, Xia T, Mädler L, Li N (2006) Toxic potential of materials at the nanolevel. Science 311(5761):622–627

Oberdorster G, Maynard A, Donaldson K, Castranova V, Fitzpatrick J, Ausman K, Carter J, Karn B, Kreyling W, Lai D (2005) Principles for characterizing the potential human health effects from exposure to nanomaterials: elements of a screening strategy. Part Fibre Toxicol 2:8

Palza H, Gutiérrez S, Delgado K, Salazar O, Fuenzalida V, Avila JI, Figueroa G, Quijada R (2010) Toward tailor-made biocide materials based on poly(propylene)/copper nanoparticles. Macromol Rapid Commun 31(6):563–567

Perelshtein I, Applerot G, Perkas N, Wehrschuetz-Sigl E, Hasmann A, Guebitz G, Gedanken A (2009) CuO–cotton nanocomposite: formation, morphology, and antibacterial activity. Surf Coat Technol 204(1–2):54–57

Poland CA, Duffin R, Kinloch I, Maynard A, Wallace WAH, Seaton A, Stone V, Brown S, MacNee W, Donaldson K (2008) Carbon nanotubes introduced into the abdominal cavity of mice show asbestos-like pathogenicity in a pilot study. Nat Nanotechnol 3(7):423–428

Powers KW, Brown SC, Krishna VB, Wasdo SC, Moudgil BM, Roberts SM (2006) Research strategies for safety evaluation of nanomaterials. Part VI: Characterization of nanoscale particles for toxicological evaluation. Toxicol Sci 90(2):296–303

Powers KW, Moudgil BM, Roberts SM (2007) Characterization of the size, shape and state of dispersion of nanoparticles for toxicological studies. Nanotoxicology 1(1):42–51

Priester JH, Stoimenov PK, Mielke RE, Webb SM, Ehrhardt C, Zhang JP, Stucky GD, Holden PA (2009) Effects of soluble cadmium salts versus CdSe quantum dots on the growth of planktonic Pseudomonas aeruginosa. Environ Sci Technol 43(7):2589–2594

Qi L, Xu Z, Jiang X, Hu C, Zou X (2004) Preparation and antibacterial activity of chitosan nanoparticles. Carbohydr Res 339(16):2693–2700

Ren G, Hu DH, Cheng EWC, Vargas-Reus MA, Reip P, Allaker RP (2009) Characterization of copper oxide nanoparticles for antimicrobial applications. Int J Antimicrob Agents 33:587–590

Rispoli F, Angelov A, Badia D, Kumar A, Seal S, Shah V (2010) Understanding the toxicity of aggregated zero valent copper nanoparticles against Escherichia coli. J Hazard Mater 180:212

Roach S (2006) Nanotechnology passes first toxicity hurdle. http://www.foodproductiondaily-usa.com/news/ng.asp?id=69557

Robertson JMC, Robertson PKJ, Lawton LA (2005) A comparison of the effectiveness of TiO2 photocatalysis and UVA photolysis for the destruction of three pathogenic micro-organisms. J Photochem Photobiol, A 175(1):51–56

SCENIHR (2007a) Scientific Committee on Emerging and Newly Identified Health Risk SCENIHR. Opinion on: the appropriateness of the risk assessment methodology in accordance with the technical guidance documents for new and existing substances for assessing the risks of nanomaterials European Commission Health and Consumer Protection Directorate-General. Directorate C—Public Health and Risk Assessment C7—Risk Assessment

SCENIHR (2007b) Scientific Committee on Emerging and Newly Identified Health Risk SCENIHR. Opinion on: the scientific aspects of the existing and proposed definitions relating to products of nanoscience and nanotechnologies European Commission Health and Consumer Protection Directorate-General. Directorate C—Public Health and Risk Assessment C7—Risk Assessment

Sharifi S, Behzadi S, Laurent S, Forrest ML, Stroeve P, Mahmoudi M (2012) Toxicity of nanomaterials. Chem Soc Rev 41:2323–2343

Sharma C, Dhiman R, Rokana N, Panwar H (2017) Nanotechnology: an untapped resource for food packaging. Nanotechnol Food Packag 8(1735):1–22

Sevinç BA, Hanley L (2010) Antibacterial activity of dental composites containing zinc oxide nanoparticles. J Biomed Mater Res, Part B 94(1):22–31

Silvestre C, Duraccio D, Cimmino S (2011) Food packaging based on polymer nanomaterials. Prog Polym Sci 36:1766–1782

Singh T, Shukla S, Kumar P, Wahla V, Bajpai VK, Rather IA (2017) Application of nanotechnology in food science: perception and overview. Front Microbiol 8:1501

Shintani H, Kurosu S, Miki A, Hayashi F, Kato S (2006) Sterilization efficiency of the photocatalyst against environmental microorganisms in a health care facility. Biocontrol Sci 11(1):17–26

Stoimenov PK, Klinger RL, Marchin GL, Klabunde KJ (2002) Metal oxide nanoparticles as bactericidal agents. Langmuir 18(17):6679–6686

Takagi A, Hirose A, Nishimura T, Fukumori N, Ogata A, Ohashi N, Kitajima S, Kanno J (2008) Induction of mesothelioma in p53± mouse by intraperitoneal application of multi-wall carbon nanotube. J Toxicol Sci 33(1):105–116

Tam KH, Djurišic AB, Chan CMN, Xi YY, Tse CW, Leung YH, Chan WK, Leung FCC, Au DWT (2008) Antibacterial activity of ZnO nanorods prepared by a hydrothermal method. Thin Solid Films 516(18):6167–6174

The Royal Society and the Royal Academy of Engineering (2004) The Royal Society and the Royal Academy of Engineering. Nanoscience and nanotechnologies: opportunities and uncertainties. London, UK

Thomas T, Thomas K, Sadrieh N, Savage N, Adair P, Bronaugh R (2006) Research strategies for safety evaluation of nanomaterials, Part VII: Evaluating consumer exposure to nanoscale materials. Toxicol Sci 91(1):14–19

US Food and Drug Administration (2009) Food additives permitted for direct addition to food for human nutrition. Fed. Regist. 74:11476

Warheit DB (2008) How meaningful are the results of nanotoxicity studies in the absence of adequate material characterization? Toxicol Sci 101(2):183–185

Xie Y, He Y, Irwin PL, Jin T, Shi X (2011) Antibacterial activity and mechanism of action of zinc oxide nanoparticles against Campylobacter jejuni. Appl Environ Microbiol 77(7):2325–2331

Xing K, Chen XG, Kong M, Liu CS, Cha DS, Park HJ (2009) Effect of oleoyl-chitosan nanoparticles as a novel antibacterial dispersion system on viability, membrane permeability, and cell morphology of *Escherichia coli* and *Staphylococcus aureus*. Carbohydr Polym 76(1):17–22

Yang FM, Li HM, Li F, Xin ZH, Zhao LY, Zheng YH, Hu QH (2010) Effect of nano-packing on preservation quality of fresh strawberry (Fragaria ananassa Duch. Cv Fengxiang) during storage at 4 °C. J Food Chem 75:C236–C240

Yoon K-Y, Byeon JH, Park J-H, Hwang J (2007) Susceptibility constants of Escherichia coli and Bacillus subtilis to silver and copper nanoparticles. Sci Total Environ 373(2–3):572–575

Yuvaraj D, Kaushik R, Rao KN (2010) Optical, field-emission, and antimicrobial properties of ZnO nanostructured films deposited at room temperature by activated reactive evaporation. ACS Appl Mater Int 2(4):1019–1024

Zeng H, Gao C, Wang Y, Watts PCP, Kong H, Cui X (2006) In situ polymerization approach to multiwalled carbon nanotubes-reinforced nylon 1010 composites: mechanical properties and crystallization behavior. Polymer 47:113–122

Zhang Q, Kusaka Y, Zhu X, Sato K, Mo Y, Kluz T et al (2003) Comparative toxicity of standard nickel and ultrafine nickel in lung after intratracheal instillation. J Occup Health 45:23–30

Chapter 4
Toxicity and Safety Evaluation of Nanoclays

Adriano Brandelli

Abstract The term nanoclay is often associated to the natural clay minerals that contain silica and alumina as the main components. Nanoclays have been frequently incorporated into polymeric systems, resulting in nanocomposites that display a significant enhancement in barrier properties besides improved thermal and mechanical resistance. Recent studies describe the potential use of nanoclays as drug carriers and as modulators of drug release. Because of the increase in manufacturing of nanoclay-containing products, including an extensive array of industrial applications, information on the toxicological and health effects of nanoclay exposure is necessary. Methods conventionally used for evaluation of nanoparticles can be employed or adapted for toxicity evaluation of nanoclay materials. In vitro systems based on biochemical and cell culture methods and in vivo evaluation using lower organisms and animal models are used to access the potential toxicity of nanoclays. Most studies indicate that nanoclays are relatively safe and toxicity is often associated to nanoclay functionalized with organic modifiers.

Keywords Montmorillonite · Halloysite · Toxicity evaluation · Food packaging

4.1 Introduction

Nanoscience and nanotechnology are supposed to provide answers and the essential tools to improve the complex set of contemporary engineering and scientific challenges. Nanotechnology involves the manufacture, characterization and/or manipulation of structures with at least one dimension in the nanometer range, in most cases considered as less than 100 nm. The physical, chemical, and biological characteristics of nanometric systems and structures are significantly different as compared to their equivalents at micro or macro-scale. Thus, unique properties and unusual

A. Brandelli (✉)
Centro de Nanociência e Nanotecnologia, Universidade Federal do Rio Grande do Sul, Porto Alegre, Brazil
e-mail: abrand@ufrgs.br

M. Rai and J. K. Biswas (eds.), *Nanomaterials: Ecotoxicity, Safety, and Public Perception*, https://doi.org/10.1007/978-3-030-05144-0_4

functional applications, in most cases offering better performances, are often expected for nanomaterials (Duncan 2011; Neethirajan and Jayas 2011).

Clay minerals are very promising materials with a great diversity of applications owing their large availability, natural occurrence, low cost, and ecological appeal (Pavlidou and Papaspyrides 2008; Meira et al. 2014). Several nanoclays have been currently used in the industry to improve mechanical and barrier properties of packaging materials, as they produce an increased tortuosity of the diffusive route for a penetrant molecule, constraining them to travel a longer way to diffuse through the matrix (Azeredo 2013). Moreover, nanoclays can be used as carriers of several substances, including natural antimicrobials in active packaging (Meira et al. 2016). Montmorillonite (MMT) clays are among the most frequently used solid fillers for obtaining novel polymer composites, named nanocomposites. MMT is a type of nanoclay that originates from the clay fraction of the soil and has been incorporated into polymers to form nanocomposites with improved mechanical strength, barrier, and flammability properties for use in food packaging, automotive, and medical devices (Brandelli and Meira 2017).

The incorporation of clay minerals into polymers can result in non-intentional exposure to the consumer due to the possible migration from the packaging material. This can result in direct contact and possible skin absorption or even inhalation of the nanoparticle depending on the formulation. In the particular case of food packaging, the migration of nanoclays to the food product would result in ingestion of these nanomaterials by the consumer. Although nanoclays have been usually considered safe to humans (Ibarguren et al. 2014), several reports have already pointed on some toxic effects in different experimental models (Lordan et al. 2011; Baek et al. 2012; Houtman et al. 2014; Maisanaba et al. 2014).

Nanoclays have found increasingly uses in materials science. Currently, attempts have been developed to incorporate nanoclays into different materials and food packaging, and therefore systematic toxicological evaluation is needed. In this chapter, some properties of nanoclays and their use in packaging technology are presented, with a discussion on methods for safety evaluation and examples of toxicity assessment.

4.2 Nanoclay Chemistry and Structure

Mineral clays employed in the manufacture of nanocomposite materials are natural or synthetic layered aluminum phyllosilicates, often hydrated, containing neutral or negative charged layers. The building units of clays are comprised by tetrahedral and octahedral sheets where silicon is surrounded by four oxygen atoms, and octahedral sheets where a metal like aluminum is surrounded by eight oxygen atoms (Fig. 4.1). The tetrahedral (T) and octahedral (O) sheets can be superimposed to originate both T–O–T (2:1 T:O ratio) and T–O (1:1 ratio) layered structures, in which the quantity of each sheet makes a layer known as a platelet (Pavlidou and Papaspyrides 2008).

Montmorillonite (MMT) belongs to the smectite group, a member of the structural family of the 2:1 phyllosilicates. The general formula $M_x(Al_{4-x}Mg_x)Si_8O_{20}(OH)_4$ is used to represents MMT, where M is a monovalent cation and x is the degree of isomorphous substitution, ranging from 0.5 to 1.3 (Pavlidou and Papaspyrides 2008; Azeredo 2013). MMT layers are maintained together by relatively weak forces. Polar molecules, including water, can enter into the layers causing the lattice expansion (Unalan et al. 2014). These nanoclays are considered environmentally friendly, naturally abundant and inexpensive (Hatzigrigoriou and Papaspyrides 2011). Thus, MMTs are attracting substantial attention due to their strong adsorptive capacity, chemical inertia, high structural stability, strong capacity to form stable suspensions and ease with which their surfaces can be modified (Baek et al. 2012). In this way, surface modification of MMT nanoclays has been frequently used to attain a better compatibility of the clay mineral with the polymer, subsequently reach higher exfoliation degrees, and therefore better properties (Wagner et al. 2017a). Consequently, MMTs are one of the most extensively used natural clays, especially for the fabrication of polymer/clay nanocomposites.

Halloysite is an important member of the kaolin group of clay minerals and has recently gained attention. It is a 1:1 two-layered aluminum hydrated phyllosilicate with a composition of $Al_2Si_2O_5(OH)_4.2H_2O$ showing a predominantly hollow tubular structure in the nanometric range (Fig. 4.1b). These structures show as mean values an outer diameter of 40–70 nm, an inner diameter of 10–20 nm and a length of 500–1500 nm (Rawtani and Agrawal 2012). Cylindrical forms of halloysite are typically smaller in both diameter and length than prismatic forms, which present mean values of 60–200 nm diameter. Prismatic tubes are formed by continued growth of cylindrical forms, and large prismatic tubes often show hexagonal outlines on their surfaces (Hillier et al. 2016). This peculiar tubular shape originates to the frequently used term “halloysite nanotubes” (HNTs). Halloysite nanotubes are able to entrap

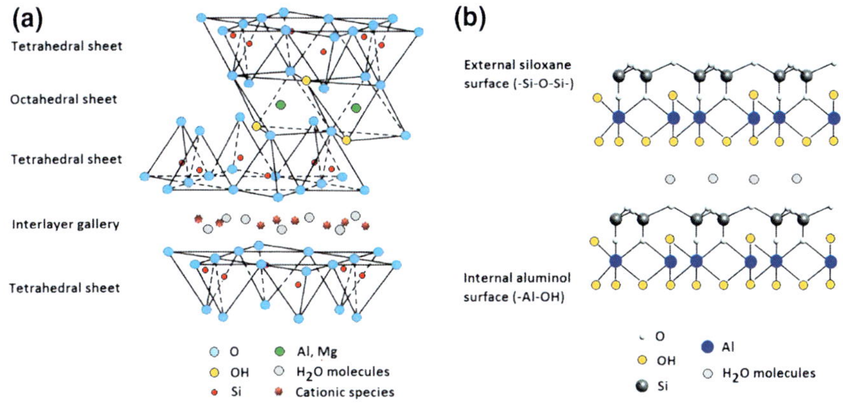

Fig. 4.1 Schematic representation of the structural features of montmorollonite clay (**a**) and halloysite nanotubes (**b**)

diverse molecules with specific sizes and can serve as a viable nano-cage for active molecules, owing to the empty space inside the nanotubes (Lvov and Abdullayev 2013). Like MMT nanosheets, halloysite nanotubes have been also incorporated into polymers to improve their functional properties (Meira et al. 2014; Unalan et al. 2014).

4.3 Nanoclays in Packaging Technology

The four basic functions of traditional packaging materials are to provide containment, protection, convenience and communications. However, the modern concept also include an adequate physicochemical condition for products that are essential for obtaining a satisfactory shelf-life and maintaining quality and safety during transportation and storage (Vanderroost et al. 2014). In addition to the basic mechanical, optical, and thermal properties of conventional packaging materials, food package should prevent microbial contamination, delay moisture loss or gain, avoid loss of nutrients and act as a barrier against gas, flavors and water vapor permeation (Marsh and Bugusu 2007; Rhim et al. 2013).

Metal, glass and paper (cardboard) are customary materials for packaging. Although these materials are still employed, plastics are constantly growing as packaging materials since they present many advantages such as low price, ease of processing and formability, low weight, and flexibility in thermal and mechanical properties (Arora and Padua 2010). Indeed, food industry appears as the largest plastics demanding application in the form of films, bottles, cups, sheets, trays, among other (Lagaron and Lopez-Rubio 2011; Rhim et al. 2013). Polymers normally used in this segment are petroleum-based materials such as polypropylene, polyethylene, polyethylene terephthalate (PET), polyvinyl chloride (PVC) and polystyrene (Marsh and Bugusu 2007). While packaging can help reduce organic waste by preserving foods, the increased utilization of synthetic plastics has also raised serious environmental concerns from a waste management point of view (Lagaron and Lopez-Rubio 2011). As a result, increased attention has be devoted to the development and use of biopolymer-based biodegradable materials (Brandelli et al. 2017a).

Unfortunately, biopolymers have limited industrial utilization because of their moderately weak mechanical and poor barrier properties for packaging applications. A particular challenge is the improvement of moisture barrier properties due to the inherent hydrophilic nature of most biopolymers (Arora and Padua 2010). The functional properties of these materials have been upgraded by reinforcement of the biopolymer matrix with nanosized inorganic or organic fillers, resulting in a new class of materials named bio-nanocomposites (Zhao et al. 2008). These materials exhibit improved performance as compared to the original biopolymers due to the high aspect ratio and high surface area of nanoparticles (Bordes et al. 2009).

Materials used in nanoreinforcement include clay and silicate nanoplatelets, silica nanoparticles, graphene, carbon nanotubes, starch nanocrystals, cellulose nanofibers or nanowhiskers, chitosan or chitin nanoparticles and other inorganics. Among these nanomaterials, only the layered inorganic solids like clays have attracted some attention by the packaging industry to date. This is essentially attributable to their large availability, low cost, relative simple processability, and substantial enhancements of mechanical and barrier properties (Brandelli and Meira 2017). Clay minerals are considered as safe food additives according to the US Food and Drug Administration and European Food Safety Authority (Ibarguren et al. 2014).

4.3.1 Nanoclay Reinforcement

A good interaction between polymer matrix and filler is strongly desired for reinforcing purposes. This is often one of the main challenges for developing novel nanocomposite materials. Considering that both phases are compatible and the filler is appropriately dispersed, the matrix-filler interactions can be significantly enhanced when decreasing the size of the reinforcing agent (Azeredo 2009). Three different types of nanoclay morphologies can be found in the polymer/clay nanocomposites (Fig. 4.2). The first is a phase-separated state that occurs as a microcomposite of layered silicate and the polymer. In this case, a poor physical attraction between them leads to frail mechanical properties. The second type corresponds to an intercalated state occurring in the form of a nanocomposite between the silicate layers and one or more polymeric chains. The third corresponds to an exfoliated nanocomposite made by extensive polymer penetration, with the nanoclay layers delaminated and randomly dispersed in the polymer matrix, resulting the best film properties due to the optimal interaction between polymer and clay (Alexandre and Dubois 2000; Pavlidou and Papaspyrides 2008; Azeredo 2009).

The superior barrier properties of polymer-clay nanocomposites appear to be associated to an increased tortuosity of the diffusive pathway for a penetrant molecule, forcing them to move a longer way to diffuse through the polymeric matrix. The increase in the distance is dependent on the properties of the clay and the volume fraction of the nanofiller in the composite (Azeredo 2009; Ranjan et al. 2014).

Montmorillonite layers were efficiently dispersed and distributed within protein-based films prepared with soy protein (Echeverría et al. 2014), chicken feather protein (Song et al. 2013) and sesame seed meal protein (Lee et al. 2014). The resulting films showed a significant strengthening leading to resistance to breakage, modulation of elasticity, and decrease in extension, moisture content, solubility, and permeability to water vapor. These protein-based nanocomposite films can be prepared with improved properties by the addition of nanoclays and could be employed as a packaging material in the food industry. Biodegradable starch/MMT nanocomposite films intended for food packaging were prepared via polymer melt processing techniques. The resulting film showed enhanced mechanical properties and a complete intercalation of the polymeric matrix into clay interlayer's. The migration tests performed with vegetables and stimulants showed that the dried films fulfilled the overall migration limit of 60 mg/kg required for plastics in European regulations (Avella et al. 2005).

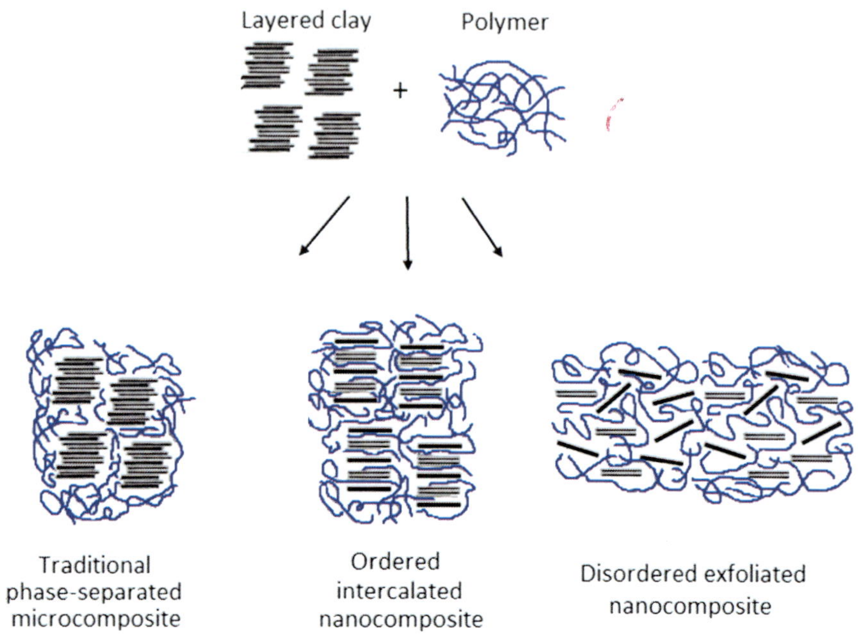

Fig. 4.2 Schematic representation of different morphologies of polymer-clay nanocomposites

Although the inclusion of halloysite nanotubes in natural polymeric films has been less reported, the resulting nanocomposites often display upgraded mechanical and barrier characteristics. Composite films prepared with potato starch (He et al. 2012) or wheat starch (Schimitt et al. 2012) showed improved thermal and tensile mechanical properties with addition of halloysite. Nanocomposite films prepared by introducing halloysite nanotubes into chitosan/poly vinyl alcohol matrix using the casting method showed increased tensile strength and elongation at break in comparison with the pristine film (Huang et al. 2012).

4.3.2 Commercial Nanoclay-Based Composites

Some food packaging materials incorporating nanoclays are currently commercialized. These materials are essentially based on the combination of nanoclays with synthetic polymers. Imperm® (Nanocor Inc.) is prepared by dispersing nanoclays into Nylon MXD6 as base resin. This material is used in multi-layer PET bottles used to ship beer, fruit juice, and soft drinks maintaining freshness and extending shelf-life by minimizing the loss of CO_2 from the beverage and the entrancement of O_2 into the bottle. An example of nanocomposite film prepared with high barrier nylon for similar purpose is Aegis™ OX (Honeywell Inc.). The hybrid plastic film

Durethan® (Bayer Polymers) is another example, comprising polyamide and layered silicate particles with increased barrier properties, enhanced brightness and rigidity. This material is intended for applications where conventional polyamide is more permeable and expensive, for example, paperboard juice vessels (Ranjan et al. 2014). A cooperative development among the U.S. military, NASA and Triton Systems Inc., resulted in a long-shelf-life packaging intended for non-refrigerated food based on nanoclays as barrier enhancers for ethylene vinyl alcohol (EVOH). This collaboration led to an experimental thermoformed food tray from EVOH loaded with 3% of Southern Clay's Cloisite® in a layer sandwiched between two PP layers (Unalan et al. 2014).

Some biodegradable nanocomposites have been also developed. For example, Nanolok™ (InMat Inc.) is a water based environmentally friendly nanocomposite coating with high oxygen barrier intended for transparent packaging applications. Likewise, a water-based coating made of PVA and nanoclay vermiculite is commercialized under the trade name Nano-Seal™ (NanoPack Inc.). This film is specifically proposed for food packaging applications where protection of food against oxygen and taints is required. This material can be industrially composted, and the manufacturer guarantee that the film deteriorates and reverts to carbon dioxide and water in less than 90 days. Plantic® biodegradable packaging (Plantic Technologies Ltd.) is a biodegradable and completely compostable bioplastics packaging, prepared from organic corn starch using nanotechnology (Neethirajan and Jayas 2011). These biopolymer-based nanocomposites are supplied to 80% of the Australian chocolate tray market.

4.4 Safety Evaluation of Nanomaterials

Nanotoxicology is a specific scientific subject that studies the interactions between nanomaterials and biological systems. The properties of nanoscale-range materials are different as compared with the same material on bulk state (Dhawan et al. 2011). Considering that a substance can be beneficial or harmful to a living organism, it is essential to understand its behavior in the biological environment and its mechanism of action. Toxicological evaluation is important to avoid possible adverse effects of nanomaterials on humans and the environment. Toxicological experiments often assess the features of health hazard, in vitro and in vivo dose-responses, the exposure time of the organism to the nanoparticle, and risk identification (Alves et al. 2014).

A new formulation or process involving nanomaterials must be safe to be approved for use. One of the utmost challenges in the nanotoxicology field is to warrant that engineered nanomaterials are not harmful to human health and the environment. Nanomaterials can be complex, and therefore it is extremely important to characterize their physicochemical properties, such as particle size and shape, surface area, dispersion, solubility, composition, purity and surface charges (Fröhlich 2012). In general, nanomaterials are firstly tested in vitro, using bacteria or cell cultures, which provides information about their interaction with the biological environment. In vivo assessment should involves a dose-response test, to determine the antagonistic effect caused in a population, in a certain period and amount of sample. Subsequently, the exposure assessment is determined, considering the mode the nanoparticle may contact de organism (oral, dermal or inhalation), its concentration and the exposure time. The process is concluded with the risk classification, where the probability of health risk in the organism/environment is calculated based on in vitro, in vivo and exposure assessment data (Oberdörster et al. 2005).

Several methods have been proposed to assess the potential toxicological effects of nanomaterials and a combination of them is frequently adopted since there is no perfect procedure to predict the behavior of these materials in a human population and the environment. Nanomaterials can entry into the environment through the release of materials originated from the recycling or deterioration of complex nanocomposites. The environmental consequences are frequently overlooked and are still poorly understood (Turco et al. 2011). The study on the effect of polymeric nanocomposites and metallic nanoparticles on the structural diversity of soil microbiota revealed that polymers, TiO_2 and Au nanoparticles affected significantly the microbial diversity (Nogueira et al. 2012).

In vivo animal studies traditionally used for toxicity assessments have been substituted by in vitro assays, in vivo tests with lower organisms and in silico computational modeling (Table 4.1). Information about integrated biological effects of nanoparticles can be achieved by in vivo studies, including the identification of the accumulation site and toxicological profile within a specific organ. After the identification of the tissue affected by nanoparticle, the research on the toxicity mechanism should be performed by in vitro studies. Despite in vivo models present some limitations such as time-consuming, high costs, and fraught ethical issues, they enable toxicokinetic studies, chronic evaluations and the analysis of transport mechanisms. Although in vitro models are faster, less expensive, without ethical issues, they do not allow all the possible outcomes reached in vivo (Garcia et al. 2014; Brandelli et al. 2017b).

In vitro testing of nanoparticles is conducted within a limited space, whereas in vivo the nanoparticles seem to be constantly transported from organ to organ. Consequently, the differential toxicity between in vivo and in vitro assays may be at least associated to exposure concentration and time. Many questions about in vivo behavior of nanoparticles remain unanswered, such as how they are metabolized, what is their long term destiny, how their particular physicochemical properties affect in vivo distribution, and the toxicity of derivatives.

The lack of correlation between in vivo and in vitro effects has been reported, suggesting that in vitro experiments should be validated in vivo (Sayes et al. 2007). A critical point of in vitro studies is checking for interferences in analytical methods, such as spectroscopy techniques or fluorescence quenching by nanoparticles (Liu et al. 2012). Thus, depending on the nanomaterial, in vitro interferences may cause inaccurate responses for cell viability, inflammation and oxidative stress (Kroll et al. 2011). In toxicity evaluation, the primary goal of in vitro models is to predict the in vivo toxicity. However, it is relevant to consider how the particle works in the cellular environment containing multiple biomolecules, since the media composition may have influence on the cytotoxicity of nanomaterials. Because their large surface

Table 4.1 Toxicological evaluation of nanomaterials. Different in vivo, in vitro and in silico methods can be applied to assess the toxicology of nanomaterials

System	Remarks
In vitro	
Cell viability assays	Provides information on cell death under controlled reproducible conditions
DNA damage assays	Assess the mechanisms underlying DNA damage and repair
Oxidative stress	Good correlation with damage to cellular macromolecules
In vivo	
Higher animals	Ethical issues, expensive, time-consuming
C. elegans	Provides data from a whole animal with intact and metabolically active systems
G. mellonela	Many larvae can be used in each experiment making pharmacological data easy to obtain
D. rerio	Zebrafish develop most of the major organ systems present in mammals
In silico	
Transcriptome analysis	Provides a global view of gene expression under exposure to different stressors
Metabolomics	Provides detailed information on the biochemical status of a biological system, and about the changes caused by chemicals
Proteomics	Identification of toxicity biomarkers and mechanisms

area, nanoparticles could absorb essential nutrients in cell culture medium, making it difficult to conclude from some cytotoxicity results (Guo et al. 2008). In this regard, the absorption of peptides and fatty acids by MMT clays at physiological pH and temperature conditions has been reported (Kalra et al. 2003; Lu et al. 2010). Another challenge is the standardization of nanoparticle concentration. Although it is normally expressed and mass to volume, it is difficult to compare because of variations in size, shape and surface chemistry. As an alternative, the number of particles per millimeter can be used.

4.4.1 *In Vitro Testing*

After nanoparticle characterization, the initial step for a toxicological study often involves a cell viability assay, also called cytotoxicity assay. Cytotoxicity can be defined as the antagonistic effects resulting from reactions with structures or processes that are essential for maintenance of cell proliferation, survival, and normal biochemistry and physiology. These effects may involve undesirable changes in the cellular metabolism, degradation or release of cellular constituents, or damage to the integrity of membranes and the cytoskeleton.

Assays with cell cultures are often useful because they are performed under a controlled physicochemical environment (pH, temperature, O_2 and CO_2 tension, osmotic pressure) and physiological conditions (defined concentrations of hormones and nutrients, cell line homogeneity, assays replicable). Diverse characteristics of the nanoparticles such as morphology, size and surface charge have great influence on their cytotoxicity, and some parameters like hydrophobicity and porosity induce the production of reactive oxygen species and determine binding sites for cell receptors contributing to the toxic effects (Fröhlich 2012). Alterations in cell morphology are easy to observe, and can be considered as a toxicity marker. However, the extrapolation of cytotoxicity results obtained with cell cultures to the real toxicity to an organism is a rather controversial, since the cell capacity to differentiate inside a population can result in variability from one passage to the next. In addition, parameters like cell type, cell density, nanoparticle concentration and medium composition may interfere in the cytotoxicity. Some studies demonstrated that the classic cytotoxicity assays may not be adequate to investigate nanomaterials toxicity since they can interfere with reagents inducing false-positive or false-negative results, such as inducing changes in redox processes and/or changing membrane permeability, which are essential properties to determine cell viability using MTT reduction and neutral red uptake tests (Kroll et al. 2011).

The MTT assay measure cell proliferation and cytotoxicity based on mitochondrial activity. The yellow water-soluble tetrazolium salt is reduced to an aqueous insoluble purple product by living cells. The assay is relatively rapid, sensitive and inexpensive and the use of microplate reader allows the handling of great number of samples. The results are sensitive to environmental conditions and the assay depends on cell metabolism for the generation of formazan. The background is usually variable due to protein precipitation on adding organic solvent to dissolve the blue formazan product. Nanoparticles can interfere in the results of MTT, thus testing cytotoxicity in 2D cell culture is not an accurate method to evaluate real toxicity of nanoparticles (Hernández-Ortiz et al. 2012).

The neutral red assay measures cell viability and cytotoxicity, determining the accumulation of neutral red dye in lysosomes of viable cells. Thus, the loss of viable cells is correlated with a reduction in the absorbance. The test is inexpensive and more sensitive than other cytotoxicity tests, but prolonged exposure of the cells to the fixative can result in discharge of the dye into the solution. Precipitation of neutral red medium and compounds with maximum absorbance wavelength near 540 nm may interfere the results (Stern et al. 2012).

The lactate dehydrogenase (LDH) assay is based on the measurement of LDH activity in the extracellular medium. LDH converts lactate is converted into pyruvate, with reduction of NAD^+ to NADH, resulting in a change of absorbance at 340 nm. Thus, the LDH released into the media acts like a marker of cell membrane integrity. The method is fast and simple, a nondestructive assay used for short-term exposure studies, but it is not reliable at lower compound concentrations.

Conventional methods used to assess DNA damage are the comet assay and the micronucleus test. The comet assay is a simple, sensitive and rapid method for measuring single and double DNA strand breaks in each cell. This method detects DNA disruptions, but does not permit inspection of any repair process. These disruptions may result in broken chromosomes, which can be assessed by the micronucleus test. The micronucleus test can be performed in vivo or in vitro and allows the observation of disruptions that may not be fixed during cell division. The method is fast and simple but validation requires analysis of a large number of cells.

Extensive oxidative stress has been proposed as an important molecular mechanism of nanotoxicity (Shvedova et al. 2008). The association between DNA damage and oxidative stress has been largely recognized and reactive oxygen species are highly reactive molecules that can disturb the intracellular homeostasis by attacking cellular macromolecules, such as lipids, proteins and DNA (Møller et al. 2012). The most popular assay for reactive oxygen species (ROS) involves the use of Dichlorofluorescin diacetate (DCFDA) in vivo and in vitro. This method involves the action of endogenous esterases that deacetylates DCFDA to dichorofluorescein (DCFH), which later reacts with reactive oxygen species to form the fluorophore DCF.

4.4.2 Animal Models

The use of animals, in particular mammals, in scientific research has been the subject of many ethical debate, because of the number of required individuals and the suffering caused during experiments. In addition, the high cost of maintaining laboratory animals is another concern in animal experimentation. In this context, the scientific community has devoted efforts on development of methods or alternative models to reconsider the use of mammalian models, through a program known as the 3Rs (Reduction, Refinement, Replacement). The 3Rs program aims to reduce the number of animals and their suffering, and develop new methods to replace animal experimentation (Cazarin et al. 2004). Thus, lower animals are being used as models for nanotoxicity evaluation, including the nematode *Caenorhabditis elegans*, the moth *Galleria mellonella,* and the zebrafish *Danio rerio*.

In this regard, *C. elegans*, one of the best characterized animals at the genetic, physiological, molecular and developmental levels, has been successfully used in toxicological evaluations of different substances (Zhang et al. 2013; Charão et al. 2015). Zebrafish-based assays are gaining interest for toxicity evaluation, since they allow the screening of a large number of chemicals for many different and significant toxic endpoints (Rubinstein 2006). Zebrafish embryos develop most of the key organ systems existing in mammals, such as the nervous, digestive and cardiovascular systems, in less than 1 week. Additionally, the use of antisense technology for gene function analysis is now routine procedure (Raldúa and Piña 2014). Therefore, it is relatively easy to check whether compounds or gene knockdowns cause toxic effects in zebrafish.

4.4.3 In Silico Evaluation

Traditional in vivo and in vitro methods normally provide restricted mechanistic information while novel high-throughput screening methodologies could reveal the interactions of chemicals with biochemical pathways that control cell function, communication and adaptation to environmental changes. Isolated in vitro tests or in silico approaches will not be capable to reveal the complex responses of the whole organism and its imbalance by toxic substances. Multi-omics research, comprising data integration and modeling, using both in vivo and cell culture-based assays, has gaining attention as a feasible approach in the development of a systematic view of chemically induced biological activities (Brandelli et al. 2017b).

Transcriptomics studies global mRNA expression, based on the fact that gene expression can rapidly change in response to chemical exposure providing a sensitive marker for toxicity. Advances in microarray technology include the capability to survey the entire transcriptome, providing a global evaluation of cellular changes. Measures at different dose and time are performed for toxicology applications. In general, the changes observed in transcriptomic studies are complemented by proteomics and metabolomics approaches for validation. The use of proteomics in toxicological studies has been dedicated to the identification of biomarkers and the understanding of molecular mechanisms of toxicity (Kennedy 2002). The sensitivity of proteomics methods for identification of single proteins is lower than immunological methods. However, proteomic approaches permit the detection of changes in unexpected and unknown proteins and often allows subsequent identification with mass spectrometry. Typically, only water-soluble proteins within a specified mass and isoelectric point range will be analyzed. Metabolomics emphases on high-throughput identification and characterization of metabolites and small molecules in biological matrices (German et al. 2005). Metabolomics can be considered a more friendly procedure than proteomics, particularly in deciphering results to phenotypic alterations, since the metabolic profiling of a biological system can be investigated under different exposure condition. The key technologies for metabolomics studies are NMR spectroscopy and mass spectrometry.

Recently, the possibility of analyzing the effects of nanomaterials on a large number of genes has led to toxicogenomic studies in nanotechnology. Microarrays and serial analysis of gene expression (SAGE) analysis are some of the tests utilized. The microarray technology is a hybridization-based process, which allows measuring mRNA concentration based on the intensity of fluorescence. Specific sequences are selected and immobilized in a DNA chip. Then, gene expression is detected by hybridization with complementary DNA (cDNA) synthetized by reverse transcription of mRNA obtained from biological samples and labeled with fluorophores. SAGE is a cDNA sequence-based test, where eukaryotic mRNA is isolated by a poly-thymidine oligonucleotide column and cDNA is synthesized by reverse transcription (Saha et al. 2002). The sequence analysis allows quantification of each specific label, which delineates the identification and abundance of the transcripts. Microarray and SAGE technologies can be useful to study the effects of different stressors, such as temperature, pH, nutrient deprivation, oxidation and metals on genes associated with oxidative stress, apoptosis, cell development, metabolic process, DNA repair, among other.

4.5 Toxicological Evaluation of Nanoclays

The toxicological evaluation of nanoclays has been performed in different cell lines and animal models. Although increasing research has been devoted to this subject, it is a point of conflicting description since some studies indicate the absence of toxicity while other reports describe the toxic effects of nanoclays. In vitro studies often indicate that nanoclays induce cytotoxicity, depending on the clay, concentration, experimental method. Different primary mechanisms such as necrosis, apoptosis, oxidative stress or genotoxicity have been implied in the observed effects. On contrary, most in vivo trials performed in rodents revealed no clear indications of systemic toxicity at doses as high as 5 g/kg (Maisanaba et al. 2015a). Some examples of studies on the toxicity evaluation of nanoclays are summarized in Table 4.2.

The cytotoxic effects of non-modified nanoclay Cloisite $Na^{+®}$ and the organoclay Cloisite 93A® in human hepatoma HepG2 cells was investigated. A significant

Table 4.2 Toxicity evaluation of nanoclays

Nanoclay	Evaluation system	Major outcome	Reference
Montmorilloninte, halloysite, kaolin, bentonite	*P. caudatum*	Very low or no toxicity	Kryuchkova et al. (2016)
Cloisite Na^+ and Cloisite 30B	Lung epithelial cells	Toxic effects with decrease in cell viability	Wagner et al. (2017a, b)
Cloisite Na^+ and Cloisite 93A	HepG2 cells	Significant cell death	Lordan et al. (2011)
Nanoclay	Mice	Potential hepato and nephrotoxicity	Isoda et al. (2017)
Cloisite Na^+ and Cloisite 30B	HepG2 cells	Genotoxic effects and morphological alterations by Cloisite 30B	Maisanaba et al. (2013)
Cloisite Na^+ and Cloisite 30B	Caco-2 cells	Oxidative stress and morphological alterations by Cloisite 30B	Maisanaba et al. (2014)
Halloysite	HepG2 and HCT116 cells	General safe at practical doses	Ahmed et al. (2015)
Halloysite	*C. elegans*	Absence of toxic effects	Fakhrullin et al. (2015)

decrease in cell viability was observed after 24 h exposure to nanoclays. Intracellular reactive oxygen species (ROS) formation, coinciding with increased membrane damage, was induced by Cloisite Na^+ but not in Cloisite 93A-induced cell death. The nanoclays aggregated differently in the cell culture medium and this appeared to have an effect on their mechanisms of toxicity (Lordan et al. 2011). The same cell line was used to investigate the cytotoxicity of Cloisite Na^+ and the organically modified clay Cloisite 30B®. Only Cloisite 30B induced toxic effects, and the study on mechanisms underlying cytotoxicity showed genotoxic effects and glutathione decrease. The most relevant ultrastructural changes were mitochondrial deterioration, dilated endomembrane systems, formation of heterophagosomes, and presence of fat droplets and nuclear lipid inclusions (Maisanaba et al. 2013). A similar study using the intestinal cell line Caco-2 also showed that only Cloisite 30B caused cytotoxicity (Maisanaba et al. 2014). Results showed increased intracellular ROS production and glutathione content, suggesting that oxidative stress may be associated with toxicity. DNA damage was not observed and morphological alterations included dilated cisternae edge in Golgi apparatus and nucleolar segregation, which suggest deficiency in secretory functions.

The potential of pristine Cloisite Na^+ and Cloisite 30B and their thermally degraded byproducts to induce toxicity in model human lung epithelial cells was investigated. The experimental set-up mimicked biological exposure in manufacturing and disposal areas and employed cellular treatments with occupationally relevant doses of nanoclays. The analyses of exposed cellular systems confirmed that nanoclay induced significant toxic effects, with Cloisite 30B showing time-dependent decreases in live cell count and cellular viability relative to control and pristine nanoclay, respectively. Byproducts produced less toxic effects; all treatments caused alterations in the cell morphology upon exposure (Wagner et al. 2017a). In another study, Maisanaba et al. (2015b) evaluated the toxicity of four MMT-based nanoclays: the unmodified clay Cloisite Na^+, the commercial clay Cloisite 30B and two novel experimental modified organoclays named Clay 1 and Clay 2. The cytotoxic effects were assessed in the Human Umbilical Vein Endothelial Cells (HUVEC) and the potential mutagenicity was evaluated by the Ames test. Cloisite Na^+ showed no cytotoxicity neither mutagenicity, while Cloisite 30B showed positive results in both assays. Clay 1 did not caused cytotoxic effects, but exhibited potential mutagenicity. In contrast, Clay 2 induced cytotoxicity in endothelial cells, however no mutagenicity was observed. Based on this study, the authors suggest that clay minerals have a different toxicity profile and evaluation should be performed case by case (Maisanaba et al. 2015b). Houtman et al. (2014) also investigated the potential cytotoxic effect of commercial organoclay Cloisite® 20B in comparison with experimental Clay 1 and Clay 2. The results showed that Clay 2 induced cytotoxicity in both HepG2 and Caco-2 cell lineages, where DNA damage was induced as determined by comet assay. Production of intracellular ROS and leakage of interleukin were not detected, while glutathione content was decreased in HepG2. Toxicity tests of nanoclay-embedded low density polyethylene composites, including reactive oxygen species generation and cell viability were performed on A594 adenocarcinomic human alveolar basal epithelial cells (Han et al. 2018). Essentially, the released nanoclays showed no toxicity, which can help the understanding of risk based-formulations of exposure.

The cytotoxicity evaluation of a series of pristine and organo-modified nanoclays in different cell lines revealed that IC_{50} values for cell viability ranged from 1.4 to 47 μg/mL for the six organoclays tested and above 100 μg/mL for the pristine nanoclays. The amount and structure of the quaternary ammonium compound used as surface modifier determined the IC_{50} values of the organoclays. Despite pristine nanoclays showed lower toxicity, they induced apoptosis and were found in cytoplasmic vesicles of treated cells. Organoclays were also found in cytoplasmic vesicles, but larger agglomerates were observed (Janer et al. 2014). The systematic analysis of one pristine and three organoclays to induce deleterious effect in primary and immortalized human lung epithelial cell lines was developed by Wagner et al. (2017b). The study was conducted with both as-received and thermally degraded clays. Overall, the cells exposed to organoclays showed reduced viability and elongated cell morphologies in comparison to pristine nanoclay. In addition, thermal degradation caused loss of organic modifiers leading to changes in toxicity to the cells. In a recent study, Stueckle et al. (2018) investigated the toxicity of pre- and post-incinerated organoclay in the mice model. The results showed that pulmonary toxicity and inflammatory potential relies on presence of coating and incineration status in that coated and incinerated nanoclay exhibited less inflammation and granuloma formation than pristine montmorillonite. High doses of both pre- and post-incinerated organoclay may cause potential pulmonary health hazards over long-term occupational exposures.

Although most toxicity studies on nanoclays have been developed with MMTs, the potential cytotoxic effects of halloysite nanotubes has investigated as well. A monolayer co-culture of Caco-2 and HT29-MTX cells was exposed to halloysite nanotubes as an in vitro model of the large intestine (Lai et al. 2013). Despite of elevated pro-inflammatory cytokine release, the absence of cytotoxic effect suggest a high degree of biocompatibility. Proteomic analysis of the differentially expressed protein profiles suggest that halloysite stimulates cellular processes associated to cell growth and proliferation, responses to cell infection, irritation and injury, and improved antioxidant ability (Lai et al. 2013). The in vivo toxicity of halloysite nanotubes was investigated using the nematode *Caenorhabditis elegans* as a model organism. *C. elegans* is one of the first organisms that may encounter these nanoclays in a polluted soil. Halloysite nanotubes were found only in the alimentary system and caused no severe toxic effects on nematodes. Halloysite was harmless for *C. elegans* at up to 1 mg/mL, which is about 1000 times higher than the possible soil contamination concentrations (Fakhrullin et al. 2015). The in vitro cytotoxicity of halloysite was also evaluated against HCT116 (colorectal carcinoma) and HepG2 (hepatocellular carcinoma) cells, which denote the primary entry point and the first accumulating organ, respectively, for nanoparticles in route to systemic circulation after oral delivery (Ahmed et al. 2015). In addition, halloysite was tested against human peripheral blood lymphocytes. Collectively, the results suggested that halloysite nanotubes are generally safe at practical concentrations of excipients for oral dosage forms.

The in vivo effects and drug interactions of nanoclays have been poorly investigated. Nanoclays can absorb diverse molecules and the resulting effects on the organisms is not well understood. In mice, administration of nanoclay particles via the tail vein led to acute liver injury (Isoda et al. 2017). Co-administration of nanoclay and carbon tetrachloride, paraquat, or cisplatin resulted in both liver and kidney injury, suggesting that nanoclay particles can potentially induce hepato- and nephrotoxicity. MMT and halloysite nanoclays can absorb the antimicrobial peptides nisin and pediocin, without losing their antimicrobial activities (Meira et al. 2015).

4.6 Conclusions

The toxicity evaluation of nanoparticles is a topic of growing interest as they are currently included in different applications. Information about toxicity of nanoclays is still limited in relation to their specific interaction with different organisms, food, and the environment. The use of nanoclays for development of nanocomposites raises concern about possible migration to food matrices and potential changes of these nanoparticles in the food or digestive tract play an essential role in the assessment of risk to the consumer. The main questions are related to their biological reactivity, ability to cross biological barriers and accumulation in target organs, their biopersistence, and how they induce dose-response biological effects. However, due to the complexity of the biological systems, it is important that such characterizations are performed under the conditions in which they could be absorbed. Therefore, it seems necessary to perform additional toxicological studies, in order to evaluate the real possibility of application of nanoclays without causing adverse effects to the consumer. The combination of both in vitro and in vivo models should be used to achieve more reliable results. Despite the increasing number of studies available, there is also a need of systematic in vitro-in vivo extrapolation studies, since detailed data on toxicity biomarkers linked to immunomodulatory effects or alteration of the genetic expression are still scarce. In conclusion, a case by case toxicological evaluation is required considering that different clays have their own toxicological profiles, their modification can change this profile, and the probable increase of the human and environmental exposure to clay minerals due to their novel applications.

References

Ahmed FR, Shoaib MH, Azhar M, Um SH, Yousuf RI, Hashmi S, Dar A (2015) *In vitro* assessment of cytotoxicity of halloysite nanotubes against HepG2, HCT116 and human peripheral blood lymphocytes. Colloid Surf B 135:50–55

Alexandre M, Dubois P (2000) Polymer-layered silicate nanocomposites: preparation, properties and uses of a new class of materials. Mater Sci Eng 28:1–63

Alves OL, Moraes ACM, Simões MB, Fonseca LC, Nascimento RO, Holtz RD, Faria AF (2014) Nanomaterials. In: Durán N, Guterres SS, Alves OL (eds) Nanotoxicology—materials, methodologies and assessments. Springer Science+Business Media, New York, pp 1–29
Arora A, Padua GW (2010) Review: nanocomposites in food packaging. J Food Sci 75:R43–R49
Avella M, de Vlieger JJ, Errico ME, Fischer S, Vacca P, Volpe MV (2005) Biodegradable starch/clay nanocomposite films for food packaging applications. Food Chem 93:467–474
Azeredo HMC (2009) Nanocomposites for food packaging applications. Food Res Int 42:1240–1253
Azeredo HMC (2013) Antimicrobial nanostructures in food packaging. Trends Food Sci Technol 30:56–69
Baek M, Lee AJ, Choi SJ (2012) Toxicological effects of a cationic clay, montmorillonite *in vitro* and *in vivo*. Mol Cell Toxicol 8:95–101
Bordes P, Pollet E, Avérous L (2009) Nano-biocomposites: biodegradable polyester/nanoclay systems. Progr Polym Sci 20:125–155
Brandelli A, Meira SMM (2017) Food packaging applications of renewable polymers incorporating nanocomposites. In: Inamuddin (ed) Green polymer composites technology. CRC Press, Boca Raton, pp 149–162
Brandelli A, Brum LFW, Santos JHZ (2017a) Nanostructured bioactive compounds for ecological food packaging. Environ Chem Lett 15:193–204
Brandelli A, Lopes NA, Boelter JF (2017b) Food applications of nanostructured antimicrobials. In: Grumezescu AM (ed) Food preservation—Nanotechnology in the agri-food industry. Elsevier, Londres, pp 35–74
Cazarin KCC, Corrêa CL, Zambrone FAD (2004) Redução, refinamento e substituição do uso de animais em estudos toxicológicos: uma abordagem atual. Braz J Pharm Sci 40:289–299
Charão MF, Souto C, Brucker N, Barth A, Jornada DS, Fagundez D, Ávila DS, Eifler-Lima VL, Guterres SS, Pohlmann AR, Garcia SC (2015) *Caenorhabditis elegans* as an alternative *in vivo* model to determine oral uptake, nanotoxicity, and efficacy of melatonin-loaded lipid-core nanocapsules on paraquat damage. Int J Nanomed 10:5093–5106
Dhawan A, Shanker R, Das M, Gupta KC (2011) Guidance for safe handling of nanomaterials. J Biomed Nanotechnol 7:218–224
Duncan TV (2011) Applications of nanotechnology in food packaging and food safety: barrier materials, antimicrobials and sensors. J Colloid Interface Sci 363:1–24
Echeverría I, Eisenberg P, Mauri AN (2014) Nanocomposites films based on soy proteins and montmorillonite processed by casting. J Membr Sci 449:15–26
Fakhrullin GI, Akhatova FS, Lvov YM, Fakhrullin RF (2015) Toxicity of halloysite clay nanotubes *in vivo*: a *Caenorhabditis elegans* study. Environ Sci Nano 2:54–59
Fröhlich E (2012) The role of surface charge in cellular uptake and cytotoxicity of medical nanoparticles. Int J Nanomed 7:5577–5591
Garcia SC, Guterres SS, Bubols GB, Bulcão RP, Charão MF, Pohlmann AR (2014) Polymeric nanoparticles: toxicological evaluation, cardiotoxicity and hepatotoxicity. In: Durán N, Guterres SS, Alves OL (eds) Nanotoxicology—materials, methodologies and assessments. Springer Science+Business Media, New York, pp 299–324
German JB, Hammock BD, Watkins SM (2005) Metabolomics: building a century of biochemistry to guide human health. Metabolomics 1:3–9
Guo L, von dem Bussche A, Buechner M, Yan A, Kane AB, Hurt RH (2008) Adsorption of essential micronutrients by carbon nanotubes and the implications for nanotoxicity testing. Small 4:721–727
Han C, Zhao A, Varughese E, Sahle-Demessie E (2018) Evaluating weathering of food packaging polyethylene-nanoclay composites: release of nanoparticles and their impacts. NanoImpact 9:61–71
Hatzigrigoriou NB, Papaspyrides CD (2011) Nanotechnology in plastic food-contact materials. J Appl Polym Sci 122:3720–3739
He Y, Kong W, Wang W, Liu T, Liu Y, Gong Q, Gao J (2012) Modified natural halloysite/potato starch composite films. Carbohydr Polym 87:2706–2711

Hernández-Ortiz M, Acosta-Torres LS, Hernández-Padrón G, Mendieta AI, Bernal R, Cruz-Vázquez C, Castaño VM (2012) Biocompatibility of crystalline opal nanoparticles. Biomed Eng Online 11:78

Hillier S, Brydson R, Delbos E, Fraser T, Gray N, Pendlowski H, Phillips I, Robertson J, Wilson I (2016) Correlations among the mineralogical and physical properties of halloysite nanotubes (HNTs). Clay Miner 51:325–350

Houtman J, Maisanaba S, Puerto M, Gutiérrez-Praena D, Jordá M, Aucejo S, Jos A (2014) Toxicity assessment of organomodified clays used in food contact materials on human target cell lines. Appl Clay Sci 90:150–158

Huang DJ, Wang WB, Kang YR, Wang AQ (2012) A chitosan/poly(vinyl alcohol) nanocomposite film reinforced with natural halloysite nanotubes. Polym Compos 33:1693–1699

Ibarguren C, Naranjo PM, Stötzel C, Audisio MC, Sham EL, Torres EMF, Müller FA (2014) Adsorption of nisin on raw montmorillonite. Appl Clay Sci 90:88–95

Isoda K, Nagata R, Hasegawa T, Taira Y, Taira I (2017) Hepatotoxicity and drug/chemical interaction toxicity of nanoclay particles in mice. Nanoscale Res Lett 12:199

Janer G, Fernández-Rosas E, Molino EM, González-Gálvez D, Villar G, López-Iglesias C, Ermini V, Vázquez-Campos S (2014) *In vitro* toxicity of functionalised nanoclays is mainly driven by the presence of organic modifiers. Nanotoxicology 8:279–294

Kalra S, Pant CK, Pathak HD, Mehata MS (2003) Studies on the absorption of peptides of glycine/alanine on montmorillonite clay with or without co-ordinated divalet cations. Colloid Surfaces A 212:43–50

Kennedy S (2002) The role of proteomics in toxicology: identification of biomarkers of toxicity by protein expression analysis. Biomarkers 7:269–290

Kroll A, Dierker C, Rommel C, Hahn D, Wohlleben W, Schulze-Isfort C, Göbbert C, Voetz M, Hardinghaus F, Schnekenburger J (2011) Cytotoxicity screening of 23 engineered nanomaterials using a test matrix of ten cell lines and three different assays. Part Fibre Toxicol 8:9

Kryuchkova M, Danilushkina A, Lvov Y, Fakhrullin R (2016) Evaluation of toxicity of nanoclays and graphene oxide *in vivo*: a *Paramecium caudatum* study. Environm Sci Nano 3:442–452

Lagaron JM, Lopez-Rubio A (2011) Nanotechnology for bioplastics: opportunities, challenges and strategies. Trends Food Sci Technol 22:611–617

Lai X, Agarwal M, Lvoy YM, Pachpande C, Varahamyan K, Witzmann FA (2013) Proteomic profiling of halloysite clay nanotube exposure in intestinal cell co-culture. J Appl Toxicol 33:1316–1329

Lee JH, Song NB Jo WS, Song KB (2014) Effects of nano-clay type and content on the physical properties of sesame seed meal protein composite films. Int J Food Sci Technol 49:1869–1875

Liu Q, Li R, Zhu Z, Qian X, Guan W, Yu L, Yang M, Jiang X, Liu B (2012) Enhanced antitumor efficacy, biodistribution and penetration of docetaxel-loaded biodegradable nanoparticles. Int J Pharm 430:350–358

Lordan S, Kennedy JE, Higginbotham CL (2011) Cytotoxic effects induced by unmodified and organically modified nanoclays in the human hepatic HepG2 cell line. J Appl Toxicol 31:27–35

Lu L, Cai J, Frost RL (2010) Near infrared spectroscopy of stearic acid adsorbed on montmorillonite. Spectrochim Acta A 75:960–963

Lvov Y, Abdullayev E (2013) Functional polymer-clay nanotube composites with sustained release of chemical agents. Prog Polym Sci 38:1690–1719

Maisanaba S, Puerto M, Pichardo S, Jordá M, Moreno FJ, Aucejo S, Jos A (2013) *In vitro* toxicological assessment of clays for their use in food packaging applications. Food Chem Toxicol 57:266–275

Maisanaba S, Gutiérrez-Praena D, Pichardo S, Moreno FJ, Jordá M, Cameán AM, Aucejo S, Jos A (2014) Toxic effects of a modified montmorillonite clay on the human intestinal cell line Caco-2. J Appl Toxicol 34:714–725

Maisanaba S, Pichardo S, Puerto M, Gutiérrez-Praena D, Cameán AM, Jos A (2015a) Toxicological evaluation of clay minerals and derived nanocomposites: a review. Environ Res 138:233–254

Maisanaba S, Prieto AI, Pichardo S, Jordá-Beneyto M, Aucejo S, Jos A (2015b) Cytotoxicity and mutagenicity assessment of organomodified clays potentially used in food packaging. Toxicol In Vitro 29:1222–1230

Marsh K, Bugusu B (2007) Food packaging—roles, materials, and environmental issues. J Food Sci 72:38–55

Meira SMM, Zehetmeyer G, Jardim AI, Scheibel JM, Oliveira RVB, Brandelli A (2014) Polypropylene/montmorillonite nanocomposites containing nisin as antimicrobial food packaging. Food Bioprocess Technol 7:3349–3357

Meira SMM, Jardim AI, Brandelli A (2015) Adsorption of nisin and pediocin on nanoclays. Food Chem 188:161–169

Meira SMM, Zehetmeyer G, Scheibel JM, Werner JO, Brandelli A (2016) Starch-halloysite nanocomposites containing nisin: characterization and inhibition of *Listeria monocytogenes* in soft cheese. LWT Food Sci Technol 68:226–234

Møller P, Folkmann JK, Danielsen PH, Jantzen K, Loft S (2012) Oxidative stress generated damage to DNA by gastrointestinal exposure to insoluble particles. Curr Mol Med 11:732–745

Neethirajan S, Jayas DS (2011) Nanotechnology for the food and bioprocessing industries. Food Bioprocess Technol 4:39–47

Nogueira V, Lopes I, Rocha-Santos T, Santos AL, Rasteiro GM, Antunes F, Gonçalves F, Soares AM, Cunha A, Almeida A, Gomes NC, Pereira R (2012) Impact of organic and inorganic nanomaterials in the soil microbial community structure. Sci Total Environ 424:344–350

Oberdörster G, Oberdörster E, Oberdörster J (2005) Nanotoxicology: an emerging discipline evolving from studies of ultrafine particles. Environ Health Perspect 113:823–839

Pavlidou S, Papaspyrides CD (2008) A review on polymer-layered silicate nanocomposites. Progr Polym Sci 33:1119–1198

Raldúa D, Piña B (2014) *In vivo* zebrafish assays for analyzing drug toxicity. Expert Opin Drug Metabol Toxicol 10:685–697

Rawtani D, Agrawal YK (2012) Multifarious applications of halloysite nanotubes: a review. Rev Adv Mater Sci 30:282–295

Ranjan S, Dasgupta N, Chakraborty AR, Samuel SM, Ramalingam C, Shanker R, Kumar A (2014) Nanoscience and nanotechnologies in food industries: opportunities and research trends. J Nanoparticle Res 16:2464

Rhim JW, Park HM, Ha CS (2013) Bio-nanocomposites for food packaging applications. Progr Polym Sci 38:1629–1652

Rubenstein AL (2006) Zebrafish assays for drug toxicity screening. Expert Opin Drug Metabol Toxicol 2:231–240

Saha S, Sparks AB, Rago C, Akmaev V, Wang CJ, Vogelstein B, Kinzler KW, Velculescu VE (2002) Using the transcriptome to annotate the genome. Nat Biotechnol 20:508–512

Schmitt H, Prashantha K, Soulestin J, Lacrampe MF, Krawczak P (2012) Preparation and properties of novel melt-blended halloysite nanotubes/wheat starch nanocomposites. Carbohydr Polym 89:920–927

Sayes CM, Reed KL, Warheit DB (2007) Assessing toxicity of fine and nanoparticles: comparing *in vitro* measurements to *in vivo* pulmonary toxicity profiles. Toxicol Sci 97:163–180

Shvedova AA, Kisin E, Murray AR, Johnson VJ, Gorelik O, Arepalli S, Hubbs AF, Mercer RR, Keohavong P, Sussman N, Jin J, Yin J, Stone S, Chen BT, Deye G, Maynard A, Castranova V, Baron PA, Kagan VE (2008) Inhalation vs. aspiration of single-walled carbon nanotubes in C57BL/6 mice: inflammation, fibrosis, oxidative stress, and mutagenesis. Am J Physiol Lung Cell Mol Physiol 295:552–565

Song NB, Jo WS, Song HY, Chung KS Won M, Song KB (2013) Effects of plasticizers and nano-clay content on the physical properties of chicken feather protein composite films. Food Hydrocoll 31:340–345

Stern ST, Adiseshaiah PP, Crist RM (2012) Autophagy and lysosomal dysfunction as emerging mechanisms of nanomaterial toxicity. Part Fibre Toxicol 14:9–20

Stueckle TA, Davidson DC, Derk R, Kornberg TG, Battelli L, Friend S, Orandle M, Wagner A, Dinu CZ, Sierros KA, Agarwal S, Gupta RK, Rojanasakul Y, Porter DW, Rojanasakul L (2018) Short-term pulmonary toxicity assessment of pre- and post-incinerated organomodified nanoclay in mice. ACS Nano 12:2292–2310

Turco RF, Bischoff M, Tong ZH, Nies L (2011) Environmental implications of nanomaterials: are we studying the right thing? Curr Opin Biotechnol 22:527–532

Unalan UI, Cerri G, Marcuzzo E, Cozzolino CA, Farris S (2014) Nanocomposite films and coatings using inorganic nanobuilding blocks (NBB): current applications and future opportunities in the food packaging sector. RSC Adv 4:29393–29428

Vanderroost M, Ragaert P, Devlieghere P, de Meulenaer B (2014) Intelligent food packaging: the next generation. Trends Food Sci Technol 39:47–62

Wagner A, Eldawud R, White A, Agarwal S, Stueckle TA, Sierros KA, Rojanasakul Y, Gupta RK, Dinu CZ (2017a) Toxicity evaluations of nanoclays and thermally degraded byproducts through spectroscopical and microscopical approaches. Biochim Biophys Acta 1861:3406–3415

Wagner A, White AP, Stueckle TA, Banerjee D, Sierros KA, Tojanasakul Y, Agarwal S, Gupta RK, Dinu CZ (2017b) Early assessment and correlations of nanoclay's toxicity to their physical and chemical properties. ACS Appl Mater Interfaces 20:32323–32335

Zhang Y, Chen D, Ennis AC, Polli JR, Xiao P, Zhang B, Stellwag EJ, Overton A, Pan X (2013) Chemical dispersant potentiates crude oil impacts on growth, reproduction, and gene expression in *Caenorhabditis elegans*. Arch Toxicol 87:371–382

Zhao R, Torley P, Halley PJ (2008) Emerging biodegradable materials: starch- and protein-based bio-nanocomposites. J Mater Sci 43:3058–3071

Chapter 5
Ecotoxicity of Metal Nanoparticles on Microorganisms

Patrycja Golinska, Magdalena Świecimska and Magdalena Wypij

Abstract Metal nanoparticles are commonly used in various products including appliances, electronics and computers, automotive, health and fitness, food and beverage, cross cutting, goods for children, home and garden, among which they are used most frequently in personal care, clothing and cosmetic sunscreen. Developments in nanotechnology are leading to a rapid increase of new nanomaterials that are likely to become a source of such materials to the environment, where their possible ecotoxicological impacts remain unknown. The surface properties of metal nanoparticles are of essential importance for their behavior and toxicity in the environment. Physico-chemical properties of metal nanoparticles affect their mobility in aquatic and terrestrial systems and their interactions with microorganisms. Metal nanoparticles present in the environment may impact on microbial activity, biomass and diversity. Thus, for risk assessment of metal nanoparticles on health and environment comprehensive study of metal NP properties as well as their quantities and concentrations in environmental systems are required.

Keywords Metal nanoparticles · Ecotoxicity · Microorganisms

5.1 Introduction

Nanoparticles (NPs), which measure from 1 to 100 nm in diameter, have a greater surface area than bulk materials, yielding a higher proportion of atoms on the surface relative to the interior and resulting in a higher surface reactivity (Auffan et al. 2009). Due to their unique physicochemical properties, nanoparticles have been increasingly

P. Golinska (✉) · M. Świecimska · M. Wypij
Department of Microbiology, Nicolaus Copernicus University, Lwowska 1, 87-100 Toruń, Poland
e-mail: golinska@umk.pl

M. Świecimska
e-mail: magda.swiecimska@gmail.com

M. Wypij
e-mail: mwypij@umk.pl

M. Rai and J. K. Biswas (eds.), *Nanomaterials: Ecotoxicity, Safety, and Public Perception*, https://doi.org/10.1007/978-3-030-05144-0_5

used in a wide range of technical applications and consumer products in recent years (Shen et al. 2015).

Metal nanoparticles, namely silver (Ag), zinc oxide (ZnO), gold (Au), titanium dioxide (TiO_2), iron (Fe) etc. are common and widely used nanoparticles in many nanoproducts and biological applications, including electronic products, cosmetics, pigments, semiconductors, textiles, sunscreen products, medical appliances, as well as industrial coatings, plastic additives, antibacterial agents and many more (Ju-Nam and Lead 2008; Tolaymat et al. 2010). The wide applications of metal nanoparticles affect their inflow into the environment where their potential fate and effects have remained unclear (Siani et al. 2017) (Fig. 5.1). Some metal nanoparticles such as silver or zinc oxide NPs are among the 14 representative engineered nanoparticles (ENPs) recommended for testing by the Organization for Economic Cooperation and Development (OECD) for their potential toxicity (Kahru and Dubourguier 2010; OECD 2010). Direct measurement of NP concentration in the environment is difficult because of technical limitations (Cornelis et al. 2014), and current amount of nanoparticles in the environment is only anticipated (Sun et al. 2014). Exposure modeling strongly suggests that soil could be a major location of NPs compared to air and water ecosystems (Gottschalk et al. 2009; Keller et al. 2013; Sun et al. 2014). Nanoparticles in soil may adversely affect the beneficial activities of natural microor-

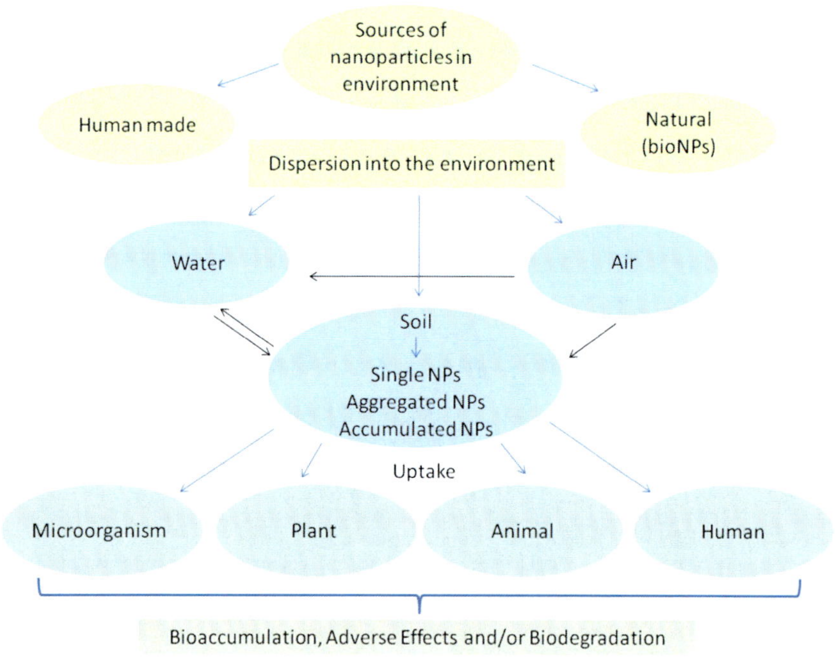

Fig. 5.1 Sources of metal nanoparticle released into the environment and their potential ecological risk

ganisms within environment (Gajjar et al. 2009; Dimkpa et al. 2013). Studies have shown that these materials can have significant negative effects on both plants and soil microbes in agro-ecosystems but the effects of specific metal NP types and the conditions under which negative effects may be observed have still not been studied well (Ma et al. 2010; Priester et al. 2012).

To date, most studies have been focused mainly on toxicity of AgNPs in relation to its well-known potency as a broad spectrum antimicrobial agent effective against many Gram-positive and Gram-negative bacteria (Morones et al. 2005; Kim et al. 2007), as well as against fungal species (Panacek et al. 2009). As both bacteria and fungi are present in the rhizosphere of plants, the antimicrobial effects from AgNP exposure may have significant implications for the host plant health (Sillen et al. 2015). However, the amount of research investigating the implications of rhizosphere-based nanoparticle exposure on both the microbes present and the supporting host plant is scarce. The influence of metal nanoparticles on soil microorganisms, namely microbial activity, biomass and diversity will be detailed in this chapter.

5.2 Sources of Metal Nanoparticles in the Environment

Nanoparticles can originate from natural (biogenic, geogenic, atmospheric, pyrogenic) and anthropogenic (by-products, manufactured, engineered) sources (Renzi and Guerranti 2015). In aquatic systems, NPs include metal sulfide nanoclusters emanating from hydrothermal systems (Luther and Rickard 2005). Naturally produced nanoparticles may include humic and fulvic acids, fullerenes, organic acids, carbon nanotubes, nanospheres and metals (e.g. Ag, Au, Fe-oxides) while manmade nanoparticles would include carbon black, fullerenes, functionalized fullerenes, polyethylene glycol, platinum, TiO_2, SiO_2, metal phosphates, zeolites, and ceramics (Nowak and Bucheli 2007; Renzi and Guerranti 2015) (Fig. 5.1).

Investments in nanotechnology are thus increasing rapidly worldwide. The total global investment in nanotechnologies was around \$10 billion in 2005 (Harrison 2007) and it is estimated that the annual value for all nanotechnology-related products will be \$1 trillion by 2011–2015 (Roco 2005).

It was estimated that annual global production of nanomaterials will be 58,000 metric tonnes between 2011 and 2020 in comparison with 2000 metric tonnes in 2004 (Maynard et al. 2006; O'Brien and Cummins 2016). This production includes ENPs made for use in agricultural applications (Kuzma and Ver Hage 2006) and environmental remediation, such as zero-valent iron (Zhang 2003). Such widespread and expanding production and use of metal nanoparticles increases the potential for their release to the environment.

Nanoparticles during the different phases of their life cycle, from production to disposal, can be released to the environment (aquatic, terrestrial, and atmospheric environments) (Nowack and Bucheli 2007), raising great concerns about potential ecological risks (Fig. 5.1). NPs can enter soil through various pathways, such as agricultural amendments of sewage sludge, atmospheric deposition, landfills, or accidental spills during industrial production (Simonin and Richaume 2015).

The current models estimate that in sewage sludge-treated soil, TiO_2NP concentrations increase between 0.94 and 3.6 mg kg^{-1} per year, whereas for AgNPs and fullerenes, the yearly predicted increases are more than 1000-fold lower. In this case, concentration increases are between 0.09 and 0.65 μg kg^{-1} and between 0.38 and 1.5 μg kg^{-1}, respectively (Sun et al. 2014). Current estimates of ZnONP concentrations in the UK environment range from less than 100 μg L^{-1} (in water) to a few mg kg^{-1} (in soil) (Boxall et al. 2007), and a more recent study by Gottschalk et al. (2009) reported modeled ZnONP concentrations of 10 ng L^{-1} in natural surface water and 430 ng L^{-1} in treated wastewater in Europe (Gottschalk et al. 2009; Ma et al. 2013). It is suggested that ZnONPs could also be added to formulations as a nanofertilizer (Milani et al. 2012). Another anticipated formulation of ZnONPs for agricultural applications could be as a pesticide (He et al. 2011; Kairyte et al. 2013). This aspect is important because nontarget effects from the large scale applications of fertilizers/pesticides occur in agriculture (Li and Schuster 2014). The use of products such as fertilizers and pesticides could lead to direct releases into the soil (Ma et al. 2013). Thus, formulations containing ZnONPs in the form of nano-fertilizer, nano-pesticides, contaminated sludge or biosolids applied in agriculture could, however, potentially release ZnONPs as emerging toxicant leading to exposure to various ecological organisms including the microorganisms and crop plants (Ma et al. 2013; Pokhrel and Dubey 2013, 2015).

Therefore, there is a demand to assess the risks associated to NPs in soils, in order to preserve the soil capacity to fulfill essential ecosystem services.

5.3 Behavior of Metal Nanoparticles in the Environment

Soil environment is a main reservoir for nanoparticles (Joner et al. 2007; Gottschalk et al. 2009; Keller et al. 2013; Sun et al. 2014). However, the fate and behaviour of nanoparticles in this ecosystem is not fully explained. The important questions are whether nanoparticles are subject of transformation or retain their normal nanoscale size, original structure and activity in the environmental systems? (Petersen et al. 2008; Ray et al. 2009). The transformation of NPs may alter the chemical and/or physical properties of NPs and consequently affect their fate (Petersen et al. 2008). The crucial information is lacking regarding mobility, transfer, and uptake as affected by environmental matrices. Thus it is expected that the exposure of organisms to ENPs will be strongly determined by the environmental conditions prevailing in each ecosystem (Joner et al. 2007).

The surface properties of NPs are one of the most important factors that determine their stability and mobility as colloidal suspensions or their aggregation into larger particles and deposition in the environment (Gimbert et al. 2007). Nanoparticle surface properties are mainly dependent on parameters such as temperature, ionic strength, pH, particle concentration and size, etc. (Lecoanet et al. 2004; Guzman et al. 2006a; Khan et al. 2017). Surface properties of metal oxide NPs are also determined by their acidity constants and zero-point of charge (Giammar et al. 2007; Hristovski et al. 2007). pH changes affect charge of metal nanoparticles (Guzman et al. 2006b; Ridley et al. 2006).

Metallic ENPs are usually coated with inorganic or organic compounds, such as citrate, cysteine, carbonate or surfactants such as sodium dodecyl sulfate to maintain the stability of the colloidal suspension (Mafune et al. 2000). Therefore, any destabilization of NPs may lead to their aggregation (Mafune et al. 2000; Mandal et al. 2001). Both, environmental factors (e.g. pH, ionic strength) and physical-chemical properties of NPs may determine whether they are bound within or transported out of soils and sediments (Rana and Kalaichelvan 2013). However, it is claimed that nanoparticles are more likely to accumulate in the environment than to disappear (Rana and Kalaichelvan 2013).

Therefore, a better understanding of the NPs' aggregation, deposition, and mobilization will help us to better predict the NPs' fate in the environment as well as their biological effects.

Due to chemical composition, surface charge and high specific surface area, as well as potentially high sorption capacities for other compounds dissolved in water some nanoparticles may bind with both, natural organic matter (NOM) as well as inorganic and organic pollutants (Fukushi and Sato 2005; Brant et al. 2007; Joner et al. 2007). NOM may influence the surface speciation and charge of NPs, and thus affect their aggregation/deposition properties. These organic matter compounds may be bonded to the surfaces of ENPs by various types of interactions, including electrostatic, hydrogen bonding, and hydrophobic interactions (Ojamae et al. 2006; Luo et al. 2018). The bond of NPs with pollutants might change the transport and bioavailability of both compounds in natural systems and alter their toxic effects. Trace metal ion speciation might be altered by NPs (especially oxide and oxide-coated NPs), therefore altering their bioavailability and potential toxicity (Chen et al. 2007; Navarro et al. 2008).

5.4 Impact of Nanoparticles on Soil Microorganisms

Soil microorganisms are indispensable constituents of soil ecosystem and play a crucial role in transformation and degradation of a wide variety of pollutants, thereby sustaining soil productivity and ecological functions. They also can influence plant growth and survival (Smith and Read 2008). Terrestrial plant species are colonized by a large number of microorganisms in several morphological regions, including interior tissues (endophytes), on their leaves (epiphytes) and in their rhizosphere. Rhizosphere microorganisms can increase nutrient uptake, protect against pathogens, enhance abiotic stress tolerance and promote the development of the plant (Berendsen et al. 2012). However, some rhizosphere microorganisms are also capable of causing plant disease or impeding plant growth. Thus, metal nanoparticles present in soil may adversely affect microbial pathogens but also beneficial activities of natural microorganisms within environment (exposure for non-target species) (Gajjar et al. 2009; Dimkpa et al. 2013). The important plant symbionts, such as arbuscular mycorrhizal fungi (AMF) that colonize plant roots and enhance nutrient uptake have also been found to be affected by the application of various metal nanoparticles such as TiO_2, ZnO or Ag NPs (Burke et al. 2014; Wang et al. 2014), which could have significant long-term effects on plant nutrient uptake and productivity in agro-ecosystems. On the other hand, they can ameliorate metal toxicity to host plants by mediating interactions between metals and plant roots (Zhu et al. 2001; Chen et al. 2004). It was found that AMF inoculation to corn alleviated ZnO-induced toxicity, suggesting its protective role (Chen et al. 2004). Similarly, AMF alleviated nanotoxicity of AgNPs as their increased mycorrhizal colonization was observed in increasing concentration of AgNPs (Feng et al. 2013). Arbuscular mycorrhizal fungi are characterized by high tolerance to heavy metals, and their uptake and accumulation (Gaur and Adholeya 2004; Zarei et al. 2010; Wu et al. 2010; Miransari 2011b). Moreover, AMF produce an alkali-soluble glycoprotein (glomalin) extracted from soil (Singh et al. 2013), which was found to play a key role in metal chelation (Gonzalez-Chavez et al. 2004; Xu et al. 2012).

In soils, the mobility of NPs in pore water is an essential condition for interactions with plant roots or fungi hyphae (Navarro et al. 2008).

It is well known that toxic effect of nanoparticles to prokaryotic or eukaryotic cells depends on their size and shape (Pal et al. 2007; Sharma et al. 2009; Stensberg et al. 2011) as well as on their stability (Shrivastava et al. 2007; Sharma 2013). Destabilization of NPs affect their aggregation in the bigger forms. Smaller nanoparticles cause higher toxicity to organisms, as these nanoparticles probably diffused more easily than the larger ones (Panacek et al. 2006; Mohan et al. 2007).

Effect of metal nanoparticles on microbial cells has been known mainly for AgNPs, to date. Silver nanoparticle exposure to microorganisms cause adhesion of NPs onto the cell wall and the membrane (Abbaszadegan et al. 2015). Antimicrobial activity of biogenic AgNPs results from the interaction of AgNPs with the sulfur-containing proteins present in the cell wall. It causes irreversible changes in cell wall structure resulting in its disruption (Ghosh et al. 2012). This in turn affects the integrity of lipid bilayer and permeability of the cell membrane, which becomes unable to proper regulation of transport activity through the plasma (Panacek et al. 2006; Ghosh et al. 2012; Chauhan et al. 2013). Moreover, silver ions, as a secondary oxidation process of silver nanoparticles, contribute to the toxic properties of silver nanoparticles (Morones et al. 2005; Manivasagan et al. 2013). Silver ions can affect transport and the release of potassium (K^+) ions from the microbial cells, which results in the increase in membrane permeability leading to more pronounced effects such as leakage and loss of cellular contents, including ions, proteins, reducing sugars and sometimes cellular energy reservoir (ATP) (Lok et al. 2006; Kim et al. 2011; Li et al. 2013). AgNPs can penetrate inside the microbial cell and interact with cellular structures and biomolecules such as proteins, lipids, and DNA. Interactions with cellular structures and biomolecules have damaging effect on microbes. In particular, AgNPs interaction with ribosomes lead to their denaturation causing inhibition of translation and protein synthesis (Morones et al. 2005; Jung et al. 2008; Rai and

Ingle 2012). Silver ions can have adverse effects on DNA which losses its replication ability as well as on cellular proteins which become inactivated (Chauhan et al. 2013).

5.4.1 *Effect of Metal Nanoparticles on Microbial Activity*

Microbial activity is a potential indicator of soil quality and has the ability to predict changes in soil properties as soil microorganisms control the transformation and mineralization of organic compounds and xenobiotics (Schloter et al. 2018). The influence of various agents on microbial activities is estimated by study of soil respiration or enzyme activities such as β-glucosidase, urease, and dehydrogenase, nitrification in nitrogen cycle or the ability to degrade specific pollutants (Chen et al. 2003; Schimel and Schaeffer 2012). Microbial activity is one of the parameters to assess the effect of metal or metal oxide NPs on microbial cells. However, the measured results depend on the type of NPs, the concentration, the exposure time, and the measured activity (Hänsch and Emmerling 2010; He et al. 2011; Shin et al. 2012; Colman et al. 2013).

Colman and colleagues (2013) studied effect of time and concentration of AgNPs on substrate-induced respiration and enzymatic activity of the soil microbial extracellular enzymes, namely leucine amino peptidase and phosphatase in an experimental system of sewage sludge. They found that such activities were reduced after 50 days of exposure to a low concentration of nanoparticles (0.14 mg kg^{-1} soil). The adverse effect of silver nanoparticles on six microbial exoenzymes related to nutrient cycles (urease, acid phosphatase, arylsulfatase, β-glucosidase) and the overall microbial activity (dehydrogenase, fluorescein diacetate hydrolase) were tested in soils treated with AgNPs (1, 10, 100 and 1000 μg g^{-1}) (Shin et al. 2012). AgNPs were capable of inhibiting the activities of all the exoenzymes tested in this study. Especially, the urease and dehydrogenase activities were significantly related to the presence of AgNPs. However, high reduction of enzymatic activity was found for much higher concentrations (100 and 1000 mg kg^{-1} soil) of AgNPs (Shin et al. 2012).

In contrast, Hänsch and Emmerling (2010) made an assessment of Ag nanoparticle toxicity using six soil enzymes assay (leucine-aminopeptidase, β-cellobiohydrolase, acid phosphatase, β-glucosidase, chitinase, and β-xylosidase), and found no significant changes in enzymatic activity in presence of Ag-NPs in concentration of 0.0032, 0.032, and 0.32 mg kg^{-1} soil. However, they found a reduction of the net nitrogen mineralization and an increase of the metabolic quotient which suggest a reduction in substrate use efficiency. The decrease in respiration rate and enzymatic activities in response to very low concentrations of AgNPs indicates the high toxic potential of such metal NPs (Simonin and Richaume 2015).

Similarly, other metal nanoparticles such as Fe_2O_3, Fe_3O_4, TiO_2 and ZnONPs have diverse effects on microbial activities. Higher activities of urease and invertase were recorded in the presence of high concentrations of Fe_2O_3-NPs (420, 840, and 1260 mg kg^{-1} soil), while the presence of Fe_3O_4-NPs did not affect any changes in enzymatic activity (He et al. 2011). TiO_2-NPs (1 and 500 mg kg^{-1}) impact on substrate-induced respiration was also assessed in six diverse agricultural soils, and a significant decrease of this activity was recorded only in one soil presenting a silty clay texture and a high organic matter content (Simonin et al. 2014). Du and coauthors (2011) found that TiO_2-NPs at a concentration of 91 mg kg^{-1}soil stimulated urease activity but decreased catalase and peroxidase activities after9 months of incubation, while ZnONPs (45 mg kg^{-1} soil) reduced protease, catalase, and peroxidase activities. Similarly, Ge et al. (2011) studied the impact of TiO_2 and ZnONPs on substrate induced respiration and found a reduction after 15 and 60 days with 500, 1000, and 2000 mg kg^{-1} soil.

These results indicate that metal and metal oxides can induce modifications of microbial activities in soil and consequently affect biogeochemical cycles (Simonin and Richaume 2015).

5.4.2 *Effect of Metal Nanoparticles on Microbial Biomass*

Microbial biomass is a sensitive indicator of pollution in soils (Brookes et al. 1985), which can be used as an indicator of the impact of metal NPs in assessment of their ecotoxicity. Evaluation of the total microbial biomass (bacteria, fungi and archea) content can be performed by using different methods, such as fumigation–extraction or extractable soil DNA (Hänsch and Emmerling 2010; Ge et al. 2011). Moreover, content of specific groups of microorganisms targeting universal genes or functional genes can be assessed by quantitative PCR (Fajardo et al. 2012). Many authors studied impact of metal nanoparticles on microbial biomass (Hänsch and Emmerling 2010; Ge et al. 2011; Kumar et al. 2011; Antisari et al. 2013; Simonin et al. 2014).

Hänsch and Emmerling (2010) found that different concentrations of AgNPs (0.0032, 0.032, 0.32 mg kg^{-1} soil) induced a decrease in microbial biomass in a dose–response manner. However, soil contamination with iron oxide magnetic NPs (Fe_2O_3 and Fe_3O_4) had no effect on microbial biomass with10 and 100 mg kg^{-1} soil (Antisari et al. 2013) nor on bacterial abundance with 420, 840, and 1260 mg kg^{-1} soil (He et al. 2011). Similarly, Simonin et al. (2014) found no significant effect of TiO_2-NPs on bacterial abundances in the six soils tested. Ge and colleagues (2011) observed that TiO_2 and ZnO NPs in concentrations of 500, 1000, and 2000 mg kg^{-1} soil reduced extractable soil DNA. The impact of a combination of Ag, Cu and Si NPswas investigated in an arctic soil (Kumar et al. 2011), where a significant decrease of microbial biomass was observed after 6 months of incubation fora mixture concentration of 660 mg kg^{-1} soil (220 mg kg^{-1} of each NP).

The impact of metal NPs on microbial biomass may be dependent on the organic matter content of soil, which has been found to enhance NPs mobility in porous media (Ben-Moshe et al. 2010; Thio et al. 2011; Wang et al. 2012). It is supposed that organic matter may be an important factor favoring NPs bioavailability for soil microorganisms (Fang et al. 2009; Cornelis et al. 2013; Pawlett et al. 2013; Peyrot et al. 2014; Simonin et al. 2014).

5.4.3 Effect of Metal Nanoparticles on Microbial Diversity

Soil is a source of an immense diversity of microorganisms (bacteria, fungi, and archaea) which may be affected by metal nanoparticles (Torsvik and Øvreås 2002). The pyrosequencing data indicated that some of the declining taxa are known to be associated with nitrogen fixation (Rhizobiales, *Bradyrhizobiaceae* and *Bradyrhizobium*) and methane oxidation (*Methylobacteriaceae*), while some positively impacted taxa are known to be associated with the decomposition of recalcitrant organic pollutants (*Sphingomonadaceae*) and biopolymers including protein (*Streptomycetaceae* and *Streptomyces*). The roles of these taxa in soil functioning suggest potential consequences on ecosystem-scale processes. That is why, investigating the impact of NPs on microbial diversity is crucial to provide information on how and why soil ecosystem functioning is affected (Simonin and Richaume 2015).

To date, a panel of techniques, such as PLFA profiles, FISH, PCR-DGGE, T-RFLP analysis, or next generation sequencing, has been used to evaluate the impact of NPs on microbial diversity (Fajardo et al. 2012; Simonin and Richaume 2015).

Effect of metal nanoparticles, mainly of silver NPs, was reported by many authors (Kumar et al. 2011; Colman et al. 2013). Kumar and colleagues (2011) observed that AgNPs modified bacterial community in an arctic soil but after higher concentration used (660 mg kg^{-1}). However, other authors reported modification of bacterial community structure after shortterm exposure of sewage sludge containing much lower concentration of AgNPs (0.14 mg kg^{-1}) (Colman et al. 2013). Authors also reported that zero valent Cu and ZnONPs had no effect on PLFA profiles and on microbial community composition as determined by pyrosequencing (Collins et al. 2012). Changes in soil bacterial community composition were also observed due to the presence of Fe_3O_4-NPs after a short incubation (24 or 48 h) as well as after 15 and 30 days (He et al. 2011; Ben-Moshe et al. 2013).

Cloning–sequencing of DGGE bands indicated a stimulation of specific groups of *Actinobacteria, Duganella, Streptomycetaceae* and *Nocardioides* (He et al. 2011). These groups facilitate the decomposition of organic matter, which could explain the concomitant soil invertase and urease increases measured during this experiment. Different concentrations of TiO_2 and ZnONPs decreased soil bacterial diversity after 60 days of incubation (Ge et al. 2011, 2012). Nogueira et al. (2012) assessed the effect of five inorganic nanomaterials (TiO_2, $TiSiO_4$, CdSe/ZnS quantum dots, gold nanorods, and Fe/Co magnetic fluid) on soil bacterial community structure using DGGE. After 30 days of soil exposure, TiO_2 and gold nanorods induced the highest changes in the structural diversity of bacterial community. The limited effect of $TiSiO_4$, CdSe/ZnS quantum dots, and Fe/Co magnetic fluid NPs on DGGE profiles was attributed to their zeta potential values reflecting an unstable state (Nogueira et al. 2012).

Based on above reports it is concluded that metal or metal oxide NPs maybe responsible for a biodiversity loss and a modification of soil microbial community composition. Although many researches have been focused on the effect of metal NPs on bacterial communities, more work is still required to assess their impact on fungal and archaeal communities.

5.5 Environmental Risk Assessment of Metal Nanoparticles

Nanomaterials are currently utilized in various products and applications for a broad range of purposes. Based on the utilization of metal nanoparticles in various commercial goods, various extents of their release from products are expected and diverse routes of entry into the environment need to be considered. The differences in release and entry will result in differing environmental consequences, i.e. with respect to the extent of accumulation of metal NPs in freshwater, sediments and soils (Voelker et al. 2015). For environmental exposure, it is important to have empirical data or procedures to predict the persistence and mobility in air, soil, and water (Joner et al. 2007).

Direct assessment of NP concentration in the environment (e.g. soil) remains difficult due to current technical limitations (Cornelis et al. 2014). Prior to studying the impact of nanoparticles on the environment and health, minimal characterization of the engineered nanoparticles has to be carried out. The physicochemical properties of the nanoparticles which affect their behaviour and toxicity must be understood in detail for a realistic risk assessment of NPs in the environment (Love et al. 2012). Moreover, an evaluation of the expected quantities and concentrations of NP in environmental systems is also required. To date nothing is known about this issue, neither from an analytical point of view (e.g. actual measurements of NPs in the environment) nor with respect to theoretical or modelling studies.

Currently, the only way to get information on existing levels of NPs in the environment is to model predicted environmental concentrations (Sun et al. 2014). Exposure modelling strongly suggests that soil could be a major sink of NPs compared to air and water ecosystems (Gottschalk et al. 2009; Keller et al. 2013; Sun et al. 2014).

The first step in risk characterisation is to identify and characterise the source of nanoparticles in the environment, assess the hazards, establish the relationship between dose and response for various endpoints, and then predict the likelihood of exposure. When both dose-response and exposure is quantified, they are compared to characterise the relative risk (Rana and Kalaichelvan, 2013).

Hazards, such as toxicity and ecotoxicity, can be measured using different end points, including physiological, genetic, or functional effects, either acute or chronic. But there may also be other potentially negative environmental effects apart from toxicity that need to be considered in a risk assessment paradigm adapted to nanoparticles like impact on atmospheric/stratospheric processes, stability of soil, effects on bioavailability of mineral nutrients, and so forth (Joner et al. 2007).

Dose-response assessments follow the hazard identification in the risk assessment process. The establishment of dose-response relationships may involve performing experiments in the laboratory or using mathematical models. However, dose-response relationships may not be straight forward for nanoparticles, as dose based on mass concentration may be less relevant than dose based on surface area, and as different nanoparticle preparations may result in differences in surface reactivity, and thus toxicity (Joner et al. 2007).

5.6 Conclusion and Future Perspectives

It is evident from this chapter that the nanotechnology has a significant potential and is used in almost all branches of industry from electronic industry to pharmaceuticals and everyday consumer products. With the increase of technological development, the number of nanoparticles produced increases. As a result, metal NP contamination also increases in soil, water and air with gradual build up in the environment.. Their level is still unknown but it is considered that the most nanoparticles are deposited in the soil. Accumulation of nanoparticles in environment has negative effect on host organisms, including soil microbes (bacteria and fungi). Nanoparticles may impact either microbial activities or their biomass and diversity. Hence, judicious applications of nanoparticles are warranted instead of their widespread applications.

It is expected that in 2020 the production of nanoparticles will increase to 58,000 tons and over the years it will increase even more (Roco 2005). There are still considerable gaps in knowledge about the potential risks of accumulation of nanoparticles in the environment (Ray et al. 2009). Due to this problem the nature of interaction of nanoparticles with the environmental system represents a new challenge for current ecotoxicological research (Handy 2008). To determine the environmental performance of nanoparticle contaminations it is crucial to establish their concentration, the degree of their environmental mobility and bioavailability (Hristozov and Malsch 2009). Future research also should address the effects of accumulation of nanoparticles in the environment, their fate and potential toxicological risks.

Currently available technologies can't quantify environmental levels of nanoparticles (Nowack 2009). It is necessary to develop new research techniques for more precise determination of the level of nanoparticle contamination and toxicity. Attention should be paid to the use of modern technologies to design sensors detecting the concentration of nanoparticles, which would allow for more accurate tracking of nanoparticles accumulation (Ray et al. 2009).

On the other hand, further development and application of greener nanomaterials across the compositions, sizes, shapes, and functionality will provide synthesizing less toxic nanoparticles and avert release of these toxicants into the environment. To implement safety will need to develop validated models capable of predicting accumulation, transport, transformation, and toxicity on organisms. These models is necessary to design less toxic nanoparticles (Ray et al. 2009).

The international cooperation is important for evaluation and remediation of ecotoxicity as well as reduction of the use of toxic nanoparticles and development of less toxic nanoparticles (Ray et al. 2009).

Acknowledgements The grant Nos. 2016/23/N/NZ9/00247 and 2017/01/X/NZ8/00140 from National Science Centre (NCN) in Poland are acknowledged.

References

Abbaszadegan A, Ghahramani Y, Gholami A, Hemmateenejad B, Dorostkar S, Nabavizadeh M, Sharghi H (2015) The effect of charge at the surface of silver nanoparticles on antimicrobial activity against Gram-positive and Gram-negative bacteria: a preliminary study. J Nanomater 2015:720654

Antisari LV, Carbone S, Gatti A, Vianello G, Nannipieri P (2013) Toxicity of metal oxide (CeO_2, Fe_3O_4, SnO_2) engineered nanoparticles on soil microbial biomass and their distribution in soil. Soil Biol Biochem 60:87–94

Auffan M, Rose J, Wiesner MR, Bottero JY (2009) Chemical stability of metallic nanoparticles a parameter controlling their potential cellular toxicity in vitro. Environ Pollut 157:1127–1133

Ben-Moshe T, Dror I, Berkowitz B (2010) Transport of metal oxide nanoparticles in saturated porous media. Chemosphere 81:387–393

Ben-Moshe T, Frenk S, Dror I, Minz D, Berkowitz B (2013) Effects of metal oxide nanoparticles on soil properties. Chemosphere 90:640–646

Berendsen RL, Pieterse CMJ, Bakker PAHM (2012) The rhizosphere microbiome and plant health. Trends Plant Sci 17:478–486

Boxall A, Chaudhry Q, Sinclair C, Jones A, Aitken R, Jefferson B, Watts C (2007) Current and future predicted environmental exposure to engineered nanoparticles. Central Science Laboratory, York, UK

Brant JA, Labille J, Bottero JY, Wiesner MR (2007) Nanoparticle transport, aggregation and deposition. In: Wiesner MR, Bottero JY (eds) Environmental nanotechnology applications and impacts of nanomaterials. McGraw Hill, New York, USA, pp 231–294

Brookes PC, Landman A, Pruden G, Jenkison DS (1985) Chloroform fumigation and the release of soil nitrogen: a rapid direct extraction method for measuring microbial biomass nitrogen in soil. Soil Biol Biochem 17:837–842

Burke DJ, Zhu S, Pablico-Lansigan MP, Hewins CR, Samia ACS (2014) Titanium oxide nanoparticle effect on the composition of soil microbial communities and plant performance. Biol Fertil Soils 50:1169–1173

Chauhan R, Kumar A, Abraham J (2013) Biological approach to the synthesis of silver nanoparticles with *Streptomyces* sp JAR1 and its antimicrobial activity. Sci Pharm 81:607–621

Chen BD, Li XL, Tao HQ, Christie P, Wong MH (2003) The role of arbuscular mycorrhiza in zinc uptake by red clover growing in a calcareous soil spiked with various quantities of zinc. Chemosphere 50:839–846

Chen HT, Neerman MF, Parrish AR, Simanek EE (2004) Cytotoxicity, hemolysis, and acute *in vivo* toxicity of dendrimers based on melamine, candidate vehicles for drug delivery. J Am Chem Soc 126:10044–10048

Chen W, Duan L, Zhu D (2007) Adsorption of polar and nonpolar organic chemicals to carbon nanotubes. Environ Sci Technol 41:8295–8300

Collins D, Luxton T, Kumar N, Shah S, Walker VK, Shah V (2012) Assessing the impact of copper and zinc oxide nanoparticles on soil: a field study. PLoS ONE 7:e42663

Colman BP, Arnaout CL, Anciaux S, Gunsch CK, Hochella MF Jr, Kim B, Lowry GV, McGill BM, Reinsch BC, Richardson CJ, Unrine JM, Wright JP, Yin L, Bernhardt ES (2013) Low concentrations of silver nanoparticles in sewage sludge cause adverse ecosystem responses under realistic field scenario. PLoS ONE 8:e57189

Cornelis G, Hund-Rinke K, Kuhlbusch T, Van den Brink N, Nickel C (2014) Fate and bioavailability of engineered nanoparticles in soils: a review. Crit Rev Environ Sci Technol 44:2720–2764

Cornelis G, Pang L, Doolette C, Kirby JK, McLaughlin MJ (2013) Transport of silver nanoparticles in saturated columns of natural soils. Sci Total Environ 463:120–130

Dimkpa CO, Latta DE, McLean JE, Britt DW, Boyanov MI, Anderson AJ (2013) Fate of CuO and ZnO nano- and microparticles in the plant environment. Environ Sci Technol 47:4734–4742

Du W, Sun Y, Ji R, Zhu J, Wu J, Guo H (2011) TiO_2 and ZnO nanoparticles negatively affect wheat growth and soil enzyme activities in agricultural soil. J Environ Monitor 13:822–828

Fajardo C, Ortíz LT, Rodríguez-Membibre ML, Nande M, Lobo MC, Martin M (2012) Assessing the impact of zero-valent iron (ZVI) nanotechnology on soil microbial structure and functionality: a molecular approach. Chemosphere 86:802–808

Fang J, Shan XQ, Wen B, Lin JM, Owens G (2009) Stability of titania nanoparticles in soil suspensions and transport in saturated homogeneous soil columns. Environ Pollut 157:1101–1109

Feng Y, Cui X, He S, Dong G, Chen M, Wang J, Lin X (2013) The role of metal nanoparticles in influencing arbuscular mycorrhizal fungi effects on plant growth. Environ Sci Technol 47:9496–9504

Fukushi K, Sato T (2005) Using a surface complexation model to predict the nature and stability of nanoparticles. Environ Sci Technol 39:1250–1256

Gajjar P, Pettee B, Britt D, Huang W, Johnson W, Anderson A (2009) Antimicrobial activities of commercial nanoparticles against an environmental soil microbe, *Pseudomonas putida* KT2440. J BiolEng 3:9–21

Gaur A, Adholeya A (2004) Prospects of arbuscular mycorrhizal fungi in phytoremediation of heavy metal contaminated soils. Curr Sci 86:528–534

Ge Y, Schimel JP, Holden PA (2012) Identification of soil bacteria susceptible to TiO_2 and ZnO nanoparticles. Appl Environ Microb 78:6749–6758

Ge YG, Schimel JP, Holden PA (2011) Evidence for negative effects of TiO_2 and ZnO nanoparticles on soil bacterial communities. Environ Sci Technol 45:1659–1664

Ghosh S, Patil S, Ahire M, Kitture R, Kale S, Pardesi K, Cameotra SS, Bellare J, Dhavale DD, Jabgunde A, Chopade BA (2012) Synthesis of silver nanoparticles using Dioscorea bulbifera tuber extract and evaluation of its synergistic potential in combination with antimicrobial agents. Int J Nanomed 7:483–496

Giammar DE, Maus CJ, Xie LY (2007) Effects of particle size and crystalline phase on lead adsorption to titanium dioxide nanoparticles. Environ Eng Sci 24:85–95

Gimbert LJ, Hamon RE, Casey PS, Worsfold PJ (2007) Partitioning and stability of engineered ZnO nanoparticles in soil suspensions using flow field-flow fractionation. Environ Chem 4:8–10

Gonzalez-Chavez MC, Carrillo-Gonzalez R, Wright SF, Nichols KA (2004) The role of glomalin, a protein produced by arbuscular mycorrhizal fungi, in sequestering potentially toxic elements. Environ Pollut 130:317–323

Gottschalk F, Sonderer T, Scholz RW, Nowack B (2009) Modeled environmental concentrations of engineered nanomaterials (TiO2, ZnO, Ag, CNT, Fullerenes) for different regions. Environ Sci Technol 43:9216–9222

Guzman KA, Finnegan MP, Banfield JF (2006a) Influence of surface potential on aggregation and transport of titania nanoparticles. Environ Sci Technol 40:7688–7693

Guzman KA, Taylor MR, Banfield JF (2006b) Environmental risks of nanotechnology: national nanotechnology initiative funding, 2000–2004. Environ Sci Technol 40:1401–1407

Handy RD, Owen R, Valsami-Jones E (2008) The ecotoxicology of nanoparticles and nanomaterials: current status, knowledge gaps, challenges, and future needs. Ecotoxicology 17:315–325

Hänsch M, Emmerling C (2010) Effects of silver nanoparticles on the microbiota and enzyme activity in soil. J Plant Nutr Soil Sci 173:554–558

Harrison P (ed) (2007) Emerging challenges: nanotechnology and the environment. GEO Year Book 2007. United Nations Environment Programme (UNEP), Nairobi, Kenya, pp 61–68

He L, Liu Y, Mustapha A, Lin M (2011) Antifungal activity of zinc oxide nanoparticles against *Botrytis cinerea* and *Penicillium expansum*. Microbiol Res 166:207–215

Hristovski K, Baumgardner A, Westerhoff P (2007) Selecting metal oxide nanomaterials for arsenic removal in fixed bed columns: from nanopowders to aggregated nanoparticle media. J Hazard Mater 147:265–274

Hristozov D, Malsch I (2009) Hazards and risks of engineered nanoparticles for the environment and human health. Sustainability 1:1161–1194

Joner EJ, Hartnik T, Amundsen CE (2007) Environmental fate and ecotoxicity of engineered nanoparticles. Norwegian Pollution Control Authority Report no. TA 2304/2007, Bioforsk, As, Norway, pp 1–64

Ju-Nam Y, Lead R (2008) Manufactured nanoparticles: an overview of their chemistry, interactions and potential environmental implications. Sci Total Environ 400:396–414
Jung J, Ahn YJ, Kang LW (2008) A novel approach to investigating protein/protein interactions and their functions by TAP-tagged yeast strains and its application to examine yeast transcription machinery. J Microbiol Biotechnol 18:631–638
Kahru A, Dubourguier HC (2010) From ecotoxicology to nanoecotoxicology. Toxicology 269:105–119
Kairyte K, Kadys A, Luksiene Z (2013) Antibacterial and antifungal activity of photo activated ZnO nanoparticles in suspension. J Photochem Photobiol 5:78–84
Keller AA, McFerran S, Lazareva A, Suh S (2013) Global life cycle releases of engineered nanomaterials. J Nanopart Res 15:1692
Khan I, Saeed K, Khan I (2017) Nanoparticles: properties, applications and toxicities. Arabian J Chem. https://doi.org/10.1016/j.arabjc.2017.05.011
Kim E, Kim S-H, Kim H-C, Lee SG, Lee SJ, Jeong SW (2011) Growth inhibition of aquatic plant caused by silver and titanium oxide nanoparticles. ToxEHS 3:1–6
Kim JS, Kuk E, Yu KN, Kim JH, Park SJ, Lee HJ, Kim SH, Park YK, Park YH, Hwang CY, Kim YK, Lee YS, Jeong DH, Cho MH (2007) Antimicrobial effects of silver nanoparticles. Nanomed Nanotech Biol Med 3:95–101
Kumar N, Shah V, Walker VK (2011) Perturbation of an arctic soil microbial community by metal nanoparticles. J Hazard Mater 190:816–822
Kuzma J, VerHage P (2006) Nanotechnology in agriculture and food production—anticipated applications. Woodrow Wilson International Center for Scholars, Project on Emerging Nanotechnologies, Washington, DC
Lecoanet HF, Bottero JY, Wiesner MR (2004) Laboratory assessment of the mobility of nanomaterials in porous media. Environ Sci Technol 38:5164–5169
Li L, Schuster M (2014) Influence of phosphate and solution pH on the mobility of ZnO nanoparticles in saturated sand. Sci Total Environ 472:971–978
Li M, Lin D, Zhu L (2013) Effects of water chemistry on the dissolution of ZnO nanoparticles and their toxicity to Escherichia coli. Environ Pollut 173:97–102
Lok CN, Ho CM, Chen R, He QY, Yu WY, Sun HZ, Tam PKH, Chiu JF, Che CM (2006) Proteomic analysis of the mode of antibacterial action of silver nanoparticles. J Proteome Res 5:916–924
Love SA, Maurer-Jones MA, Thompson JW, Lin Y-S, Haynes C-L (2012) Assessing nanoparticle toxicity. Annu Rev Analytical Chem 5:181–205
Luo M, Huang Y, Zhu M, Tang Y, Ren T, Ren J, Wang H, Li F (2018) Properties of different natural organic matter influence the adsorption and aggregation behavior of TiO_2 nanoparticles. J Saudi Chem Soc 22:146–154
Luther GW, Rickard DT (2005) Metal sulfide cluster complexes and their biogeochemical importance in the environment. J Nanopart Res 7:389–407
Ma H, Williams PL, Diamond SA (2013) Ecotoxicity of manufactured ZnO nanoparticles—a review. Environ Pollut 172:76–85
Ma X, Geisler-Lee J, Deng Y, Kolmakov A (2010) Interactions between engineered nanoparticles (ENPs) and plants: phytotoxicity, uptake and accumulation. Sci Total Environ 408:3053–3061
Mafune F, Kohno J, Takeda Y, Kondow T, Sawabe H (2000) Structure and stability of silver nanoparticles in aqueous solution produced by laser ablation. J Phys Chem B 104:8333–8337
Mandal S, Gole A, Lala N, Gonnade R, Ganvir V, Sastry M (2001) Studies on the reversible aggregation of cysteine-capped colloidal silver particles interconnected via hydrogen bonds. Langmuir 17:6262–6268
Manivasagan P, Venkatesan J, Sivakumar K, Kim SK (2013) Production, characterization and antioxidant potential of protease from *Streptomyces* sp. MAB18 using poultry wastes. BioMed Res Int 2013:496–586
Maynard AD, Aitken RJ, Butz T, Colvin V, Donaldson K, Oberdörster G, Philbert MA, Ryan J, Seaton A, Stone V, Tinkle SS, Tran L, Walker NJ, Warheit DB (2006) Safe handling of nanotechnology. Nature 444(7117):267–269

Milani N, McLaughlin MJ, Stacey SP, Kirby JK, Hettiarachchi GM, Beak DG, Cornelis G (2012) Dissolution kinetics of macronutrient fertilizers coated with manufactured zinc oxide nanoparticles. J Agri Food Chem 60:3991–3998

Miransari M (2011) Hyperaccumulators, arbuscular mycorrhizal fungi and stress of heavy metals. Biotechnol Adv 29:645–653

Mohan YM, Lee K, Premkumar T, Geckeler KE (2007) Hydrogel networks as nanoreactors: a novel approach to silver nanoparticles for antibacterial applications. Polymer 48:158–164

Morones JR, Elechiguerra JL, Camacho A, Holt K, Kouri JB, Ramirez JT, Yacaman MJ (2005) The bactericidal effect of silver nanoparticles. Nanotechnology 16:2346–2353

Navarro E, Baun A, Behra R, Hartmann NB, Filser J, Miao A-J, Quigg A, Santschi PH, Sigg L (2008) Environmental behavior and ecotoxicity of engineered nanoparticles to algae, plants, and fungi. Ecotoxicology 17:372–386

NogueiraV Lopes I, Rocha-Santos T, Santos AL, Rasteiro GM, Antunes F, Gonçalves F, Soares AMVM, Cunha A, Almeida A, Gomes NNCM, Pereira R (2012) Impact of organic and inorganic nanomaterials in the soil microbial community structure. Sci Total Environ 424:344–350

Nowack B (2009) The behavior and effects of nanoparticles in the environment. Environ Pollut 157:1063–1064

Nowack B, Bucheli TD (2007) Occurrence, behavior and effects of nanoparticles in the environment. Environ Pollut 150:5–22

O'Brien NJ, Cummins EJ (2016) Environmental exposure modeling methods for engineered nanomaterials. In: Xing B, Vecitis CD, Sensi N (eds.) Engineered nanoparticles and the environment: biophysicochemical processes and toxicity. Wiley, Haboken, New Jersey, USA, pp 118–138

OECD (2010) Publications in the series on the safety of manufactured nanomaterials No. 27: List of manufactured nanomaterials and list of endpoints for phase one of the sponsorship programme for the testing of manufactured nanomaterials. OECD Publishing, Paris, France, p 13

Ojamae L, Aulin C, Pedersen H, Kall PO (2006) IR and quantum chemical studies of carboxylic acid and glycine adsorption on rutile TiO_2 nanoparticles. J Colloid Interface Sci 296:71–78

Pal S, Tak YK, Song JM (2007) Does the antibacterial activity of silver nanoparticles depend on the shape of the nanoparticle? A study of the gram-negative bacterium Escherichia coli. Appl Environ Microbiol 73:1712–1720

Panacek A, Kolar M, Vecerova R, Prucek R, Soukupova J, Krystof V, Hamal P, Zboril R, Kvitek L (2009) Antifungal activity of silver nanoparticles against Candida spp. Biomaterials 30:6333–6340

Panacek A, Kvitek L, Prucek R, Kolar M, Vecerova R, Pizurova N, Sharma VK, Nevecna T, Zboril R (2006) Silver colloid nanoparticles: synthesis, characterization, and their antibacterial activity. J Phys Chem B 110:16248–16253

Pawlett M, Ritz K, Dorey RA, Rocks S, Ramsden J, Harris JA (2013) The impact of zero-valent iron nanoparticles upon soil microbial communities is context dependent. Environ Sci Pollut Res 20:1041–1049

Petersen EJ, Huang Q, Weber WJ (2008) Bioaccumulation of radio-labeled carbon nanotubes by *Eisenia foetida*. Environ Sci Technol 42:3090–3095

Peyrot C, Wilkinson KJ, Desrosiers M, Sauvé S (2014) Effects of silver nanoparticles on soil enzyme activities with and without added organic matter. Environ Toxicol Chem 33:115–125

Pokhrel LR, Dubey B (2013) Evaluation of developmental responses of two crop plants exposed to silver and zinc oxide nanoparticles. Sci Total Environ 452–453:321–332

Pokhrel LR, Dubey B (2015) Early developmental responses of plants exposed to metals and oxides nanomaterials. In: Siddiqui MH, Al-Whaibi MH, Mohammad F (eds) Nanotechnology and plant sciences: nanoparticles and their impact on plants. Springer-Verlag, Berlin, Heidelberg, Germany, pp 153–164

Priester JH, Ge Y, Mielke RE, Horst AM, Moritz SC, Espinosa K, Gelb J, Walker SL, Nisbet RM, An YJ, Schimel JP, Palmer RG, Hernandez-Viezcas JA, Zhao LJ, Gardea-Torresdey JL, Holden PA (2012) Soybean susceptibility to manufactured nanomaterials with evidence for food quality and soil fertility interruption. PNAS 109:2451–2456

Rai M, Ingle A (2012) Role of nanotechnology in agriculture with special reference to management of insect pests. Appl Microbiol Biotechnol 94:287–293

Rana S, Kalaichelva PT (2013) Ecotoxicity of nanoparticles. ISRN Toxicology. http://dx.doi.org/10.1155/2013/574648

Ray PC, Yu H, Fu PP (2009) Toxicity and environmental risks of nanomaterials: challenges and future needs. J Environ Sci Health C Environ Carcinog Ecotoxicol Rev 27:1–35

Renzi M, Guerranti C (2015) Ecotoxicity of nanoparticles in aquatic environments: a review based on multivariate statistics of meta-data. J Environ Anal Chem 2:149

Ridley MK, Hackley VA, Machesky ML (2006) Characterization and surface-reactivity of nanocrystalline anatase in aqueous solutions. Langmuir 22:10972–10982

Roco MC (2005) Environmentally responsible development of nanotechnology. Environ Sci Technol 39:106A–112A

Schimel JP, Schaeffer SM (2012) Microbial control over carbon cycling in soil. Front Microbiol. https://doi.org/10.3389/fmicb.2012.00348

Schloter M, Nannipieri D, Sorensen S, van Elsas JD (2018) Microbial indicators for soil quality. Biol Fertil Soils 54:1–10

Sharma VK (2013) Silver nanoparticles in aquatic environment: a review. In: Shamim N, Sharma VK (eds) Sustainable nanotechnology and the environment: advances and achievements. ACS symposium series, vol 1124. American Chemical Society, Washington, DC, USA, pp 165–179

Sharma VK, Yngard RA, Lin Y (2009) Silver nanoparticles: green synthesis and antimicrobial activities. Adv Colloid Interface Sci 145:83–96

Shen Z, Chen Z, Hou Z, Li T, Lu X (2015) Ecotoxicological effect of zinc oxide nanoparticles on soil microorganisms. Front Environ Sci Eng 9:912–918

Shin YJ, Kwak JI, An YJ (2012) Evidence for the inhibitory effects of silver nanoparticles on the activities of soil exoenzymes. Chemosphere 88:524–529

Shrivastava S, Bera T, Roy A, Singh G, Ramachandra Rao P, Dash D (2007) Characterization of enhanced antibacterial effects of novel silver nanoparticles. Nanotechnology 18:225103

Siani NG, Fallah S, Pokhrel LR, Rostamnejadi A (2017) Natural amelioration of zinc oxide nanoparticle toxicity in fenugreek (Trigonella foenum-gracum) by arbuscular mycorrhizal (*Glomus intraradices*) secretion of glomalin. Plant Physiol Biochem 112:227–238

Sillen W, Thijs S, Abbamondi GR, Janssen J, Weyens N, White JC, Vangronsveld J (2015) Effects of silver nanoparticles on soil microorganisms and maize biomass are linked in the rhizosphere. Soil Biol Biochem 91:14–22

Simonin M, Guyonnet JP, Martins JMF, Ginot M, Richaume A (2014) Influence of soil properties on the toxicity of TiO_2 nanoparticles on carbon mineralization and bacterial abundance. J Hazard Mater. https://doi.org/10.1016/j.jhazmat.2014.10.004

Simonin M, Richaume A (2015) Impact of engineered nanoparticles on the activity, abundance, and diversity of soil microbial communities: a review. Environ Sci Pollut Res 1–14

Singh PK, Singh M, Tripathi BN (2013) Glomalin: an arbuscular mycorrhizal fungal soil protein. Protoplasma 250:663–669

Smith SE, Read DJ (2008) Mycorrhizal symbiosis. Academic Press, London, UK

Stensberg MC, Wei Q, McLamore ES, Marshall D, Porterfield DM, Wei A, Sepúlveda MS (2011) Toxicological studies on silver nanoparticles: challenges and opportunities in assessment, monitoring and imaging. Nanomedicine 6:879–898

Sun TY, Gottschalk F, Hungerbühler K, Nowack B (2014) Comprehensive probabilistic modelling of environmental emissions of engineered nanomaterials. Environ Pollut 185:69–76

Thio BJR, Zhou D, Keller AA (2011) Influence of natural organic matter on the aggregation and deposition of titanium dioxide nanoparticles. J Hazard Mater 189:556–563

Tolaymat TM, El Badawy AM, Genaidy A, Scheckel KG, Luxton TP, Suidan M (2010) An evidence-based environmental perspective of manufactured silver nanoparticle in syntheses and applications: a systematic review and critical appraisal of peer-reviewed scientific papers. Sci Total Environ 408:999–1006

Torsvik V, Øvreås L (2002) Microbial diversity and function in soil: from genes to ecosystems. Curr Opin Microbiol 5:240–245

Voelker D, Schlich K, Hohndorf L, Koch W, Kuehnen U, Polleichtner C, Kussatz C, Hund-Rinke K (2015) Approach on environmental risk assessment of nanosilver released from textiles. Environ Res 140:661–672

Wang Q, Yang Z, Yang Y, Long C, Li H (2014) A bibliometric analysis of research on the risk of engineering nanomaterials during 1999–2012. Sci Total Environ 473–474:483–489

Wang ZY, Xie XY, Zhao J, Liu XY, Feng WQ, White JC, Xing BS (2012) Xylem- and phloem-based transport of CuO nanoparticles in maize (*Zea mays* L.). Environ Sci Technol 46:4434–4441

Wu B, Wang Y, Lee YH, Horst A, Wang Z, Chen DR, Sureshkumar R, Tang YJ (2010) Comparative eco-toxicities of nano-ZnO particles under aquatic and aerosol exposure modes. Environl Sci Technol 44:1484–1489

Xu ZY, Tang M, Chen H, Ban YH, Zhang HH (2012) Microbial community structure in the rhizosphere of *Sophora viciifolia* grown at a lead and zinc mine of northwest China. Sci Total Environ 453–464

Zarei M, Hempel S, Wubet T, Schafer T, Savaghebi G, Jouzani GS, Nekouei MK, Buscot F (2010) Molecular diversity of arbuscular mycorrhizal fungi in relation to soil chemical properties and heavy metal contamination. Environ Pollut 158:2757–2765

Zhang WX (2003) Nanoscale iron particles for environmental remediation: an overview. J Nanopart Res 5:323–332

Zhu YG, Christie P, Scott Laidlaw A (2001) Uptake of Zn by arbuscular mycorrhizal white clover from Zn-contaminated soil. Chemosphere 42:193–199

Chapter 6
Ecotoxicity of Nanometals: The Problems and Solutions

Irina A. Shurygina, Larisa M. Sosedova, Mikhail A. Novikov, Eugeniy A. Titov and Michael G. Shurygin

Abstract The chapter deals with the ecotoxicity of nanometals. It presents the results of studying the effects of metal nanoparticles on environmental objects such as plants, animals, soil, water and microorganisms. The study of metal nanoparticles effects on the algae showed that it has pronounced dose-dependent effects and the level of influence depends on the type of nanoparticle, its size and the type of algae. Some nanocomposites, such as cerium oxide, do not lose their toxic properties even in a bound state. It is demonstrated that Fe_2O_3 has the lowest destabilization risk for aquatic ecosystems. Soil pollution with metal nanocomposites has negative effects resulting in death of its inhabitants and decrease in their reproduction. These changes have a dose-dependent effects. Moreover, the severity of negative effects depends on the type of nanocomposite and the composition of the soil fauna. Copper and silver nanocomposites have the most pronounced toxicity for earthworms. Phytotoxicity of metal nanocomposites directly depends on the type of nanoparticles and concentration of accumulated matter in plant tissues. The ability of plants to accumulate metal nanoparticles in tissues is quite promising which can be used as utilization instrument of nanoparticles from environment. Nowadays, there is growing knowledge about ecological hazards and potential risk of exposure to nanomaterials entering into objects of the environment. While the next step will be protection of plants,

I. A. Shurygina (✉) · M. G. Shurygin
Irkutsk Scientific Center of Surgery and Traumatology,
1, Bortsov Revolutsii street, Irkutsk 664003, Russia
e-mail: irinashurygina@gmail.com

M. G. Shurygin
e-mail: shurygin@rambler.ru

L. M. Sosedova · M. A. Novikov · E. A. Titov
East-Siberian Institution of Medical and Ecological Research,
PO Box 1170, 3, district 12a, Angarsk, Irkutsk region 665827, Russia
e-mail: sosedlar@mail.ru

M. A. Novikov
e-mail: novik375@bk.ru

E. A. Titov
e-mail: g57097@yandex.ru

M. Rai and J. K. Biswas (eds.), *Nanomaterials: Ecotoxicity, Safety, and Public Perception*, https://doi.org/10.1007/978-3-030-05144-0_6

animals, and primarily, the human being from the negative effects and addressing the issue of environmentally safe usage of nanometals.

Keywords Nanocomposite · Ecosystem · Ecotoxicity · Aquatic biocenosis Soil biocenosis

Nomenclature

$AgNO_3$	Silver nitrate
Al_2O_3	Aluminum oxide
ATP	Adenosine Tri-Phosphate
bcl-2	B-cell lymphoma 2
Cd	Cadmium
CuO	Copper oxide
$CuSO_4$	Copper(II) sulfate
DNA	Deoxyribonucleic acid
Fe_2O_3	Iron(III) oxide
γ-GT	Gamma-glutamyltransferase
hsp	Heat shock protein
H_2O_2	Hydrogen peroxide
HTS	Highthroughput screening
nCuO	Nano-copper oxide
nZnO	Nano-zinc oxide
OECD	Organization for Economic Co-operation and Development
StAR	Steroidogenic acute regulatory protein
TiO_2	Titanium dioxideTitanium dioxide
ZnO	Zinc oxide
$ZnSO_4$	Zinc sulfate

6.1 Introduction

Nanotechnological industry is growing at a rapid pace which leads to concerns about possible ecological consequences of nanomaterials spreading and accumulating in the environment. The problems of assessing the ecological risks of nanotechnology arise because of incredible variety of forms, chemical composition, and the dimension of nanoparticles (George et al. 2011; Thomas et al. 2011).

Recent research has revealed 1,317 nanotechnological consumer products which unite 587 companies in 30 countries around the world (Vance et al. 2015).

With the laboratory and industrial production moving higher, nanoparticles penetrate water and earth where their life and behavior have yet to be studied. The report presented at Organization for Economic Co-operation and Development (OECD)

expert meeting on environmental safety and ecotoxicity (2014) dealt with the issues of chemical agents testing, including nanoparticles and nanomedicines, in terms of environmental safety. It was mentioned that there is a need to make one universal approach and unified research protocols of studying the toxicity properties of nanomaterials for scientists from different countries (Vance et al. 2015).

At the same time, when registering new contaminants, including those in nanoform, chemical industry faces a severe problem: the need of toxicological profile and regulation of more than 100,000 industrial chemical agents and mixtures. However, the system of making safety data sheets, classifications of chemical hazards for warning about possible negative effects on the environment and finding measures for their elimination has neither high-throughput methods nor alternatives to animal testing to solve this problem. The lack of screening, objective and economically efficient approach for assessing the environmental safety of innovative nanomaterials and nanomedicines led to new dangers and the issue of their safe usage became global (Kahru and Savolainen 2010; Holden et al. 2013).

6.2 Effect of Metal Nanoparticles on Aquatic Biocenosis

Behavior and transformation of nanoparticles in different water mediums (ground, open and sea waters) are under researched as well as their potential toxicity. The scientists are concerned about negative effects of nanoparticles on aquatic ecosystems (Wiesner et al. 2006; Navarro et al. 2008). Mostly, the characteristics of influence depend on the size of nanoparticles and their stability in dispersion phase. While studying the aquatic ecotoxicity, multi-layered approach is very important; in this approach main characteristics of the studied materials are taken into consideration such as solubility, agglomeration, deterioration and transformation in water medium. Unicellular organisms such as bacteria and/or algae are considered more resistant to toxic effects of nanoparticles than protozoa and multicellular organisms which have more developed system for its absorption and biotransformation (Bondarenko et al. 2013).

Daphnia magna is recognized as a key organism of the freshwater ecosystem, being an important phytoplankton. It is used to assess the state of the environment in ecotoxicological experiments due to its high sensitivity to environmental pollution, small body size and short life expectancy (Clément et al. 2013). Consequently, a lot of studies dedicated to the effects of nanoparticles on the state of aquatic ecosystem have been carried out on *Daphnia magna*.

Methodological approaches are developed to test the toxicity of nanoparticles in aqueous medium and to assess their stability using *Daphnia magna*. Also, the agglomeration process of particles in water is very important because their instability can lead to a change in bioavailability and toxicity. The degree of agglomeration and solubility comparison of ZnO, TiO_2 and Ag nanoparticles showed significant dependence on pH of the medium and ionic force of solution. The degree of toxicity was assessed by the immobilization of *Daphnia magna*. When studying silver nanoparticles, which form large stable agglomerates in a medium, high toxicity is shown at increasing solubility. Titanium dioxide caused the formation of small stable agglomerates (200 nm) in a medium with lower ionic strength at pH 7 whereas in a solution with high ionic strength the diameters of the agglomerates reached the micrometer range. Stable suspensions of TiO_2 caused immobilization of *Daphnia magna* for 48 h. While there was no toxicity in unstable solutions with high ionic strength containing highly agglomerated nanoparticles. Small agglomerates of the suspension for zinc oxide (ZnO) were obtained in the medium with low ionic strength at pH 7, but an increase in solubility was detected which increased the toxicity of solution (Damoiseaux et al. 2011; Nel et al. 2013; Cupi et al. 2016).

As described above, the agglomeration can change bioavailability and toxicity of nanoparticles as well as stability of suspensions, and consequently, affects the reproducibility of test results (González-Sánchez et al. 2013).

In studies on acute and chronic toxicity assessment of chromium (III) oxide nanoparticles the effect on viability of *Daphnia magna* was shown. When assessing acute toxicity, the value obtained while testing on *Daphnia magna* was 6.79 mg/l. In long-time trials in 0.50 mg/l dose of nanoparticles the effect on growth, reproduction, and life expectancy of *Daphnia magna* was noted (Puerari et al. 2016).

Researchers witnessed high significance of toxicity studies of nanoparticles for minimizing the risk of environmental pollution. Based on the results of experiments in testing three different metal nanoparticles, the dependence of their toxicity for *Ceriodaphnia dubia* on dispersion and solubility was shown (Kennedy et al. 2017). It was found that powder of silver nanoparticles was not dispersible, nano-TiO_2 powder was dispersible but unstable, inclusion of silver nanoparticles into the matrix of polyvinylpyrrolidone increases its stability in the test environment. Based on the performed functional results, mathematical indicators were estimated for calculating and expressing toxicity: nominal, arithmetic average, geometric average, and time-weighted average. The results showed that while arithmetic means were effective for expressing toxicity of more stable materials, time-weighted average was suitable for unstable nanoparticles of nano-TiO_2. The authors presented an additional approach for monitoring the effect of nanoparticles on bioassays while solving the ecotoxicity problems (Kennedy et al. 2017).

The objective of the study by Tomacheski et al. (2017) was to evaluate the impact of shape, surface, size and matrix containing silver and titanium dioxide (TiO_2) on survival of *Daphnia magna* in suspension. *Daphnia magna* demonstrated high susceptibility to all the silver-based composites with 100% death rate after 24 h of exposure. While incubating with TiO_2 a linear concentration-response versus mortality was not achieved. It was concluded that silver toxicity does not depend on its form (nano or ion) or matrix properties (silicon dioxide, phosphate glass or bentonite). TiO_2 demonstrated low acute toxicity against *Daphnia magna* (Tomacheski et al. 2017).

Special attention should be given to time duration while studying ecotoxicity of metal oxides. Usually, short duration impact is tested whereas it is necessary to study an entire life cycle of nanomaterials. In course of time configuration, chemical and physical characteristics of nanoparticles undergo transformation which leads to different bioavailability and routes of exposure (Salieri et al. 2015).

Nowadays, the central mechanism of nanoparticles toxicity is their ability to generate reactive oxygen species capable of oxidizing biomolecules (Maness et al. 1999; Shurygina et al. 2015; Jukapli and Bagheri 2016). Thus, the development of sensitive instruments for quantitative evaluation of reactive oxygen species and oxidative stress generation is highly valued (Zhang et al. 2013; Yang et al. 2016).

To search for new quick ways to assess nanoparticles biotoxicity the researchers offer to register reactive oxygen species, forming under the influence of nanoparticles. The study was focused on copper, zinc and titanium oxides, silver and fullerenes. It was suggested to use the combination of highly-productive luminescent bacteria tests using *Escherichia coli* strains. The authors compared toxicity of TiO_2, CuO, ZnO (25, 30, 70 nm, respectively) nanoparticles and soluble salts ($ZnSO_4$, $CuSO_4$ and $AgNO_3$) against the studied strains. Bacterial toxicity tests showed that metals in nanoform were more toxic. According to authors, screening of chemical agents forming reactive oxygen species plays a significant role in different scopes of inquiry and gives new opportunities for fast screening of ecotoxicity (Ivask et al. 2010). The studies continue the works where summary review on toxicity of nanoparticles is provided for aquatic microorganisms and based on paradigm of oxidative stress. The researchers developed optical biosensor based on absorption measurement of cytochrome C for sensitive, non-invasive and continuous measurement of H_2O_2 which in real-time allows to determine quickly the rate and amount of H_2O_2 in extracellular medium in response to an impact of nanoparticles on cells of aquatic microorganisms (Santschi et al. 2017). In numerous studies the antibacterial properties of silver nanoparticles were shown. However, their antibacterial mechanism remains unclear (Oberdörster et al. 2005; Shurygina et al. 2011b; Fadeeva et al. 2015; Kon and Rai 2016; Ganenko et al. 2017; Lesnichaya et al. 2017).

With small diameter silver particles penetrate microorganisms easily, bind with thiol-containing groups of proteins and ferments. Metabolic disturbance in bacteria can lead to their quick death. Moreover, penetration of metal particles disturbs oxidation-reduction potential which leads to an increase in permeability of the cell membranes and decrease in the intracellular level of ATP (Cao et al. 2016). Mi et al. (2002) found out that nano-silver can destroy DNA structure of bacteria at the same time disturbing the cloning process. Moreover, silver can interact directly with bacteria causing transformational changes and functional damage in cell membrane, and these effects can lead to cytoplasmic leakage and bacteriolysis (Shurygina et al. 2011a; Mosselhy et al. 2015).

Considering that organisms, including aquatic ones, are rarely exposed to certain chemical agents, overall estimate should be done judging by the content of contaminants in water with different toxic potentials (Altenburger et al. 2004). Consequently, raising awareness about the degree of mix of pollutants should increase the degree of credibility while estimating the toxicity of systems containing several nanoparticles (Song et al. 2014; Tong et al. 2015; Sanchís et al. 2016). Ye et al. (2017) carried out combined ecotoxicity study with nano-zinc oxide (nZnO) and nano-copper oxide (nCuO) to freshwater algae *Scenedesmus obliquus* (Ye et al. 2017). The study showed that influence of nZnO and nCuO mixture can be predicted according to the effects of certain mixture components. At the same time, the effects of saluted fractions of metal oxides mixture are not identical to the effects of isolated and combined toxicity. Contribution of nCuO (for metal mass) into combined toxicity was more than that of nZnO. The authors offer to use suggested mathematic model at screening assessment of mixture toxicity of contaminants in nanoform.

Shang with co-authors (2015) showed that to understand the ecological consequences of accumulation of copper oxide nanoparticles in water, the characteristics of water medium has crucial importance. The phototoxicity of copper oxide nanoparticles varied depending on the chemical composition of water. The phototoxicity of CuO nanoparticles against *Escherichia coli* was the highest in deionized water. The concentration of copper oxide nanoparticles in water and their antibacterial activity correlate among themselves which gives the authors an opportunity to speak of its prognostic significance of copper oxide when assessing toxicity in ecosystem.

A promising direction is the use of quantum dots as new materials for photovoltaic technologies, light-emitting devices and biomedical applications. Zhou et al. (2016) studied the effect of CdSe/ZnS quantum dots on the growth rate of four microalgae: *Phaeodactylum tricornutum*, *Rhinomonas reticulata*, *Isochrysis galbana*, and *Dunaliella tertiolecta*. Furthermore, the influence of quantum dots on copepod *Acartiatonsa* was analyzed. They estimated survival rate, fertility and heat shock protein expression hsp70. Quantum dots with the size from 0.15 to 1.5 nm were studied. In acute experiments several factors were revealed: high mortality rate of copepods, decrease in the growth rate of algae and increase in expression of hsp70 in adult specimens. In general, the results indicate that species unable to swim in deep water such as *P. tricornutum* and early hatched copepods may be more prone to toxic effects of quantum dots that tend to aggregate and settle in seawater (Zhou et al. 2016).

Viability tests of bacterial isolate *Bacillus licheniformis* from fresh lake water are of interest because most of studies were carried out in experiments in vitro. It is found that influence of Al_2O_3 nanoparticles in 1 μg/ml concentration for 2 h leads to decrease in cell viability by 17%. According to authors, the stability of nanoparticles is determined by the presence of nanosized natural colloids in lake water (Pakrashi et al. 2011).

When analyzing the toxicity of ZnO and TiO_2 nanoparticles for *Escherichia coli*, it is demonstrated that its viability varied linearly with the concentration of nanoparticles (Kumar et al. 2011). The pathogenetic mechanism of the toxic effect of nanoparticles ZnO and TiO_2 is described. Decrease in lactate dehydrogenase activity, glutathione content, lipid hydroperoxides, malondialdehyde, and reactive oxygen species was found. The authors assume that oxidative stress is the main cytotoxicity mechanism of the studied nanoparticles and leads to DNA damage. Thus, nanoparticles of ZnO and TiO_2 are genotoxic for *Escherichia coli*. The researchers pay special attention to the necessity of detailed evaluation of safety/toxicity for metal oxides (Kumar et al. 2011).

High throughput screening (HTS) is a popular approach to cytotoxicity assessment which offers opportunity to test quickly the effect of nanoparticles on bacteria in several concentrations simultaneously. However, its drawback is that it is possible to test only one bacterial species which limits the ability to predict reactions of complex microbial communities (Jin et al. 2010; Tong et al. 2013). It is shown that entry of nano-TiO_2 into the environment can change the composition of aqueous bacterial communities, which has significantly affect the stability and functioning of the aquatic ecosystems. HTS usage demonstrated that response reaction of the four bacterial species, widespread in aquatic habitat, was different in short-term exposure of TiO_2 nanoparticles (Binh et al. 2014a). The authors supposed that the responses of bacterial communities to nanoparticles are complex, and they clearly demonstrated it in their further studies using HTS test modifications. Bacterial community compositions of Lake Michigan and the Chicago River were changed significantly under the influence of TiO_2 nanoparticles characterized by quantitative decrease of Actinomycetales, Sphingobacteriales, Limnohabitans, Flavobacterium and significant increase of Limnobacter (Binh et al. 2014b).

Algae are one of the most important participants of ecosystem and one of the main components of food chain, which allows to recommend them as ecological indicator when conducting the monitoring of environmental pollution with metal nanoparticles. The study of copper nanoparticles influence on unicellular algae *Chlamydomonas reinhardtii* showed that short exposure of algae to copper nanoparticles even with low doses caused changes in cell size and led to discoloration of chlorophyll i.e. destroyed its protein structure and led to initiation and development of the oxidative stress in a cell (von Moos et al. 2015). In return, the influence of zinc oxide nanoparticles was characterized by growth retardation of algae. Also, the dose-dependent character of zinc oxide nanoparticles influence was clearly evidenced; with escalating the dose lethal effect was observed. Freshwater algae turned out to be more stable to zinc nanoparticles influence than marine ones (Aravantinou et al. 2015). Whereas ceric oxide nanoparticles had expressed toxic properties even in bound state and at the same time the toxicity of nanoparticle depended on their size. Ceric oxide, as well as other metal nanocomposites, inhibits the growth of algae by binding nanoparticles of ceric oxide with components of cellular membranes and, consequently, disturbing transport of nutrients to the cell and withdrawal of some products of life from the cell. Ceric nanocomposite can damage cell membranes also by causing oxidative stress (Manier et al. 2013). As well as cerium, barium titanate nanoparticles had toxic effect on algae even in small doses. Titanate nanoparticles bind themselves to cellular membrane components disturbing the transport process of nutrients and products of life withdrawal. These nanoparticles also disturbed cell proliferation of algae by decreasing photosynthetic activity of a cell and ATP level (which indicates depression of mitochondrial cell activity), increasing the content of superoxide dismutase in a cell and development of oxidative stress (Polonini et al. 2015).

Convincing evidences about abilities of nano-CuO and nano-TiO_2 to generate reactive oxygen species and oxidative stress in unicellular green algae *Chlamydomonas reinhardtii* were demonstrated. Superoxides were also obtained by influence of synthetic metal-oxide nanoparticles on marine bacteria *Photobacterium phosphoreum* (Gomes et al. 2005; González-Sánchez et al. 2013; Wang et al. 2017). Influence of TiO_2 nanoparticles on growth, oxidative stress and chlorophyll accumulation in green alga *Dunaliella tertiolecta* was studied in sea water (Callegaro et al. 2015; Morelli et al. 2018). In the studied algae early (after 6 h of exposure) oxidative stress was observed, which subsequently decreased with longer exposure. At the same time, the greatest number of active forms of oxygen was detected in extracellular spaces. There was no decrease in growth of green alga *Dunaliella tertiolecta* as well as the chlorophyll synthesis despite the fact that in other studies there was a decrease in the content of chlorophyll in cells (Metzler et al. 2012; Ayatallahzadeh Shirazi et al. 2015).

In general, the research of metal nanoparticles influence on algae showed that it has expressed dose-dependent effect, but level of exposure depends on nanoparticle type, its size and algae species. Some nanocomposites, such as ceric oxide, do not lose their toxic properties even in bound state (Fig. 6.1).

Using nanomaterials as sorbents to clean Cd-contaminated water and soil is a promising strategy. Hence, there arises the question about the influence of nanomaterials on the ecological state of these biotopes. The model shows the toxicity of Cd nanoparticles, which form in water at its adsorption, towards *Daphnia magna* (Gao et al. 2018). Concentration increase of Cd nanoparticle showed oxidative damage with increase in activity of peroxidases, superoxide dismutase, catalase and anti-superoxide anion. The damage to cell membranes was indicated by a gradual increase in the malondialdehyde content. Therefore, the use of nanomaterials for Cd-contamination adsorption is associated with potential risk to aquatic organisms (Gao et al. 2018).

Some studies are dedicated to positive effects of zerovalent iron nanoparticles on remediation processes (Cullen et al. 2011; Semerád and Cajthaml 2016). There were described cases of artificial introduction of some chemical agents of human origin into the environment, which were later proved as toxic against biota, and it was not anticipated earlier. Thus, certain concern about the use of zerovalent iron particles is quite reasonable. Many consider nano-iron with zero valency as a promising chemical agent with the possibility to use in the aqueous phase including groundwater. Its effectiveness against various types of contaminants, including chlorinated derivatives, pesticides and heavy metals has been demonstrated (Barrera-Díaz et al. 2012; Gomes et al. 2013; Jang et al. 2014; Tesh and Scott 2014).

However, the use of zerovalent iron requires a thorough study because of contradicting results. It is shown that low concentration of zerovalent iron nanoparticles (5.1 mg/l) can increase accumulation of lipids significantly, decrease the content of saturated and monounsaturated fatty acids excluding palmitoleic acid, and increase the content of polyunsaturated fatty acids in cells of microorganisms used in the study (Pádrová et al. 2016). The authors even assumed that adding zerovalent iron nanoparticles can contribute to optimization of economic production of oil from oil-containing organisms.

The importance of the effect of zerovalent iron concentration was well demonstrated by Jang et al. (2014), who studied its influence on the growth of *Paracoccus* sp. The authors found out that low concentration (50 mg/l) can contribute to cell growth and nitrate biodegradation; very high concentration (1000 mg/l) causes toxicity towards the studied culture (Jang et al. 2014).

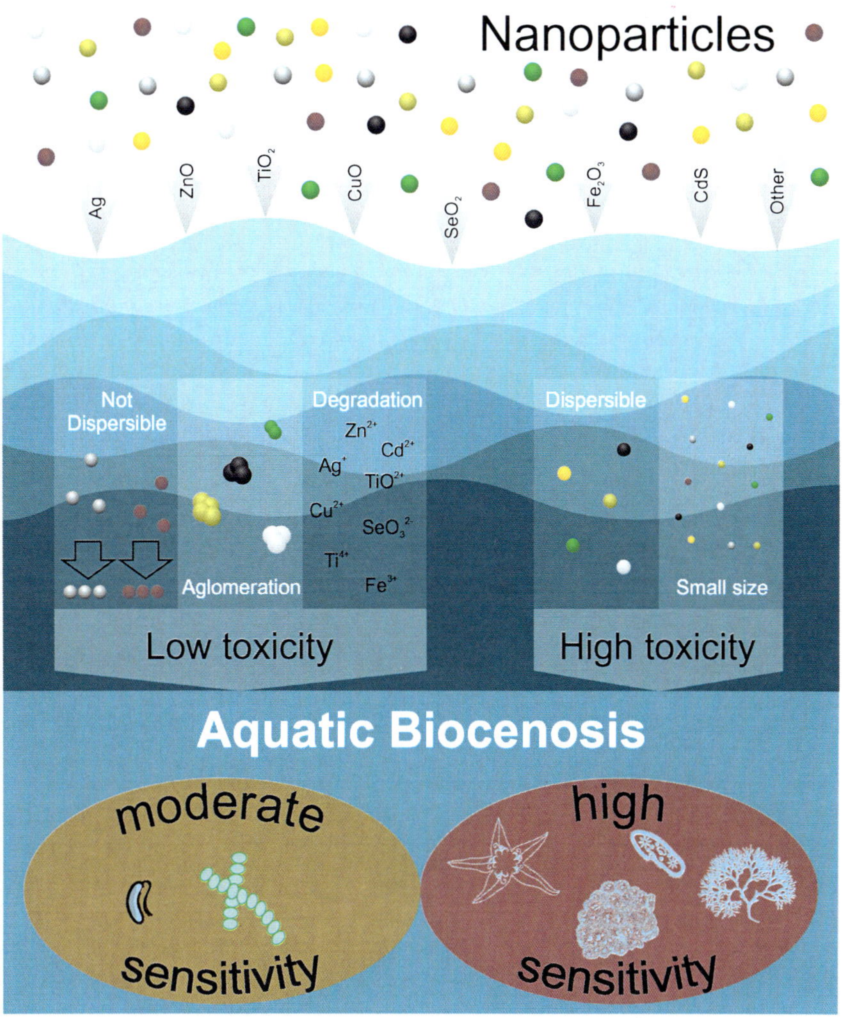

Fig. 6.1 Effect of metal nanoparticles on aquatic biocenosis

In general, it is considered that zerovalent iron nanoparticles participate in the transformation of contaminants, sustain microbial community in biotope and reconstruct contaminated sites decreasing general toxicity (Mitrano et al. 2015).[2] In this case, the originated iron oxides are insoluble and non-toxic. Thus, we should not overestimate the danger of using zerovalent iron nanoparticles for the environment which can be considered as decontaminant of a specific contaminated site.

Environmental problems of water pollution with wastewater effluents are global. Adsorption is one of the most effective methods for removing toxic heavy metals from wastewater effluents. Nowadays, there is a growing tendency to use nanostructured materials such as fullerenes (Darwish 2012), carbon nanotubes in particular (Burakov et al. 2014) and graphene (Melezhyk et al. 2016) for removing heavy metals from wastewater effluents thanks to unique physico-chemical and mechanical properties (Nogueira et al. 2015). Nanomaterials have very high strength, endurance, conductivity, and thermal stability. Moreover, high surface area, combined with various types of intermolecular interactions, makes it possible to effectively use them in different adsorption systems (Burakov et al. 2018).

When choosing the most suitable adsorbents for wastewater treatment, the potential toxicity of NiO (100 and 10–20 nm), Fe_2O_3($\approx$85 × 425 nm) and TiO_2 (<25 nm) was assessed against a group of aquatic organisms: *Vibrio fischeri*, *Raphidocelis subcapitata*, *Brachionus plicatilis*, and *Daphnia magna* (Nogueira et al. 2012). Based on the results of this study, tested metal nanoparticles can be classified from highest to lowest toxicity of NiO (100 nm) > NiO (10–20 nm) > TiO_2 (<25 nm) > Fe_2O_3, which allows to conclude that Fe_2O_3 has the least risk for stability disturbance of aquatic ecosystems.

6.3 Effect of Metal Nanoparticles on Soil Biocenosis

Nowadays, metal nanocomposites are becoming increasingly widespread in industry which leads to an increase in concentrations of these elements in soils. Secondary studies on soil contamination with metal nanoparticles and ecotoxicity showed that these nanoparticles have negative effects on the soil fauna. The studies are often carried out with different types of earthworms. The choice of model organism is conditioned by the ability of earthworms to accumulate and biotransform metal ions, which makes them suitable biological indicators for biomonitoring of soil contamination. Also, during life processes worms ingests large amount of soil and as a result the contaminants have great toxic effect on them. It is assumed that the toxicity of nanometals in the soil depends on the concentration of their dissolved fraction (Fig. 6.2).

Nanocomposites of bismuth are widely used in the production of low-melting alloys, cosmetic products and pharmacology. For several years bismuth was widely used as a non-toxic alternative to lead which resulted in a sharp increase in its production. It has increased the concentration of bismuth and its derivatives in soil. The studies showed that the content of bismuth up to 300 mg/kg of soil does not have lethal effect on earthworms nor does it lead to their weight loss. At the same time soil contamination with bismuth nanoparticles led to a decrease in the reproduction of worms. It was also shown that the presence of worms in the soil increased the bioavailability of bismuth compared to soil in which there were no worms. It is possibly due to the ability of earthworms to release bismuth from the bound organic compounds to the soil as well as their ability to change the soil pH. Due to the feeding habits worms can accumulate high concentration of bismuth nanoparticles in their body, but the character of accumulation depends on the soil type, its moisture content and exposure time (Omouri et al. 2018).

Fig. 6.2 Effect of metal nanoparticles on soil biocenosis

Iron nanoparticles are used in soil remediation processes in the form of chemical adsorbents of contaminants of soil. Therefore, the assessment of the toxic effect of iron nanoparticles on soil fauna is a question of present interest. Characteristic of iron nanocomposites is their rapid oxidation and the formation of iron oxides which are already widespread in soil. It is shown that the effect of iron nanoparticles caused weight loss dysfunction in reproduction in earthworms. When the iron nanoparticles dose was 300 mg/kg of soil the death of the soil fauna was noted. Also, the mortality of worms depended on the type of soil. More significant changes were noted at the dose of 500 mg/kg. At the dose of 1000 mg/kg 100% cases of animal death were noted. In a long-term period, the toxic potentials of iron nanoparticles decreased. In addition, one of the favorable features of toxic effects of iron nanocomposites is the reduction in the severity of process in the distant period (El-Temsah and Joner 2012; Brami et al. 2017).

Nanoparticles of silver make up about 25% of nanomaterials used in commercial products. Due to the proven antimicrobial characteristics, silver nanocomposites are a part of food, cosmetic and medical products. As a result, silver nanocomposites get into wastewaters and industrial wastes in enormous quantities, which causes the accumulation of these nanocomposites in water and soil medium. Also, silver ions are one of the most toxic forms of metals. Study on the effects of silver nanocomposite on earthworms showed that the death of animals occurred at a concentration of 100 mg/kg of soil. Silver nanoparticles also caused the weight loss and dysfunction in reproduction of earthworms (Brami et al. 2017).

Copper nanoparticles, having antimicrobial characteristics, are widely used in manufacturing, for example, in the form of a preservative in the wood-working industry, food products and medical supplies. Copper nanoparticles can disturb the immunological status of earthworms causing the development of a pronounced phagocytic reaction, dysfunction in reproduction and, as a result, they can lead to decrease in earthworm population. Copper nanoparticles at a concentration of 50 mg/kg of soil caused the death of soil fauna. Also, other nanocomposites caused a decrease in reproductive function (Gomes et al. 2015; Gautam et al. 2018).

Therefore, soil contamination with metal nanocomposites has a pronounced negative impact involving death of its inhabitants and decrease in their reproduction. These changes show dose-dependent nature. What is more, the severity of the negative effect depends on the type of nanocomposite and the composition of soil fauna. Silver and copper nanoparticles have the most pronounced toxicity for earthworms.

6.4 Effect of Metal Nanoparticles on Plants as a Component of Biocenosis

Nanocomposites of metals find expanding applications in agriculture, for example, for chemical disinfection of seeds to reduce the losses of the seed fund. Nanosized metals have high bactericidal activity.

One of the benefits of nano-silver is its ability to catalyze metabolic reactions. Several studies have shown that seed treatment with silver nanocomposite at low concentrations does not affect their germination ability and can be used as an antiseptic in storing them. What is more, low concentrations of silver nanoparticles caused increase in growth of plant root system (Barabanov et al. 2018). When increasing the concentration, silver nanocomposite conversely acquires phytotoxic properties: inhibits seed sprouting and slows down the sprouts growth. Perhaps it is because the liquid medium of the seed coat parenchyma helps silver nanocomposite to enter the brood case. At seed sprouting the main silver concentration is seen in root system of a sprout. When increasing the concentration, dysfunction in fibrils development of the suction zone is seen, root cap breaks, cells of epidermis and root cortex are vacuolated (Amooaghaie et al. 2015).

It is assumed that the phytotoxic properties of silver nanoparticles are connected either directly with the toxic effect of nanocomposite or indirectly by altering the metabolic activity of a root or developing oxidative stress in tissue. Also, phytotoxicity depends on nanoparticle size which is, apparently, because smaller particles penetrate the plant tissues easier (Cvjetko et al. 2017). One of research trends of metal nanoparticles, for example iron (Manquián-Cerda et al. 2017) and silver (Tripathi et al. 2017), is method for production by accumulation in plant leaves because this method is more beneficial compared to the "chemical" one. The main advantage of the "green" method is the decrease of toxic properties of nano-silver in comparison to the "chemical" analogues. Another advantage is collective effect of silver nanoparticles and vegetable matter. The studies conducted in *Mukia maderaspatana* showed that the phenolic groups contained in the leaves of this plant act as stabilizers and restorator for silver nanoparticles, and together they enhance their antioxidant properties (Harshiny et al. 2015). Supposedly, such conjugates will become highly-demanded in pharmacology (Harshiny et al. 2015). Similar studies were carried out in *Pongamia pinnata* (Priya et al. 2016) and *Carica papaya* (Kokila et al. 2016). These plants also can accumulate silver nanoparticles and the extracts contained in them significantly enhance antimicrobial and antioxidant characteristics of nanoparticles. Such conjugates of silver nanoparticles and plant components got the name "plant antibiotics" (Kokila et al. 2016).

The studies on the effects of titanium oxide nanoparticles on plants are of particular interest due to their extremely wide application in various aspects of human life. The studies show that these nanoparticles can accumulate in plants at very high concentrations mainly in the root system, but no toxic effect of titanium oxide on plants was observed (Foltête et al. 2011). When treating the radish seeds with titanium oxide nanoparticles, it did not affect their germinating ability. Moreover, the presence of these nanoparticles in seeds decreases the level of toxic effects of cadmium on them (Manesh et al. 2018).

Plants can regulate the content of metal nanoparticles in them. A study carried out in rice showed that a number of amino acids, for example aspartate and lysine, can regulate the accumulation of gold nanoparticles in the seed coat and plant roots. Unfortunately, the mechanism of such regulation needs further investigation (Ye et al. 2018).

Therefore, metal nanocomposites can have positive and negative effects on plants. Phytotoxicity of metal nanocomposites directly depends on the type of nanoparticles and the concentration of accumulated matter in plants tissues. The ability of plants to accumulate metal nanoparticles in tissues is very promising, and can be used in development of medicines of plant origin and as an instrument for nanoparticles utilization from the environment. The discovery about the plants regulation mechanisms of metal nanoparticles content in them is also quite interesting.

6.5 Effect of Metal Nanoparticles on the Animal World as a Component of Biocenosis

Metal nanoparticles refer to prioritized materials; negative effects of inhalation of certain nanomaterials by living organism are thoroughly studied. The most reliable method of studying the effects of nanoparticles, as well as any other substances, on living systems as components of biocenosis is in vivo test method (Rothman 2002). Various living organisms can act as models for studies of nanoparticles. However, the most developed studies in science are the ones on animal models: either vertebrate or invertebrate. The studies in animal models have found out that nanoparticles can penetrate the organism by inhalation, per os, through the damaged skin, and by medicine injections with nanocarriers (Oberdörster et al. 2005). What is more, a lot of metals in macroform are completely harmless to an organism and are already widely used in industry and at home. For example, titanium oxide in its usual state is a nontoxic, indifferent compound, and it is widely used in the pharmaceutical and cosmetic industry. However, exposure to titanium oxide nanoparticles leads to the stimulation of free radicals production and causes a pronounced prooxidant effect; inflammation and even pulmonary fibrosis can also be observed with inhalation exposure (Warheit et al. 2003). It is noted that inflammations of pulmonary tissue are mainly due to the prooxidant and genotoxic effect of nanoparticles (Elder et al. 2000). At inhalational exposure to TiO_2 ultrafine particles (0.8 μm, 10 mg/m^3) for a year caused many factors: decrease in the life duration of animals, accumulation of TiO_2 nanoparticles in the organism, weight loss in rats, increase in the number of neutrophils and phagocytes in bronchoalveolar lavage, development of inflammatory changes in lungs and other organs, induction of epithelial proliferation and fibroproliferative lung injury with the development of bronchus obliteration (Bermudez et al. 2002). When administered orally, titanium oxide nanoparticles are characterized by even greater toxicity: more intensive lungs, liver, spleen, and kidneys penetration, considering that nanoparticles cannot be removed from the kidneys (Wang et al. 2007).

The inhalation effect of even low concentrations of copper nanoparticles leads to a potentiation of toxicity towards lung tissue i.e. other substances in the presence of copper nanoparticles will have more pronounced negative effect (Meng et al. 2007). The per oral effect of copper nanoparticles disrupts the acid-base balance of blood and causes the development of metabolic alkalosis, degenerative changes of liver and signs of glomerulonephritis (Williams 1998; Galla 2000).

The reproductive system is one of the points of metal nanoparticles application. For example, bismuth nanoparticles have been proved to be toxic for earthworms *Eisenia andrei*, reducing their reproduction at concentration of more than 116 mg/kg (Omouri et al. 2018). The effect of titanium oxide leads to endocrine profile change (an increase in the activity of gamma-glutamyl transferase (γ-GT) of testes and a decrease in steroidogenic acute regulatory protein (StAR)) in rats (Shahin and Mohamed 2017). It also affects the expression of the c-kit gene, testosterone level in serum and quantity of spermatozoids (Shahin and Mohamed 2017). Also, toxic effects on the spermatogenesis of experimental animals were detected while exposing to silver and cadmium nanoparticles (Braydich-Stolle et al. 2005). The studies of Gao et al. (2017) proved that silver nanoparticles cause transcriptomic changes which potentially can have harmful effect on fetation (Gao et al. 2017). In the nerve tissue cells of white rats, when exposed to nano-silver, an increase of DNA damage was noted according to the results of the DNA comet test which indicates possible development of genotoxic effect (Sosedova et al. 2015). Great emphasis was put on the toxicity studies of silver nanoparticles because it has been applied in medicine for 15 years in bone grafting, dressing materials and other products for bactericidal purposes (Alt et al. 2004; Pozdnyakov et al. 2016). Also, silver nanoparticles can settle in lungs, liver and kidneys. As a result of axonal transport, they can penetrate the olfactory bulb and other structures of the brain through hematoencephalic barrier as well (Andrusishina et al. 2011).

To prove the penetration of silver nanoparticles into the brain tissue of white rats, the elemental composition of tissue samples was examined using a FEI Techni2G transmission microscope with a field emitter and an energy dispersive detector (Sosedova et al. 2016). In animals, exposed to silver nanoparticles enclosed in matrix of natural polymer arabinogalactan, silver was found in the brain tissue, which indicated the ability of nanoparticles to penetrate the brain through hematoencephalic barrier (Fig. 6.3).

Morphological study on the nerve tissue of white rats receiving silver nanoparticles enclosed in a matrix of a natural polymer–arabinogalactan which indicated significant changes in the structure of the nerve tissue. There was degenerative changes in neurons, karyorexis, an expansion of the perivascular spaces, neuronophagy, thickening of the arterial wall with swelling of epitheliocytes and myocytes (Figs. 6.4 and 6.5) (Novikov et al. 2014; Titov et al. 2015a).

Similar problems were observed when gold nanoparticles were administrated in similar doses (Abdelhalim and Jarrar 2012).

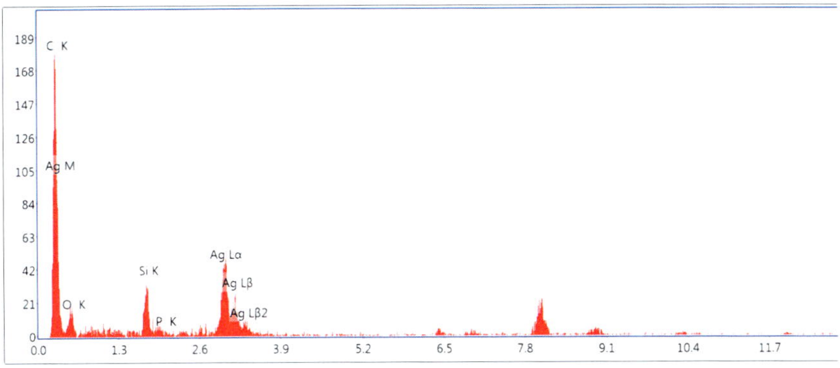

Fig. 6.3 The spectrometry results of an animal brain tissue section exposed to silver nanoparticles, enclosed in a matrix of a natural polymer—arabinogalactan

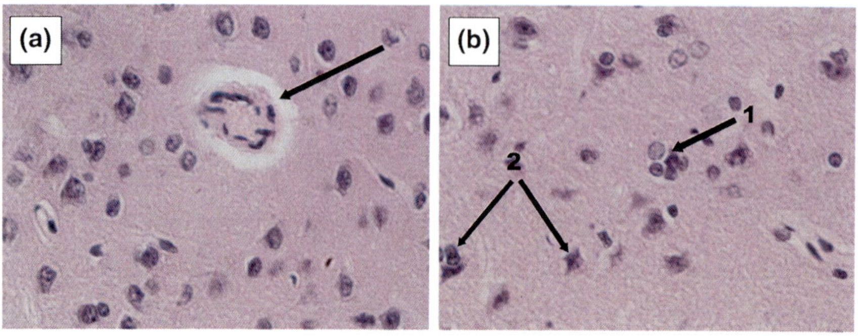

Fig. 6.4 Results of a histology of the parietotemporal lobe of cerebral cortex in white rats which received silver nanoparticles, enclosed in a matrix of a natural polymer—arabinogalactan **a** expansion of perivascular spaces, artery wall thickened, spongy, myocytes and endotheliocytes swollen (↑); **b** 1—neuronophagy, 2—degenerately changed neurons. Hematoxylin-eosin.400x

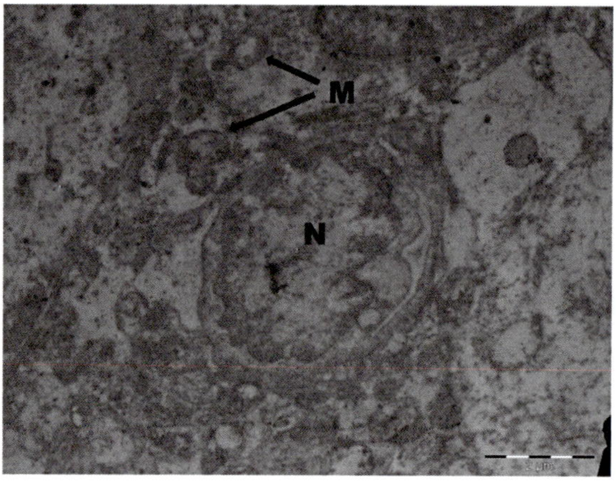

Fig. 6.5 Results of electron microscopy of the parietotemporal lobe of cerebral cortex in white rats which received silver nanoparticles, enclosed in a matrix of natural polymer N-nucleus; M-mitochondria. Nucleus deformation of a neuron. 10000x

Besides, silver nanoparticles cause damage to the nuclear apparatus of neurons, the development of an apoptotic process which is indicated by an increase extension of mitochondria space and activation of apoptosis protein modulators caspase-3 and bcl-2 persisting for a long time in the post-exposure period (Sosedova et al. 2014; Titov and Novikov 2014; Titov et al. 2016a).

There were reports of over stability of nanoparticles in the environment and the ability to preserve toxic properties both in an extended period of exposure (Novikov et al. 2014; Titov et al. 2015b; Sosedova and Filippova 2017), and during several years (Sahoo et al. 2007).

Studies on implantation of metal discs in rats and mice were carried out; the discs consisted of cobalt, nickel, titanium and silicon dioxides in the form of usual samples and discs containing nanomaterials. Histologic examination after 6 and 12 months showed that discs, containing cobalt and nickel in nanoform, induced rhabdomyosarcoma development in muscles of rats while in place of usual metal implants of all tested metals only fibrosis of the capsule was observed (Gatti 2005). Studies in laboratory animals allow to model different exposure routes of metal nanoparticles at different doses. Also, it allows to undertake studies in different time intervals (Rukavishnikov et al. 2017). At the same time, insufficient methodological inventory on safety of nanomaterials assessment, which some researchers recognize, dictates the need to create new approaches for toxicity assessment, including the use of intracellular proteomics evaluation methods (Titov et al. 2016b).

6.6 Conclusion

As can be seen from the information presented in the chapter, metal nanoparticles play a significant role as pollutants for the environment. In this case, the potentially negative impact of nanoparticles can have on both the animal and plant worlds. First of all, water and soil biocenoses are under the threat of negative impact.

Therefore, in today's world there is more and more knowledge about environmental hazards and the potential risk of exposure to nanomaterials entering environmental objects. Whereas, the next step will be protection of plants, animals, and, primarily, human beings from its negative effects and solving problems of nanomaterials in the environment ensuring their safe usage.

References

Abdelhalim MA, Jarrar BM (2012) Histological alterations in the liver of rats induced by different gold nanoparticle sizes, doses and exposure duration. J Nanobiotechnol 10:5

Alt V, Bechert T, Steinrücke P, Wagener M, Seidel P, Dingeldein E et al (2004) An in vitro assessment of the antibacterial properties and cytotoxicity of nanoparticulate silver bone cement. Biomaterials 25(18):4383–4391

Altenburger R, Walter H, Grote M (2004) What contributes to the combined effect of a complex mixture? Environ Sci Technol 38(23):6353–6362

Amooaghaie R, Saeri MR, Azizi M (2015) Synthesis, characterization and biocompatibility of silver nanoparticles synthesized from Nigella sativa leaf extract in comparison with chemical silver nanoparticles. Ecotoxicol Environ Saf 120:400–408

Andrusishina IN, Golub IA, Didikin GG, Litvin SE, Gromovoy TY, Gorchev VF, Movchan VA (2011) Structure, properties and toxicity of nanoparticles of silver and copper oxides. Biotechnology (Russian) 4(6):51–59

Aravantinou AF, Tsarpali V, Dailianis S, Manariotis ID (2015) Effect of cultivation media on the toxicity of ZnO nanoparticles to freshwater and marine microalgae. Ecotoxicol Environ Saf 114:109–116

Ayatallahzadeh Shirazi M, Shariati F, Keshavarz AK, Ramezanpour Z (2015) Toxic effect of aluminium oxide nanoparticles on green micro-algae Dunaliella salina. Int J Environ Res 9(2):585–594

Barabanov PV, Gerasimov AV, Blinov AV, Kravtsov AA, Kravtsov VA (2018) Influence of nanosilver on the efficiency of Pisum sativum crops germination. Ecotoxicol Environ Saf 147:715–719

Barrera-Díaz CE, Lugo-Lugo V, Bilyeu B (2012) A review of chemical, electrochemical and biological methods for aqueous Cr(VI) reduction. J Hazard Mater 223–224:1–12

Bermudez E, Mangum JB, Asgharian B, Wong BA, Reverdy EE, Janszen DB et al (2002) Long-term pulmonary responses of three laboratory rodent species to subchronic inhalation of pigmentary titanium dioxide particles. Toxicol Sci 70(1):86–97

Binh CT, Tong T, Gaillard JF, Gray KA, Kelly JJ (2014a) Common freshwater bacteria vary in their responses to short-term exposure to nano-TiO_2. Environ Toxicol Chem 33(2):317–327

Binh CT, Tong T, Gaillard JF, Gray KA, Kelly JJ (2014b) Acute effects of TiO_2 nanomaterials on the viability and taxonomic composition of aquatic bacterial communities assessed via high-throughput screening and next generation sequencing. PLoS ONE 9(8):e106280

Bondarenko O, Juganson K, Ivask A, Kasemets K, Mortimer M, Kahru A (2013) Toxicity of Ag, CuO and ZnO nanoparticles to selected environmentally relevant test organisms and mammalian cells in vitro: a critical review. Arch Toxicol 87(7):1181–1200

Brami C, Glover AR, Butt KR, Lowe CN (2017) Effects of silver nanoparticles on survival, biomass change and avoidance behaviour of the endogeic earthworm Allolobophora chlorotica. Ecotoxicol Environ Saf 141:64–69

Braydich-Stolle L, Hussain S, Schlager JJ, Hofmann MC (2005) In vitro cytotoxicity of nanoparticles in mammalian germline stem cells. Toxicol Sci 88(2):412–419

Burakov A, Romantsova I, Kucherova A, Tkachev A (2014) Removal of heavy-metal ions from aqueous solutions using activated carbons: effect of adsorbent surface modification with carbon nanotubes. Adsorpt Sci Technol 32(9):737–747

Burakov AE, Galunin EV, Burakova IV, Kucherova AE, Agarwal S, Tkachev AG, Gupta VK (2018) Adsorption of heavy metals on conventional and nanostructured materials for wastewater treatment purposes: a review. Ecotoxicol Environ Saf 148:702–712

Callegaro S, Minetto D, Pojana G, Bilanicová D, Libralato G, Volpi Ghirardini A et al (2015) Effects of alginate on stability and ecotoxicity of nano-TiO_2 in artificial seawater. Ecotoxicol Environ Saf 117:107–114

Cao H, Meng F, Liu X (2016) Antimicrobial activity of tantalum oxide coatings decorated with Ag nanoparticles. J Vac Sci Technol A 34(4):102

Clément L, Hurel C, Marmier N (2013) Toxicity of TiO(2) nanoparticles to cladocerans, algae, rotifers and plants—effects of size and crystalline structure. Chemosphere 90(3):1083–1090

Cullen LG, Tilston EL, Mitchell GR, Collins CD, Shaw LJ (2011) Assessing the impact of nano- and micro-scale zerovalent iron particles on soil microbial activities: particle reactivity interferes with assay conditions and interpretation of genuine microbial effects. Chemosphere 82(11):1675–1682

Cupi D, Hartmann NB, Baun A (2016) Influence of pH and media composition on suspension stability of silver, zinc oxide, and titanium dioxide nanoparticles and immobilization of Daphnia magna under guideline testing conditions. Ecotoxicol Environ Saf 127:144–152

Cvjetko P, Milošić A, Domijan AM, VinkovićVrček I, Tolić S, Peharec Štefanić P et al (2017) Toxicity of silver ions and differently coated silver nanoparticles in Allium cepa roots. Ecotoxicol Environ Saf 137:18–28

Damoiseaux R, George S, Li M, Pokhrel S, Ji Z, France B et al (2011) No time to lose–high throughput screening to assess nanomaterial safety. Nanoscale 3(4):1345–1360

Darwish AD (2012) Fullerenes. Annu Rep Prog Chem Sect A Inorg Chem 108:464–477

Elder AC, Gelein R, Finkelstein JN, Cox C, Oberdörster G (2000) Pulmonary inflammatory response to inhaled ultrafine particles is modified by age, ozone exposure, and bacterial toxin. Inhal Toxicol 12(4):227–246

El-Temsah YS, Joner EJ (2012) Ecotoxicological effects on earthworms of fresh and aged nano-sized zero-valent iron (nZVI) in soil. Chemosphere 89(1):76–82

Fadeeva TV, Shurygina IA, Sukhov BG, Rai MK, Shurygin MG, Umanets VA et al (2015) Relationship between the structures and antimicrobial activities of argentic nanocomposites. Bull Russ Acad Sci Phys 79(2):273–275

Foltête AS, Masfaraud JF, Bigorgne E, Nahmani J, Chaurand P, Botta C et al (2011) Environmental impact of sunscreen nanomaterials: ecotoxicity and genotoxicity of altered TiO_2 nanocomposites on Viciafaba. Environ Pollut 159(10):2515–2522

Galla JH (2000) Metabolic alkalosis. J Am Soc Nephrol 11(2):369–375

Ganenko TV, Tantsyrev AP, Fadeeva TV, Shurygina IA, Shurygin MG, Sukhov BG, Trofimov BA (2017) Antimicrobial activity of silver nanocomposites of arabinogalactan and its sulfo derivatives. J Infectol (Russian) 9(1):51–52

Gao M, Zhang Z, Lv M, Song W, Lv Y (2018) Toxic effects of nanomaterial-adsorbed cadmium on Daphnia magna. Ecotoxicol Environ Saf 148:261–268

Gao X, Topping VD, Keltner Z, Sprando RL, Yourick JJ (2017) Toxicity of nano- and ionic silver to embryonic stem cells: a comparative toxicogenomic study. J Nanobiotechnol 15(1):31

Gatti AM (2005) Risk assessment of micro and nanoparticles and the human health. In: Handbook of nanostructured biomaterials and their applications. American Scientific Publisher, USA, pp 347–369

Gautam A, Ray A, Mukherjee S, Das S, Pal K, Das S et al (2018) Immunotoxicity of copper nanoparticle and copper sulfate in a common Indian earthworm. Ecotoxicol Environ Saf 148:620–631

George S, Xia T, Rallo R, Zhao Y, Ji Z, Lin S et al (2011) Use of a high-throughput screening approach coupled with in vivo zebrafish embryo screening to develop hazard ranking for engineered nanomaterials. ACS Nano 5(3):1805–1817

Gomes A, Fernandes E, Lima JL (2005) Fluorescence probes used for detection of reactive oxygen species. J Biochem Biophys Methods 65(2–3):45–80

Gomes HI, Dias-Ferreira C, Ribeiro AB (2013) Overview of in situ and ex situ remediation technologies for PCB-contaminated soils and sediments and obstacles for full-scale application. Sci Total Environ 445–446:237–260

Gomes SI, Murphy M, Nielsen MT, Kristiansen SM, Amorim MJ, Scott-Fordsmand JJ (2015) Cu-nanoparticles ecotoxicity–explored and explained? Chemosphere 139:240–245

González-Sánchez MI, González-Macia L, Pérez-Prior MT, Valero E, Hancock J, Killard AJ (2013) Electrochemical detection of extracellular hydrogen peroxide in Arabidopsis thaliana: a real-time marker of oxidative stress. Plant, Cell Environ 36(4):869–878

Harshiny M, Matheswaran M, Arthanareeswaran G, Kumaran S, Rajasree S (2015) Enhancement of antibacterial properties of silver nanoparticles—ceftriaxone conjugate through Mukia maderaspatana leaf extract mediated synthesis. Ecotoxicol Environ Saf 121:135–141

Holden PA, Nisbet RM, Lenihan HS, Miller RJ, Cherr GN, Schimel JP, Gardea-Torresdey JL (2013) Ecological nanotoxicology: integrating nanomaterial hazard considerations across the subcellular, population, community, and ecosystems levels. Acc Chem Res 46(3):813–822

Ivask A, Bondarenko O, Jepihhina N, Kahru A (2010) Profiling of the reactive oxygen species-related ecotoxicity of CuO, ZnO, TiO_2, silver and fullerene nanoparticles using a set of recombinant luminescent Escherichia coli strains: differentiating the impact of particles and solubilised metals. Anal Bioanal Chem 398(2):701–716

Jang MH, Lim M, Hwang YS (2014) Potential environmental implications of nanoscale zero-valent iron particles for environmental remediation. Environ Health Toxicol 29:e2014022

Jin X, Li M, Wang J, Marambio-Jones C, Peng F, Huang X et al (2010) High-throughput screening of silver nanoparticle stability and bacterial inactivation in aquatic media: influence of specific ions. Environ Sci Technol 44(19):7321–7328

Jukapli NM, Bagheri S (2016) Recent developments on titania nanoparticle as photocatalytic cancer cells treatment. J Photochem Photobiol B 163:421–430

Kahru A, Savolainen K (2010) Potential hazard of nanoparticles: from properties to biological and environmental effects. Toxicology 269(2–3):89–91

Kennedy AJ, Coleman JG, Diamond SA, Melby NL, Bednar AJ, Harmon A et al (2017) Assessing nanomaterial exposures in aquatic ecotoxicological testing: framework and case studies based on dispersion and dissolution. Nanotoxicology 11(4):546–557

Kokila T, Ramesh PS, Geetha D (2016) Biosynthesis of AgNPs using Carica Papaya peel extract and evaluation of its antioxidant and antimicrobial activities. Ecotoxicol Environ Saf 134(Pt 2):467–473

Kon K, Rai M (eds) (2016) Antibiotic resistance: Mechanisms and new antimicrobial approaches. Academic Press, Amsterdam

Kumar A, Pandey AK, Singh SS, Shanker R, Dhawan A (2011) Engineered ZnO and TiO(2) nanoparticles induce oxidative stress and DNA damage leading to reduced viability of Escherichia coli. Free Radic Biol Med 51(10):1872–1881

Lesnichaya MV, Alexandrova GP, Fadeeva TV, Shurygina IA, Shurygin MG, Sukhov BG and Trofimov BA (2017) Preparation of new antimicrobial agents based on silver nanocomposites in matrices of natural polysaccharides. J Infectol (Russian) 9(11):92

Manesh RR, Grassi G, Bergami E, Marques-Santos LF, Faleri C, Liberatori G, Corsi I (2018) Co-exposure to titanium dioxide nanoparticles does not affect cadmium toxicity in radish seeds (Raphanus sativus). Ecotoxicol Environ Saf 148:359–366

Maness PC, Smolinski S, Blake DM, Huang Z, Wolfrum EJ, Jacoby WA (1999) Bactericidal activity of photocatalytic TiO(2) reaction: toward an understanding of its killing mechanism. Appl Environ Microbiol 65(9):4094–4098

Manier N, Bado-Nilles A, Delalain P, Aguerre-Chariol O, Pandard P (2013) Ecotoxicity of non-aged and aged CeO_2 nanomaterials towards freshwater microalgae. Environ Pollut 180:63–70

Manquián-Cerda K, Cruces E, Angélica Rubio M, Reyes C, Arancibia-Miranda N (2017) Preparation of nanoscale iron (oxide, oxyhydroxides and zero-valent) particles derived from blueberries: reactivity, characterization and removal mechanism of arsenate. Ecotoxicol Environ Saf 145:69–77

Melezhyk AV, Kotov VA, Tkachev AG (2016) Optical properties and aggregation of graphene nanoplatelets. J Nanosci Nanotechnol 16(1):1067–1075

Meng H, Chen Z, Xing G, Yuan H, Chen C, Zhao F et al (2007) Ultrahigh reactivity provokes nanotoxicity: explanation of oral toxicity of nano-copper particles. Toxicol Lett 175(1–3):102–110

Metzler DM, Erdem A, Tseng YH and Huang CP (2012) Responses of algal cells to engineered nanoparticles measured as algal cell population, chlorophyll a, and lipid peroxidation: effect of particle size and type. J Nanotechnol 2012. Article ID 237284

Mi FL, Wu YB, Shyu SS, Schoung JY, Huang YB, Tsai YH et al (2002) Control of wound infections using a bilayer chitosan wound dressing with sustainable antibiotic delivery. J Biomed Mater Res 59:438–449

Mitrano DM, Motellier S, Clavaguera S, Nowack B (2015) Review of nanomaterial aging and transformations through the life cycle of nano-enhanced products. Environ Int 77:132–147

Morelli E, Gabellieri E, Bonomini A, Tognotti D, Grassi G, Corsi I (2018) TiO_2 nanoparticles in seawater: aggregation and interactions with the green alga Dunaliella tertiolecta. Ecotoxicol Environ Saf 148:184–193

Mosselhy DA, El-Aziz MA, Hanna M, Ahmed MA, Husien MM, Feng QL (2015) Comparative synthesis and antimicrobial action of silver nanoparticles and silver nitrate. J Nanopart Res 17:1–10

Navarro E, Baun A, Behra R, Hartmann NB, Filser J, Miao AJ et al (2008) Environmental behavior and ecotoxicity of engineered nanoparticles to algae, plants, and fungi. Ecotoxicology 17(5):372–386

Nel A, Xia T, Meng H, Wang X, Lin S, Ji Z, Zhang H (2013) Nanomaterial toxicity testing in the 21st century: use of a predictive toxicological approach and high-throughput screening. Acc Chem Res 46(3):607–621

NogueiraV Lopes I, Rocha-Santos TA, Rasteiro MG, Abrantes N, Gonçalves F et al (2015) Assessing the ecotoxicity of metal nano-oxides with potential for wastewater treatment. Environ Sci Pollut Res Int 22(17):13212–13224

NogueiraV Lopes I, Rocha-Santos T, Santos AL, Rasteiro GM, Antunes F et al (2012) Impact of organic and inorganic nanomaterials in the soil microbial community structure. Sci Total Environ 424:344–350

Novikov MA, Titov EA, Sosedova LM, OstroukhovaLA Trofimova NN, Babkin VA (2014) Biochemical and morphological changes in white rats after intragastric injection of a synthetic nanobiocomposite based on silver nanoparticles and arabinogalactan. Pharm Chem J 48(6):387–390

Oberdörster G, Oberdörster E, Oberdörster J (2005) Nanotoxicology: an emerging discipline evolving from studies of ultrafine particles. Environ Health Perspect 113(7):823–839

Omouri Z, Hawari J, Fournier M, Robidoux PY (2018) Bioavailability and chronic toxicity of bismuth citrate to earthworm Eisenia andrei exposed to natural sandy soil. Ecotoxicol Environ Saf 147:1–8

Pádrová K, Čejková A, Cajthaml T, Kolouchová I, Vítová M, Sigler K, Řezanka T (2016) Enhancing the lipid productivity of yeasts with trace concentrations of iron nanoparticles. Folia Microbiol (Praha) 61(4):329–335

Pakrashi S, Dalai S, Sabat D, Singh S, Chandrasekaran N, Mukherjee A (2011) Cytotoxicity of Al_2O_3 nanoparticles at low exposure levels to a freshwater bacterial isolate. Chem Res Toxicol 24(11):1899–1904

Polonini HC, Brandão HM, Raposo NR, Brandão MA, Mouton L, Couté A et al (2015) Size-dependent ecotoxicity of barium titanate particles: the case of Chlorella vulgaris green algae. Ecotoxicology 24(4):938–948

Pozdnyakov AS, Emel'yanov AI, Kuznetsova NP, Ermakova TG, Fadeeva TV, Sosedova LM, Prozorova GF (2016) Nontoxic hydrophilic polymeric nanocomposites containing silver nanoparticles with strong antimicrobial activity. Int J Nanomedicine 11:1295–1304

Priya RS, Geetha D, Ramesh PS (2016) Antioxidant activity of chemically synthesized AgNPs and biosynthesized Pongamia pinnata leaf extract mediated AgNPs—A comparative study. Ecotoxicol Environ Saf 134(Pt 2):308–318

Puerari RC, da Costa CH, Vicentini DS, Fuzinatto CF, Melegari SP, Schmidt ÉC et al (2016) Synthesis, characterization and toxicological evaluation of Cr_2O_3 nanoparticles using *Daphnia magna* and *Aliivibrio fischeri*. Ecotoxicol Environ Saf 128:36–43

Rothman SS (2002) Lessons from the living cell: the culture of science and the limits of reductionism. McGraw-Hill, New York

Rukavishnikov VS, Sosedova LM, Vokina VA, Titov EA, Novikov MA, Yakimova NL (2017) Assessment of neurotoxicity of nanometals encapsulated on arabinogalactan matrix. Occup Med Ind Ecol 10:25–29. Russian

Sahoo SK, Parveen S, Panda JJ (2007) The present and future of nanotechnology in human health care. Nanomedicine 3(1):20–31

Salieri B, Righi S, Pasteris A, Olsen SI (2015) Freshwater ecotoxicity characterisation factor for metal oxide nanoparticles: a case study on titanium dioxide nanoparticle. Sci Total Environ 505:494–502

Sanchís J, Olmos M, Vincent P, Farré M, Barceló D (2016) New insights on the influence of organic co-contaminants on the aquatic toxicology of carbon nanomaterials. Environ Sci Technol 50(2):961–969

Santschi C, Von Moos N, Koman VB, Slaveykova VI, Bowen P, Martin OJ (2017) Non-invasive continuous monitoring of pro-oxidant effects of engineered nanoparticles on aquatic microorganisms. J Nanobiotechnol 15(1):19

Semerád J, Cajthaml T (2016) Ecotoxicity and environmental safety related to nano-scale zerovalent iron remediation applications. Appl Microbiol Biotechnol 100(23):9809–9819

Shahin NN, Mohamed MM (2017) Nano-sized titanium dioxide toxicity in rat prostate and testis: possible ameliorative effect of morin. Toxicol Appl Pharmacol 334:129–141

Shang E, Li Y, Niu J, Guo H, Zhou Y, Liu H, Zhang X (2015) Effect of aqueous media on the copper-ion-mediated phototoxicity of CuO nanoparticles toward green fluorescent protein-expressing Escherichia coli. Ecotoxicol Environ Saf 122:238–244

Shurygina IA, Shurygin MG, Dmitrieva LA, Fadeeva TV, Ganenko TV, Tantsyrev AP et al (2015) Bacterio- and lymphocytotoxicity of silver nanocomposite with sulfated arabinogalactan. Russ Chem Bull 64(7):1629–1632

Shurygina IA, Sukhov BG, Fadeeva TV, Umanets VA, Shurygin MG, Ganenko TV et al (2011a) Bactericidal action of Ag(0)-antithrombotic sulfated arabinogalactan nanocomposite: coevolution of initial nanocomposite and living microbial cell to a novel nonliving nanocomposite. Nanomedicine 7(6):827–833

Shurygina IA, Sukhov BG, Fadeeva TV, Umanets VA, Shurygin MG, Verecshagina SA et al (2011b) Mechanism of bactericidal action Ag(0)-nanobiocomposite: the evolution of the original composite and live microbial cells in the new nanocomposite. Russ Phys J 54(2):285–288. Russian

Song M, Wang F, Zeng L, Yin J, Wang H, Jiang G (2014) Co-exposure of carboxyl-functionalized single-walled carbon nanotubes and 17-aethinylestradiol in cultured cells: effects on bioactivity and cytotoxicity. Environ Sci Technol 48(23):13978–13984

Sosedova LM, Filippova TM (2017) The effects of nanosilver, encapsulated in a polymeric matrix, on albino rats brain tissue. Nano Hybrids and Compos 13:263–267

Sosedova LM, Kapustina EA, Novikov MA (2015) Activity of apoptosis in the nerve tissue of white rats exposed to nanosilver arabinogalactan. Toxicol Rev 6:27–31. Russian

Sosedova LM, Novikov MA, Titov EA, Rukavishnikov VS (2016) [Induction of apoptosis in neurons of white rats under exposure of nanobiocomposite based on Ag(0) nanoparticles and arabinogalactan. Gig Sanit (Russian) 95(12):1210–1213

Sosedova LM, Titov EA and Novikov MA (2014) Morphofunctional evaluation of the effects of silver nanoparticles encapsulated in a polymer matrix. Microelem Med (Russian) 15(4):39–43

Tesh SJ, Scott TB (2014) Nano-composites for water remediation: a review. Adv Mater 26(35):6056–6068

Thomas CR, George S, Horst AM, Ji Z, Miller RJ, Peralta-Videa JR et al (2011) Nanomaterials in the environment: from materials to high-throughput screening to organisms. ACS Nano 5(1):13–20

Titov EA, Novikov MA (2014) Bcl-2 expression as criteria indicator of exposure to nanobiocomposites. Toxicol Rev (Russian) 4:34–38

Titov EA, Novikov MA, Sosedova LM (2015a) Effect of silver nanoparticles encapsulated in a polymer matrix on the structure of nervous tissue and expression of caspase-3. Nanotechnol Russ 10(7–8):640–644

Titov EA, Sosedova LM and Novikov MA (2015b) Alteration of white rats brain tissue inducted by assessment of silver nanocomposite in capsulated in polymer matrix. Patol Fiziol Eksp Ter (Russian) 59(4):41–44

Titov EA, Novikov MA, Sosedova LM (2016a) Expression of caspase-3 and bcl-2 proteins as an indicator of functional state of brain tissue in white rats exposed to argentum arabinogalactan. Occup Med Ind Ecol (Russian) 5:40–43

Titov EA, Novikov MA, Sosedova LM (2016b) Method of evaluating toxic effects of silver nanoparticles encapsulated within polymer matrix of arabinogalactan on brain tissue of laboratory animals in long-term. RU patent 2578545

Tomacheski D, Pittol M, Simões DN, Ribeiro VF, Santana RMC (2017) Effects of silver adsorbed on fumed silica, silver phosphate glass, bentonite organomodified with silver and titanium dioxide in aquatic indicator organisms. J Environ Sci (China) 56:230–239

Tong T, Binh CT, Kelly JJ, Gaillard JF, Gray KA (2013) Cytotoxicity of commercial nano-TiO_2 to Escherichia coli assessed by high-throughput screening: effects of environmental factors. Water Res 47(7):2352–2362

Tong T, Wilke CM, Wu J, Binh CT, Kelly JJ, Gaillard JF, Gray KA (2015) Combined toxicity of nano-ZnO and nano-TiO_2: from single- to multinanomaterial systems. Environ Sci Technol 49(13):8113–8123

Tripathi DK, Tripathi A, Shweta Singh S, Singh Y, Vishwakarma K et al (2017) Uptake, accumulation and toxicity of silver nanoparticle in autotrophic plants, and heterotrophic microbes: a concentric review. Front Microbiol 8:07

Vance ME, Kuiken T, Vejerano EP, McGinnis SP, Hochella MF, Rejeski D, Hull MS (2015) Nanotechnology in the real world: redeveloping the nanomaterial consumer products inventory. Beilstein J Nanotechnol 6:1769–1780
von Moos N, Maillard L, Slaveykova VI (2015) Dynamics of sub-lethal effects of nano-CuO on the microalga Chlamydomonas reinhardtii during short-term exposure. Aquat Toxicol 161:267–275
Wang D, Zhao L, Ma H, Zhang H, Guo LH (2017) Quantitative analysis of reactive oxygen species photogenerated on metal oxide nanoparticles and their bacteria toxicity: the role of superoxide radicals. Environ Sci Technol 51(17):10137–10145
Wang J, Zhou G, Chen C, Yu H, Wang T, Ma Y et al (2007) Acute toxicity and biodistribution of different sized titanium dioxide particles in mice after oral administration. Toxicol Lett 168(2):176–185
Warheit DB, Webb TR, Reed KL (2003) Pulmonary toxicity studies with TiO_2 particles containing various commercial coatings. Toxicologist 72(1):298–300
Wiesner MR, Lowry GV, Alvarez P, Dionysiou D, Biswas P (2006) Assessing the risks of manufactured nanomaterials. Environ Sci Technol 40(14):4336–4345
Williams AJ (1998) ABC of oxygen: assessing and interpreting arterial blood gases and acid-base balance. BMJ 317(7167):1213–1216
Yang H, Mei S, Zhao L, Zhang Y (2016) Effects of ultraviolet irradiation on the antibacterial activity of TiO_2 nanotubes. Nanosci Nanotechnol Lett 8(6):498–504
Ye N, Wang Z, Fang H, Wang S, Zhang F (2017) Combined ecotoxicity of binary zinc oxide and copper oxide nanoparticles to Scenedesmus obliquus. J Environ Sci Health A Tox Hazard Subst Environ Eng 52(6):555–560
Ye X, Li H, Wang Q, Chai R, Ma C, Gao H, Mao J (2018) Influence of aspartic acid and lysine on the uptake of gold nanoparticles in rice. Ecotoxicol Environ Saf 148:418–425
Zhang W, Li Y, Niu J, Chen Y (2013) Photogeneration of reactive oxygen species on uncoated silver, gold, nickel, and silicon nanoparticles and their antibacterial effects. Langmuir 29(15):4647–4651
Zhou C, Vitiello V, Pellegrini D, Wu C, Morelli E, Buttino I (2016) Toxicological effects of CdSe/ZnS quantum dots on marine planktonic organisms. Ecotoxicol Environ Saf 123:26–31

Chapter 7
Bioelectrocatalytic Assessment of the Activating Effect and Toxic Interaction Between Carbon Nanomaterials and Microbial Cells

A. N. Reshetilov, S. E. Tarasov and Yu V. Plekhanova

Abstract Carbon and other types of nanomaterials have found broad application in various fields of human activities. However, their impact on living organisms, including microorganisms, still needs to be understood more profoundly. Some nanomaterials cause activating effects, others are characterized by toxicity. The action of nanomaterials on microorganisms is assessed by bringing them into mutual contact by chemical immobilization, sorption or other techniques and registering the caused effect. Changes of the bioelectrocatalytic characteristics—the main parameters of the electrodes (anodes or working cathodes) in such devices as biosensors or microbial fuel cells (MFC)—are widely used for assessment. These characteristics are studied by cyclic voltammetry, chronoamperometry and potentiometry as well as impedance spectroscopy. In this chapter, we briefly describe the effects of mainly carbon nanomaterials on microorganisms. Emphasis is made on presenting data obtained for *Gluconobacter*, which is used as the basis of biosensors and MFC and can be considered as model biomaterial.

Keywords Carbon nanomaterials · Modification of graphite electrode
Activating and toxic effects · Immobilized *Gluconobacter* cells
Bioelectrocatalytic testing · Microbial fuel cell · Biosensors

---

A. N. Reshetilov (✉) · S. E. Tarasov · Y. V. Plekhanova
FSBIS G.K. Skryabin Institute of Biochemistry and Physiology of Microorganisms, Russian Academy of Sciences, Moscow Region, Pushchino 142290, Russia
e-mail: anatol@ibpm.pushchino.ru

S. E. Tarasov
e-mail: setar25@gmail.com

Y. V. Plekhanova
e-mail: plekhanova@ibpm.pushchino.ru

M. Rai and J. K. Biswas (eds.), *Nanomaterials: Ecotoxicity, Safety, and Public Perception*, https://doi.org/10.1007/978-3-030-05144-0_7

7.1 Introduction

Nanomaterials (NM) of various nature have recently attracted much theoretical and applied attention. One of the facets of this attention is due to the assessment of their impact on biological material. The high abundance of nanoparticles in the environment causes a growing concern with respect to their impact on human health. Still, knowledge of how NM interact with cells or tissues is limited. Inasmuch as the European Union calls for a significant reduction of experiments on animals for scientific purposes (Directive 2010/63), additional efforts are required to develop in vitro and in vivo models for assessing potentially harmless and potentially hazardous substances (Dubiak-Szepietowska et al. 2016).

A significant number of works have investigated the interaction of nanomaterials with microorganisms (MO), as reviewed in detail in Ghasemi et al. (2013). The effects of NM on MO are assessed by providing for their physical contact with microbial cells. In Putzbach and Ronkainen (2013), it is shown that the contact can be performed either by simple physical adsorption or covalent binding or else by growing a microbial film on the investigated surface; in Shan et al. (2010), a microbial population is immobilized in a polymer carrier. Cultivation of MO in an NM-containing medium is also widely used (Dar et al. 2013). Of special significance are micro- and nanotechnologies used in developing microbial biosensors (Lim et al. 2015). In other words, immobilization on NM is an obligatory requirement in assessing their effects.

The spectrum of investigated nanomaterials is rather wide. The largest number of publications, probably, focus on such materials as single- (SWCNT) and multiwalled carbon nanotubes (MWCNT) (Zou et al. 2008; Murray et al. 2009; Ghasemi et al. 2013), which lead to a decrease in electrode's charge transfer resistance due to an increase of conductance; thermally expanded graphite (TEG), which enables direct electron transfer to membrane fractions of *Gluconobacter* bacteria (Reshetilov et al. 2015); highly oriented pyrolytic graphite (HOPG), the structure of which can be described as an alternation of graphene planes, and which can be considered to be nontoxic only indirectly, owing to the general inertness of the material (Luo et al. 2010); oxidized and reduced graphene oxide (Bianco 2013; Goenka et al. 2014); carbon fibres possessing a high electrical conductance, also attributed to NM based on the high dispersity of their structure (Reshetilov et al. 2017a). Regarding NM of metallic nature, Franci et al. (2015) described in their review the interaction of MO with silver nanoparticles; in Shin and Cha (2008), the interaction of MO with zero-valent iron nanoparticles is discussed. The synthesis of metallic nanoparticles by microorganisms is presented in Talebi et al. (2010).

Assessment of NM effects extensively uses changes of the bioelectrocatalytic characteristics, which are the main parameters of the electrodes (anodes/measuring electrodes) in such devices as biosensors or microbial fuel cells (MFC). The measured values are current–voltage characteristics, generated currents and voltages, impedance parameters, electric power.

This chapter gives a brief overview of the effects of mainly carbon NM on MO. Emphasis is made on the description of data obtained for *Gluconobacter*, which are used as the basis of biosensors and MFC and, in a sense, can be considered as a model biomaterial to form bioelectrodes and to assess the action of NM.

7.2 Types and Properties of Carbon Nanomaterials Used in Assessing Their Effects on Measuring Electrodes and Biological Material (Microbial Cells)

This Section briefly discusses the properties of carbon NM used frequently in studies for the assessment of their effect on biomaterials, in particular, on microbial cells. A feature of microbial cells is that, on the one hand, enzymes in them are protected from external impacts, but, on the other hand, cells themselves can be more subject to the toxic effect of external factors. In the case of NM, their toxicity can be associated with damage of microbial membranes, formation of reactive oxygen species or oxidative DNA damage (Kim et al. 2009). Besides, features of this interaction include the ability of nanomaterials to accumulate in living cells and to increase their cytotoxicity in the course of time. It is to be noted that to date all features of the interaction of nanoparticles and bacterial cells have not been entirely studied.

Modern NM are represented by substances of various chemical nature with diverse properties (de Dios and Diaz-Garcia 2010; Vashist and Luong 2015). In the general case, when choosing an NM to solve a particular problem, one should take into account its nature, production technique, availability, as well as consider its physical and chemical properties. Most nanomaterials used can be divided into three groups: carbon NM (graphene, carbon nanotubes, carbon fibres, fullerenes, etc.); metallic nanoparticles (particles of gold and silver, metal oxides, quantum dots are used most often (Su et al. 2009); as well as branched polymers (dendrimers) (Astruc et al. 2003). Each of these types of NM possesses its own unique properties, which enable perfecting the quality of bioelectrochemical devices' electrodes at the modification of their surface. Carbon NM are primarily distinguished by high electrical conductance. Nanoparticles of metals also provide for a highly developed surface and are involved in electrochemical processes (Habib et al. 2012).

Nanomaterial applications in biosensors expand year by year, and analysis of literature shows that the use of carbon nanotubes and metal nanoparticles may lead to the formation of biosensors' new useful properties. Nanomaterials are used increasingly frequently for modification of electrode surfaces in the development of modern amperometric biosensors (Pei et al. 2013; Šefčovičová and Tkac 2015). The fields of biosensors' commercial use are constantly expanding at present (chemical synthesis, quality control, ecological monitoring, biomedical research), and investigators often turn to nanomaterials when refining these devices. One of biosensors' essential parameters is the rate of the electrochemical reaction between the biological component and the electrode surface. Modification of the surface by carbon nanomaterials possessing a high electrical conductance can increase the rate of electron transfer or facilitate direct electron transfer. In some works, an opinion is expressed (Chang et al. 2013) that a large surface area of multiwalled carbon nanotubes contributes to an improved immobilization of biomolecules (e.g., enzymes or antibodies), thus enhancing biosensors' sensitivity through amplification of electric signals. Use of carbon nanotubes in biosensors for signal transfer without participation of the mediator between the electrically active component and the electrode is also possible (Jacobs et al. 2010).

7.3 Assessment of the Effect of NM: Positive and Negative Effects

Some NM cause positive effects, others are characterized by the absence of a positive effect and the occurrence of a negative effect. Let us first consider examples of positives, then negatives.

In a general case, to assess the properties of nanomaterials and their effect on bacterial cells, it is necessary to assess a number of parameters of nanoparticles themselves. In particular, an important role, as it was pointed out, is played by the size, shape, surface area and charge, chemical groups on the surface of nanomaterial, its porosity, solubility and capability of agglomeration and crystallization (Dhawan and Sharma 2010). To determine these properties, such methods of analysis are used as dynamic light scattering, electron microscopy, atomic force microscopy, capillary electrophoresis, inductively coupled plasma mass spectrometry, field flow fractionation, nanoparticle tracking analysis, mathematical modelling (Brunauer–Emmett–Teller (BET) method) (Gupta et al. 2005; Weibel et al. 2005; Powers et al. 2007). Particular methods of analysis depend on the problem set by the researcher and types of nanoparticles and investigated microorganisms. Inasmuch as studies based on bioelectrodes (biosensors and MFC) have gained momentum recently, research into the effect of nanoparticles on microbial cells is efficiently carried out by electrochemical methods. This enables following both the potential toxic effect of nanomaterials on bacterial cells and the integral positive impact on the properties of the bioelectrodes themselves.

The evident positive characteristics of NM include such properties as the large surface area, high penetrability and chemical selectivity, which can be useful in the development of bioelectrodes. A larger surface area and catalytic ability (as compared with traditional materials) enables increasing the rates of chemical reactions on the surface of modified bioelectrodes, which leads to an improvement of the electrochemical parameters (e.g., an increase of conductance). Nanoparticles are often more biologically active than traditionally used materials (metals, graphite, etc.) and can interact more efficiently with biological objects (Uskokovic 2007; Linazasoro 2008). Finally, a useful property of NM is that their use makes possible the development of nano-sized devices (Ghasemi et al. 2013).

Considering the negative (toxic) effects of NM, it should be noted that most research works deal with the antibacterial action of nanomaterials (Kang et al. 2009). It is known that MWCNT can be toxic for bacterial cells, damaging their surface (Kang et al. 2008). In Liu et al. (2009), SWCNT in solution were called "nano darts", because they constantly interacted with the surface of bacteria (*Escherichia coli*, *Pseudomonas aeruginosa*, *Staphylococcus aureus*, *Bacillus subtilis*), deteriorating its condition. The efficiency of cell surface destruction was enhanced by increasing the fraction (amount) of separate SWCNT (as compared with their aggregates) and their total concentration. The fact that the tube size is indeed of importance was supported by a research into the effect of the length of SWCNT on their antimicrobial activity. Longer SWCNT (5 μm) were shown to be more efficient in comparison with shorter tubes (<1 μm), as they have a greater chance of interaction with cell surface, whereas shorter SWCNT form aggregates not affecting microbial cells (Yang et al. 2010).

Bacterial survival in the presence of nanoparticles depend on their composition, size, surface charge and shape, and also on the investigated bacterial strain (Simon-Deckers et al. 2009). It is highly likely that the affinity of nanoparticles to microbial cell surface is a key point in induction of toxic effects, which are caused by nanoparticles' adhesion to the cell surface, as well as because they are more efficient in destroying cells as compared with their macro analogues (Jiang et al. 2009). Cytotoxicity of nanoparticles induced by ions released from nanoparticles is to be taken into account, because it has a direct effect on cell viability, as exemplified by the yeast *Saccharomyces cerevisiae* (Kasemets et al. 2009) and a number of bacteria (Dhas et al. 2014). All these factors are to be considered in experiments with NM.

7.4 Bioelectrocatalytic Parameters Used in Research

Quite a number of electrochemical methods are used to study the positive and negative effects of nanoparticles on microbial cells. In particular, special attention is paid to the mechanisms of bioelectrocatalytic oxidation of substrate by bacterial cells and on the impact of various external factors on these mechanisms (Schroder 2007). That said, studies of bacterial cells by electrochemical methods are complicated by the multi-stage microbial metabolism and the integrated nature of bacterial cells.

7.4.1 Cyclic Current–Voltage Characteristics

Cyclic voltammetry is one of the standard methods in electrochemistry (Heinze 1984) and can be used to study both the properties of biofilms and changes occurring in them (Busalmen et al. 2008). A feature of this method is that, using cyclic voltammetry, we can characterize bioelectrodes at various stages of bacterial culture development; herewith, the obtained characteristics will change depending on the stage of cell growth and development (Fricke et al. 2008). This technique can be also used in studies of the effect of nanoparticles on microbial cells. For this, two current–voltage curves should be registered—one for the bioelectrode based on pure microbial culture and the other for the bioelectrode based on bacterial culture subjected to the action of nanoparticles. In the case of a toxic effect of nanoparticles on bacterial cells we will observe a drop of oxidative currents due to a disturbed bioelectrocatalytic oxidation of substrate by bacteria. If the effect of nanoparticles is positive, the oxidation peaks will be observed to be enhanced and restored on the cyclic current–voltage curve (Zhang et al. 2011).

7.4.2 Power Characteristics

Another method for assessing the effect of NM on bacterial cells is the plotting of bioelectrodes' power density curves. The two-electrode measuring circuit is used, which models an MFC whose main task is to generate electric energy by microbial cells. In this case, one of the electrodes is coated by bacterial cells and it acts as an anode; the other electrode is used as a cathode (Logan et al. 2006). A number of external resistors is connected to the produced system, which makes it possible to obtain a dependence of power generated by the cell on current density or on voltage in the system. The power characteristics can be affected by numerous factors, such as temperature (Feng et al. 2008), pH (Rozendal et al. 2006) or the presence of nanomaterials in the system. The method is exceptionally useful in studies of NM's positive properties with respect to bacterial cells, as it enables observation of all effects from the interaction of nanomaterials with biological objects simultaneously. An increase of power generated by microbes under the influence of nanomaterials can be associated in this case with improved electron transport between bacteria and the electrode (Jiang et al. 2014), with the emergence of direct electron transfer or else with the increase of bioelectrodes' active surface (Fogel and Limson 2016).

7.4.3 Electrochemical Impedance Spectroscopy

An efficient method to study the electrochemical properties of bacterial cells is electrochemical impedance spectroscopy (EIS). It is a spectral technique, and the measurement produces a dependence characteristic of given bacterial cells under certain conditions. EIS has no serious restrictions for the size of a measured specimen, so the use of small-sized electrodes proved a convenient solution for studies of the properties of cells and cell systems (Liu et al. 2011). The principle of impedance studies of individual cells is not complex: usually two microelectrodes are used, the surface of the working electrode remains accessible for investigated cells, and the surface of the counter electrode is coated with a special passivating layer (Yang and Bashir 2008). For the electrodes to have no great effect on the cell properties, a small current of no more than 1 μA in amplitude should be passed through cells (Bard and Faulkner 2001). After the electrodes are connected, the impedance spectrum of cells is studied to determine the region subjected the most to the change of impedance during any intracellular processes or external excitation factors. Further measurements are possible without scanning the entire frequency range but only in the region that proved the most susceptible to changes. On the whole, current methods of cell diagnostics using EIS can be divided into three groups that differ by the sensor mechanism. The first group registers the change of ions' release from living cells by the change of electrolyte resistance in the system. The second category of the methods studies the insulation properties of the cell membrane. It is known that various components of cells strongly differ by their electric properties: if the cytoplasm conducts electric current well (conductivity, 0.3–0.6 S/m), the cell membrane is a high-ohmic insulator (up to 10^{-7} S/m) (Pethig and Markx 1997). These electric properties are characteristic of each type of cell, and EIS can measure any changes during cell growth or propagation. The third group of methods investigates flows of substances and spatial distribution of cells or cell fragments, which also affects the impedance of the space between two measuring electrodes. All these methods can be used to study the properties of both whole cells and their various constituents (particular enzymes, DNA, mitochondria) (Du et al. 2013; Padmaraj et al. 2014). The field has been intensively developing over the recent years.

7.5 Modification of the Properties of Graphite Electrodes Making the Basis of Biosensor Anodes and MFC

Materials based on carbon nanotubes have been actively used in microbial fuel cells in recent decades. Their use is intended to solve two tasks: to improve the electrochemical activity of bacterial communities on bioanodes and biocathodes, as well as to enhance the oxygen reduction reaction on cathodes (Yazdi et al. 2016).

Refinement of the properties of MFC using two-dimensional graphene sheets is considered in a review by Ci et al. (2015). Variants of using composite nanomaterials of the type of metal/polymer, metal/carbon and polymer/carbon in microbial fuel cells are reviewed in Antolini (2015).

Microbial biosensors based on nanomaterials have demonstrated improved detection ranges and broader substrate specificity. Possibly, this is due to the direct contact of enzymes occurring inside the cell with the NM-modified electrode surface. Positive properties of such biosensors include a reduced response due to the facilitated access of substrate to enzymes and improved selectivity (Li et al. 2013; Wang et al. 2013).

7.6 Bioelectrocatalytic Tests on Intact *Gluconobacter*

In the recent decade, NM have been actively used to develop microbial biosensors and MFC. It could also be noted that greater attention has been paid to consider the interaction of *Gluconobacter* and NM (Svitel et al. 2006; Sefcovicova et al. 2014; Bertokova et al. 2015). The results obtained to date are briefly as follows.

7.6.1 Thermally Expanded Graphite (TEG)

Electrically conductive graphene-like material, thermally expanded graphite (TEG), is among promising nanomaterials. Owing to its fibrous laminated structure, TEG is easily pressed and formed into electrodes; it also has a high specific surface reaching values of 2000 m^2/g (Gorshenev et al. 2003). As shown by the experiment, immobilization of cells on an electrode fabricated from TEG led to increase the current on the current–voltage curves and an increase of power of an MFC based on this bioelectrode and *G. oxydans* cells up to 6 $\mu W/cm^2$ as compared with 2.5 $\mu W/cm^2$ produced for the control electrode from spectral graphite. TEG electrodes enable using membrane fractions (MF) of *Gluconobacter* cells as an alternative biocatalyst (Reshetilov et al. 2015). In essence, MF are parts of the cell membrane and contain various cell membrane structures, including membrane PQQ-dependent dehydrogenases (DH). The authors of Reshetilov et al. (2015) have first shown that electrooxidation of ethanol on the bioanode containing MF immobilized on TEG can proceed by two mechanisms—direct electron transfer and mediator transfer.

An MFC based on plastic miniature electrodes from TEG and *Gluconobacter* bacteria developed by the authors (Reshetilov et al. 2016) was incorporated into the frog *Rana temporaria* and used for generation of electric energy at the oxidation of endogenous glucose.

These are examples of the positive interaction of nanomaterials and biological objects.

7.6.2 Single- and Multi-walled Carbon Nanotubes

The use of carbon nanotubes for modification of anodes in MFC show an important role of these nanoparticles in increasing electrochemically active surface area (Ghasemi et al. 2013). Thus, in Zou et al. (2008) it was shown that the anodes' charge transfer resistance after they were coated with polypyrrol–nanotubes nanomaterial reduced from 500 Ω to almost zero. These results show that nanoparticles increase the conductance due to a decrease of the electric resistance between the electrode and electrolyte by improving the diffusion of mediator to the electrode surface.

The effect of carbon nanotubes on the biosensor signal was studied using a microbial sensor based on *G. oxydans* cells and MWCNT dispersed in chitosan (Odaci et al. 2009). The biosensor enabled glucose detection was made at concentrations from 0.05 up to 1 mM over a time of 40 s.

Reshetilov et al. (2017b) reported positive effect of functionalized carbon nanotubes on the performance of bioelectrodes based on *G. oxydans* cells. Combined immobilization of bacterial cells and nanotubes in chitosan gel on the bioanode surface led to an increase in the MFC power by 25% and a decrease of internal resistance by 13% as compared with MFC based on spectral graphite electrodes.

7.6.3 Highly Oriented Pyrolytic Graphite

Highly oriented pyrolytic graphite is a graphite modification of laminated structure. This material is produced as a result of the uniaxial pressure on pyrocarbon annealed at high temperature (up to 3300 °K). The structure of HOPG can be described as an alternation of graphene planes. HOPG is often used as a support for microscopic examination and is a material for fabrication of graphene by mechanical exfoliation. HOPG is an inert material and, owing to this, it is biologically compatible with all microbial cells. Still, studies in Reshetilov et al. (2017b) showed that even a mere addition of HOPG particles to *G. oxydans*-based bioelectrodes and the formation of a bacterial cell–HOPG suspension leads to a sharp degradation of electrodes' electrochemical characteristics, and the use of electrodes from pure HOPG totally blocks electron transfer in the system. As this effect is not due to the destructive toxic effect of material on whole bacterial cells and their fragments, it can be related only to the low conductance of the suspension of material itself, which makes its use in biosensors and MFC inefficient.

7.6.4 Oxidized and Reduced Graphene Oxide

Graphene is a two-dimensional material that possesses a high electric conductance and mechanical strength, as well as a number of other properties valuable in designing electronic devices. Graphene can be used to form highly structured electrode surfaces, which, owing to a high surface area/volume ratio enable immobilization of a significant amount of biocatalyst. Still, under some conditions graphene and its oxide demonstrate a high antibacterial activity (Akhavan and Ghaderi 2010; Aslan and Anik 2016), and further research is required for developing bioelectrodes based on this nanomaterial.

Reshetilov et al. (2017b) investigated the properties of bioelectrodes based on spectral graphite and *G. oxydans* cells modified by graphene and graphene oxide. It should be noted that the resistance of the graphite bioanode modified by graphenes decreased by 34% for pure graphene and by 27% for graphene oxide, which implies a satisfactory electric conductance of this material. The total power of MFC based on these electrodes increased by only 5–10% as compared with the control MFC. This may indicate that, despite its excellent electrochemical characteristics, graphene still has some toxic activity inhibiting the metabolism of bacteria and preventing a significant enhancement of electron transfer in this system.

7.6.5 Carbon Fibres

Carbon fibre (CF) is a nanostructured organic material consisting of fine threads 5–15 μm in diameter, which contains 92–99.9% carbon and possesses high values of strength and elastic modulus. Carbon atoms are combined into microscopic crystals aligned parallel to one another. Carbon fibres are characterized by a low specific weight, low thermal expansion coefficient (Dong et al. 2018), chemical inertness and high active surface (300–1500 m^2/g) (Huang 2009). This material is inert and its toxic effect on bacterial cells has not been revealed. The electrically conductive properties of CF differ only slightly from those of graphite. For this reason, the use of CF in bioelectrodes makes sense only in the case when electrodes with certain physical properties are required, e.g., with increased flexibility and plasticity, as demonstrated by Reshetilov et al. (2017a).

7.7 Conclusion and Future Perspectives

The research carried out in the field of the effect of NM on MO are important as they make the basis of fundamental knowledge of biological science. Besides, the results are essential for implementing applied issues—development of biosensors and MFC. In this context, it can be noted that current research into the applications of carbon NM in bioelectrocatalysis is to study the properties of conducting substances for developing electrodes, to search for microbial cells capable of efficient electron transfer by the mechanism of mediator and direct bioelectrocatalysis. SWCNT, MWCNT, graphene are among the nanomaterials used both in biosensors and MFC—in this case, biosensors are associated with MFC as their precursors, because the structure and function of MFC and electrochemical type biosensors much coincide. Analysis of available literature data on the development of MFC with novel characteristics suggests that one of the tendencies is the development of miniature planar and 3D MFC. Such systems would require, correspondingly, miniature electrodes, the anode and cathode. Along with other known NM, graphene-like materials possess properties that provide for their use in constructing miniature MFC; thus, they have a high surface to volume ratio, possess a high electric conductance, high strength. Inasmuch as a considerable number of positive results have been obtained on the use of graphene-like materials in MFC, it is highly unlikely that the answer to the question whether graphene-like materials, nanotubes, TEG, metal particles and the like should be continued to be used for research into NM–MO combinations would be negative. Applications of the NM–MO complex expands the window of opportunity and enables the development of new-generation efficient devices, which can be certainly said to be intended for practice.

Acknowledgements The authors are grateful for the support by the Russian Science Foundation within the framework of the project "Design, Fabrication and Study of New Hybrid Integrated Sensors Based on Nanoelectronic, Acoustoelectronic and Electrochemical Technologies for Biological Applications" No. 18-49-08005.

References

Akhavan O, Ghaderi E (2010) Toxicity of graphene and graphene oxide nanowalls against bacteria. ACS Nano 4(10):5731–5736
Antolini E (2015) Composite materials for polymer electrolyte membrane microbial fuel cells. Biosens Bioelectron 69:54–70
Aslan S, Anik U (2016) Microbial glucose biosensors based on glassy carbon paste electrodes modified with *Gluconobacter oxydans* and graphene oxide or graphene–platinum hybrid nanoparticles. Microchim Acta 183:73–81
Astruc D, Blais JC, Daniel MC, Gatard S, Nlate S, Ruiz J (2003) Metallodendrimers and dendronized gold colloids as nanocatalysts, nanosensors and nanomaterials for molecular electronics. C R Chim 6:1117–1127
Bard A, Faulkner L (2001) Electrochemical methods. In: Fundamentals and Application, 2nd ed. Wiley, New York, pp 368–414
Bertokova A, Bertok T, Filip J, Tkáč J (2015) *Gluconobacter* sp. cells for manufacturing of effective electrochemical biosensors and biofuel cells. Chem Papers 69:27–41
Bianco A (2013) Graphene: safe or toxic? The two faces of the medal. Angew Chemie Int Ed 52:4986–4997
Busalmen JP, Esteve-Nunez A, Feliu JM (2008) Whole cell electrochemistry of electricity-producing microorganisms evidence an adaptation for optimal exocellular electron transport. Environ Sci Technol 42:2445–2450
Chang YT, Huang J-H, Tu MC, Chang P, Yew T-R (2013) Flexible direct-growth CNT biosensors. Biosens Bioelectron 41:898–902
Ci S, Cai P, Wen Z, Li J (2015) Graphene-based electrode materials for microbial fuel cells. Sci China Mater 58(6):496–509
Dar MA, Ingle A, Rai M (2013) Enhanced antimicrobial activity of silver nanoparticles synthesized by *Cryphonectria* sp. evaluated singly and in combination with antibiotics. Nanomed NBM 9:105–110
de Dios AS, Diaz-Garcia ME (2010) Multifunctional nanoparticles: analytical prospects. Anal Chim Acta 666:1–22
Dhas SP, Shiny PJ, Khan S, Mukherjee A, Chandrasekaran N (2014) Toxic behavior of silver and zinc oxide nanoparticles on environmental microorganisms. J Basic Microbiol 54(9):916–927
Dhawan A, Sharma V (2010) Toxicity assessment of nanomaterials: methods and challenges. Anal Bioanal Chem 398:589–605
European Parliament and the Council of the European Union (2010) Directive 2010/63/EU of the European Parliament and of the Council of 22 September 2010 on the protection of animals used for scientific purposes. Official Journal of the European Union L 276, vol. 53, pp 33–79
Dong C, Li K, Jiang Y, Arola D, Zhang D (2018) Evaluation of thermal expansion coefficient of carbon fiber reinforced composites using electronic speckle interferometry. Opt Express 26:531

Du E, Ha S, Diez-Silva M, Dao M, Suresh S, Chandrakasan AP (2013) Electric impedance microflow cytometry for characterization of cell disease states. Lab Chip 13(19):3903–3909

Dubiak-Szepietowska M, Karczmarczyk A, Winckler T, Feller K-H (2016) A cell-based biosensor for nanomaterials cytotoxicity assessment in three dimensional cell culture. Toxicology 370:60–69

Feng Y, Wang X, Logan BE, Lee H (2008) Brewery wastewater treatment using air-cathode microbial fuel cells. Appl Microbiol Biotechnol 78:873–880

Fogel R, Limson JL (2016) Applications of nanomaterials in microbial fuel cells. In: Ozoemena K, Chen S (eds) Nanomaterials for fuel cell catalysis. Nanostructure science and technology. Springer, Cham, pp 551–575

Franci G, Falanga A, Galdiero S, Palomba L, Rai M, Morelli G, Galdiero M (2015) Silver nanoparticles as potential antibacterial agents. Molecules 20:8856–8874

Fricke K, Harnisch F, Schroder U (2008) On the use of cyclic voltammetry for the study of anodic electron transfer in microbial fuel cells. Energy Environ Sci 1:144–147

Ghasemi M, Daud WRW, Hassan SHA, Ohc S-E, Ismail M, Rahimnejad M, Md Jahim J (2013) Nano-structured carbon as electrode material in microbial fuel cells: a comprehensive review. J Alloys Compd 580:245–255

Goenka S, Sant V, Sant S (2014) Graphene-based nanomaterials for drug delivery and tissue engineering. J Controll Release 173:75–88

Gorshenev VN, Bibikov SB, Novikov YuN (2003) Conducting materials based on thermally expanded graphite. Russ J Appl Chem 76(4):603–606

Gupta S, Brouwer P, Bandyopadhyay S, Patil S, Briggs R, Jain J, Seal S (2005) TEM/AFM investigation of size and surface properties of nanocrystalline ceria. J Nanosci Nanotechnol 5(7):1101–1107

Habib O, Demirkol DO, Timur S (2012) Sol–gel/chitosan/gold nanoparticle-modified electrode in mediated bacterial biosensor. Food Anal Methods 5:188–194

Heinze J (1984) Cyclovoltammetrie—die "Spektroskopie" des Elektrochemikers. Angew Chem 96:823–840

Huang X (2009) Fabrication and properties of carbon fibers. Materials 2:2369–2403

Jacobs CB, Peairs MJ, Venton BJ (2010) Review: carbon nanotube based electrochemical sensors for biomolecules. Anal Chim Acta 662(2):105–127

Jiang W, Mashayekhi H, Xing BS (2009) Bacterial toxicity comparison between nano- and micro-scaled oxide particles. Environ Pollution 157:1619–1625

Jiang X, Hu J, Lieber AM, Jackan CS, Biffinger JC, Fitzgerald LA, Ringeisen BR, Lieber CM (2014) Nanoparticle facilitated extracellular electron transfer in microbial fuel cells. Nano Lett 14(11):6737–6742

Kang S, Mauter MS, Elimelech M (2008) Physicochemical determinants of multiwalled carbon nanotube bacterial cytotoxicity. Environ Sci Technol 42:7528–7534

Kang S, Mauter MS, Elimelech M (2009) Microbial cytotoxicity of carbon-based nanomaterials: implications for river water and wastewater effluent. Environ Sci Technol 43:2648–2653

Kasemets K, Ivask A, Dubourguier HC, Kahru A (2009) Toxicity of nanoparticles of ZnO, CuO and TiO_2 to yeast *Saccharomyces cerevisiae*. Toxicol In Vitro 23:1116–1122

Kim S, Choi JE, Choi J, Chung KH, Park K, Yi J, Ryu DY (2009) Oxidative stress-dependent toxicity of silver nanoparticles in human hepatoma cells. Toxicol In Vitro 23(6):1076–1084

Li L, Liang B, Li F, Shi JG, Mascini M, Lang QL, Liu A (2013) Co-immobilization of glucose oxidase and xylose dehydrogenase displayed whole cell on multiwalled carbon nanotube nanocomposite films modified electrode for simultaneous voltammetric detection of D-glucose and D-xylose. Biosens Bioelectron 42:156–162

Lim JW, Ha D, Lee J, Lee SK, Kim T (2015) Review of micro/nanotechnologies for microbial biosensors. Front Bioeng Biotechnol 3:61

Linazasoro G (2008) Potential applications of nanotechnologies to Parkinson's disease therapy. Parkinsonism Relat Disord 14:383–392

Liu JJ, Li H, Zhang F, Li X, Wang L, Chen Y (2011) Online impedance monitoring of yeast cell culture behaviours. Microel Eng 88(8):1711–1713

Liu SB, Wei L, Hao L, Fang N, Chang MW, Xu R, Yang YH, Chen Y (2009) Sharper and faster "nano darts" kill more bacteria: a study of antibacterial activity of individually dispersed pristine single-walled carbon nanotube. ACS Nano 3:3891–3902

Logan BE, Hamelers B, Rozendal RA (2006) Microbial fuel cells: methodology and technology. Environ Sci Technol 17:5181–5191

Luo J, Chan WB, Wang L, Zhong CJ (2010) Probing interfacial interactions of bacteria on metal nanoparticles and substrates with different surface properties. Int J Antimicrob Agents 36(6):549–556

Murray AR, Kisin E, Leonard SS, Young SH, Kommineni C, Kagan VE, Castranova V, Shvedova AA (2009) Oxidative stress and inflammatory response in dermal toxicity of single-walled carbon nanotubes. Toxicology 257(3):161–171

Odaci D, Timur S, Telefoncu A (2009) A microbial biosensor based on bacterial cells immobilized on chitosan matrix. Bioelectrochemistry 75:77–82

Padmaraj D, Pande R, Miller JH, Wosik J, Zagozdzon-Wosik W (2014) Mitochondrial membrane studies using impedance spectroscopy with parallel pH monitoring. PLoS ONE 9(7):e101793

Pei X, Zhang B, Tang J, Liu B, Lai W, Tang D (2013) Sandwich-type immunosensors and immunoassays exploiting nanostructure labels: a review. Anal Chim Acta 758:1–18

Pethig R, Markx GH (1997) Applications of dielectrophoresis in biotechnology. Trends Biotechnol 15(10):426–432

Powers K, Palazuelos M, Moudgil B, Roberts S (2007) Characterization of the size, shape and state of dispersion of nanoparticles for toxicological studies. Nanotoxicology 1(1):42–51

Putzbach W, Ronkainen NJ (2013) Immobilization techniques in the fabrication of nanomaterial-based electrochemical biosensors: a review. Sensors 13:4811–4840

Reshetilov AN, Kitova AE, Kolesov VV, Yaropolov AI (2015) Mediator-free bioelectrocatalytic oxidation of ethanol on an electrode from thermally expanded graphite modified by *Gluconobacter oxydans* membrane fractions. Electroanalysis 27(6):1443–1448

Reshetilov AN, Plekhanova YV, Tarasov SE, Kitova AE, Uteshev VK, Vasilov RG, Kolesov VV (2016) Sposob poluchenia electricheskoi energii s pomoshiu mikrobnogo biotoplivnogo elementa implantirovannogo v organism zhivoi travyanoi lyagushki *Rana Temporaria*. RU Patent No. 2599421, 15 Sept 2016

Reshetilov AN, Plekhanova JV, Tarasov SE, Bykov AG, Gutorov MA, Alferov SV, Tenchurin TK, Chvalun SN, Orekhov AS, Shepelev AD, Gotovtsev PM, Vasilov RG (2017a) Evaluation properties of bioelectrodes based on carbon superfine materials containing model microorganisms *Gluconobacter*. Nanotechnol Russ 12:107–115

Reshetilov AN, Plekhanova YV, Tarasov SE, Arlyapov VA, Kolesov VV, Gutorov MA, Gotovtsev PM, Vasilov RG (2017b) Effect of some carbon nanomaterials on ethanol oxidation by *Gluconobacter oxydans* bacterial cells. Appl Biochem Microbiol 53:123–129

Rozendal RA, Hamelers HM, Buisman CJN (2006) Effects of membrane cation transport on pH and microbial fuel cell performance. Environ Sci Technol 40:5206–5211

Schroder U (2007) Anodic electron transfer mechanisms in microbial fuel cells and their energy efficiency. Phys Chem Chem Phys 9:2619–2629

Šefčovičová J, Tkac J (2015) Application of nanomaterials in microbial-cell biosensor constructions. Chem Papers 69:42–53

Sefcovicova J, Filip J, Gemeiner P, Tkac J (2014) Nanomaterial-based microbial biosensor for detection of ethanol in real samples. J Biotechnol 185:S21

Shan C, Yang H, Han D, Zhang Q, Ivaska A, Niu L (2010) Graphene/AuNPs/chitosan nanocomposites film for glucose biosensing. Biosens Bioelectron 25:1070–1074

Shin KH, Cha DK (2008) Microbial reduction of nitrate in the presence of nanoscale zero-valent iron. Chemosphere 72:257–262

Simon-Deckers A, Loo S, Mayne-L'hermite M, Herlin-Boime N, Menguy N, Reynaud C, Gouget B and Carriere M (2009). Size-, composition- and shape-dependent toxicological impact of metal oxide nanoparticles and carbon nanotubes toward bacteria. Environ Sci Technol 43:8423–8429

Su Y, He Y, Lu H, Sai L, Li Q, Li W, Wang L, Shen P, Huang Q, Fan C (2009) The cytotoxicity of cadmium-based, aqueous phase-synthesized quantum dots and its modulation by surface coating. Biomaterials 30(1):19–25

Svitel J, Tkac J, Vostiar I, Navratil M, Stefuca V, Bucko M, Gemeiner P (2006) *Gluconobacter* in biosensors: applications of whole cells and enzymes isolated from *Gluconobacter* and *Acetobacter* to biosensor construction. Biotechnol Lett 28:2003–2010

Talebi S, Ramezani F, Ramezani M (2010) Biosynthesis of metal nanoparticles by micro-organisms. Nanocon Olomouc 10:12–18

Uskokovic V (2007) Nanotechnologies: what we do not know. Technol Soc 29:43–61

Vashist SK, Luong JHT (2015) Recent advances in electrochemical biosensing schemes using graphene and graphene-based nanocomposites. Carbon 84:519–550

Wang HW, Lang QL, Li L, Liang B, Tang XJ, Kong LG, Mascini M, Liu A (2013) Yeast surface displaying glucose oxidase as whole-cell biocatalyst: construction, characterization, and its electrochemical glucose sensing application. Anal Chem 85:6107–6112

Weibel A, Bouchet R, Boulc'h F, Knauth P (2005) The big problem of small particles: a comparison of methods for determination of particle size in nanocrystalline anatase powders. Chem Mater 17(9):2378–2385

Yang CN, Mamouni J, Tang YG, Yang LJ (2010) Antimicrobial activity of single-walled carbon nanotubes: length effect. Langmuir 26:16013–16019

Yang L, Bashir R (2008) Electrical/electrochemical impedance for rapid detection of foodborne pathogenic bacteria. Biotechnol Adv 26(2):135–150

Yazdi AA, D'Angelo L, Omer N, Windiasti G, Lu X, Xu J (2016) Carbon nanotube modification of microbial fuel cell electrodes. Biosens Bioelectron 85:536–538

Zhang YZ, Mo GQ, Li XW, Zhang W, Zhang JQ, Ye JS, Huang XD, Yu CZ (2011) A graphene-modified anode to improve the performance of microbial fuel cells. J Power Sources 196:5402–5407

Zou Y, Xiang C, Yang L, Sun L, Xu F, Cao Z (2008) A mediatorless microbial fuel cell using polypyrrole coated carbon nanotubes composite as anode material. Int J Hydrogen Energy 33:4856–4862

Chapter 8
Nanotoxicity of Lipid-Based Nanomedicines

Maria Jose Morilla and Eder Lilia Romero

Abstract The impact of lipid-based nanomedicines on living beings widely differs from that caused by non-intentional exposition of metallic, metallic oxides, or carbon-based inhaled nanomaterials. The scarce data gathered on nanotoxicity of lipid-based nanomedicines is mostly related to intravenous and inhalation routes. This last is also the most important non-intentional exposition route. Intravenous nanomedicines mostly consist of antitumoral and antimycotic agents, both responsible for the induction of potentially lethal acute hypersensitivity reactions. Their nanotoxicity is explained according to different mechanisms. The so called double hit theory states that the clinical symptoms arise from the sum of the complement activation (Complement Activation Related Pseudo Allergy, CARPA effect), where the anaphylatoxins C3a, C4a and C5a induce the release of allergomedins by mast cells, and from the direct activation by nanomedicines binding to pattern recognizing receptors on macrophages; activation of bradykinin receptors B1 and B2 may contribute to the clinical picture. The rapid phagocytic response hypothesis instead, states that the complement activation would not be responsible for the clinical symptoms, whereas the hypersensitivity reaction occur by direct activation of IgG opsonized nanomedicines, of receptors expressed mostly in pulmonary intravascular macrophages (PIM). Activated PIMs thus, would release platelet activating factor (PAF), a highly active and very short-lived molecule, a mast cells activator responsible for the potentially lethal toxic effects. The nanotoxicity of inhaled liposomes, developed to treat bronchiectasis caused by bacterial infections on the other hand, was preclinically observed in rodents (species of unknown value as reliable predictors of the effects of inhaled medicines in humans), while in dogs was reported as transient, lymph node reactions (perivascular/peribronchiolar lymphoid cell infiltration) that are not readily reversible and lymphoid findings associated with foamy macrophages. In early clinical phase, the nanotoxicity manifested after repeated administrations

M. J. Morilla · E. L. Romero (✉)
Department of Science and Technology, Nanomedicines Research and Development Center, Quilmes National University, Buenos Aires, Argentina
e-mail: elromero@unq.edu.ar

M. J. Morilla
e-mail: jmorilla@unq.edu.ar

M. Rai and J. K. Biswas (eds.), *Nanomaterials: Ecotoxicity, Safety, and Public Perception*, https://doi.org/10.1007/978-3-030-05144-0_8

along months, as higher airway reactogenicity, higher rates of treatment-related adverse events and higher rates of discontinuations due to adverse events. In case of liposomal amikacin—a per se irritant drug—, the nanotoxicity was related to the presence of free antibiotic; in case of liposomal ciprofloxacin, it was manifested as dyspnea, bronchospasm, hemoptysis, cough, taste disorders. Not surprisingly, the nanotoxicity of nanomedicines has not been properly anticipated by preclinical studies. More accurate disease models for example based on artificial three-dimensional human cell organs, may aid preclinical studies to acquire higher significance.

Keywords Intravenous · Inhalation · Liposomes · Antitumoral · Antimycotic Antimicrobial

8.1 Introduction

It is widely accepted that bulk and nanomaterials (NMs) sharing the same chemical composition, display completely different toxicity mechanisms (Sharifi et al. 2012). The effects of NMs on living beings are complex and depend on: (1) chemical structure, nature and biological identity of the NMs, (2) type of exposure (intentional or not), (3) route of exposure, and (4) dose (Ganguly et al. 2018). However, gathered data pertaining to the interactions between NMs with living being's health are complex and difficult to classify. A search in Science Direct (accessed in July 2018), showed 32,449 publications between 1994 and 2019 for the keywords "toxicity nanomaterials" covering from in vitro to pre-clinical toxicological studies, many of them employing non-realistic routes of administration and doses, in acute or chronic regimens, most of them focused on the effects of non-intentional (environmental, occupational) contact with non-lipid NMs (Alsaleh and Brown 2018). In comparison, nearly 7695 publications were found between 1997 and 2019 for the keywords "toxicity nanomedicines" and only 3893 between 1998 and 2019 for the more specific terms "toxicity lipid nanomedicines"

The toxic effects of lipid nanomaterials are less significantly reported than those of NM having different and wide variety of chemical natures. In general terms, known is that due to their large surface area/mass ratio, NMs may possess high reactivity and toxicity, including ROS generation, inflammation, cytokine production, glutathione depletion, mitochondrial exhaustion, cellular injury, and protein and DNA damage. The low water solubility may be responsible for bioaccumulation inside lungs and have the potential to cause long-term effects. The aggregation/agglomeration may cause interruption of cellular processes and cellular injury (Bakand et al. 2012). Almost none of these considerations however, apply to the toxic effects caused by lipid nanomedicines.

The toxicity arising from environmental or occupational non-intentional exposure to NMs occurs mostly by inhalation. In the last case, the exposure is expected to be limited by maximal occupational exposure limits (OEL, defined by ISO as "maximum concentration of airborne contaminants deemed to be acceptable, as defined by the authority having jurisdiction" (ISO 16972:2010), by behaviour patterns and protective equipment. The non-intentional exposure may result in acute or chronic nanotoxicity, and mainly occurs in response to inhaled NMs of varied chemical nature, that exclude lipid-based NMs, because of the lack of non-intentional contact. Not all the information on toxicity of nanoparticulate materials thus, applies to the toxicity of nanomedicines. The nanotoxicity arising from intentional exposure instead, is a concept applying only to therapeutic/diagnostic nanomedicines. The intentional exposure of patients to nanomedicines, takes place after successfully passing at least two phase 3 clinical trials, where safety and security are evaluated at large scale along 2–4 years in thousands of patients before its commercialization is approved (FDA approvals). The intravenous administration of nanomedicines is the major employed route together with the inhalatory one. These routes are also responsible for the most pronounced toxic effects. Importantly, despite the structural characterization, manufacture at industrial scale and efficacy/toxicity are ascribed to increasingly stringent regulation by the US Food and Drug Administration (FDA) and European Medicines Agency (EMA), nanomedicines are not exempted from inducing deleterious acute effects on human health.

The aim of the chapter is to describe the deleterious effects of lipid NMs employed as nanomedicines, on the human health. Different from other NMs, the exposure to nanomedicines is studied in a much more limited scope, where doses and exposure route are restricted to those pertaining to the intended use. In such context, the deleterious effects of nanomedicines will be described as a function of clinical immunological manifestations, that can be grouped in: (1) complement activation, (2) inflammatory response, and (3) classical hypersensitivity activation (Alsaleh and Brown 2018).

We will start by acknowledging that, despite the number of consumer products that utilize NMs is exponentially increasing (Vance et al. 2015), the NMs still lack of a universally accepted definition; the same occurs with the nature and field of action of the nanomedicines. Because of those reasons, as a kind of introduction, we will briefly discuss each of the most relevant current definitions.

8.2 There Is no Single Definition Neither of Nanomaterials nor of Nanomedicine

The definitions of nanomaterial employed by the European Union (EU), United States Environmental Protection Agency (EPA) and International Organization for Standardization (ISO) differ among themselves, as follows:

(1) The EU adopted a definition of a nanomaterial in 2011 (Recommendation on the definition of a nanomaterial (2011/696/EU) *as* "a natural, incidental or manufactured material containing particles, in an unbound state or as an aggregate or as an agglomerate and where, for 50% or more of the particles in the number size distribution, one or more external dimensions is in the size range 1–100 nm.

In specific cases and where warranted by concerns for the environment, health, safety or competitiveness the number size distribution threshold of 50% may be replaced by a threshold between 1 and 50%. A material can also be considered as nanomaterial if its volume-specific surface area is larger than 60 m^2/cm^3.

(2) The EPA defines a nanomaterial as "chemical substance that is solid at 25 °C and standard atmospheric pressure, that is manufactured or processed in a form where the primary particles, including aggregates and agglomerates, are in the size range of 1–100 nm in at least one dimension, and that is intentionally manufactured or processed to exhibit unique and novel characteristics or properties because of their size. A reportable chemical substance does not include a chemical substance that only has trace amounts of primary particles, aggregates, or agglomerates in the size range of 1–100 nm, such that the chemical substance does not exhibit the unique and novel properties because of its size.

(3) The International Organization for Standardization (ISO) defines a nanomaterial as a "material with any external nanoscale dimension or having internal structure or surface structure in the nanoscale" (ISO/TS 80004-1:2015) where the nanoscale is the length range approximately from 1 to 100 nm. The lower limit (approximately 1 nm) in the definition of nanoscale is introduced to avoid single and small groups of atoms, as well as individual molecules, from being designated as *nano-objects* (see below) or elements of *nanostructures* (see below), which might be implied by the absence of a lower limit. Further, size-dependent biological effects, specifically particle-cell interactions, and environmental interactions related to nanotechnology, involve structures below 1 nm and above 100 nm. In addition to size, the complex interplay of parameters such as aspect ratio, core chemistry, agglomeration state, physical state, surface properties and others will influence biological and environmental interactions associated with nanostructured materials.

We observe that according to each organism, the scope of the nanomaterial definition can be more or less comprehensive. For instance, the EPA definition only includes manufactured nanomaterials (intentionally produced to have selected properties or composition), while the EU definition includes manufactured, natural and incidental ones (generated as an unintentional by-product of a process). On the other hand, the EPA defines materials by size (1–100 nm) and requires the exhibition of some novel property (like catalytic activity, chemical reactivity, or electrical conductivity), while the EU definition is limited to fix the upper size. And it goes on with differences in composition requirements, aggregate thresholds, and other characteristics. The more open NM definition is that given by ISO, since it does not include the source of the material (therefore including all the sources), fixes the lower limit as 1 nm, and states the upper limit as approximately 100 nm, clarifying the upper limit can be shifted beyond the 100 nm, especially in the field of biological uses of NM.

Besides, according to their dimensions, ISO classifies the NMs into two big categories as follows: (1) nano-object as discrete piece of material with one, two or three external dimensions in the nanoscale, and (2) nanostructured material as material having internal nanostructure or surface nanostructure (ISO/TS 80004-1:2015). According to the shape, the nano-objects are classified into three categories: (1) nanoplate with one external dimension in the nanoscale (1D) (for instance graphene sheets); (2) nanofibers having 2 external dimensions in the nanoscale (2D) (for instance Carbon nanotubes); and (3) nanoparticles (NPs) with three external dimensions in the nanoscale (3D) (such as liposomes, micelles or polymeric nanoparticles) (ISO/TS 80004-2:2015).

Finally, according to its chemical composition, the NMs can be classified into 4 categories: (i) Carbon-based nanomaterials (e.g., single- and multi-walled carbon nanotubes, fullerenes); (ii) Inorganic-based nanomaterials: including metal and metal oxide NPs including saline quantum dots (QDs); (iii) Organic-based nanomaterials: that by noncovalent (weak) interactions self-assembled into desired structures such as micelles, liposomes and polymeric NPs; (iv) Composite-based nanomaterials: multiphase NPs with one phase on the nanoscale dimension that can either combine NPs with other NPs or NPs combined with larger or with bulk-type materials (Jeevanandam et al. 2018). This is a key classification since living beings react differently to NMs displaying equal size, shape, and topology according to their chemical nature (Sharifia et al. 2012).

The medical applications of NMs are more specifically defined by the Center for Drug Evaluation Research (CDER) compared to that of FDA and EMA.

The CDER (FDA) defines NM in its Manual of Policies and Procedures (MAPP) as "any material with at least one dimension smaller than 1,000 nm" (OSHA). From a strict point of view, the FDA considers a product incorporate nanotechnology if they contain or are manufactured with: nanoobjects that range from 1 to 100 nm, which due to their small size and high surface area exhibit key differences in comparison to bulk materials; or materials outside of this range that exhibit related dimension dependent properties or phenomena up to 1000 nm (FDA 2014). FDA has stated that the agency does not adhere to a specific definition of nanotechnology, but rather focuses on material behaviour (Ramachandran et al. 2012).

The EMA employs the EU definition, added with recommendations such as "acknowledged that an upper limit of 100 nm is not scientifically justified across the whole range of nanomaterials" (Commission Recommendation 2011) and 'special circumstances prevailing in the pharmaceutical sector' and stated that the Recommendation should 'not prejudice the use of the term "nano" when defining certain pharmaceuticals and medical devices' (European Union 2010). Remarkably, there is a need to include sizes higher than 100 nm according to the object of application (EU Scientific Committee 2011).

Just as with nanomaterials, there are multiple definitions of Nanomedicine as well. The European Science Foundation defined Nanomedicine as the science and technology of diagnosing, treating, and preventing disease and traumatic injury, of relieving pain, and of preserving and improving human health, using molecular tools and molecular knowledge of the human body (European Science Foundation 2005). In simpler terms, the European Technology Platform (ETP) defines Nanomedicine as the use of nanotechnologies in medicine. It exploits the improved and novel physical, chemical, and biological properties of materials at the nanoscale to prevent early detected and treated diseases and traumatic injury (European Technology Platform on Nanomedicine 2005).

The US NIH defines Nanomedicine as highly specific medical intervention at the molecular scale for curing disease or repairing damaged tissues, such as bone, muscle, or nerve (NIH 2008). The goal of nanomedicine is developed drugs, imaging agents, vaccines and tissue engineering with higher efficacy and improved safety and toxicological profiles. Nanomedicine has a significant impact on many different medical developments in the three main areas: therapeutics, diagnostics/imaging, and regenerative medicine (Nanomedicine 2013).

EMA is working on a definition of Nanomedicines and it should contain the following items: purposefully designed systems for clinical applications, at least one component at nano-scale size, resulting in definable specific properties and characteristics-related to the specific nanotechnology application and characteristics for the intended use (route of administration, dose) and associated with the expected clinical advantages of the nano-engineering (e.g., preferential organ/tissue distribution) (Pérez de la Ossa 2014).

In this Chapter we will apply the term nano-object in reference to engineered NMs (including nanomedicines) and reserve the term nanomaterial to make reference to engineered nanoparticles, nanotubes or nanoplates, from natural or incidental sources.

We will define a therapeutic nanomedicine as a nano-object associated with a given amount of active pharmaceutical ingredient (API). This type of non monodrug entity, with a difficulty to completely characterize structure, known as non-biological complex drug (NBCD) (Crommelin et al. 2015), acts as a nano-drug delivery system. The API´s pharmacokinetics, biodistribution and intracellular traffic no longer depends on its chemical structure but on the structural features of the nano-object.

8.3 One Single Definition for Nanotoxicology

Currently, there is a single definition for Nanotoxicology provided by Oberdörster (Oberdörster 2010). Nanotoxicology is defined as "the study of adverse effects of the engineered nanoobjects on the organisms and ecosystems. It also includes the prevention and amelioration of such adverse effects". This definition implies that, as the new properties arising from reducing the size to the nanoscale constitute useful features of new products, materials that are not harmful in their bulk form are also likely to be associated with unwanted greater biological/toxicological reactivity. Bulk gold, for example, is normally inert but Au NPs that are useful as image agents are not inert at all. Besides, NPs are also more likely to react with cells and biological components such as proteins, and to travel through organisms (translocation), which increases their chances of entering various organs and activating inflammatory and immunological responses. Concepts of nanotoxicology are aimed at identifying potential hazards are also applicable for safety evaluation in nanomedicine for applications of nanotechnology-based therapeutics and diagnostics.

The size of typical nanomedicines is in the order of that of macromolecules and cell ultrastructure. The hydrodynamic radii of a typical protein ranges between 3 and 10 nm, the size of viruses between 50 and 150 nm; accordingly, the size of therapeutic liposomes is between 80 and 120 nm in diameter, considerably smaller than 6000–8000 nm (6–8 μm), the size of a red cell, well beyond the nanoscale

At the systemic level, certain nanomedicines of less than 100 nm in mean diameter, because of their high surface area, increase the bioavailability of oral drugs (nanocrystal technology). Other intravenously injected nanomedicines may remain in blood circulation for long time, because of the reduced renal clearance. At the tissue level, certain nanomedicines may experience convective extravasation in areas of increased vascular permeability and accumulate in the vicinities of solid tumors aided by the low lymphatic drainage (enhanced permeability and retention (EPR) effect); this may also occur in inflamed tissues. If properly designed, the nanomedicines may cross anatomo-patho/physiological barriers in a controlled fashion, to get therapeutic targets of hidden accessibility. This potential has to be clearly distinguished from unwanted translocation to secondary organs observed for certain inhaled NM. At the cellular level, nanomedicines are recognized and captured by different types of endocytic mechanisms, to be intracellularly submitted to a destructive process that can be exploited to provide massive drug delivery to target organelles or to trigger protective immune reactions.

Currently, the FDA has approved several types of nanomedicines e.g., liposomes, micelles, nanocrystals, polymeric nanoparticles, microemulsions, and nanoemulsions. Although these nanomedicine products do not fulfil the US National Nanotechnology Initiative's (NNI) size definition of nanotechnology, these are still considered nanotechnology-derived medicine, due to their unique features and advantages when compared to their micro-sized counterpart drug products (Weissig et al. 2014). Nearly 50 FDA-approved nanomedicines are available for clinical use (Bobo et al. 2016; Ventola and Lee 2017) and 60 in clinical trials (Caster et al. 2017). Nearly 34% of the products in the market are polymer-drug conjugates (including PEGylated biologic drugs), 20% are liposomes, 30% are nanocrystals, 10% inorganic nanoparticles, and the remainder micelles (1) and proteins (2). Accordingly, 56% of the nanomedicines in clinical trials are liposomes, 18% polymeric nanoparticles, 15% micelles. Nanomedicines are typically administered intravenously (≈78%), followed by orally (17%) or less frequently pulmonary (2%) or topical (3%).

8.4 Common Step of NMs Biokinetics: The Biocorona Formation

Whereas the large surface-to-mass ratio of NM makes them biologically more reactive than larger particles of identical material and shape (Duffin et al. 2002; Napierska et al. 2009), the biokinetics of NMs presented in Fig. 8.1 cannot be generalized: certain NMs will be rapidly degraded along the first minutes of contact

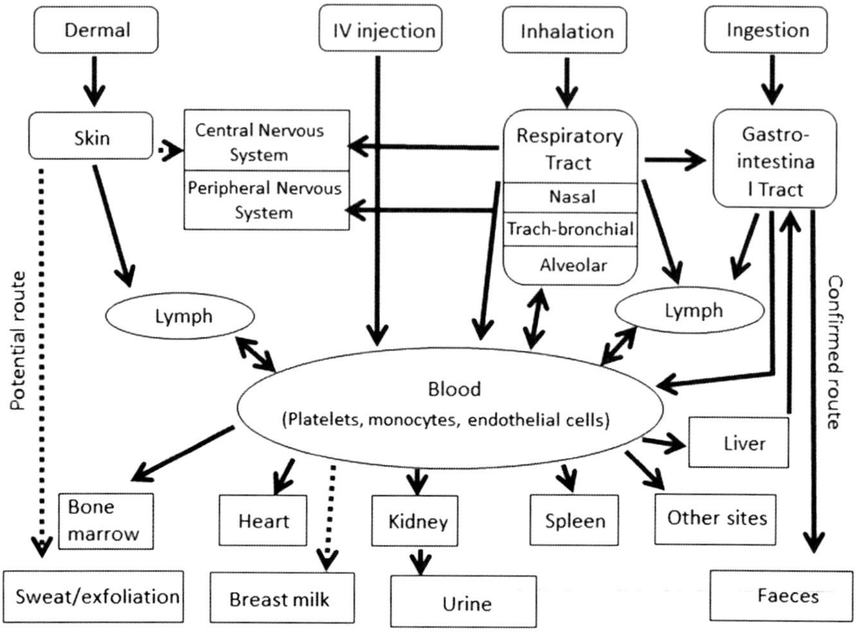

Fig. 8.1 Biokinetics of nanoparticles. Solid arrows: confirmed routes; dashed arrows: potential routes (modified from Oberdorster et al. 2013)

and internalization by biological barriers (epithelia/endothelia/blood/mucus/gastrointestinal media/solid organs), reducing their chances of translocation to secondary organs/compartments beyond the administration site. The chance of translocation of a NM depends on its chemical/structural nature; lipid-based nano-objects for instance, do not translocate. Certain other NMs however, may remain structurally stable by weeks or months, being capable of translocation and constituting a source of chronic toxicity. The chemical structure, together with a highly varied structural nature (geometry, topology, inner structure) of NM, will be responsible for the transformations that will undergo before and along their contact with the biointerface, defining the solubility, susceptibility to oxidation or to hydrolysis, biodegradability of biopersistence (Nel et al. 2009).

The sequence of events occurred upon contact between NM and biointerface may be depicted as follows:

(1) Formation of the biocorona
(2) Interaction with blood cells (only for intravenous (IV) route)
(3) Or interaction with mucus (gastrointestinal route or big sized NM administered by inhalation route)
(4) Interaction with cells/internalization/degradation/biopersistence
(5) Renal clearance (only for <5 nm NM administered by IV route).

In the following section, only the formation of the biocorona will be addressed in detail, since this step is common to all the contact routes.

8.4.1 Formation of the Biocorona

Upon contact with biological and/or abiotic environments, due to their high surface free energy, the NMs acquire a protein (that may be accompanied by nucleic acids, sugar polymers and lipids) known as biocorona, that will change its original chemical identity to a new biological identity (Monopoli et al. 2011; Tenzer et al. 2013). The corona composition depends on NM size, surface curvature, charge and functionalization, relative ratio of media to NM concentration, topology exposure temperature, exposure time (Docter et al. 2015) and the biological/abiotic environment (Lundqvist et al. 2008; Zhu et al. 2013). For instance, hydrophobic nanoparticles, e.g., carbon nanotubes, usually attract proteins with several hydrophobic residues (Ge et al. 2011). In addition, enhanced protein adsorption is observed with increased nanoparticle size (reduced curvature, larger surface area) (Cedervall et al. 2007); positive surface charge is typically correlated with increased protein adsorption.

The protein/biomolecule corona is dynamic in nature. The amount of protein/biomolecules in the corona changes over time, while the types of bound protein continually adsorbing and desorbing remaining relatively constant (Vroman 1962; Vogler 2012; Tenzer et al. 2013). The time-dependence of these processes, i.e., the rates of association and dissociation of a protein/biomolecule, are described by the parameters k_{on} and k_{off}, respectively. The balance between k_{on} and k_{off} determines the affinity of a protein/biomolecule to the NM and is defined by the dissociation constant (K_d). The value of k_{on} depends on the contact frequency between the protein/biomolecule and the NM, along with the probability that such a contact results in an adsorption event. Protein/biomolecules that are present at high concentrations, and that diffuse rapidly mostly show high k_{on} values. The value of k_{off} depends on the binding energy of the protein/biomolecule–NM complex. Protein/biomolecules that adsorb with a large binding energy thus have a lower k_{off} value (Vogler 2012).

In complex environments, protein/biomolecules compete for binding sites on the (nano)material surface during adsorption, with their characteristic individual Kd. Notably, Kds determined for the adsorption of certain proteins in isolation to NPs were reported to range between 10^{-4} and 10^{-9} M—very similar to what is measured for antibody–antigen interactions. In contrast to bulk materials, a certain binding specificity of nano-sized objects for a specific (class of) protein(s), and other biomolecules may exist (Docter et al. 2015).

The plasma protein corona is highly complex, containing over 200 different proteins, and is established in less than one minute (Tenzer et al. 2013). In an early phase, a highly complex corona is established already in 30 s, which may be composed of multiple core–shell structures ('new') (Docter et al. 2015). In the late phase, the corona composition ex situ remains stable, and changes predominantly quantitatively rather than qualitatively over time. The Vroman-effect describes a time-dependent composition of the corona, in which highly abundant proteins adsorbing only weakly dominate the early state (Vroman 1962). Subsequently, these adsorbed proteins are replaced by less abundant proteins, which, however, bind with a higher affinity, resulting in a complex series of adsorption and displacement steps. In the late phase, Vroman effect dependent and independent binding kinetics contribute to the corona account (Docter et al. 2015). Over time, new proteins will be added to the original surface, by protein-protein interaction. The initial set of proteins may vary depending on the chances of accessing the NM surface.

The biocorona may modify the original surface of a given NM by hiding ligands or granting a new biological identity by exhibiting new ligands that may increase or decrease their nanotoxicity (Wolfram et al. 2014).

8.4.2 Biocorona and Nanotoxicity

The acquisition of a biocorona may trigger nanotoxicity for instance, if during adsorption to NM surface the endogenous proteins experience misfolding (Pelaz et al. 2013; Zuo et al. 2013). For instance, nanoparticle-induced conformational changes in fibrinogen were shown to activate inflammatory signalling pathways (Deng et al. 2011).

On the contrary, the formation of a protein corona can also mitigate the cytotoxicity of NPs. Especially in the case of NP–cell membrane interactions, the presence of a protein coat can decrease cell damage (Hu et al. 2011; Lesniak et al. 2012). Indeed, nanoparticles that are not pre-coated with a protein layer may instead interact with cell membrane proteins, causing damage to them. For instance, it has been shown that the toxicity of carbon nanotubes (Ge et al. 2011) and gold nanorods (Wang et al. 2013a, b) decreases upon coating with plasma proteins. A positive correlation also exists between protein density and decreased toxicity. In addition, whereas pristine silica NPs have been shown to activate thrombocytes and cause hemolysis, these adverse events can be prevented in the presence of a protein coating (Tenzer et al. 2013). In conclusion, the protein corona can reduce the overall toxicity of NPs, while simultaneously impairing the function of a subset of plasma proteins.

8.5 Biokinetics of Respirable NM

Upon the biocorona formation, the NM will interact with biological media according to the specific features of each microenvironment, as described above. Since the inhalation pathway most likely results in larger systemic doses of natural and incidental NM displaying different chemical nature, the inhalation route constitutes the main non-intentional contact route, followed by the oral and the topical route in third place (Oberdorster et al. 2005).

The respiratory system receives the entire cardiac output, meaning that the respiratory system can be exposed secondary to a systemic passage of NM. The inhalation route also becomes an important for intentional exposure (administration) to lipid NM aerosolized by nebulization. The nanotoxicity resulted from the two types of exposure can be described in common. Importantly, whereas NMs of varied chemical nature of size ranging from ultrafine to less than 5 mm diameter, may have access to the alveoli, water drops produced by vibrating mesh nebulization and having only ~2 μm diameter aerodynamic size can carry tens of nm sized liposomes to the conducting airways and the respiratory zone (Cipolla et al. 2014). Given the important physical barrier that limits the biodistribution of inhaled NMs, we will briefly address the main anatomical and functional features of the lung. Then the factors that may trigger nanotoxicity by inhalation of NMs will be revised, to finally make a focus on the controversial approval pathway currently followed by inhalable nanomedicines. The last section of the Chapter will specifically address the nanotoxicity of antitumoral and anti-infective lipid-based nanomedicines already in the market.

8.5.1 Inhalatory Route: An Overview of the Conducting and the Respiratory Regions

The *extrathoracic (or upper) airways* are composed of the nose, mouth, pharynx and larynx, while intra-thoracic (or lower) airways are represented by the conducting airways (trachea, bronchi), and the respiratory airways, with pulmonary alveoli. The lung resembles an inverted tree, where the trachea or trunk subdivides into two main bronchi and these latter successively branch into increasingly narrow and short bronchioles. In total, the trachea undergoes 23 bifurcations before it reaches the alveolar sacs. In the conducting region the air is filtered, warmed, humidified and conducted to the respiratory region.

The *proximal conducting airways* contain ciliated columnar cells, and goblet or mucus secreting cells (Weibel 1963). The mucus is produced by goblet cells and sub-mucosa glands. It covers the entire airway surface, and its thickness ranges from 5 to 55 μm. It consists of an upper gel phase made of 95% water, 2% mucin, a highly glycosylated and entangled polymer, as well as salts, proteins, and lipids. A periciliary liquid layer underlies the mucus gel and its low viscosity allows effective beating of the cilia. Mucociliary clearance is one of the most important defence mechanisms to trap and eliminate the higher inhaled microparticles and microorganisms in the lungs. The mucus (together with trapped NM) is transported by the coordinated beating of the cilia and by expiratory airflow towards the oropharynx at an average flow rate of 5 mm per minute. Mucus, cells, and debris coming from the nasal cavities and from the lung meet in the pharynx, are mixed with saliva and are swallowed.

The *deeper respiratory region* comprises the respiratory bronchioles, the alveolar ducts and the alveolar sacs, where the gas exchange between airspaces and blood capillaries occurs. The alveolar surface is 95% covered by squamous type I pneumocytes of large apical surface and thinness (0.05 μm). Cuboidal type II pneumocytes produce the *lung surfactant* and are located in the corners of the alveoli. (approximately 300 million alveoli in the adult lung). Inhaled NM accessing the alveoli is trapped by alveolar macrophages (AM). Besides of offering a very thin air-blood barrier (<1 μm), the surface area of the alveolar epithelium reaches 100 m^2, which is enormous as compared to the 0.25 m^2 surface area of the airways (Gehr et al. 1978; Patton and Byron 2007).

8.5.2 Pulmonary Surfactant (PS)

The lung is ceaselessly submitted to compression–expansion cycles (~15–20 cycles per minute) and undergoes a typical total volume variation of~20% (Krafft 2015); the pulmonary surfactant (PS) lowers surface tension at the air–liquid interface to prevent alveolar collapse and enabling muscular expansion work (Rugonyi et al. 2008; Lopez-Rodriguez and Pérez-Gil 2014). PS is a complicated mixture of~80 wt% phospholipids (PL), 5–10% neutral lipids (primarily cholesterol), and 5–10% proteins (SP)-A (~28–36 kD), B (~8 kD), C (~4 kD), and D (~43 kD) (Possmayer 2004). The phospholipids together with the lipophilic SP-B and SP-C reduce the surface tension. The hydrophilic SP-A and SP-D play an important role in the immune system due to enhanced clearance of pathogens as well as specific modulation of the inflammatory immune response (Wright 2005). The PS is secreted by pulmonary type II epithelial cells, and plays a crucial role in maintaining normal respiratory mechanics. At least three biophysical properties of PS are essential for normal respiratory physiology, especially in the neonatal period. They are rapid adsorption, very low surface tension on film compression, and effective film replenishment on film expansion (Possmayer 2004). PS films can lower the alveolar surface tension to near-zero values at end expiration and are able to maintain these low values for prolonged periods (Schurch 1982). In other words, functional PS films should be capable of increasing the surface pressure (π) to a value close to 70 mN/m at 37 °C. This π is substantially higher than the equilibrium spreading pressure (πe) of PL vesicles, i.e., ~45–50 mN/m (Possmayer 2004).

The function of PS is essential. Ventilation of lungs that lack adequate surfactant, both in premature babies (Robertson 1984) and in experimental animals subjected to repeated lavage (Lachmann et al. 1980) damages the alveolocapillary barrier, leading to pulmonary edema, respiratory failure, and death.

In children and adults, PS inactivation often occurs as a consequence of acute lung injury (ALI), and is a primary cause of acute respiratory distress syndrome (ARDS). ALI/ARDS may result from both direct lung injury (pneumonia, aspiration, drowning or toxic inhalation) and indirect lung injury (sepsis, nonthoracic trauma, multiple blood transfusions, major surgery, acute pancreatitis or shock) (Dushianthan et al. 2012).

By 2030, the annual incidence of ALI/ARDS is predicted to be 335,000, with 147,000 deaths per year (Rubenfeld et al. 2005). Although PS therapy has demonstrated some improvements in gas exchange and survival in animal ARDS models, its efficacy in human clinical trials is limited, and mortality remains high (Dushianthan et al. 2012). It has been recognised that PS inactivation is a major factor in ARDS, and that the lack of clinical efficacy of the exogenous PS preparations results from such inactivation.

The PS interposes between respirable nanomedicines and target cells in the alveoli. Inhalable nanomedicines have to grant safety for the respiratory function. As depicted above, PS inactivation is life-threatening, and refers to all processes that interfere with and decrease surface activity of surfactant. PS inactivation may occur because of external [leakage of surface active serum proteins (albumin, fibrinogen monomer, fibrinogen, immunoglobulin, C-reactive protein), meconium, fatty acids, lysoderivatives, pro-inflammatory cytokines (IL-1, TNF), lipopolysaccharides (LPS), and certain lipids such as cholesterol (a major component of meconium), bile salts, protein C reactive, red blood cell components (haemoglobin)] or endogenous (altered secretion by alveolar type II cells) factors. The PS components can also be hydrolysed by phospholipases or proteases (the concentration of which can increase strongly during microbial infection), or peroxidised by reactive oxygen species (ROS), or altered by various pollutants, including acids, fumes, and volatile particles (Lopez-Rodriguez and Pérez-Gil 2014). The advent of new techniques for detection, production, and structural characterization of respirable nanoparticles, enabled to survey their impact on respiratory function, specifically of their interaction with PS (Schleh and Hohlfeld 2009; Harishchandra et al. 2010).

An ample evidence has emerged supporting the fact however, that translocation of polymeric, silicon oxide pristine or methylated titania oxide (Chhoden et al. 2015), negatively- and positively-charged poly(styrene) (Beck-Broichsitter et al. 2014) or metallic nanoparticles, induces PS inactivation. Hydroxyapatite nanoparticles injected to PS Infasurf subphase for instance, cause a significant time dependent shift of compression isotherms to the left, i.e., more area reduction is required to increasing surface pressure, indicating surfactant inactivation (Fan et al. 2011). Nanoparticles in contact with lung surfactant have to prove therefore, not to induce the surfactant inactivation (Altube et al. 2017).

8.5.3 *The Alveolar Macrophages (AMs)*

In conjunction with mucociliary transport, phagocytosis by AMs contributes to removal of inhaled particles from the lungs. The AM resides on the air-blood barrier in the fluid bathing the epithelial layer, just below the surfactant layer, and are specialized in eliminating respirable particulate material (Martin and Frevert 2005). AMs are the major antigen-presenting cells in the alveolar space and serve as dominant phagocytes in mediating innate immune and inflammatory responses against inhaled foreign matters (Rubins 2003). Although they occupy only 1% of the alveolar surface, they are capable to clean particles from the entire alveolar surface due to amoeboid movements (Geiser 2010). AMs are highly adapted to the unique environment of the lung, and at steady state their ability to generate inflammatory responses is regulated tightly to ensure that lung injury is kept to a minimum, thus preventing precarious alveolar physiology and preserving gas exchange (Piantadosi and Schwartz 2004). AMs with M2 polarization play a key role in lung development (Jones et al. 2013). When challenged by inhaled pathogenic organisms however, AMs synthesize and secret a series of inflammatory mediators including cytokines, chemokines, arachidonic acid metabolites, lysozymes and antimicrobial proteins to potentiate inflammatory and phagocytic responses and recruit activated neutrophils into the alveolar region.

8.5.4 Inhalable NMs and Distribution Sites in the Lungs

The site of deposition of inhaled NMs within the respiratory tract depends on the aerodynamic diameter of the aerosol particles (d_{aer}). The aerodynamic diameter of a particle (d_{aer}), on the other hand, that can be calculate from the following formula, is equivalent to the diameter of a unit density (ρ0) sphere that has the same terminal velocity in still air as the particle, where d is the geometric diameter of the particle, ρ is the particle density and χ is the particle dynamic shape factor denoting deviation of shape from sphericity (Hinds 1999).

$$d_{aer} = d\sqrt{\rho/\rho 0\chi}$$

The inhaled NMs interact with the complex lung geometry in such a way that the large particles ($d_{aer} < 5$ μm) are retained in upper airways (mouth, trachea and main bronchi) by inertial impaction. One to 5 μm d_{aer} particles deposit by gravitational settling in the central and distal tract. Particles with $d_{aer} < 1$ μm remain suspended in the air and are mostly exhaled. Ultrafine particles (<100 nm) generated in abundance by the most significant pollution sources, which are those related to combustion processes (Morawska et al. 2005) can largely deposit in the respiratory tract by random Brownian motion. Particles<100 nm reaches the alveolar region while particles<10 nm already deposit in the tracheo-bronchial region due to their high diffusion coefficients (Heyder et al. 1986). Particulate matter (PM) is categorized into particles<10 μm (PM_{10}), <2.5 μm ($PM_{2.5}$) and<0.1 μm ($PM_{0.1}$) in diameter ultrafine particles, the size fitting in the definition of NPs and considered NMs (Roller et al. 2011). Ultrafine particles exhibit a high number/volume and surface area/volume, hence tend to agglomerate when suspended in air and in liquids, and to adsorb proteins and lipids that lately will influence their biokinetics (Oberdörster 2010). The size (airborne, hydrodynamic), shape, the tendency to agglomeration/aggregation, the surface properties such as area, electrical charge, presence of coating or contaminants, defects, solubility and crystallinity have an impact on the biological/toxic effects of ultrafine NPs. All these properties may also be modified according to the production method of NPs, during storage, and the contact with the biointerface. In particular, agglomeration (held together by weak van der Waals forces) and aggregation (held together by strong chemical bonds or sintered) both in liquids or in air

Table 8.1 Particle mass, number, and surface area: comparing different particle sizes of airborne concentration of 10 pg/cm (unit density particles)

Particle diameter (nm)	Particle number (N/cm^3)	Particle surface area ($\mu m^2/cm^3$)
5	153,000,000	12,000
20	2,400,000	3016

determine the actual particle size for ultrafine-biointeractions or the deposition-site and -efficiency in the respiratory tract upon inhalation. Both in dry dusts as in aqueous dispersions, ultrafine NPs may be found as isolated NPs, agglomerated NPs, aggregated NPs or agglomerated aggregated NPs. The media of the actual size of NPs in different environments is essential to properly understand the toxic effects and predict its further fate since size is a key factor with respect to translocation across cell barriers (Oberdörster 2010) (see Table 8.1).

8.6 Nanotoxicity by Inhalatory Route

Many different inhaled agents may damage the lung. However, the pattern of cellular injury and toxic responses may often be mediated by a few biochemical events such as oxidative injury, signalling pathways and genotoxicity. Imbalanced ROS generation may lead to lipid peroxidation, membrane damage and DNA damage. Alteration in growth factors and/or cytokine signalling molecules may induce inflammation, proliferative injuries or apoptosis. Acute inflammatory, cellular, and molecular responses of the respiratory tract to inhaled materials can be characterised using human lung cells in vitro (Hayes and Bakand 2014).

The inhalation of ultrafine NM, initially known as poorly soluble low toxicity dusts (Dankovic et al. 2007), poorly soluble particles (PSPs) (Borm et al. 2015) or respirable granular biopersistent particles (GBP) without known significant specific toxicity at concentrations below existing exposure limits (Roller and Pott 2006), is related with the development of lung diseases (Kuempel et al. 2014), including exacerbation of lung diseases (Li et al. 2008), airway inflammation (Donaldson et al. 2005), fibrosis (Churg and Wright 2002; Churg et al. 2003), cancer (Dankovicet al. 2007) and myocardial infarction (Peters et al. 2004). GBP comprise various materials such as minerals, metals, metal oxides, or polymers that show a negligible solubility in lung fluids (extracellular lung lining fluid, intracellular lysosomal fluid), and are known to induce very low inflammogenicity (granular biopersistent). The deleterious effect of GBP would be owed not only to their size, small enough to access the alveoli and to be internalized by AM, but also to their own chemical nature (namely: TiO_2, C black, $BaSO_4$, AlOOH, CeO_2). The AM overloading is a phenomenon associated to inhalation of GBP and occurrence of inflammation and translocation. The phenomenon consists in the progressive reduction in NM clearance from the deep lung and occurs because of the failure of AM to remove NM due to the loss of AM motility (Morrow 1988). In rats, it has been observed that overloading starts when the particulate volume of NM in an AM exceeded 6% of the AM volume, and the complete cessation of AM-mediated clearance occurs when the phagocytosed particle volume reaches about 60% of the AM volume, as demonstrated in a subsequent study with 3 and 10 μm polystyrene particles (Oberdörster et al. 1992).

The volumetric overload hypothesis specifies that it is not the material density but the packing density that determines the volume. According to this, the void spaces between the packed particles inside AMs are also responsible for the induced toxicity. Those GBP that exhibit a higher portion of void spaces in their agglomerates thus possess a higher displacement volume in AM when comparing identical mass concentrations, leading to a higher inflammatory potency. Therefore, the inflammation caused by repeated dosing is explained by particle agglomerate volume (Pauluhn 2011; MAK Commission 2014). Particle agglomerate volume is thus the best dose metric explaining the chronic inflammation of nanoscale GBP (Committee on Hazardous Substances 2015).

The biologically effective dose of instilled or inhaled GBP that causes acute effects, on the other hand, is estimated employing surface area metrics (Donaldson et al. 2013; Pauluhn 2014; Schmid and Stoeger 2016). For instance, from sub chronic inhalation testing of TiO_2 and $BaSO_4$ particles in Wistar rats, a threshold of 1 cm^2 particle surface area per cm^2 proximal alveolar region was concluded for the onset of inflammation based on PMN activation in the BALF (Donaldson et al. 2013). The facts about dose metrics apply mostly to the inhalation of GBP, nanomaterials with a strong tendency to agglomeration or aggregation, but is less applicable to nanomedicines, as we will see below.

Innate immune cells express receptors that can recognize certain molecular patterns (e.g., pathogen-associated molecular patterns, PAMP) through pattern recognition receptors (PRR) such as the Toll-like receptors (TLR) and NOD-like receptors (NLR). These receptors can recognize certain molecular patterns of microbial origin such as lipopolysaccharides and peptidoglycan. Innate immune cells also recognize danger- or damage associated molecular patterns (DAMPs) (also called 'alarmins') such as interleukin-1 (IL-1). These are endogenous biomolecules that trigger a sterile inflammatory response. It has been proposed that GBP may act as danger signals termed nanomaterial-associated molecular patterns (NAMPs), which could provoke cellular stress responses, cell injury and inflammation (Fadeel 2012). TiO_2 NP is an agonist of NALP3 (also known as NLRP3 and cryopyrin), a NLRs that mediates the formation of caspase-1-activating multiprotein complex known as the inflammasome, which processes proinflammatory caspases and cytokines (Davis et al. 2011). Ag NP are also capable of activating the inflammasome through degradation of ATF-6 (activating transcription factor-6), an endoplasmic reticulum stress sensor (Simard et al. 2015) triggering subsequent cellular stress, injury, and inflammation. Additionally, NP different to the GBP, for instance, rigid biopersistent respirable fibrous NMs with a specific geometry and high aspect ratio [(>5 μm length, diameter<3 μm, aspect ratio>3:1) so-called WHO fibres] such as silica micro and NP and MWCNT

(Gebel et al. 2014) are known to induce the so called 'alarmins' IL-1a and IL-33, responsible for acute lung inflammation (Wang et al. 2013a, b; Rabolli et al. 2014).

The frustrated phagocytosis of nano-objects (Champion and Mitragotri 2006; Frohlich 2012) is another example of inflammation arisen upon interaction between macrophages and fibres or nanofibers of high aspect ratio. Frustrated phagocytosis is the failure to engulf after attaching, spreading and manipulation of fibres by macrophages. The extent to which high aspect ratio materials evade clearance is strongly length-dependent. High aspect ratio materials such as asbestos fibres or carbon nanotubes can be cleared by phagocytosis or persist and induce frustrated phagocytic interactions leading to chronic inflammation, oxidative stress, direct cell injury, and chromosomal abnormalities. This evasion can lead to diseases such as fibrosis, asbestosis, lung cancer, mesothelioma for asbestos-like materials; carbon nanotubes have recently been shown to induce asbestos-like chronic inflammation. Uptake of short fibres occurred more readily than for long, but long fibres are more potent stimulators of inflammatory molecules. Stimulation resulted in dose-dependent secretion of inflammatory biomolecules but no cytotoxicity or strong ROS production (Padmore et al. 2017).

Overall, particle clearance from the lung decreases in a linear fashion with increasing NM load (Roller 2003). The consequence of higher NM loading is chronic inflammation mediated by AMs and neutrophils in the deeper airways. Lung carcinogenicity and other lung abnormalities are the consequence of chronic inflammation, which was assumed by some researchers to be species-specific for the rat as hamsters and mice did not show lung tumors after GBP exposure. Others argued that the latter species may not be adequate indicators for human lung carcinogenicity.

The impact caused by inhaled lipid-based NM or nanomedicines on the other hand, is pursued along the time range (usually months) during which the medication is administered, in a very well-defined dosage regimen. It is not accidental that liposomes are the most extensively investigated nanomedicines for drug delivery to the lungs (Mansour et al. 2009). This is because upon intravenous administration, a route allowing direct contact with mediators of innate and acquired immunity, certain NM may elicit life threatening acute reactions, as we will see below. The lipid-based nano-objects have repeatedly shown to display high biocompatibility in comparison to nano-objects of other than lipid chemical nature (Mallick and Choi 2014), and do not fit neither into the WHO nanofibrer not into the GBP definition. The soft and biodegradable lipid-based nano-objects induce minimal perturbations in terms of biocorona formation, immune recognition and impact on cell signalling pathways. At the lung level however, the inhaled NM have not only to exhibit higher efficacy than pre-existent products, but also to prove not to perturb the gaseous exchange at alveolar level. Once in the deep lung, this may happen by different mechanisms: by triggering inflammatory mediators in AM, or decreasing the tensioactive function of the PS. Eventually, considerable amounts of liposomes may induce AM overloading, allowing the access of inhaled material to interstitial places and from there to systemic circulation, a fact potentially deleterious. In this scenario, we will describe in more detail the implications of the development of the two unique nebulizable liposomal formulations in advanced clinical phases, aimed to treat lung infections: liposomal amikacin and liposomal ciprofloxacin

8.6.1 Liposomal Amikacin

Amikacin Liposomal Inhalation Suspension (ALIS) previously known as Arikace, Arikayce and LAI is amikacin encapsulated in liposomes for treatment of non-tuberculous mycobacteria lung disease (NTM-LD) (Phase 3) and chronic *Pseudomonas aeruginosa* in non-cystic fibrosis (CF) bronchiectasis (Phase 2) developed by Insmed (Insmed Company website). Previously ALIS was evaluated in a Phase 3 clinical trial in CF (Bilton et al. 2013) and in a Phase 2 trial in non-CF bronchiectasis (O'Donnell et al. 2009).

The liposomal formulation of amikacin was developed specifically for inhalation with the aim of overcoming the rapid clearance of drug from the lung tissue, so allowing once daily dosing, shortening the duration of inhalation and achieving improved penetration of biofilms without significant systemic exposures. Liposomes are composed of dipalmitoylphosphatidylcholine (DPPC) and cholesterol, at a 2:1 weight ratio, with 70 mg/mL of amikacin and 47 mg/mL of lipid (DPPC and cholesterol combined). The liposomes are relatively small (<300 nm) and release about 30% of the initial amikacin load during nebulization using an eFlow Nebuliser System. Approximately 67% of the aerosol droplets produced are within the respirable range (<5 μm), which enables 43% of the nominal liposome dose to be deposited in the lungs after nebulization (Olivier et al. 2016).

Since liposomal components of ALIS are found in lung surfactant, which is normally cleared rapidly by several lung cell types, (type II pneumocytes, Clara cells, and macrophages) the liposomes would lack of toxicity (Chinoy et al. 1994). Turnover rates in the adult human lung for DPPC and cholesterol are approximately 79 mg/h and 24 mg/day, respectively (Martini et al. 1999). Following a single 560 mg ALIS dose, approximately 73 mg of DPPC (29 mg to lung periphery) and 36 mg of cholesterol (14 mg to lung periphery) are delivered to the lung. Repeated dosing over prolonged periods in animal models has little lasting effect on lung histology or pulmonary macrophage function (Neville et al. 2009). If well these data may be realistic, the drug toxicity is not accounted, the same as the lung functionalism of the patients, that can be very different from that of healthy individuals.

Subacute (1 month) and chronic (3 months) inhalation studies in rats showed inflammation and alveolar epithelial hyperplasia that accompanied macrophage infiltration. The NOAEL (no observed adverse effect level) for changes in the lung was<10 mg/kg/day, while the NOAEL for nephropathy was 30 mg/kg/day. All changes were reversed or showed evidence of reversibility following a 3-month recovery period. In 6-month rat study foamy macrophage accumulation was associated with inflammation, that was accompanied by regenerative hyperplasia of the alveolar epithelium (EMA 2016). Inhalation studies in dogs (9 months) did not show such effects; whereas some of them were transient, lymph node reactions (perivascular/peribronchiolar lymphoid cell infiltration) that are not readily reversible and lymphoid findings associated with foamy macrophages were observed, that can indicate an adverse effect (EMA 2016). The EMA concluded that the findings of the rat studies with ALIS were species-specific and that (1) the cause could be due to lung burden and (2) a threshold effect could exist (these findings were not seen in the dog). However, the relevance of the rodent data to humans is still unknown. Therefore, the decision on the duration of therapy should take into account the nonclinical findings as well as the safety profile observed in patients in the Phase 3 study. In addition, findings should be discussed considering the disease to be treated itself; i.e., does the disease itself affect the sensitivity of the human lung (EMA 2016)

Across all clinical studies ALIS has been administered to 519 individuals, including 6 healthy subjects, 383 patients with CF, 43 patients with bronchiectasis and 89 patients with NTM-LD. The safety database relevant to the final proposed daily dose includes 519 patients exposed to 560 mg or 590 mg amikacin. Studies provide a comparison with saline, with empty liposomes and to tobramycin solution for inhalation in a population in which the majority had received tobramycin previously. In each comparison made ALIS was associated with higher airway reactogenicity, higher rates of treatment-related adverse events and higher rates of discontinuations due to adverse events. However, the comparison with empty liposomes indicates that amikacin is per se an irritant. ALIS is clearly less well tolerated than empty liposomes and less well tolerated than tobramycin. These results however, would not be an impediment for an approval. However, since efficacy is not currently supported in either indication the comparative safety profile of ALIS adds to the overall conclusions on the lack of a favourable benefit-risk relationship. In June 2016, the Committee for Medicinal Products for Human Use of EMA did not approve the application of ALIS for the treatment of *Mycobacterium avium complex (MAC)* lung disease in adult patients who have persistent positive sputum cultures despite the use of medically appropriate first-line therapy (EMA 2016). In May 2018, FDA accepted the New Drug Application (NDA) for ALIS for adult patients with NTM-LD caused by MAC.

8.6.2 Liposomal Ciprofloxacin

Liposomal ciprofloxacin formulations (Lipoquin®) are unilamellar liposomes between ~50 and 100 nm in diameter composed by hydrogenated soy phosphatidylcholine (HSPC) and cholesterol. Linhaliq, previously known as Pulmaquin®, a 1:1 vol:vol mixture of liposomal ciprofloxacin (Lipoquin®, 45 mg/ml) and free ciprofloxacin (18 mg/ml) (in two separate vials that are mixed at the moment of nebulization) (Cipolla et al. 2016), is an inhalation formulation of liposome encapsulated and unencapsulated ciprofloxacin in development for the treatment of non-cystic fibrosis bronchiectasis (NCFBE) patients with chronic lung infections with *Pseudomonas aeruginosa* developed by Aradigm. Inhaled liposomes could extend the residence time of ciprofloxacin in the lung, may allow for a reduction in the dosing frequency while still continuously maintaining high drug concentrations in the lung, above the MIC, and the free ciprofloxacin is available to act rapidly (Cipolla et al. 2013). Liposomes, in addition can also deliver the drug to macrophages via phagocytosis, improving the treatment of intracellular infections such as tuberculosis and non-tuberculous mycobacteria, as well as respiratory infections that are a bioterrorism threat (Conley et al. 1997; Bermudez et al. 2015). The liposomes retain the amount of encapsulated drug and mean vesicle size in response to storage and nebulization.

The clinical safety was evaluated in one Phase 2 study, and two Phase 3 studies, 389 patients with NCFBE with chronic *P. aeruginosa* lung infections received at least one dose of Linhaliq (inhaled once daily (189 mg ciprofloxacin, 135 mg liposomal + 54 mg free) in 28-day on/off 6 cycles, followed by an open-label extension, in which all patients received Linhaliq for another 28 days), 193 patients received at least one dose of placebo liposomes and saline (Smith 2018).

The FDA conclusions on the safety were that at similar rates of common treatment-emergent adverse events, adverse events leading to withdrawal, serious adverse events, and adverse events leading to death were found in all groups. Most treatment-emergent adverse events appeared to be related to local effects of Linhaliq: dyspnea, bronchospasm, hemoptysis, cough, taste disorders. Patients treated with Linhaliq more likely to have treatment-emergent ciprofloxacin-resistant *P. aeruginosa* cultured at any point post-baseline. Unknown whether exposure beyond one year may lead to additional safety concerns, further increase in resistance to fluoroquinolones, or reduced treatment effect.

In January 2018, the FDA denied the New Drug Application for Linhaliq for treatment of NCFB patients with chronic lung infections with *P. aeruginosa*. The areas of concern include clinical data, human factors validation study and product quality. The recommendations included additional Phase 3 clinical trial that demonstrates a significant treatment effect on clinically meaningful endpoints which could evaluate the co-primary endpoints of frequency and severity of exacerbations to assess for durable evidence of efficacy over a period of two years. In March 2018, Aradigm submitted a Marketing Authorisation Application to EMA for Linhaliq for the treatment of NCFBE patients with chronic lung infections with *P. aeruginosa*.

8.6.3 A Summary and a Comparison Between the Inhalatory and IV Route

In summary, once inhaled the NMs access the mucus of the airways where can be trapped and removed towards the gastrointestinal tract without contacting the underlying cells. If the aerodynamic size is small enough, the inhaled NMs can access the deeper respiratory zone of the alveoli, where after interacting with the PS and gaining a biocorona, can be internalized by AMs. In case of AM overloading, the NMs can cross the alveolar epithelia, gain access to the lung interstice, cross the endothelia, and access the blood compartment. An associated phenomena, the translocation, is a phenomenon linked to the inhalation of ultrafine NPs (of non-lipid composition) and is estimated to occur for only 1% or 2% of the NMs deposited in the lung (Kreyling et al. 2009). Additionally, hard NMs may impair the tensioactive function of the PS. The PS inhibition may cause respiratory distress and be fatal.

When NMs are administered either into the respiratory tract or intravenously, fundamental differences in NM transfer routes to blood and body organs are reported, tough in both cases, the NM gains access to the blood compartment. Whereas dose and dose rates associated with NM administration by direct intravenous injection are obviously very high, both dose and entry rate of NPs from lung deposits into the blood compartment (arterial) are low. Thus, NM delivered to the respiratory tract enter the blood compartment not only at different dose levels and different dose rates than directly IV injected NM, but also into blood of different oxygenation states (arterial versus venous). Furthermore, NM will adsorb proteins and/or lipids at the site of entry, i.e., constituents of the epithelial lining fluid (ELF) in the respiratory tract and different plasma proteins in the blood compartment. Such secondary coating of NM occurring at several stages of entry into and biodistribution to secondary body compartments will affect their kinetics (Rinderknecht et al. 2009).

8.7 Biokinetics of Lipid Intravenous Nanomedicines

After intravenous administration, nanomedicines distribute directly to the heart, and from there to the lungs, to be pumped within oxygenated blood at 5.8 l/min to the different organs: 700 ml/min (13%) to the CNS, 1,1 l/min (20%) to the GUT; 77 ml/min (1.4%) spleen; 300 ml/min (5%) liver; 1.24 l/min (22%) kidneys and 1.310 l/min (23%) muscle, skin and adipose tissue (Bertrand and Leroux 2012).

8.7.1 Toxicity of Lipid Intravenous Nanomedicines

From nearly 40 years ago, the intravenous infusion of certain micellar formulations of *paclitaxel* Taxol®, docetaxel (Taxotere), cyclosporine (Neoral, Sandimmune), etoposide (Etopophos) and different iodinated contrast media, are known to induce idiopathic hypersensitivity (HSR) (O'Dwyer and Weiss 1984; Szebeni et al. 1998, 2001).

HSR responses arise at first exposure without prior sensitization. They lessen, resolve, or disappear on later treatment. Some individuals experience a delayed onset of symptoms. Typical symptoms of HSR include facial swelling, flushing, rash, cough, chill, fever, and shortness of breath, tachypnea, hypertension, hypotension, chest pain, and back pain. These symptoms differ between individuals and vary from light to severe. Pre-existing allergen-reactive immunoglobulins (Ig), such as IgEs, do not initiate these reactions (Moghimi et al. 2012). Fatal reactions have been reported during Taxol® infusion, and these have been linked to cardiac arrest (Rowinsky et al. 1993). HSRs affects up to some half-million patients in the USA every year (Lazarou et al. 1998). Such reactions can be caused by almost all types of drugs, irrespective of their chemical composition, complexity, or route of administration.

HSR is, according to the severity of symptoms, graded in categories I IV, from mild to severe, to life-threatening reactions (Ring and Messmer 1977; Ring et al. 2010). Expressed as the percentage of patients displaying HSRs among all those treated, the prevalence range spans 5 orders of magnitude (0.001–70%), and the % values were clustered into five groups (very high, high, moderate, low and very low) with numerical ranges coinciding with the orders of magnitude in the 10^{-3}–10^{2} range (Szebeni et al. 2015). The highest rate of HSRs (>10%) occur with certain monoclonal antibodies (mAbs; rituximab, infliximab), liposomal drugs [Doxil (Caelyx), AmB (AmBisome)] and anticancer agents (taxanes: (paclitaxel, docetaxel) which are delivered in micellar solvents, such as Cremophor EL (CrEL). In comparison, the frequency of HSRs to penicillin, the textbook example of drug-induced allergic reactions, falls at the moderate/high borderline (1%).

Specifically, liposome reactions were reported as early as 1986, the first human study when large doses of liposomes were infused in cancer patients (Sculier et al. 1986). Since then, all kinds of liposomal drugs (examples: Doxil® (or Caeylx®, and the follow-on product Lipodox®) and Ambisome® (liposomal-based formulations of doxorubicin and amphotericin B, respectively), were shown to cause HSRs (Money-Kyrle et al. 1993; Cabriales et al. 1998). In particular HSR (also known as infusion reactions) to NBCDs remain unpredictable by standard allergy tests (Moghimi et al. 2004, 2010; Hamad et al. 2008, 2010).

Currently, the origin of the HSR is explained by two main contraposed theories, whose fundamentals are worth to be briefly reviewed.

According to the *double hit theory*, HSR occur by the contribution of the CARPA (complement activation-related pseudoallergy) and of the CIPA (complement independent pseudoallergy) effects. *Carpa effect:* The complement system comprises at least 30 soluble and membrane-bound proteins and/or glycoproteins, and functions as an efficient nonspecific defence system in the body (Holers 2014). NBCD formulations, depending on their size, morphology, and surface characteristics, can trigger complement activation through any of three pathways (i.e., classical, lectin, and alternative), or combinations thereof. Complement activation also proceeds with stealth entities, such as PEGylated liposomes (e.g., Doxil®), and various polymer-grafted drug carriers (Moghimi et al. 2006, 2011; Hamad et al. 2010, 2013; Pedersen et al. 2010; Farhangrazi and Moghimi 2016; Wibroe et al. 2016; Chen et al. 2017). Complement activation is an enzymatic cascade and, apart from its opsonization function, another important consequence is the sequential proteolytic cleavage of C3, C4, and C5, resulting in the release of anaphylatoxins C4a, C3a, and C5a from their respective parent molecule (Holers 2014). Anaphylatoxins, depending on their level and circulation half-life, can activate mast cells (as well as other immune cells, such as macrophages, basophils, and eosinophils) on binding to their corresponding signalling receptors (e.g., C5aR1 and C3aR1) molecule (Moghimi et al. 2012; Holers 2014). Accordingly, overactivation of mast cells in response organs, such as heart, lungs, and skin, could account for some of the observed adverse injection reactions in pigs, and sensitive individuals. For instance, activation of mast cells in the coronary arterial intima and perivascularly near myocytes can induce cardiac anaphylaxis. Indeed, activation of mast cells induces the release of histamine, a variety of inflammatory mediators, and vasoactive molecules, such as tumor necrosis factor-a (TNF-α), prostaglandins, and leukotrienes (from both mast cells and other immune cells) (Moghimi et al. 2012). Histamine has a range of cellular receptors (e.g., H1 and H2 receptors). H1 receptor activation could lead to the contraction of smooth muscle cells and vascular leakage, whereas H2 receptor activation could increase cellular cAMP levels and account for increased heart rate and pulse pressure.

CARPA is described as a novel subcategory of acute (type I) hypersensitivity reactions instead of IgE binding, which is mostly mild, transient, and preventable by appropriate precautions but that occasionally it can be severe or even lethal (Szebeniet al. 1999, 2011). According to this theory, the induction of pulmonary hypertension typical of HSR, would result from the release of mediators by cells such as pulmonary intravascular macrophage (PIM), after complement activation. The PIM are ATR-positive allergomedin-releasing cells that would be indirectly activated after binding to C3a and/or C5a anaphylotoxins.

CIPA on the other hand, refers to the activation of effector cells by direct interaction with nanomedicines, in a complement-independent fashion. According to this theory, the pulmonary intravascular macrophages (PIM) may be activated upon binding of reactogenic nanomedicines to pattern recognizing receptors (PRRs). The PRRs can include pathogen- and/or damage-associated molecular pattern receptors (PAMP-R, DAMP-R) that cross-react with surface polymers or other repetitive elements of the nanomedicine coats. Further receptors that can mediate the second 'hit' include Toll-like receptors, those that recognize nanomedicine corona proteins, FcRs recognizing nanomedicine-bound antibodies or CR1 and CR3 that recognize particle-bound Complement (C) protein fragments. Some of these receptors also mediate phagocytosis.

The double-hit theory is proposed to be amplified to a 'multi-hit' theory, since in addition to CARPA and CIPA, other additional different mechanisms may induce symptomatically similar toxicities. For example, contact system activation leading to bradykinin, another demonstrated mediator of HSRs during hemodialysis recognized, by bradykinin receptors B1 and B2, represents an additional pathway of HSRs (Verresen et al. 1994; Coppo and Amore 2000), (Szebeni 2018).

The *rapid phagocytic response hypothesis* (RPR) instead, states that the complement activation would play a minor or perhaps none role, in the clinical symptoms of HSR (Moghimi 2018). Such statements arise from the fact that complement activation registered upon Doxil (a nanomedicine known to induce HSR), infusion, occurs in the absence of clinical symptoms of HSR. This suggests that complement activation would not always be direct responsible of anaphylaxis. According to the RPR hypothesis, in mice and in humans the generation of intermediate responsible for HSR would occur after the phagocytosis of IgG-mediated opsonization of particulate materials, followed by recognition by low affinity FcRIII in macrophages (in lower extent by monocytes, dendritic cells and basophiles), plus the induction of PAF, a mastocytes activator (Strait 2002). PAF has a short half-life, but it is active at concentrations as low as 10^{-12} M. PAF can activate mast cells and might further induce platelet aggregation and activation, thus resulting in an increase in vascular permeability, circulatory collapse, and decreased cardiac output. It has been observed that

PIMs rapidly intercept intravenously injected particulate materials and subsequently release mediators, such as thromboxane A2, prostaglandins, and leukotrienes, which correlate with periods of peak vasoconstriction, bronchoconstriction, and pulmonary hypertension (Warner 1996; Schneberger et al. 2012).

Anaphylaxis would be triggered in patients possessing pre-existent antiphospholipid antibodies, opsonizing the lipid-based nanomedicines, a step leading to their extensive phagocytosis by PIM (or related macrophages), a process that would be linked to the release of mediators such as PAF. The robust ingestion of particles by macrophages appears, according to this viewpoint, to initiate cardiopulmonary distress, whereas the C5a factor may exacerbate the symptomatology but would not be its direct cause.

CrEL is a non-ionic solubilizer and/or emulsifier obtained by reacting castor oil with ethylene oxide. The product is complex and comprises a mixture of unmodified castor oil, ethoxylated glycerol, and polyethoxylate glycerol moieties (e.g., with one, two, or all polyethylene resides esterified by ricinoleic acid) (Szebeni et al. 2001). CrEL forms 10–25-nm micelles at a concentration>60 mg/ml that in contact with plasma, become into 100–300-nm droplets (Szebeni et al. 2001). The droplets formation result from the interaction of micelles with plasma lipoproteins, that sequester the more hydrophobic components modifying the final geometry of the aggregate. The anaphylaxis generated by CrEL is explained according to the RPR because the small micelles would not be recognised by macrophages, but the droplets might be susceptible to robust clearance by the responsive macrophages through a signalling receptor. The individual variations in adverse reactions to Taxol® could further relate to differences in the blood concentration of LDL and HDL (and their subtypes), which could limit the rate of droplet formation.

According to the RPR, new strategies other than PEGylation, aimed to delay particle recognition by the responsible macrophage subpopulations within the first few minutes of injection may overcome adverse reactions. One example is designing nano-objects exhibiting high adherence to erythrocytes, which prolongs their circulation time in the blood by avoiding macrophage availability (Chambers and Mitragotri 2004). Another example is designing nano-objects with geometries different to the sphere such as rods and disks, that can difficult the phagocytoses (Wibro et al. 2017). It has been observed that the slower clearance of rods and disks (and even spheres attached to erythrocytes) might trigger a desensitization process and prevent mediator release from macrophages. This is in agreement with the fact that in clinical practice, slowing the infusion rate of nanomedicines (and, therefore their rate of presentation to macrophages) was observed to decrease the magnitude of adverse reactions in patients (Szebeni et al. 2011; Moghimi et al. 2012). The above-mentioned strategies are more difficult to be introduced in liposomal nanomedicines where macrophage responses could be governed by the extent of vesicular deformability and bilayer fluidity and/or rigidity, which might in turn regulate macrophage receptor binding, signalling, and internalization processes.

The activation of complement can also generate opsonic intermediates such as C3b, for which most of macrophages possess receptors. There is evidence that clustering of CR3, at least on monocytic cells, recruits a Toll/IL-1R family-like signalling cascade, interacting with IL-1R-associated kinase 1 to induce NF-kB activity (Shi et al. 2001), which is a prototypical proinflammatory signalling pathway resulting in the expression of proinflammatory genes, including those encoding cytokines, chemokines, and adhesion molecules.

Remarkably, normal human lung appears to lack constitutive PIMs, but several studies have suggested the presence of induced pulmonary macrophages in patients with liver dysfunction and disease, hepatopulmonary syndrome, as well as inflammatory lung disease (Warner 1996). Hence, it is plausible that individuals who adversely respond to injection/infusion of nanomedicines either have a form of a hepatic and/or lung disorder or a predisposed liver or lung disease with induced (or responsive) macrophages in pulmonary circulation (or in other sites, such as the periportal regions of the liver and the splenic marginal zones).

According to (Moghimi 2018), the extent and frequency of adverse injection reactions to nanomedicines, including the positioning and number of reactive macrophages (e.g., in pulmonary vasculature or in the spleen) and the patient's medical history (e.g., patients with pulmonary and/or hepatic diseases or who have a predisposed disease with induced macrophages in pulmonary circulation), as well as a direct role for particle uptake by dendritic (Griffin et al. 2017) and mast cells, account for the HSR triggering. Overall, slowing the infusion rate of nanomedicines, to slower the presentation rate to reactive macrophages (whether in lungs or elsewhere) can overcome cardiopulmonary distress in responding patients.

8.8 Conclusion

The nanotoxicity of lipid-based nanomedicines is observed upon intentional administration of therapeutic doses. For intravenous nanomedicines, the nanotoxicity occurs after regulatory organisms' approval and commercialization occurred. The regulatory approval of inhaled nanomedicines instead, has been delayed around clinical Phase 3, despite of promising a rapid onset of action, a high concentration of drugs in the vicinity of the relevant anatomical region while limiting systemic exposure and reducing the likelihood of side-effects. Data gathered in preclinical and early clinical stages together with the lack of significant improvements compared to free antibiotics or conventional medication, seem to contradict former expectations. Nanomedicines are administered to patients suffering underlying diseases, exhibiting anatomo pathological scenarios different from those of healthy individuals. Because of that, assessing nanomedicines nanotoxicity is far more complex than determining the effects of non-intentional exposition, where the experimental reproduction of a human disease in, for instance, an animal model is unnecessary. Nanotoxicity of parenteral nanomedicines for instance, would be better assessed on models having PIM. In inhalation route, models mimicking the anatomical characteristics of human airways, instead of rodents, would lead to more accurate assays of regional deposition. Accordingly, heavier non-rodent animal models-more expensive and limited by ethics concerns—capable also of reproducing the microenvironment of the human disease, would led to a more accurate prediction of nanotoxicity. Animal tests however, are costly (considering the costs and duration as well as costs following wrong decisions in product development) and resource intensive. As stated before, involve ethical aspects, apart from being not sufficiently reproducible, not well enough reported, lacking inter-species predictivity. Finally, animal experiments do not reflect human diversity, exposure, and treatment. Not even the reproducibility under the optimal conditions of regulatory guideline studies does say that the results are meaningful for humans. In such landscape, options different to the use of animals, such us the development of artificial three-dimensional tissues or organs reproducing the architecture, microenvironment and allowing the normal cross talk between different human cell types, would undoubtedly led to more predictive preclinical results. This in turn would reduce the chances of discovering nanotoxicity or low efficacy once during the expensive clinical trials.

References

Alsaleh NB, Brown JM (2018) Immune responses to engineered nanomaterials: current understanding and challenges. Curr Opin Toxicol 10:8–14

Altube MJ, Cutro A, Bakas L, Morilla MJ, Disalvo EA, Romero EL (2017) Nebulizing novel multifunctional nanovesicles: the impact of macrophage-targeted-pH-sensitive archaeosomes on pulmonary surfactant. J Mater Chem B 5:8083–8095

Bakand S, Hayes A, Dechsakulthorn F (2012) Nanoparticles: a review of particle toxicology following inhalation exposure. Inhal Toxicol 24(2):125–135

Beck-Broichsitter M, Ruppert C, Schmehl T, Günther A, Seeger W (2014) Biophysical inhibition of synthetic vs. naturally-derived pulmonary surfactant preparations by polymeric nanoparticles. Biochim Biophys Acta-Biomembr 1838(1):474–481

Bermudez LE, Blanchard JD, Hauck L, Gonda I (2015) Treatment of *Mycobacterium avium* sub-sphominissuis (MAH) lung infection with liposome-encapsulated ciprofloxacin resulted in significant decrease in bacterial load in the lung. Am J Respir Crit Care Med 191:A6293

Bertrand N, Leroux J-C (2012) The journey of a drug-carrier in the body: an anatomo-physiological perspective. J Control Rel 161:152–163

Bilton D, Pressler T, Fajac I, Clancy JP, Sands D, Minic P, Cipolli M, LaRosa M, Galeva I, Sole A, Staab D, Dupont L, Goss CH, Hamblett N, Quittner A, Ramsey B, Gupta R, Konstan M (2013) Phase 3 efficacy and safety data from randomized, multicenter study of liposomal amikacin for inhalation (Arikace) compared with TOBI in cystic fibrosis patients with chronic infection due to *Pseudomonas aeruginosa*. North American Cystic Fibrosis Conference; Salt Lake City, Utah. Pediatr Pulmonol 36:290

Bobo D, Robinson KJ, Islam J, Thurecht KJ, Corrie SR (2016) Nanoparticle-based medicines: a review of FDA-approved materials and clinical trials to date. Pharm Res 33(10):2373–2387

Borm P, Cassee FR, Oberdörster G (2015) Lung particle overload: old school—new insights? Part Fibre Toxicol 12:10

Cabriales S, Bresnahan J, Testa D, Espina BM, Scadden DT, Ross M, Gill PS (1998) Extravasation of liposomal daunorubicin in patients with AIDS-associated Kaposi's sarcoma: a report of four cases. Oncol Nurs Forum 25:67–70

Caster JM, Patel AN, Zhang T, Wang A (2017) Investigational nanomedicines in 2016: a review of nanotherapeutics currently undergoing clinical trials. Wiley Interdiscip Rev Nanomed Nanobiotechnol 9(1)

Cedervall T, Lynch I, Lindman S, Berggard T, Thulin E, Nilsson H, Dawson KA, Linse S (2007) Understanding the nanoparticle–protein corona using methods to quantify exchange rates and affinities of proteins for nanoparticles. Proc Natl Acad Sci USA 104:2050–2055
Chambers E, Mitragotri S (2004) Prolonged circulation of large polymeric nanoparticles by non-covalent adsorption on erythrocytes. J Control Release 100:111–119
Champion JA, Mitragotri S (2006) Role of target geometry in phagocytosis. Proc Natl Acad Sci USA 103(13):4930–4934
Chen F, Wang G, Griffin JI, Brenneman B, Banda NK, Holers VM, Backos DS, Wu L, Moghimi SM, Simberg D (2017) Complement proteins bind to nanoparticle protein corona and undergo dynamic exchange *in vivo*. Nat Nanotechnol 12:387–393
Chhoden T, Clausen PA, Larsen ST, Nørgaard AW, Lauritsen FR (2015) Interactions between nanoparticles and lung surfactant investigated by matrix-assisted laser desorption/ionization time-of-flight mass spectrometry. Rapid Commun Mass Spectrom 29(11):1080–1086
Chinoy MR, Fisher AB, Shuman H (1994) Confocal imaging of time-dependent internalization and localization of NBD-PC in intact rat lungs. Am J Physiol 266(6 Pt 1):L713–L721
Churg A, Brauer M, del Carmen Avila-Casado M, Fortoul TI, Wright JL (2003) Chronic exposure to high levels of particulate air pollution and small airway remodeling. Environ Health Perspect 111:714–718
Churg A, Wright JL (2002) Airway wall remodelling induced by occupational mineral dusts and air pollutant particles. Chest 122:306–309
Cipolla D, Blanchard J, Gonda I (2016) Development of liposomal ciprofloxacin to treat lung infections. Pharmaceutics 8:6
Cipolla D, Gonda I, Chan H-K (2013) Liposomal Formulations for Inhalation. Ther Deliv 4:1047–1072
Cipolla D, Shekunov B, Blanchard J, Hickey A (2014) Lipid-based carriers for pulmonary products: preclinical development and case studies in humans. Adv Drug Del Rev 75:53–80
Commission Recommendation of 18 October 2011 on the definition of nanomaterial. http://eurlex.europa.eu/LexUriServ/LexUriServ.do?uri=OJ:L:2011:275:0038:0040
Committee on Hazardous Substances, BAuA (2015) Assessment Criterion (Reference Value) for Granular Biopersistent Particles Without Known Significant Specific Toxicity (Nanoscaled GBP) (Respirable Dust) Generated From Manufactured Ultrafine Particles. http://www.baua.de/en/Topics-from-A-to-Z/Hazardous-Substances/TRGS/pdf/ 910/nanoscaled-GBP.pdf
Conley J, Yang H, Wilson T, Blasetti K, Di Ninno V, Schnell G, Wong JP (1997) Aerosol delivery of liposome-encapsulated ciprofloxacin: aerosol characterization and efficacy against *Francisella tularensis* infection in mice. Antimicrob Agents Chemother 41(6):1288–1292
Coppo R, Amore A (2000) Importance of the bradykinin-nitric oxide synthase system in the hypersensitivity reactions of chronic haemodialysis patients. Nephrol Dial Transpl 15:1288–1290
Crommelin DJA, Shah VP, Klebovich I, McNeil SE, Weinstein V, Flühmann B, Mühlebach S, de Vlieger JS (2015) The similarity question for biological and non-biological complex drugs. Eur J Pharm Sci 76:10–17
Dankovic D, Kuempel E, Wheeler M (2007) An approach to risk assessment for TiO_2. Inhal Toxicol 19:205–212
Davis BK, Wen H, Ting JP (2011) The inflammasome NLRs in immunity, inflammation, and associated diseases. Annu Rev Immunol 29:707–735
Deng ZJ, Liang M, Monteiro M, Toth I, Minchin RF (2011) Nanoparticle-induced unfolding of fibrinogen promotes Mac-1 receptor activation and inflammation. Nat Nanotechnol 6:39–44
Docter D, Westmeier D, Markiewicz M, Stolte S, Knauer SK, Stauber RH (2015) The nanoparticle biomolecule corona: lessons learned—challenge accepted? Chem Soc Rev 44:6094–6121
Donaldson K, Tran L, Jimenez LA, Duffin R, Newby DE, Mills N, MacNee W, Stone V (2005) Combustion-derived nanoparticles: a review of their toxicology following inhalation exposures. Part Fibre Toxicol 2:10
Donaldson K, Schinwald A, Murphy F, Cho WS, Duffin R, Tran L, Poland C (2013) The biologically effective dose in inhalation nanotoxicology. Acc Chem Res 46(3):723–732

Duffin R, Tran CL, Clouter A, Brown DM, Macnee W, Stone V, Donaldson K (2002) The Importance of surface area and specific reactivity in the acute pulmonary inflammatory response to particles. Ann Occup Hyg 46:242–245
Dushianthan A, Cusack R, Goss V, Postle AD, Grocott MPW (2012) Clinical review: Exogenous surfactant therapy for acute lung injury/acute respiratory distress syndrome–where do we go from here? Crit Care 16:238–248
EMA. Withdrawal assessment report EMA/493973/2016
EU Scientific Committee Recommendation on the definition of a nanomaterial (2011/696/EU). http://ec.europa.eu/environment/chemicals/nanotech/faq/definition_en.htm
European Science Foundation (2005) Nanomedicine, an ESF–European Medical Research Councils (EMRC) forward look report
European Technology Platform on Nanomedicine, Nanotechnology for Health, Vision paper and basis for a strategic research agenda for nanomedicine, EC Publication Office, September 2005
European Union 004-EN-N (2010) Scientific Committee on Emerging and Newly Identified Health Risks. Scientific Basis for the Definition of the Term "nanomaterial"; BN 978-92-79-12757-1; https://doi.org/10.2772/39703 ND-AS-09
European Food Safety Agency (EFSA), EU Regulation No. 1169/2011. European Commission 2011
Fadeel B (2012) Clear and present danger? Engineered nanoparticles and the immune system. Swiss Med Wkly 142:w13609
Fan Q, Wang YE, Zhao X, Loo JS, Zuo YY (2011) Adverse biophysical effects of hydroxyapatite nanoparticles on natural pulmonary surfactant. ACS Nano 5(8):6410–6416
Farhangrazi ZS, Moghimi SM (2016) Materials etiquette and complement responses. Curr Bionanotechnol 2:6–10
FDA for patients. https://www.fda.gov/ForPatients/Approvals/Drugs/ucm405622.htm
FDA (2014) Considering whether an FDA-regulated product involves the application of nanotechnology Guidance for Industry. http://www.fda.gov/RegulatoryInformation/Guidances/ucm257698.htm
Frohlich E (2012) The role of surface charge in cellular uptake and cytotoxicity of medical nanoparticles. Int J Nanomedicine 7:5577–5591
Ganguly P, Breen A, Pillai SC (2018) Toxicity of nanomaterials: exposure, pathways, assessment and recent advances. ACS Biomater Sci Eng. https://doi.org/10.1021/acsbiomaterials.8b00068
Ge C, Du J, Zhao L, Wang L, Liu Y, Li D, Yang Y, Zhou R, Zhao Y, Chai Z, Chen C (2011) Binding of blood proteins to carbon nanotubes reduces cytotoxicity. Proc Natl Acad Sci USA 108:16968–16973
Gebel T, Foth H, Damm G, Freyberger A, Kramer P-J, Lilienblum W, Röhl C, Schupp T, Weiss C, Wollin C-M, Georg J (2014) Manufactured nanomaterials: categorization and approaches to hazard assessment Hengstler. Arch Toxicol 88:2191–2211
Gehr P, Bachofen M, Weibel ER (1978) The normal human lung: ultrastructure and morphometric estimation of diffusion capacity. Respir Physiol 32:121–140
Geiser M (2010) Update on macrophage clearance of inhaled micro- and nanoparticles. J Aerosol Med Pulm Drug Deliv 23:207–217
Griffin JI, Wang G, Smith WJ, Vu VP, Scheinman R, Stitch D, Moldovan R, Moghimi SM, Simberg D (2017) Revealing dynamics of accumulation of systemically injected liposomes in the skin by intravital microscopy. ASC Nano 11(11):11584. https://doi.org/10.1021/acsnano.7b06524
Granular biopersitent. http://nano.dguv.de/en/news/news-detail-english/new-baua-report-on-granular-biopersistent-particles-gbp-at-workplaces
Hamad I, Al-Hanbali O, Hunter AC, Rutt KJ, Andresen TL, Moghimi SM (2010) Distinct polymer architecture mediates switching of complement activation pathways at the nanosphere-serum interface: implications for stealth nanoparticle engineering. ACS Nano 4:6629–6638
Hamad I, Hunter AC, Szebeni J, Moghimi SM (2008) Poly(ethylene glycol)s generate complement activation products in human serum through increased alternative pathway turnover and a MASP-2-dependent process. Mol Immunol 46:225–232

Hamad I, Hunter AC, Moghimi SM (2013) Complement monitoring of Pluronic 127 gel and micelles: suppression of copolymer-mediated complement activation by elevated serum levels of HDL, LDL, and apolipoproteins A1 and B-100. J Control Release 170:167–174

Harishchandra RK, Saleem M, Galla H-J (2010) Nanoparticle interaction with model lung surfactant monolayers. J R Soc Interface 7(Suppl 1):S15–S26

Hayes AJ, Bakand S (2014) Toxicological perspectives of inhaled therapeutics and nanoparticles. Expert Opin Drug Metab Toxicol 10(7):933–947

Heyder J, Gebhart J, Rudolf G, Schiller CF, Stahlhofen W (1986) Deposition of particles in the human respiratory tract in the size range 0.005–15 μm. J Aerosol Sci 17:811–825

Hinds WC (1999) Uniform particle motion. In: Aerosol technology: properties, behavior and measurement of airborne particles, 2nd edn, p. 53–55

Holers VM (2014) Complement and its receptors: new insights into human disease. Annu Rev Immunol 32:433–459

Hu W, Peng C, Lv M, Li X, Zhang Y, Chen N, Fan C, Huang Q (2011) Protein corona-mediated mitigation of cytotoxicity of graphene oxide. ACS Nano 5:3693–3700

Insmed Company website. Accessed on 24 July 2018

ISO/TS 80004-1:2015 (2015) Nanotechnology—Vocabulary—Part 1: core terms. International Organization for Standardization: Geneva, Switzerland

ISO/TS 80004-2:2015 Nanotechnologies—Vocabulary—Part 2: nano-objects

Jeevanandam J, Barhoum A, Chan YS, Dufresne A, Danquah MK (2018) Review on nanoparticles and nanostructured materials: history, sources, toxicity and regulations. Beilstein J Nanotechnol 9:1050–1074

Jones CV, Williams TM, Walker KA, Dickinson H, Sakkal S, Rumballe BA, Little MH, Jenkin G, Ricardo SD (2013) M2 macrophage polarisation is associated with alveolar formation during postnatal lung development. Respir Res 14:41

Krafft MP (2015) Overcoming inactivation of the lung surfactant by serum proteins: a potential role for fluorocarbons? Soft Matter 11:5982–5994

Kreyling WG, Semmler-Behnke M, Seitz J, Scymczak W, Wenk A, Mayer P, Takenaka S, Oberdörster G (2009) Size dependence of the translocation of inhaled iridium and carbon nanoparticle aggregates from the lung of rats to the blood and secondary target organs. Inhal Toxicol 21(Sl):55–60

Kuempel ED, Attfield MD, Stayner LT, Castranova V (2014) Human and animal evidence supports lower occupational exposure limits for poorly-soluble respirable particles: letter to the Editor re: 'Low-toxicity dusts: current exposure guidelines are not sufficiently protective' by Cherrie, Brosseau, Hay and Donaldson. Ann Occup Hyg 58(9):1205–1208

Lachmann B, Robertson B. Vogel J (1980) *In vivo* lung lavage as an experimental model of the respiratory distress syndrome. Acta Anaesth Scand 24:231–236

Lazarou J, Pomeranz BH, Corey PN (1998) JAMA. Incidence of adverse drug reactions in hospitalized patients: a meta-analysis of prospective studies 15;279(15):1200–1205

Lesniak A, Fenaroli F, Monopoli MP, Aberg C, Dawson KA, Salvati A (2012) Effects of the presence or absence of a protein corona on silica nanoparticle uptake and impact on cells. ACS Nano 6:5845–5857

Li N, Xia T, Nel AE (2008) The role of oxidative stress in ambient particulate matter-induced lung diseases and its implications in the toxicity of engineered nanoparticles. Free Radic Biol Med 44:1689–1699

Lopez-Rodriguez E, Pérez-Gil J (2014) Structure-function relationships in pulmonary surfactant membranes: from biophysics to therapy. Biochim Biophys Acta-Biomembr 1838(6):1568–1585

Lundqvist M, Stigler J, Elia G, Lynch I, Cedervall T, Dawson KA (2008) Nanoparticlesize and surface properties determine the protein corona with possible implications for biological impacts. Proc Natl Acad Sci USA 105:14265–14270

MAK Commission (2014) The MAK-collection part I, MAK value documentations 2014. In: Deutsche (ed) Forschungsgemeinschaft. The MAK collection for occupational health and safety. Wiley-VCH Verlag GmbH & Co, KGaA, p. 320

Mallick S, Choi JS (2014) Liposomes: versatile and biocompatible nanovesicles for efficient biomolecules delivery. J Nanosci Nanotechnol 14(1):755–765
Mansour HM, Rhee YS, Wu X (2009) Nanomedicine in pulmonary delivery. Int J Nanomedicine 4:299–319
Martin TR, Frevert CW (2005) Innate immunity in the lungs. Proc Am Thorac Soc 2:403–411
Martini WZ, Chinkes DL, Barrow RE, Murphey ED, Wolfe RR (1999) Lung surfactant kinetics in conscious pigs. Am J Physiol 277(1 Pt 1):E187–E195
Moghimi M, Hunter C, Dadswell CM, Savay S, Alving C, Szebeni J (2004) Causative factors behind poloxamer 188 (Pluronic F68, Flocor trade mark)-induced complement activation in human sera; A protective role against poloxamer mediated complement activation by elevated serum lipoprotein levels. Biochim Biophys Acta 1689:103–113
Moghimi SM (2018) Nanomedicine safety in preclinical and clinical development: focus on idiosyncratic injection/infusion reactions. Drug Discov Today 23(5):1034–1042
Moghimi SM, Andersen AJ, Hashem SH, Lettiero B, Ahmadvand D, Hunter AC, Andresen TL, Hamad I, Szebeni J (2010) Complement activation cascade triggered by PEG-PL engineered nanomedicines and carbon nanotubes: the challenges ahead. J Control Release 146:175–181
Moghimi SM, Hamad I, Andresen TL, Jørgensen K, Szebeni J (2006) Methylation of the phosphate oxygen moiety of phospholipid-methoxy(polyethylene glycol) conjugate prevents PEGylated liposome-mediated complement activation and anaphylatoxin production. FASEB J 20:2591–2593
Moghimi SM, Andersen AJ, Ahmadvand D, Wibroe PP, Andresen TL, Hunter AC (2011) Material properties in complement activation. Adv Drug Deliv Rev 63:1000–1007
Moghimi SM, Wibroe PP, Helvig SY, Farhangrazi ZS, Hunter AC (2012) Genomic perspectives in inter-individual adverse responses following nanomedicine administration: the way forward. Adv Drug Deliver Rev 64:1385–1393
Money-Kyrle JF, Bates F, Ready J, Gazzard BG, Phillips RH, Boag FC (1993) Liposomal daunorubicin in advanced Kaposi's sarcoma: a phase II study. Clin Oncol (R Coll Radiol) 5:367–371
Monopoli MP, Bombelli FB, Dawson KA (2011) Nanobiotechnology: nanoparticle coronas take shape. Nat Nanotechnol 6:11–12
Morawska L, Hofmann W, Hitchins-Loveday J, Swanson C, Mengersen K (2005) Experimental study of the deposition of combustion aerosols in the human respiratory tract. J Aerosol Sci 36:939–957
Morrow PE (1988) Possible mechanisms to explain dust overloading of the lungs. Fundam Appl Toxicol 10(3):369–384
Nanomedicine 2020 (May 2013) Contribution of nanomedicine to horizon 2020. ETP Nanomedicine—NANOMED2020
Napierska D, Thomassen LC, Rabolli V, Lison D, Gonzalez L, Kirsch-Volders M, Martens JA, Hoet PH (2009) Size-dependent cytotoxicity of monodisperse silica nanoparticles in human endothelial cells. Small 5:846–853
Neville M, Liu S, Artis C, Dadey E, Gupta R (2009) Functionality of foamy alveolar macrophages after inhalation of aerosolized liposomal Amikacin (ArikaceTM). Pediatr Pulmonol 44:109–212
NIH Roadmap for Medical Research (2008) http://pubs.niaaa.nih.gov/publications/arh311/12-13.pdf
Nel AE, Mädler L, Velegol D, Xia T, Hoek Van, Somasundaran S, Klaessig F, Castranova V, Thompson M (2009) Understanding biophysicochemical interactions at the nano–bio interface. Nat Mater 8:543–557
O'Donnell A, Swarnakar R, Yashina L, Nikolova P, Marinov R, Waghray P (2009) A placebo-controlled study of liposomal amikacin for inhalation (ARIKACE™) nebulized once-daily in the treatment of bronchiectasis patients with chronic *Pseudomonas aeruginosa* lung infection. Eur Respir J 34(Suppl. 53):1361
Oberdörster G (2010) Safety assessment for nanotechnology and nanomedicine: concepts of nanotoxicology. J Intern Med 267:89–105

Oberdörster G, Ferin J, Morrow PE (1992) Volumetric loading of alveolar macrophages (AM): a possible basis for diminished AM-mediated particle clearance. Exp Lung Res 18(1):87–104

Oberdorster G, Kane AB, Klaper RD, Hurt RH (2013) Nanotoxicology. In: Klaassen CD (ed) Casarett and Doull's toxicology-the basic science of poisons. McGraw Hill, New York, NY, USA, pp. 1189–1229

Oberdorster G, Maynard A, Donaldson K, Castranova V, Fitzpatrick J, Ausman K, Carter J, Karn B, Kreyling W, Lai D, Olin S, Monteiro-Riviere N, Warheit D, Yang H (2005) Principles for characterizing the potential human health effects from exposure to nanomaterials: elements of a screening strategy. Part Fibre Toxicol 2:8

O'Dwyer PJ, Weiss R (1984) Hypersensitivity reactions induced by etoposide. Cancer Treat Rep 68:959–961

Olivier KN, Maas-Moreno R, Whatley M, Cheng KT, Lee J-H, Fiorentino C, Shaffer R, Macdonald S, Gupta R, Corcoran TE, Malinin VS, Eagle G, Perkins W, Paik C, Chen C (2016) Airway deposition and retention of liposomal amikacin for inhalation in patients with pulmonary nontuberculous mycobacterial disease. Am J Respir Crit Care Med 193:A3732

OSHA Occupational safety and health administration. Safety and health tips: nanotechnology http://www.osha.gov/dsg/nanotechnology/nanotechnology.html

Padmore T, Stark C, Turkevich LA, Champion JA (2017) Quantitative analysis of the role of fiber length on phagocytosis and inflammatory response by alveolar macrophages. Biochim Biophys Acta 1861(2):58–67

Patton JS, Byron PR (2007) Inhaling medicines: delivering drugs to the body through the lungs. Nat Rev Drug Discov 6:67–74

Pauluhn J (2011) Poorly soluble particulates: searching for a unifying denominator of nanoparticles and fine particles for DNEL estimation. Toxicology 279(1–3):176–188

Pauluhn J (2014) Repeated inhalation exposure of rats to an anionic high molecular weight polymer aerosol: application of prediction models to better understand pulmonary effects and modes of action. Exp Toxicol Pathol 66(5–6):243–256

Pedersen MB, Zhou X, Larsen EK, Sørensen US, Kjems J, Nygaard JV, Nyengaard JR, Meyer RL, Boesen T, Vorup-Jensen T (2010) Curvature of synthetic and natural surfaces is an important target feature in classical pathway complement activation. J Immunol 184:1931–1945

Pelaz B, Charron G, Pfeiffer C, Zhao Y, de la Fuente JM, Liang XJ, Parak WJ, Del Pino P (2013) Interfacing engineered nanoparticles with biological systems: anticipating adverse nano-bio interactions. Small 9:1573–1584

Pérez de la Ossa DH (2014) Quality aspects of nano-based medicines. In: SME Workshop: Focus on quality for medicines containing chemical entities. London, 4 April 2014

Peters A, von Klot S, Heier M, Trentinaglia I, Hormann A, Wichmann HE, Löwel H (2004) Exposure to traffic and the onset of myocardial infarction. N Engl J Med 351:1721–1730

Piantadosi CA, Schwartz DA (2004) The acute respiratory distress syndrome. Ann Intern Med 141:460–470

Possmayer F (2004) Physicochemical aspects of pulmonary surfactant. In: Polin RA, Fox WW, Abman SH (eds) Fetal and neonatal physiology. WB. Saunders Company, Philadelphia, pp 1014–1034

Rabolli V, Badissi AA, Devosse R, Uwambayinema F, Yakoub Y, Palmai-Pallag M, Lebrun A, De Gussem V, Couillin I, Ryffel B, Marbaix E, Lison D, Huaux F (2014) The alarmin IL-1alpha is a master cytokine in acute lung inflammation induced by silica micro- and nanoparticles. Part Fibre Toxicol 11:69

Ramachandran G, Howard J, Maynard A, Philbert M (2012) Handling worker and third-party exposures to nanotherapeutics during clinical trials. J Law Med Ethics 40:856–864

Rinderknecht A, Oberdörster G, de Mesy Bentley K et al (2009) Serum protein coated gold nanoparticles in the perfused human term placenta. Toxicologist 108

Ring J, Messmer K (1977) Incidence and severity of anaphylactoid reactions to colloid volume substitutes. Lancet 1:466–469

Ring J, Grosber M, Mohrenschlager M, Brockow K (2010) Anaphylaxis: acute treatment and management. Chem Immunol Allergy 95:201–210
Robertson B (1984) Pathology and pathophysiology of neonatal surfactant deficiency ("respiratory distress syndrome", "hyaline membrane disease"). In: Robertson B, Van Golde LMG, Batenburg JJ (eds) Pulmonary surfactant. Elsevier Science Publishers, Amsterdam, pp 383–418
Roller M, Pott F (2006) Lung tumor risk estimates from rat studies with not specifically toxic granular dusts. Ann N Y Acad Sci 1076:266–280
Roller J, Laschke MW, Tschernig T, Schramm R, Veith NT, Thorlacius H, Menger MD (2011) How to detect a dwarf: in vivo imaging of nanoparticles in the lung. Nanomedicine 7(6):753–762
Roller M (2003) Dose-response relationships of granular bio-durable dusts in rat lungs: does a cancer threshold exist? Eur J Oncol 8(4):277–293
Rowinsky EK, Eisenhauer EA, Chaudhry V, Arbuck SG, Donehower RC (1993) Clinical toxicities encountered with paclitaxel (Taxol). Semin Oncol 20:1–15
Rubenfeld GD, Caldwell E, Peabody E, Weaver J, Martin DP, Neff M, Stern EJ, Hudson LDN (2005) Incidence and outcomes of acute lung injury. Engl J Med 353:11685–11693
Rubins JB (2003) Alveolar macrophages—wielding the double-edged sword of inflammation. Am J Respir Crit Care Med 167:103–104
Rugonyi S, Biswas SC, Hall SB (2008) The biophysical function of pulmonary surfactant. Respir Physiol Neurobiol 163(1):244–255
Schleh C, Hohlfeld JM (2009) Interaction of nanoparticles with the pulmonary surfactant system. Inhalation Toxicol 21(sup1):97–103
Schmid O, Stoeger T (2016) Surface area is the biologically most effective dosemetric for acute nanoparticle toxicity in the lung. J Aerosol Sci 99:133–143
Schneberger D, Aharonson-Raz K, Singh B (2012) Pulmonary intravascular macrophages and lung health: what are we missing? Am J Physiol Lung Cell Mol Physiol 302:L498–L503
Schurch S (1982) Surface tension at low lung volumes: dependence on time and alveolar size. Respir Physiol 48:339–355
Sculier JP, Coune A, Brassinne C, Laduron C, Atassi G, Ruysschaert JM, Fruhling J (1986) Intravenous infusion of high doses of liposomes containing NSC251635, a water-insoluble cytostatic agent. A pilot study with pharmacokinetic data. J Clin Oncol 4:789–797
Sharifi S, Behzadi S, Laurent S, Laird Forrest M, Stroeve M, Mahmoud M (2012) Toxicity of nanomaterials. Chem Soc Rev 41:2323–2343
Shi C, Zhang X, Chen Z, Robinson MK, Simon DI (2001) Leukocyte integrin Mac-1 recruits Toll/interleukin-1 receptor superfamily signalling intermediates to modulate NF-κB activity. Circ Res 89:859–865
Simard JC, Vallieres F, de Liz R, Lavastre V, Girard D (2015) Silver nanoparticles induce degradation of the endoplasmic reticulum stress sensor activating transcription factor-6 leading to activation of the NLRP-3 inflammasome. J Biol Chem 290(9):5926–5939
Smith T (2018) Clinical team leader division of anti-infective products. FDA Opening Remarks
Strait MD (2002) Pathways of anaphylaxis in the mice. J Allergy Clin Immunol 109:658–668
Szebeni J, Muggia F, Gabizon A, Barenholz Y (2011) Activation of complement by therapeutic liposomes and other lipid excipient-based therapeutic products: prediction and prevention. Adv Drug Deliv Rev 63:1020–1030
Szebeni J (2018) Mechanism of nanoparticle-induced hypersensitivity in pigs: complement or not complement? Drug Discov Today 23(3):487–492
Szebeni J, Alving CR, Savay S, Barenholz Y, Priev A, Danino D, Talmon Y (2001) Formation of complement-activating particles in aqueous solutions of Taxol: possible role in hypersensitivity reactions. Int Immunopharm 1:721–735
Szebeni J, Muggia FM, Alving CR (1998) Complement activation by Cremophor EL as a possible contributor to hypersensitivity to paclitaxel: an *in vitro* study. J Natl Cancer Inst 90:300–330
Szebeni J, Fontana JL, Wassef NM, Mongan PD, Morse DS, Dobbins DE, Stahl GL, Bünger R, Alving CR (1999) Hemodynamic changes induced by liposomes and liposome-encapsulated

hemoglobin in pigs: a model of pseudollargic cardiopulmonary reactions to liposomes. Role of complement and inhibition by soluble CR1 and anti-C5a antibody. Circulation 99:2302–2309
Szebeni J, Fishbane S, Hedenus M, Howaldt S, Locatelli F, Patni S, Rampton D, Weiss G, Folkersen J (2015) Hypersensitivity to intravenous iron: classification, terminology, mechanisms and management. Br J Pharmacol 172:5025–5036
Technical fact sheet—Nanomaterials, November 2017
Tenzer S, Docter D, Kuharev J, Musyanovych A, Fetz V, Hecht R, Schlenk F, Fischer D, Kiouptsi K, Reinhardt C, Landfester K, Schild H, Maskos M, Knauer SK, Stauber RH (2013) Rapid formation of plasma protein corona critically affects nanoparticle pathophysiology. Nat Nanotechnol 8:772–781
Vance ME, Kuiken T, Vejerano EP, McGinnis SP, Hochella MF Jr, Rejeski D, Hull MS (2015) Nanotechnology in the real world: redeveloping the nanomaterial consumer products inventory. Beilstein J Nanotechnol 6:1769–1780
Ventola C, Lee MS (2017) Progress in nanomedicine: approved and investigational nanodrugs. P&T 42:742–755
Verresen L, Fink E, Lemke HD, Vanrenterghem Y (1994) Bradykinin is a mediator of anaphylactoid reactions during hemodialysis with AN69 membranes. Kidney Int 45:1497–1503
Vogler EA (2012) Protein adsorption in three dimensions. Biomaterials 33:1201–1237
Vroman L (1962) Effect of absorbed proteins on the wettability of hydrophilic and hydrophobic solids. Nature 196:476–477
Wang L, Li J, Pan J, Jiang X, Ji Y, Li Y, Qu Y, Zhao Y, Wu X, Chen C (2013a) Revealing the binding structure of the protein corona on gold nanorods using synchrotron radiation-based techniques: understanding the reduced damage in cell membranes. J Am Chem Soc 135:17359–17368
Wang X, Podila R, Shannahan JH, Rao AM, Brown JM (2013b) Intravenously delivered graphene nanosheets and multiwalled carbon nanotubes induce site-specific Th2 inflammatory responses via the IL-33/ST2 axis. Int J Nanomedicine 8:1733–1748
Warner AE (1996) Pulmonary intravascular macrophages. Role in acute lung injury. Clin Chest Med 17:125–135
Weibel ER (1963) Morphometry of the Human Lung; Springer: New York, NY, USA
Weissig V, Pettinger TK, Murdock N (2014) Nanopharmaceuticals (part 1): products on the market. Int J Nanomed 9:4357–4373
Wibro PP, Anselmo AC, Nilsson PH, Sarode A, Gupta V, Urbanics R, Szebeni J, Hunter AC, Mitragotri S, Mollnes TE, Moghimi SM (2017) Bypassing adverse injection reactions to nanoparticles through shape modification and attachment to erythrocytes. Nat Nanotechnol 12:589–590
Wibroe PP, Ahmadvand D, Oghabian MA, Yaghmur A, Moghimi SM (2016) An integrated assessment of morphology, size, and complement activation of the PEGylated liposomal doxorubicin products Doxil®, Caelyx®, DOXOrubicin, and SinaDoxosome. J Control Release 221:1–8
Wolfram J, Yang Y, Shen J, Moten A, Chen C, Shen H, Ferrari M, Zhao Y (2014) The nano-plasma interface: implications of the protein corona. Colloids Sur B Biointerfaces 124:17–24
Wright JR (2005) Immunoregulatory functions of surfactant proteins. Nat Rev Immunol 5:58–68
Zhu M, Nie G, Meng H, Xia T, Nel A, Zhao Y (2013) Physicochemical properties determine nanomaterial cellular uptake, transport, and fate. Acc Chem Res 46:622–631
Zuo G, Kang SG, Xiu P, Zhao Y, Zhou R (2013) Interactions between proteins and carbon-based nanoparticles: exploring the origin of nanotoxicity at the molecular level. Small 9:1546–1556

Chapter 9
Zinc and Silver Nanoparticles: Properties, Applications and Impact to the Aquatic Environment

Paulo Ricardo Franco Marcelino, Mariete Barbosa Moreira, Talita Martins Lacerda and Silvio Silvério da Silva

Abstract Recent advances in the field of nanoscience and nanotechnology enabled the development of precise processes that are used across the most diverse science fields to control individual atoms and molecules. Inorganic nanomaterials, specially zinc and silver nanoparticles (Zn-NPs, Ag-NPs), have been attracting attention in the last decades due to their versatility, with applications as active componentes in eletronics, pharmaceuticals, cosmetic and agricultural products. However, safe nanotechnology unfortunatelly did not progress at the same speed, and hazardous nanomaterials are still disposed indiscriminately, crompromising the environment. Herein, the most recent and relevant contributions related to Zn-NPs and Ag-NPs, with regard to their chemical properties, utilization, disposal, interaction with biological systems and environmental impacts, are presented and discussed.

Nomenclature

Ag-NPs	Silver Nanoparticles
A-ZnO-NPs	3-Aminopropyltrimethoxysilane
D-ZnO-NPs	Dodecyltrichlorosilane

P. R. F. Marcelino (✉) · T. M. Lacerda · S. S. da Silva (✉)
Department of Biotechnology, Engineering School of Lorena (EEL), São Paulo University (USP), CEP, 12602-810 Lorena, Brazil
e-mail: paulorfm1@hotmail.com

S. S. da Silva
e-mail: silviosilverio@usp.br

T. M. Lacerda
e-mail: talitalacerda@usp.br

M. B. Moreira
Chemistry Institute, Universidade Estadual Paulista Júlio de Mesquita Filho (UNESP), CEP, 14800-900 Araraquara, Brazil
e-mail: marietequimica@yahoo.com.br

M. Rai and J. K. Biswas (eds.), *Nanomaterials: Ecotoxicity, Safety, and Public Perception*, https://doi.org/10.1007/978-3-030-05144-0_9

NOM	Natural Organic Matter
NPs	Nanoparticles
Zn-NPs	Zinc Nanoparticles

9.1 Introduction

Along with the XXI century began the nanoscale time, and the first nanomaterials/nanoparticles (NPs, with dimensions of 10^{-9} m) were produced. Such materials can be designed and synthesized in a variety of sizes (up to 100 nm), shapes (tubes, beads, rods and prisms) and compositions (materials based on metals, metal oxides, quantum dots and carbon) (Doak et al. 2009). These characteristics of NPs determine their physico-chemical and biological properties (structure-property relationship) and consequently the application in the industrial, energy and environmental sectors, and also in medicine (Nel et al. 2006).

Currently, zinc (Zn-NPs) and silver (Ag-NPs) NPs have been attracting attention due to their versatility. Scientific community and industry have applied these metallic-NPs due to their intrinsic catalytic (Hvolbaek et al. 2007), optical (Noguez and Garzon 2009) magnetic (Niemirowicz et al. 2007), electronic (Matsui 2005; Millstone et al. 2010) and antimicrobial (Mayrinck et al. 2014) properties. These characteristics allow wide applications of metallic NPs in several areas such as medical (Huang et al. 2011; Ge et al. 2014; Marassi et al. 2018), pharmaceutical (Thakur 2015), biotechnological (Tartaj et al. 2005), cosmetic (Katz et al. 2015) and agricultural (Rai and Ingle 2012). However, the environmental consequences of using and disposing such materials can not be neglected.

Recently, terms such as "nanowastes" and "nanotoxicity" have been highlighted in the scientific community because of the popularity of nanomaterials and the impact they may cause when improperly disposed. Studies show that the structure, composition, concentration, solubility and high surface area influence the toxicity of nanoparticles (Guzman et al. 2006). Moreover, when they reach the environment, some transformations that potentiate their toxic effects may occur.

Sewage is still the most common method for the disposal of domestic and industrial rejects, and its efficiency is unfortunately limited, mainly considering the elimination of nanoscaled particles. Therefore, several studies have focused on studying the behavior of nanomaterials in the aquatic environment and the impact they can cause to living organisms. In this chapter some aspects related to the importance and toxicity of Zn-NPs and Ag-NPs in aquatic environment will be discussed.

9.2 Zinc Nanoparticles (Zn-NPs)

9.2.1 Metalic Zinc: History, Occurrence, Characteristics and Properties

Zinc is a transition metal located in the 4th period and in group 12 of the Periodic Table, with electronic configuration [Ar] $3d^{10}4\ s^2$. It has silver-gray bright color, is brittle at room temperature (298 K), malleable at temperatures between 373 and 423 K, resistant to oxidation, and is naturally found only in 2+ state (Lee 1991; http://www2.fc.unesp.br/lvq/LVQ_tabela/030_zinco.html). It is one of the 50 most abundant elements of the terrestrial crust, presenting a concentration of approximately 75 ppm or 0.007% (Medeiros 2012). During the genesis of the planet Earth was deposited in the form of zinc sulfide (ZnS) (Fig. 9.1), also known as blende or sphalerite. Other important zinc ores of economic interest which are also found in the biosphere are willemita $(Zn_2SiO)_4$, smithsonite ($ZnCO_3$), calamine ($[Zn_4Si_3O(OH)_2]$), wurtzita ((Zn;Fe)S), franklinite (Mn)O.Fe_2O_3), hydrozinitite ($[2ZnO_3.3Zn(OH)_2]$) and zincite (ZnO) (Lee 1991; Medeiros 2012).

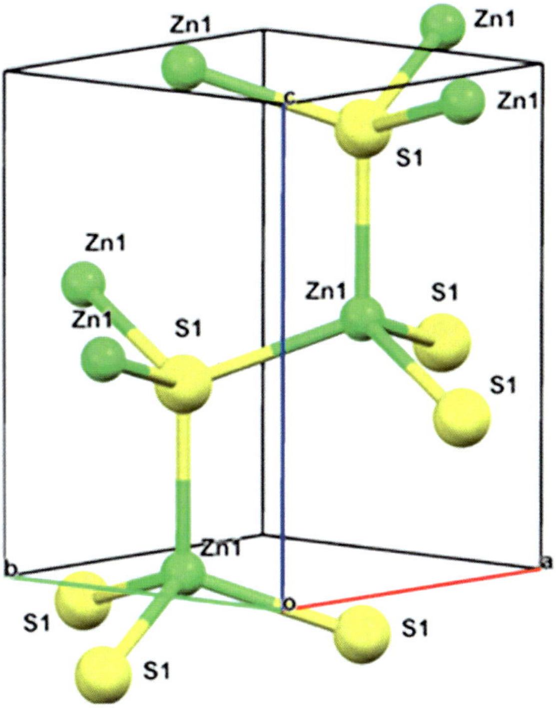

Fig. 9.1 Representation of a zinc sulphide (ZnS) unit cell—yellow spheres represent sulfur (S) atoms and green spheres represent zinc (Zn) atoms

The earliest historical accounts of zinc dates back to antiquity. In Ancient Greece, it was named *kadmeia* and in Rome *cadmia*. Afterwards, the term *cadmea* was used in latin to refer to zinc ores, while zinc oxide was called *nix alba* because of its snow-like appearance. History usually refers to zinc as an oriental metal, due to production data found in Persia, India and China between 11-7 B.C. centuries (Hanley 1933; Enghag 2004). Some reports indicate that the italian explorer Marco Polo, during expeditions to Asia between 1271 and 1295, had contact with zinc and its manufactured materials, commonly called *Tutia* (zinc oxide) and *Spodium* (zinc oxide mixed with other metal oxides) (Parks 1927; Enghag 2004; https://www.rheinzink.pt/qualidade/historia-do-zinco/).

Zinc arrived in the western civilizations through the great Portuguese navigations. Erasmus Ebener Henckel (1509) and Andreas Sigismund Marggraf (1746) worked on the isolation of metallic zinc from various oxides (Beckmann and Johnston 1814; Enghag 2004; Preedy et al. 2011; Pollard et al. 2016). The European industrial production of zinc began in Bristol in the 1740s, when John Champion developed the English method. In 1807, Abbé Dony de Liège also developed a pyrometallurgical process for zinc production, but due to financial difficulties it did not have the expected success. In 1850, John Hiltz was the first to produce metallic zinc in America (Enghag 2004).

Currently, zinc is of fundamental importance for the automotive, marine, chemical, pharmaceutical, food and construction industries. Its derivatives are mainly applied in the galvanization of steel, in the development of metal alloys and in the production of anticorrosion paints. Zinc oxide (ZnO) has also received great attention in the market, since it is used in the formulation of cements, ceramics, pharmaceuticals, cosmetics and food additives (https://www.azonano.com/article.aspx?ArticleID=3348; Mayrinck et al. 2014). Another compound of industrial relevance is zinc chloride ($ZnCl_2$) employed in the treatment of tissues, such as solder flux and killed salts dissolving metal oxides, linking the weld metal clean surface (Lee 1991).

In the current scenario, the five countries considered as the largest producers of zinc are China (5.1 million metric ton), Peru (1.4 million metric ton), India (1.3 million metric ton), and Australia (730,000 metric tons) (https://investingnews.com/daily/resource-investing/base-metals-investing/zinc-investing/top-zinc-producing-countries/). Modern industry produces metallic zinc from thermal and electrochemical treatments, being the former the most common method, in which zinc oxide reacts with carbon monoxide, at high temperatures, to form metallic zinc and carbon dioxide (Lee 1991).

Besides the economic-industrial importance, zinc is an essential metal for living beings, participating in catalytic and regulatory processes, and with structural functions in biological systems (Shriver and Atkins 2008). It is known, for example, that in the human organism, zinc is the second most abundant essential trace element and its regulation/homeostasis depends on the metallothionein/tionein protein system (Henriques et al. 2003; Tapiero and Tew 2003). It also acts as a cofactor in around 300 metalloproteins and 2000 transcription factors, with catalytic, co-catalytic and structural functions (Table 9.1) (Valee and Falchuk 1993; King 2011). Generally,

Table 9.1 Zinc metalloproteins

Protein	Role of zinc	Function
Alcohol dehydrogenase	Catalitic[a]	Dehydrogenase responsible for the oxidation of alcohols to carbonyl compounds (aldehydes/ketones) with the concomitant reduction of NAD^+ to NADH
Reverse transcriptase	Catalitic	This enzyme also called DNA-polymerase or RNA-dependent is responsible for performing the reverse transcription process by polymerizing DNA molecules from RNA molecules
Alkaline phosphatase	Catalitic	Hydrolase responsible for the reactions of dephosphorylation (removal of phosphate groups in several molecules) at alkaline pH
Carboxypeptidase A	Catalitic	Hydrolase responsible for the hydrolysis of C-terminal amino acids during the digestion process
Carbonic Anhydrase	Catalitic	Enzyme responsible for the hydration of carbon dioxide produced in the respiratory chain
β-lactamase	Catalitic	Hydrolase produced by some pathogenic microorganisms responsible for the hydrolysis of the β-lactam ring present in some drugs, thus preventing their antimicrobial action
Superoxide Dismutase Copper-Zinc	Co-catalitic[b]	Hydrolase responsible for the decomposition of hydrogen peroxide (H_2O_2) in water (H_2O) and oxygen (O_2)
Phospholipase C	Catalitic/co-catalitic	Hydrolase responsible for the cleavage of phosphate group in phospholipids
Nuclease P1	Co-catalitic	Hydrolase responsible for random nucleic acid cleavage
Leucine Aminopeptidase	Co-catalitic	Hydrolase responsible for the cleavage of N-terminal leucine residues in peptides and proteins
Zinc Finger Proteins (Zif268)	Structural[c]	Structural proteins produced by the coordination of Zn^{2+} with specific residues of histidine and cysteine forming "folds" in the protein domains that are bound to the DNA
Ferredoxin	Structural	Metalloprotein that mediates the transfer of electrons in various metabolic reactions
Protein kinase C(PKC)	Structural	Enzymes capable of phosphorylating proteins from the hydroxyl groups of serine and threonine amino acid residues

[a]The metal ion, generally present in the catalytic site of the metalloprotein, participates directly in the enzymatic catalysis and its removal may inactivate the the enzyme
[b]The presence of the metal may increase or decrease the performance of the enzymatic catalysis. However, removal of the metal does not lead to the loss of enzyme activity
[c]The presence of the metal is required only to maintain the stability of the protein conformation
Based on Henriques et al. (2003)

zinc is coordinated to cysteine, cysteine thiolate and histidine from metalloproteins (Shriver and Atkins 2008).

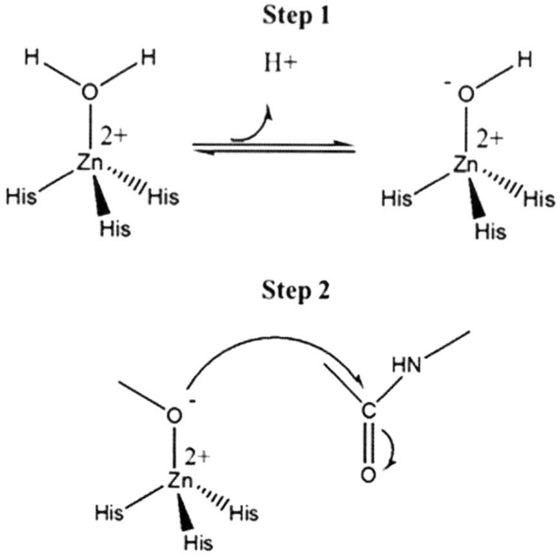

Fig. 9.2 Two-step mechanism of Zn-hydroxide catalysis

The catalytic function of zinc in metalloenzymes may occur by two mechanisms, Zn-hydroxide and Zn-carbonyl, which are described below.

Zn-hydroxide mechanism: in this case, the Zn^{2+} present in the active site of the metalloenzyme complexes with water molecules of the reaction medium. Under normal conditions, the pK_a of water is around 15, which decreases to approximatelly 7 if molecules are complexed with zinc. The change in pK_a indicates the formation of a hydroxide (OH^-) ion at the active site of the metalloenzine (Step 1), a strong nucleophile that can attack carbonyl groups (Step 2). The carbonic anhydrase enzyme is an example of metalloenzyme that follows this mechanism (Fig. 9.2) (Shriver and Atkins 2008).

Zn-carbonyl mechanism: in this case, the Zn^{2+} present in the active site of the metalloenzyme acts as a Lewis acid, receiving a pair of electrons from the oxygen atom of the carbonyl. This fact makes carbonyl polarized and susceptible to the attack of weak nucleophilic groups, such as water. The carboxypeptidase enzyme is an example of metalloenzyme that follows this mechanism (Fig. 9.3) (Shriver and Atkins 2008).

In addition to the catalytic, co-catalytic and structural functions reported in Table 9.1, Zn^{2+} also acts in humans as antioxidant agent and as cofactor of the biosynthesis, storage and secretion of insulin, a metabolite that controls the glucose levels in the blood stream (King 2011). It also influences human growth, participates in the synthesis of vitamin A and is important for the endocrine, immune and nervous systems (Fernandes Cruz and Freire Soares 2011).

In plants, Zn^{2+} is a cofactor of several enzymes and participates in the metabolism of carbohydrates, proteins, phosphates, RNA, ribosomes, auxins and phenolic com-

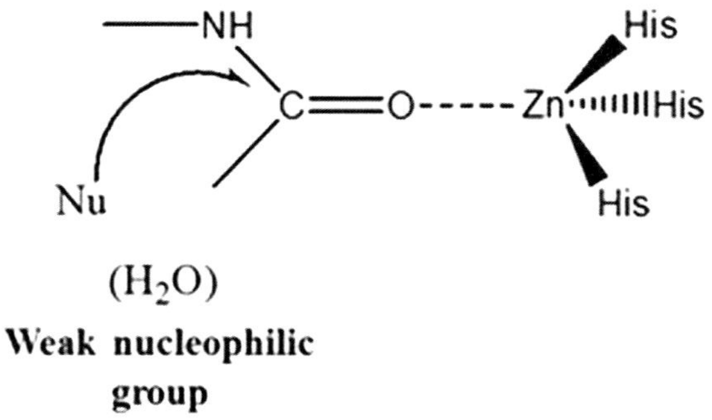

Fig. 9.3 Zn-carbonyl catalysis mechanism

pounds. It also acts to increase the size of plants, to accelerate cell multiplication and on the fertility of pollen grains (Borket 1989; Malavolta et al. 1991). In crops of economic importance, such as corn, some studies demonstrated that the addition of Zn^{2+} promotes significant increases in both dry matter and grain yield (Galrão et al. 1981; Martens and Wastermann 1991; Galrão 1995).

The importance of zinc to society and biological systems is unquestionable. With the advent and popularization of nanotechnology at the end of the XX Century, the production of its corresponding nanomaterials, also known as zinc nanoparticles (Zn-NPs), expanded its application.

9.2.2 ZnO-NPs: Properties and Applications

As previously mentioned, ZnO (Fig. 9.4) is the zinc compound with the vastest industrial applications, with researches improving considerably in the 1970s. According to the Science Direct platform (www.sciencedirect.com), between 1994 and 2017, the number of publications related to terms such as "ZnO nanostrutucture" increased approximately 100 times, while the term "ZnO nanoparticles" increased by 500 times.

The extensive use of ZnO in nanomaterials is due to its chemical, biological, optical, mechanical, electrical and magnetic properties resulting from the specific microstructure (Hong et al. 2006; Mayrinck et al. 2014). The versatility of ZnO-NPs allows applications in polymers synthesis (adhesives and rubbers), in cosmetic formulations (UV absorber and antimicrobials), in the production of paints (pigments and anticorrosive products) for ceramics, and in the development of catalysts and electronic materials (Mayrinck et al. 2014). In addition to conventional NPs, ZnO bionanoparticles are produced from its redox reactions with plant extracts. These ZnO-NPs are mainly applied as photocatalyst and antimicrobial agentes (Shah et al. 2015; Basnet et al. 2018). However, there are still few studies on their impact on the environment and toxicological effects of living organisms.

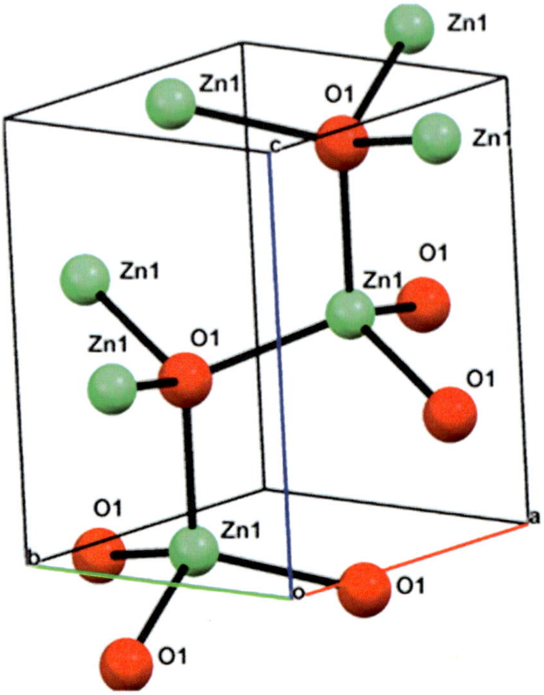

Fig. 9.4 Representation of a zinc oxyde (ZnO) unit cell—red spheres represent oxygen (O) atoms and green spheres represent zinc (Zn) atoms

ZnO-NPs have several industrial applications, which motivated countries like United States, Germany, Japan, Brazil, China and India to invest a lot in research and development. These increasing investments have, on the one hand, boosted the consumption of nanoscaled materials, products and processes but, on the other, left a gap in respect to ther toxicity, ecotoxicity and safety. The scenario here is still in the initial phase and little is known about the effects of these materials on living organisms and on the environment (http://lqes.iqm.unicamp.br/canal_cientifico/lqes_news/lqes_news_cit/lqes_news_2008/lqes_news_novidades_1143.html; Paschoalino et al. 2010).

Currently, information on the amount of discarded nanomaterials are still scarce. However, like other products, the main destination of nanoparticles are industrial or household waste, and in many regions of the world sewage treatment is ineffective. Thus, NPs can reach lakes, groundwater, rivers and oceans, threatening the potability of water and leading to environmental and public health problems (Troester et al. 2016).

9.2.3 Behaviour and Effects of ZnO-NPs in the Aquactic Environment

In the environment, NPs can be transformed by physical, chemical and biological processes that will affect their transport, bioavailability and toxicity. In the aquatic environment, ZnO-NPs are commonly converted by aggregation, deposition, dissolution and association with inorganic compounds and biological macromolecules (Fig. 9.5) (Amde et al. 2017).

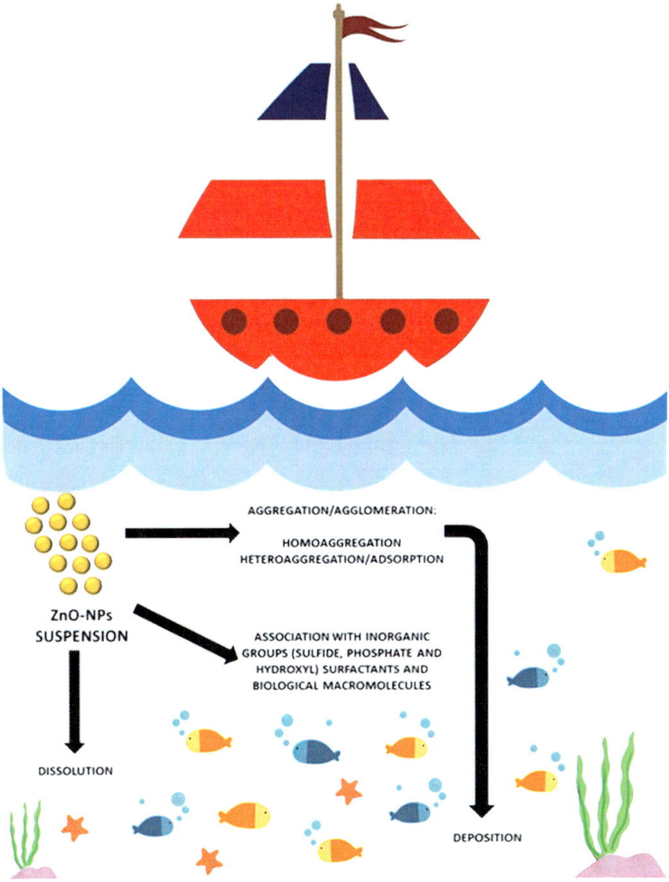

Fig. 9.5 Physical, chemical and biological processes of ZnO-NPs in the aquatic environment

9.2.3.1 Aggregation

In this process due to the effects of intermolecular interactions (permanent dipole, induced dipole and hydrogen bonds), electrostatic repulsion, hydration and magnetic mechanisms, agglomerates are easily formed in suspension, which can be of pure ZnO-NPs (homo-aggregates) or of ZnO-NPs with others metallic nanoparticles (hetero-aggregates) (Hotze et al. 2010; Lowry et al. 2012; Quik et al. 2014; Dwivedi et al. 2015; Sendra et al. 2017), the later being more common due to the complex constitution of the aquatic environment (Schultz et al. 2015). This phenomenon affects transport and favors deposition, and decrease the corresponding surface charge, reactivity, toxicity and bioavailability of NPs (Amde et al. 2017). In ZnO-NPs, aggregation and stability are influenced by factors such as shape and size of the NPs, pH (Zhou and Keller 2010; Yung et al. 2015), ionic strength (Odzak et al. 2017) and natural organic matter (NOM) (Zhou and Keller 2010; Peng et al. 2017) of the aquatic environment.

In contrast to the aggregation, the literature reports that ZnO nanoparticle agglomerates can disaggregate at low pH, ionic strength and in the presence of high concentrations of NOM (Amde et al. 2017).

9.2.3.2 Dissolution

In the dissolution of ZnO-NPs in aquatic environment, Zn^{2+} is released into the environment (dissociation of zinc oxide). The dissolution mechanism of ZnO-NPS, as well as aggregation, is directly influenced by NPs characteristics such as size and shape (Mudunkotuwa et al. 2012; Bian et al. 2011), physicochemical conditions of the aquatic environment in which it is found, such as pH (Omar et al. 2014), ionic strength (Yung et al. 2015), NOM (Omar et al. 2014; Jiang et al. 2015), temperature (Majedi et al. 2013), ultraviolet radiation (UV) (Odzak et al. 2017; Ma et al. 2014) and concentrations of other reactive species (Lv et al. 2012; Peng et al. 2017; Amde et al. 2017).

As previously reported, the ingestion of Zn^{2+} in low concentrations is fundamental for the metabolism of living organisms. However, in high doses, causes stomachache, nausea and vomiting, increases the risk of coronary heart disease, injures the pancreas and reduces the High Density Lipoproteins (HDL) content in humans. In addition, antagonistic interaction between zinc and copper may occur, reducing the absorption of copper and, consequentely, causing anemia (CETESB 2012). In the aquatic environment, the accumulation of Zn^{2+} ions cause coagulation of the mucus on the gills of fish, leading to death by suffocation (Marcantonio 2005). In plants, Zn^{2+} intoxication inhibit growth (Lenzi et al. 2011).

9.2.3.3 Association with Inorganic Compounds and Biological Macromolecules

ZnO-NPs may also be associated with inorganic compounds and biological macromolecules present in the aquatic environment, which is totally related to the physicochemical conditions of the medium (Amde et al. 2017). ZnO-NPs are commonly associated to inorganic groups such as sulfides (Lombi et al. 2012; Ma et al. 2013), phosphates (Brunetti et al. 2015), Fe-oxide and Fe-hydroxides (Lombi et al. 2012; Ma et al. 2013). In addition, biological macromolecules (proteins, carbohydrates, lipids, humic materials, NOM and others) also interact with ZnO-NPs by intermolecular and electrostatic forces, ligand exchange, bridging and complexation/chelation due presence of hydroxyl, carbonyl, amine, amide, carboxylic groups (Yang et al. 2009; Philippe and Schaumann 2014; Amde et al. 2017).

9.2.4 *Toxic Effects of ZnO-NPs to Aquatic Organisms*

The effects that ZnO-NPs can cause to living organisms and the environment are mainly influenced by the intrinsic characteristics of the NPs and how they behave in certain physical-chemical conditions. In addition, the absorption mechanisms used by living organisms and the interaction of ZnO-NPs with intracellular molecules are responsible for the corresponding bioavailability and toxicity (Amde et al. 2017). One of the main harmful effects of ZnO-NPs on living organisms is their ability to produce free radicals (ROS) due to the redox reactions that may occur in the intracellular environment. ROS produced can interact with cell membranes and biomolecules such as proteins and nucleic acids, causing oxidative stress and even genetic mutations (Chang et al. 2012). Some deleterious effects of ZnO-NPs on aquatic organisms are presented below.

9.2.4.1 Fishes

Studies were recently reported indicating that ZnO-NPs varying from 20 to 100 nm and in concentrations of 10–120 mg/L have caused genetic, morphological, immunological and serum protein concentrations in zebrafish (*Danio rerio*) embryos (Choi et al. 2016; Du et al. 2016; Zhao et al. 2016; Chun et al. 2017; Rajput et al. 2018).

A previous study with white sucker fish (*Catostomus commersonii*) indicated that ZnO-NPs caused cardiorespiratory damages and, consequently, cellular oxidative stress in the gill tissue, increasing the levels of malondialdehyde, the expression of heat shock protein (HSP) and caspase activity (Bessemer et al. 2014).

9.2.4.2 Molluscs, Algae and Microoganisms

The association of ZnO-NPs with nifedipine (used intensively in the treatment of hypertension) is a classic example of association of ZnO-NPs with macromolecules, and its consequences in aquatic environment, mainly in freshwater bivalves, were evaluated (Falfushynska et al. 2015).

According to studies of Ali et al. (2012), freshwater snail *Lymnaea luteola* L. in contact with ZnO-NPs (32 μg/ml) presented reduced levels of glutathione and malondialdehyde. Also, the activity of glutathione-S-transferase, glutathione peroxidase and catalase decreased.

In tests with mussels (*Aegyptiac coelatura*) exposed to ZnO-NPs at concentrations of 10 and 50 mg/L, increased levels of malondialdehyde and nitric oxide and a concomitant reduction in the activity of the enzymes superoxide dismutase, glutathione-S transferase and catalase were reported. Such physiological changes were probably due to oxidative stress caused by NPs (Fahmy and Sayed 2017).

The exposure of marine algae *Chlorella vulgaris*, at certain ZnO-NPs concentrations, showed reduction of LDH levels and increase in superoxide dismutase activity and lipid peroxidation (LPO) levels (Suman et al. 2015).

The marine diatom *Thalassiosira pseudonana* (phytoplankton) exposed to ZnO-NPs coated with 3-aminopropyltrimethoxysilane (A-ZnO-NPs) and dodecyltrichlorosilane (D-ZnO-NPs) resulted in a modification in expression of genes, probably due to the toxic action mechanisms of ZnO-NP coatings (Yung et al. 2017).

Zn-NPs have also been reported as toxic to some bacteria common to the aquatic environment such as *Escherichia coli* (LC_{50} 1000 mg/L) and *Staphilococcus aureus* (LC_{50} 10 mg/L) (Adams et al. 2006) and microcrustaceans such as *Daphinia magna* (LC_{50} 3 mg/L) (Heinlaan et al. 2008).

9.2.4.3 Plants

Water from rivers contaminated with ZnO-NPs, when in contact with crops through irrigation systems, caused biomass drop, damaged root surface cells and induced abnormal defense system in plants (Lee et al. 2013). Some authors have also reported the shrinking of the roots and epidermis (Lin and Xing 2008) and even inhibition in the chlorophyll biosynthesis and photosynthesis (Raskar and Laware 2014).

ZnO-NPs are of fundamental importance for the development of society. However, their use and disposal must be rational, since the damages they potentially cause to the environment are serious and often irreversible.

9.3 Silver Nanoparticles (Ag-NPs)

9.3.1 Metallic Silver: History, Occurrence, Characteristics and Properties

For centuries, silver has been one of the most known and valued metals in the world (Enghag 2004). Silver ornaments and coins are used since antiquity, as the oldest silver objects date from 5000 B.C in India. The Bible cites, in the book of Genesis, the use of silver for domestic utensils, which dates back to 3000 B.C. Silver was responsible for the wealth of many empires, such as Greece around 1000 B.C, and was considered the almost perfect metal by alchemists (Souza et al. 2013).

After 1492, world silver production increased greatly due to the discovery of silver mines in Bolivia, Peru and Mexico, and between 1500 and 1800 these countries accounted for more than 85% of the global silver production. After 1850 several other countries increased the production of silver (Enghag 2004). In 2017, the world production of silver was 26,502 tons (852.1 Moz). Currently, the main producing countries of silver are Peru and China (World Silver Survey 2018). Native silver occurs rarely, being frequently associated in minerals containing antimony (pyragyrite, Ag_3SbS_3), arsenic (proustite, Ag_3AsS_3) and chlorine (chloroarginite, AgCl). However, the main occurrence of silver is in the form of sulphide minerals such as argentite (Ag_2S), which exhibits crystalline orthorhombic system (Fig. 9.6). Silver sulfide is geologically connected with gold shafts and deposits of copper, zinc and lead (Enghag 2004).

Silver is a transition metal located in the 5th period and in group 11 of the Periodic Table, has electronic configuration [Kr] $4d^{10}5s^1$, appears as a silvery-colored solid with characteristic luster, is stable in water and atmospheric air and is resistant to corrosion. It shows oxidation states of +1 and +2, is soluble in nitric acid and insoluble in sulfuric and hydrochloric acid. Gold and silver are the metals with the highest reflectivity, thermal and electrical conductivity (Lee 1991). Silver is used in ornaments, photographic emulsions, jewelry and in the manufacture of batteries and mirrors (Souza et al. 2013). In addition, silver compounds were historically used as a drug because of the antibacterial activity of silver ions (Russell and Hugo 1994).

9.3.2 Ag-NPs: Properties, Applications and Fate in Urban Wastewater Systems

The antimicrobial properties of silver have boosted its applications since olden times, being the second most used metal for the Chaldeans (4000 B.C). In addition to domestics utensils and coins, silver has been widely used in water purification, food preservation and in the treatment of diseases for centuries (Rai et al. 2018). Before the advent of antibiotics, silver was commonly used as a bactericidal agent. Antibiotic

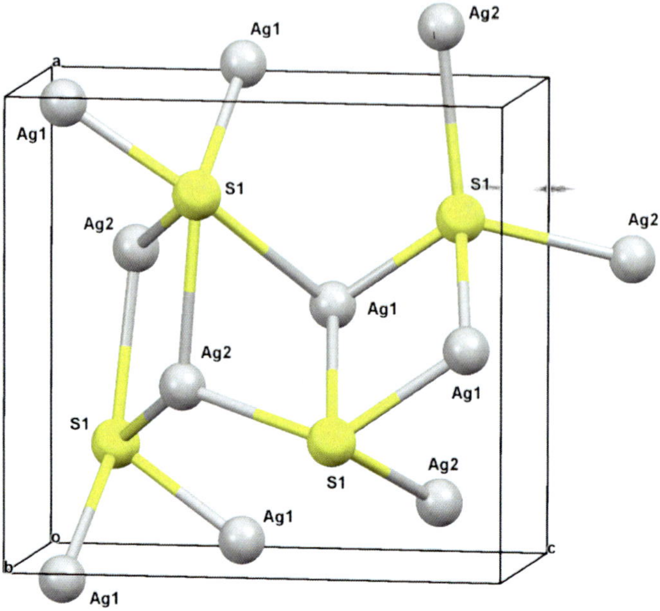

Fig. 9.6 Representation of a silver oxide (Ag_2O) unit cell—yellow spheres represent oxygen (O) atoms and silver spheres represent silver (Ag) atoms

resistance has recently motivated the return of silver as a remarkable antimicrobial agent (Rai et al. 2009; Franci et al. 2015).

The antimicrobial activity of Ag-NPs is well established in the literature, as it causes structural changes in the bacterial cell wall, resulting in increased membrane permeability and consequent cell death. The size and shape of Ag-NPs can be used to modulate antimicrobial activity, and smaller Ag-NPs promoted more intense antimicrobial activity for its higher binding surface when compared to larger Ag-NPs (Rai et al. 2018).

The biological activity of Ag-NPs is not restricted to antifungal and antibacterial activity. Recently, Ag-NPs presented promising antitumor activity (Yeasmin et al. 2017), significant activity against the parasite *Anopheles stephensi* (responsible for malaria) (Kumar et al. 2018), and potential for the treatment of drug-resistant microbes (Birla et al. 2009; Rai et al. 2012; Mudasir et al. 2013).

Ag-NPs are among the most synthesized nanostructured materials, due to its low cost and unique properties. It is possible to change the properties and reactivity of Ag-NPs by changing their sizes and morphologies (spheres, cubes, roads) (Carlson et al. 2008; Khodashenas and Ghorbani 2015). Generally, Ag-NPs are synthesized by reducing a silver salt (such as silver nitrate, silver perchlorate or silver tetrafluorborate) to Ag-NPs, using as stabilizing agent citrate or a polymer such as poly(vinyl pyrrolidone) (PVP), poly(ethylene glycol) (PEG) or poly(vinyl alcohol) (PVA) (Kapoor et al. 1994; Huang et al. 1996). Ag-NPs can also be obtained by other

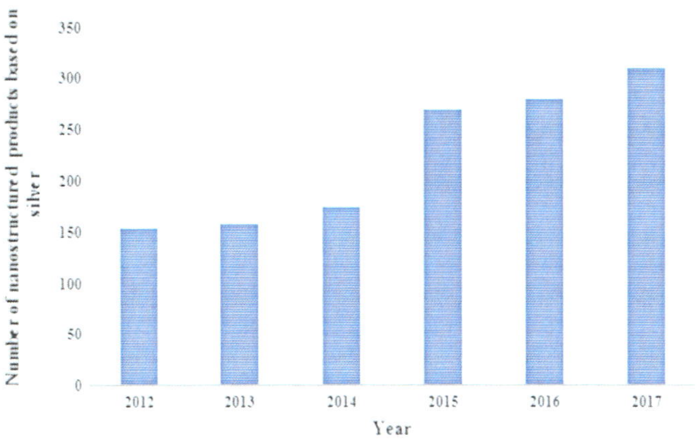

Fig. 9.7 Number of nanostructured products based on silver produced between 2012 and 2017

methods, involving biological reduction (Ahmad et al. 2014), irradiation (Bogle et al. 2006; Li et al. 2006) and electrochemistry (Rodrígues-Sánchez et al. 2000; Yin et al. 2003). Moreover, the mycosynthesis of Ag-NPs was reported, in which a solution of silver nitrate was reduced to metalic silver by the fungus *Fusarium acuminatum* cell extract (Ingle et al. 2008), *Carica papaya* callus extract (Mude et al. 2009), Saffron (*Crocus sativus* L.) extract (Bagherzade et al. 2017), *Cryphonectria* sp. (Mudasir et al. 2013) and *Phoma glomerata* (Birla et al. 2009).

Currently, according to the Danish database "The nanodatabase", which evaluates consumer products containing nanomaterials for consumption in the European market, there are a total of 3037 products, of which 309 are registered as containing silver ("The nanodatabase" 2018). In addition, it is noted that the number of registered products containing silver has increased significantly over the past five years. Figure 9.7, shows the evolution of products containing Ag-NPs from 2012 to 2017.

Several studies have boosted the consumption of Ag-NPs worldwide. According to the literature, in recent years, the average global Ag-NP consumption was 55 tons/year, with future predictions of 360–450 tons/year (Lazareva and Keller 2014). Ag-NPs are widely applied in various consumer products including paints, textiles, cosmetics, deodorants, walter filters, food packaging, medical products, biosensors, detergents and eletrical devices (Young et al. 2003; Furno et al. 2004; Kumar et al. 2008; Benn and Westerhoff 2008; Blaser et al. 2008; Roe et al. 2008; Kholoud et al. 2010; Cushen et al. 2012; Goswami et al. 2014; McGillicuddy et al. 2014).

NPs are disposed in the environment through natural sources (such as sea salts and volcanic dust) and by industrial processes. The immense global demand and use of nanoparticle-containing products, and their release in the environment, causes great concern about their corresponding fate and impact (Oberdorster et al. 2005). Nanostructured products containing silver are among the most consumed, and the main route of exposure of these nanoparticles is through wastewater. Therefore, the fate of Ag-NPs has been the target of several studies (Kaegi et al. 2013; Brunetti et al. 2015; Bruneau et al. 2016; Polesel et al. 2018).

Some years ago, the photographic industry was responsible for an important amount of all silver disposed in the aquatic environment. Nowadays, the global use and disposal of various products containing Ag-NPs such as fabrics, cosmetics, deodorants, food packaging, water filters and toys has exposed individuals and environment to Ag-NPs (Purcell and Peters 1998; Windler et al. 1998; Geranio et al. 2009; Fabrega et al. 2011). According to the literature, Ag-NPs are efficiently transported by sewage to plants and are retained in sewage sludge, which is used as fertilizer in agricultural soils. Soil leaching allows the sludge containing Ag-NPs to reach aquatic systems (Hou et al. 2005). Another destination for sewage sludge are landfills, and soil leaching causes the contamination of groundwater. (Hou et al. 2005). Kim et al. (2010) reported that in the wastewater in anaerobic medium, Ag-NPs can be converted to silver sulfide (Ag_2S), which is reasonable since silver is a mole acid and binds effectively to soft bases as sulfur atoms. In addition, the bad smell released by the sewage is an evidence of the formation of sulfides under anaerobic conditions. Brunetti et al. (2015) reported that in wastewater, Ag-NPs are converted to silver sulfide and a fraction of silver binds to cysteine and histidine residues produced by microorganisms present in the medium.

9.3.3 Behaviour and Effects of Ag-NPs in the Aquatic Environment

The behavior of Ag-NPs in the aquatic environment has been the subject of many studies (Moore 2006; Fabrega et al. 2011; Yu et al. 2014; Vale et al. 2016; Zhang et al. 2016). According to the literature, Ag-NPs may react with the species present in the aquatic environment or remain in suspension or in the form of aggregates (McGillicuddy et al. 2014). In the aquatic environment, the bivalent and trivalent species are unstable, and metallic silver and its monovalent ion, released from the Ag-NPs, react to form metallic complexes, such as Ag^0, Ag^+, Ag^{2+} and Ag^{3+} silver chloride (AgCl), silver sulphide (Ag_2S) and silver oxide (Ag_2O) (Reidy et al. 2013; McGillicuddy et al. 2014).

The speciation of Ag-NPs in the aquatic environment is dependent on the organic matter content, pH, ionic strength and concentration of ligands that bind to the silver ions, forming metal complexes (McGillicuddy et al. 2014). In a study on the fate of silver and zinc NPs, Brunetti et al. (2015) concluded that under anaerobic conditions thiolation is a dominant process, which inhibits or retards other mechanisms such as oxidative dissolution or direct sulphidation. The process of dissolution of Ag-NPs, that involves the oxidation of silver metal to monovalent silver, is very important for the understanding of the fate and behavior of Ag-NPs in aquatic environment (Liu and Hurt 2010; He et al. 2013). Recent studies demonstrated that silver dissolution can be affected by pH, temperature, oxygen levels, ionic strength, coating type present in Ag-NPs, concentration and particles size (Zhang et al. 2016). According to the authors, the aggregation of NPs, easily achieved at high concentrations, reduces the surface area, hindering the dissolution. It was also observed that smaller NPs are more soluble, and Ag-NPs coated with citrate present faster dissolution rates compared to PVP-coated Ag-NPs in fresh and saline water.

9.3.4 Toxic Effects of Ag-NP in Aquatic Organisms

The expansion of the use of products containing NPs exposes man and the environment to inevitable increasing rates of NPs. ("The nanodatabase" 2018). In aquatic biota, the uptake of NPs occurs through direct ingestion or through gills, bony plates and olfactory walls (Moore 2006). As previously reported, the most stable species of silver in aquatic environment are metallic silver and its monovalent ion (Ag^+), which can be released by Ag-NPs. In the aquatic environment, one fraction of Ag^+ remains dissolved and another predominantly reacts with silver sulphide and silver chloride. Among these species, Ag^+ is the one with the highest toxicity (Ratte et al. 1999; Tappin et al. 2010).

Studies have shown that in fish, silver ions cross the basolateral membrane of the gills, blocking Na^+/K^+ ATPase and compromising Na^+ Cl^- ion regulation. In addition, the presence of Ag-NPs compromised the precocious development of the fish, with cardiac arrhythmias and deformations in the spinal cord (Grosell et al. 1999). The toxicity of Ag-NPs may be influenced by some factors such as speciation, NPs coating, size and surface area (Burchardt et al. 2012).

The toxicity of the Ag-NPs is influenced by their type of coating, and the organic compounds used to stabilize the NPs may allow the release of Ag^+ in the aquatic environment (Kwok et al. 2012). Studies with zebrafish (fresh water fish) and Japanese medaka (saline) showed that citrate-coated Ag-NPs were accumulated in zebrafish embryos (Park et al. 2013). In addition, it has been recently reported that Ag-NPs coated with PVC and citrate exhibited different toxicity in the presence of *Japanese medaka* (Kwok et al. 2012). The toxicity of Ag-NPs is higher for smaller NPs because of higher the surface area and consequente higher release of Ag^+. In addition, studies shows that juvenile trout rainbow were exposed to Ag-NPs and morphological modifications and bioaccumulation were observed (Bruneau et al. 2016). Recent studies showed that the exposure of neotropical fish *Prochilidus lineatus* to Ag-NPs resulted in harmful effects such as bioaccumlation and damages to lipids and histophalogies (Ale et al. 2018).

Algae present an important role in the aquatic environment, as they are responsible for photosynthesis and also correspond to an importante source of food for other organisms (Miao et al. 2009). The exposure of the marine diatom *Thalassiosira weissflogiis* to Ag-NPs was studied by Miao et al. (2009), who observed that the Ag-NPs have toxic effects on phytoplankton, due to the release of Ag^+ in the medium.

9.4 Conclusion

The toxicity of nanoparticles is an issue not yet surpassed by the scientific community. In this way, efforts are necessary for the understanding of the fate and possible ecotoxicological risks that NPs present to the environment.

References

Adams LK, Lyon DY, Alvarez PJJ (2006) Comparative eco-toxicity of nanoscale TiO_2, SiO_2, and ZnO water suspensions. Water Res 40:3527–3532

Ahmad N, Shree K, Srivastava M, Dutra R (2014) Novel rapid biological approach for synthesis of silver nanoparticles and it characterization. Int J Pharmacol 1:28–31

Ale A, Barcchetta C, Rossi AS, Desimone MF, Torre FR, Gervasio S, Cazenave J (2018) Nanosilver toxicity in gills of a neotropical fish metal accumulation oxidative stress histopathology and other physiological effects. Ecotoxicol Environ Saf 148:976–984

Ali D, Alarifi S, Kumar S (2012) Oxidative stress and genotoxic effect of zinc oxide nanoparticles in freshwater snail *Lymnaea luteola* L. Aquat Toxicol 124–125:83–90

Amde M, Liu JF, Tan ZQ, Bekana D (2017) Transformation and bioavailability of metal oxide nanoparticles in aquatic and terrestrial environments. A review. Environ Pollut 230:250–267

Bagherzade G, Tavakolo MM, Namaei MH (2017) Green synthesis of silver nanoparticles using aqueous extrat of saffron (*Crocus sativus L.*) wastages and its antibacterial activity against six bacteria. Asian Pac J Trop Biomed 7:227–233

Basnet P, Chanu TI, Samanta D, Chatterjee S (2018) A review on bio-synthesized zinc oxide nanoparticles using plant extracts as reductants and stabilizing agentes. J Photochem Photobiol B 183:201–221

Beckmann J, Johnston W (1814) A history of inventions and discoveries, vol 3, 2nd edn. Walker

Benn TM, Westerhoff P (2008) Nanoparticle silver released into water from commercially available sock fabrics. Environ Sci Technol 42:4133–4139

Bessemer RA, Butler KMA, Tunnah L, Callaghan NI, Rundle A, Currie S, Dieni CA, MacCormack TJ (2014) Cardiorespiratory toxicity of environmentally relevant zinc oxide nanoparticles in the freshwater fish *Catostomus commersonii*. Nanotoxicology 9:861–870

Bian SW, Mudunkotuwa IA, Rupasinghe T, Grassian VH (2011) Aggregation and dissolution of 4 nm ZnO nanoparticles in aqueous environments: influence of pH, ionic strength, size, and adsorption of humic acid. Langmuir 27:6059–6068

Birla SS, Tiwari VV, Gade AK, Ingle AP, Yadav AP, Rai MK (2009) Fabrication of silver nanoparticle by Phoma glomerata and its combined effect against *Escherichia coli Pseudomonas aeruginosa* and *Staphylococcus aureus*. Lett Appl Microbiol 48:173–179

Blaser SA, Scheringer M, MacLeod M, Hungerbuhler (2008) Estimation of cumulative aquatic exposure and risk due to silver: Contribution of nano-functionalized plastic and textiles. Sci Total Environ 390:396–409

Bogle K, Dhole S, Bhoraskar V (2006) Silver nanoparticles: Synthesis and size control by electron irradiation. Nanotechnology 17:3204–3208

Borkert CM (1989) Micronutrientes na planta. In: Büll LT, Rosolem CA. Interpretação de análise química de solo e planta para fins de adubação, p 309–329. Fundação de Estudos e Pesquisas Agrícolas e Florestais

Bruneau A, Turcotte P, Pilote M, Gagné F, Gagnon C (2016) Fate of silver nanoparticles in wastewater and immunotoxic effects on rainbow trout. Aquat Toxicol 174:70–81

Brunetti G, Donner E, Laera G, Sekine R, Scheckel KG, Khaksar M, Vasilev K, De Mastro G, Lombi E (2015) Fate of zinc and silver engineered nanoparticles in sewerage networks. Water Res 77:72–84

Burchardt AD, Carvalho RN, Valente A, Nativo P, Gilliland D, Garcia CP, Passarella R, Pedroni V, Rossi F, Litieri T (2012) Effects of silver nanoparticles in diatom *Thalssiosira pseudonana* and *Cyanobacterium synechococcus* sp. Environ Sci Technol 46:11336–11344

Carlson C, Hussain SM, Schlager JJ (2008) Unique cellular interaction of silver nanoparticles: size dependent generation of reactive oxygen species. J Phys Chem B 112:13608–13619

CETESB—Companhia de Tecnologia de Saneamento Ambiental. Divisão de Toxicologia, Genotoxicidade e Microbiologia Ambiental. Ficha de Informações Toxicológicas do zinco (2012). http://www.cetesb.sp.gov.br/userfiles/file/laboratorios/fit/zinco.pdf. Accessed on 18 June 18)

Chang Y, Zhang M, Xia L, Zhang J, Xing G (2012) The toxic effects and mechanisms of CuO and ZnO nanoparticles. Materials 5:2850–2871

Choi JS, Kim RO, Yoon S, Kim WK (2016) Developmental toxicity of zinc oxide nanoparticles to zebrafish (*Danio rerio*): a transcriptomic analysis. PLoS ONE 11:e0160763

Chun HS, Park D, Eun Lim S, Jeong KH, Park JS, Park HJ, Kang S (2017) Two zinc-aminoclays' In-vitro cytotoxicity assessment in hela cells and In-vivo embryotoxicity assay in zebrafish. Ecotoxicol Environ Saf 137:103–112

Cushen M, Kerry J, Morris M, Cruz-Romero M, Cummins E (2012) Nanotechnologies in the food industry recent developments risks and regulation. Trends Food Sci Technol 24:30–46

Doak SH, Griffiths SM, Manshian B, Singh N, Williams PM, Brown AP, Jenkins GJS (2009) Review: confounding experimental considerations in nanogenotoxicology. Mutagenesis 24:285–293

Du J, Wang S, You H, Liu Z (2016) Effects of ZnO nanoparticles on perfluorooctane sulfonate induced thyroid-disrupting on zebrafish larvae. J Environ Sci 47:153–164

Dwivedi AD, Dubey SP, Sillanpaa M, Kwon YN, Lee C, Varma RS (2015) Fate of engineered nanoparticles: implications in the environment. Coord Chem Rev 287:64–78

Enghag P (2004) Encyclopedia of the elements, 1rd, 2nd edn. Wiley-VCH, Gothenburg

Fabrega J, Luoma SN, Tyler CR, Galloway TS, Lead JR (2011) Silver nanoparticles: behaviour and effects in the aquatic environment. Environ Int 37:517–531

Fahmy SR, Sayed DA (2017) Toxicological perturbations of zinc oxide nanoparticles in the *Coelatura aegyptiaca* mussel. Toxicol Ind Health 33:564–575

Falfushynska H, Gnatyshyna L, Yurchak I, Sokolova I, Stoliar O (2015) The effects of zinc nanooxide on cellular stress responses of the freshwater mussels *Unio tumidus* are modulated by elevated temperature and organic pollutants. Aquat Toxicol 162:82–93

Fernandes Cruz JB, Freire Soares H (2011) Uma revisão sobre o zinco. Ensaios e Ciência: Ciências Biológicas, Agrárias e da Saúde 15:207–222

Franci G, Falanga A, Galdiero S, Palomba L, Rai M, Morelli G, Galdieiro M (2015) Silver nanoparticles as potential antibacterial agentes. Molecules 20:8856–8874

Furno F, Morley KS, Wong B, Sharp BL, Arnold PL, Howdle SM, Bayston R, Reid HJ (2004) Silver nanoparticles and polymeric medical devices: a new approach to prevention of infection? J Antimicrob Chemother 54:1019–1024

Galrão EZ (1995) Níveis críticos de zinco para o milho cultivado em latossolo vermelho amarelo, fase cerrado. Revista Brasileira de Ciência do Solo 19:255–260

Galrão EZ, Mesquita Filho MV (1981) Efeito de fontes de zinco na produção de matéria seca do milho em um solo sob cerrado. Revista Brasileira de Ciência do Solo 5:167–170

Ge L, Li Q, Wang M, Ouyang J, Li X, Xing MMQ (2014) Nanosilver particles in medical applications: synthesis, performance and toxicity. Int J Nanomed 9:2399–2407

Geranio L, Heuberger M, Nowack B (2009) The behavior of silver nanotextiles during washing. Environ Sci Technol 43:8113–8118

Goswami P, Borah N, Das DK (2014) Size and shape controlled synthesis of aqueous silver nanoparticles and a comparative study of their fluorescence and eletrochemical responses toward cholesterol. J Chem Pharm Res 6:697–704

Grosell M, Boeck G, Johannson O, Wood CM (1999) The effects of silver on intestinal ion and acid-base regulation in the marine teleost fish *Papophrys vertulus*. Comp Biochem Physiol C: Pharmacol Toxicol Endocrinol 124:259–270

Guzmán KAD, Taylor MR, Banfield JF (2006) Environmental risks of nanotechnology: national nanotechnology initiative funding, 2000–2004. Environ Sci Technol 40:1401–1407

Hanley HR (1933) The story of zinc. J Chem Educ Part I:600–604; Part II:682–688

He D, Bligh MW, Waite TD (2013) Effects of aggregate structure on the dissolution kinetics of citrate stabilized silver nanoparticles. Environ Sci Technol 44:2169–2175

Heinlaan M, Ivask A, Blinova I, Dubourguier HC, Kahru A (2008) Toxicity of nanosized and bulk ZnO, CuO and TiO_2 to bacteria *Vibrio fischeri* and crustaceans *Daphnia magna* and *Thamnocephalus platyurus*. Chemosphere 71:1308–1316

Henriques GS, Hirata MH, Cozzolino SMF (2003) Aspectos recentes da absorção e biodisponibilidade do zinco e suas correlações com a fisiologia da isoforma testicular da Enzima Conversora de Angiotensina. Revista de Nutrição 16:333–345

Hong R, Pan T, Qian J, Li H (2006) Synthesis and surface modification of ZnO nanoparticles. Chem Eng J 119:71–81

Hotze EM, Phenrat T, Lowry GV (2010) Nanoparticle aggregation: challenges to understanding transport and reactivity in the environment. J Environ Qual 39:1909–1924

Hou H, Takamatsu T, Koshikawa MK, Hosomi M (2005) Migration of silver, incluim antimony and bismuth and variations in their chemical fractions on addition to uncontaminated soils. Soil Sci 170:624–639

Huang HH, Ni XP, Loy GL, Chew CH, Tan KL, Loh FC (1996) Photochemical formation of silver nanoparticles in poly (n-vinylpyrrolidone). Langmuir 12:909–912

Huang Z, Jiang X, Guo D, Gu D, Gu N (2011) Controllable synthesis and biomedical applications of silver nanomaterials. J Nanosci Nanotechnol 11:9395–9408

Hvolbaek B, Janssens VTW, Clausen SB, Falsig H, Christensen HC, Norskov KJ (2007) Catalytic activity of Au nanoparticles. Nanotoday 2:14–18

Ingle A, Gade A, Pierrat S, Sonnichsen C, Rai M (2008) Mycosynthesis of silver nanoparticles using the fungus *Fusarium acuminatum* and its activity against some human pathogenic bateria. Curr Nanosci 4:141–144

Jiang C, Aiken GR, Hsu-Kim H (2015) Effects of natural organic matter properties on the dissolution kinetics of zinc oxide nanoparticles. Environ Sci Technol 49:11476–11484

Kaegi R, Voegelin A, Ort C, Innet B, Thalmann B, Krimer J, Hagedorfer H, Elumelu M, Muller E (2013) Fate and transformation of silver nanoparticles in urban wastewater systems. Water Res 47:3866–3877

Kappor S, Lawless D, Kennepohl P, Meisel D, Serperone N (1994) Reduction and aggregation of silver ions in aqueous gelatins solutions. Langmuir 10:3018–3022

Katz LM, Dewan K, Bronaugh RL (2015) Nanotechnology in cosmetic. Food Chem Toxicol 85:127–137

Khodashenas B, Ghorbani HR (2015) Synthesis of silver nanoparticles with diferentes shapes. Arab J Chem 6:1–16

Kholoud MM, Abou E, Eftaiha A, Al-warthan A, Ammar RAA (2010) Synthesis and applications of silver nanoparticles. Arab J Chem 3:135–140

Kim B, Park CS, Murayama M, Hochella MF (2010) Discovery and characterization of silver sulfide nanoparticles in final sewage sludge products. Environ Sci Technol 44:7509–7514

King JC (2011) Zinc: an essential but elusive nutrient. Am J Clin Nutr 94:679–684

Kumar A, Vemula PK, Ajayan PM, John G (2008) Silver-nanoparticle-embedded antimicrobial paints based on vegetable oil. Nat Mater 7:236–241

Kumar D, Kumar G, Das R, Agrawal V (2018) Strong larvicidal potential of silver nanoparticles (AgNPs) synthezed using *Holarrhena antidysenterica* (L) wall bark extrat against malarial vector *Anopheles slephensi* Liston. Process Saf Environ Prot 116:137–148

Kwok KWH, Auffan M, Badireddy AR, Nelson CM, Wiesner MR, Chilkoti A, Liu J, Marinakos SM, Hinton DE (2012) Uptake of silver nanoparticles and toxicity to early life stage of japanese medaka (*Oryzias latipes*) effects of coating materials. Aquat Toxicol 120–121:59–66
Laboratório de Química do Estado Sólido (Universidade Estadual de Campinas). Riscos potenciais ligados às nanotecnologias: financiamentos para pesquisa, Europa está na frente dos Estados Unidos. 2008. Disponível em: http://lqes.iqm.unicamp.br/canal_cientifico/lqes_news/lqes_news_cit/lqes_news_2008/lqes_news_novidades_1143.html. Accessed on 7 June 18
Laboratório Virtual de Química (Universidade Estadual Paulista—Campus Bauru). Tabela Periódica—Zinco (2018). Disponível em: http://www2.fc.unesp.br/lvq/LVQ_tabela/030_zinco.html. Accessed on 7 June 18
Lazareva A, Keller AA (2014) Estimating potential life cycle releases of emgineered nanomaterials from wastewater treatment plants. ACS Sustain Chem Eng 2:1656–1665
Lee JD (1991) Concise inorganic chemistry, 4th edn. Chapman & Hall, London, UK
Lee S, Chung H, Kim S, Lee I (2013) The genotoxic effect of ZnO and CuO nanoparticles on early growth of buckwheat, *Fagopyrum esculentum*. Water Air Soil Pollut 224:1668–1678
Lenzi E, Almeida VC, Favero LOB, Becker JB (2011) Detalhes da utilização do íon Hidróxido, HO^-, no tratamento de efluentes contaminados com metal pesado zinco. Acta Scientiarum. Biol Sci 33:313–322
Li Y, Kim YN, Lee EJ, Cai WP, Cho SO (2006) Synthesis of silver nanoparticles by eléctron irradiation of silver acetate. Nucl Instrum Methods Phys Res, Sect B 251:425–428
Lin D, Xing B (2008) Root uptake and phytotoxicity of ZnO nanoparticles. Environ Sci Technol 42:5580–5585
Liu JY, Hurt RH (2010) Iron release kinetics and particle persistence in aqueous nano silver colloids. Environ Sci Technol 44:2169–2175
Lombi E, Donner E, Tavakkoli E, Turney TW, Naidu R, Miller BW, Scheckel KG (2012) Fate of zinc oxide nanoparticles during anaerobic digestion of wastewater and post-treatment processing of sewage sludge. Environ Sci Technol 46:9089–9096
Lowry GV, Gregory KB, Apte SC, Lead JR (2012) Transformations of nanomaterials in the environment. Environ Sci Technol 46:6893–6899
Lv J, Zhang S, Luo L, Han W, Zhang J, Yang K, Christie P (2012) Dissolution and microstructural transformation of ZnO nanoparticles under the influence of phosphate. Environ Sci Technol 46:7215–7221
Ma H, Wallis LK, Diamond S, Li S, Canas-Carrell J, Parra A (2014) Impact of solar UV radiation on toxicity of ZnO nanoparticles through photocatalytic reactive oxygen species (ROS) generation and photo-induced dissolution. Environ Pollut 193:165–172
Ma R, Levard C, Michel FM, Brown GE Jr, Lowry GV (2013) Sulfidation mechanism for zinc oxide nanoparticles and the effect of sulfidation on their solubility. Environ Sci Technol 47:2527–2534
Majedi SM, Lee HK, Kelly BC (2013) Role of water temperature in the fate and transport of zinc oxide nanoparticles in aquatic environment. J Phys: Conf Ser 429:012039
Malavolta E, Boaretto AE, Paulino VT (1991) Micronutrientes: uma visão geral. In: Ferreira ME, Cruz MCP. Micronutrientes na agricultura, p 1–34. POTAFOS
Marassi V, Di Cristo L, Smith S, Ortelli S, Blosi M, Costa AL, Reschiglian P, Volkov Y, Prina-Mello A (2018) Silver nanoparticles as a medical device in healthcare settings: a five-step approach for candidate screening of coating agents. Roy Soc Open Sci 5(1):171113
Marcantonio AS (2005) Toxicidade do sulfato de cobre e do sulfato de zinco para rã touro, *rana catesbeiana shaw*, 1802: Toxicidade aguda e crônica e parâmetros Hematológicos. 107 f. Tese (Doutorado em Aquicultura de Águas Continentais)—Faculdade de Ciências Agrárias e Veterinárias, Universidade Estadual Paulista, Jaboticabal (2005)
Martens DC, Wastermann DT (1991) Fertilizer applications for correcting micronutrient deficiencies. In: Mortvedt JJ, Cox FR, Shuman LM, Welch RM eds. Micronutrients in agriculture. 2nd edn. Madison, Soil Science Society of America
Matsui I (2005) Nanoparticles for eletronic devices applications: a brief review. J Chem Eng Jpn 38:535–546

Mayrinck C, Raphael E, Ferrari JL, Schiavon MA (2014) Síntese, propriedades e aplicações de óxido de zinco nanoestruturado. Revista Virtual de Química 6:1185–1204
McGillicuddy P, Borah N, Das DK (2014) Size and shape controlled synthesis of aqueous silver nanoparticles and a comparative study of their fluorescence and eletrochemical responses toward cholesterol. J Chem Pharm Res 6:697–704
Medeiros MA (2012) Elemento Químico: Zinco. Química Nova na Escola. 34:159–160
Miao A, Schwehr KA, Xu C, Zhang S, Luo Z, Quigg A, Santschi PH (2009) The algal toxicity of silver engeneered nanoparticles and detoxification by exopolymeric substances. Environ Pollut 157:3034–3041
Millstone EJ, KAvulak FJD, Wo HC, Holcombe WT, Westling JE, Briseno AL, Toney MF, Frechet JMJ (2010) Synthesis, proprieties and eletronic applications of size-controlled poly(3-hexylthiophene) nanoparticles. Langmuir 16:13056–13061
Moore MN (2006) Do nanoparticles presente ecotoxicological risks for the heath of the aquatic environment? Environ Int 32:967–976
Mudasir A, Ingle A, Rai M (2013) Enhanced antimicrobial activity of silver nanoparticles synthesized by *Cryphonectria* sp evaluated singly and in combination with antibiotics. Nanomed Nanotechnol Biol Med 9:105–110
Mude N, Ingle A, Gade A, Rai M (2009) Synthesis of silver nanoparticles using callus extract of *Carica papaya*—a first report. J Plant Biochem Biotechnol 18:83–86
Mudunkotuwa IA, Rupasinghe T, Wu CM, Grassian VH (2012) Dissolution of ZnO nanoparticles at circumneutral pH: a study of size effects in the presence and absence of citric acid. Langmuir 28:396–403
Nel A, Xia T, Madler L, Li N (2006) Toxic potential of materials at the nanolevel. Science 311:622–627
Niemirowicz K, Durmas B, Tokajuk G, Piktel E, Michalak G, Gu X, Kulakowska A, Savage BP, Burcki R (2007) Sporicidal activity of ceragenin CSA-13 against *Bacillus subtilis*. Sci Rep 7:1–12
Noguez C, Garzon L (2009) Optically active metal nanoparticles. Nanotoday 2:14–18
Oberdorster G, Oberdorster E, Oberdorster J (2005) Nanotoxicology: an emerging discipline evolving from studies of ultrafine particles. Environ Health Perspect 113:823–839
Odzak N, Kistler D, Sigg L (2017) Influence of daylight on the fate of silver and zinc oxide nanoparticles in natural aquatic environments. Environ Pollut 226:1–11
Omar FM, Abdul Aziz H, Stoll S (2014) Aggregation and disaggregation of ZnO nanoparticles: influence of pH and adsorption of Suwannee River humic acid. Sci Total Environ 468–469:195–201
Park K, Tutle G, Sinche F, Harper SL (2013) Stability of citrate-capped silver nanoparticle in exposure media na their affects on the development of embryonic zebrafish (*Dario rerio*). Arch Pharmacal Res 36:125–133
Parks GB (1927) The book of Ser Marco Polo, the Venetian, concerning the Kingdoms and Marvels of the East, Book 1, p 48, Macmillan Company
Paschoalino MP, Marcone GPS, Jardim WF (2010) Os nanomateriais e a questão ambiental. Quim Nova 33:421–430
Peng YH, Tsai YC, Hsiung CE, Lin YH, Shih YH (2017) Influence of water chemistry on the environmental behaviors of commercial ZnO nanoparticles in various water and wastewater samples. J Hazard Mater 322:348–356
Philippe A, Schaumann GE (2014) Interactions of dissolved organic matter with natural and engineered inorganic colloids: a review. Environ Sci Technol 48:8946–8962
Polesel F, Farkas J, Kjos M, Almeida CP, Flores-Alsina X, Gernaey KV, Hansen SF, Plósz BG, Booth AM (2018) Occurrence, characterization and fate of (nano) particulate Ti and Ag in two norwegian wastewater treatment plants. Water Res 141:19–31
Pollard AM, Heron C, Armitage RA (2016) Archaeological chemistry, 3rd edn. Royal Society of Chemistry (RSC)
Preedy VR, Watson RR, Martin CR (2011) Handbook of behavior, food and nutrition, 1st edn. Springer
Purcell TW, Peters JJ (1998) Sources of silver in environment. Environ Toxicol Chem 17:539–546

Quik JT, Velzeboer I, Wouterse M, Koelmans AA, van de Meent D (2014) Heteroaggregation and sedimentation rates for nanomaterials in natural waters. Water Res 48:269–279
Rai M, Deshmukh SD, Ingle AP, Gade AK (2012) Silver nanoparticles: the powerful nanoweapon against multidrug-resitant bacteria. J Appl Microbiol 12:841–852
Rai M, Ingle A (2012) Role of nanotechnology in agriculture with special reference to management of insect pest. Appl Microbiol Biotechnol 94:287–293
Rai M, Ingle A, Medici S (2018) Biomedical applications of metals, 1st edn. Springer
Rai M, Yadao A, Gade A (2009) Silver nanoparticles as a new generation of antimicrobials. Biotechnol Adv 27:76–83
Rajput VD, Minkina TM, Behal A, Sushkova SN, Mandzhieva S, Singh R, Gorovtsov A, Tsitsuashvili VS, Purvis WO, Ghazaryan KA, Movsesyan HS (2018) Effects of zinc-oxide nanoparticles on soil, plants, animals and soil organisms: a review. Environ Nanotechnol Monit Manage 9:76–84
Raskar SV, Laware SL (2014) Effect of zinc oxide nanoparticles on cytology and seed germination in onion. Int J Curr Microbiol Appl Sci 3:467–473
Ratte HT (1999) Bioacumulation and toxicity of silver compounds: a review. Environ Toxicol Chem 18:89–108
Reidy B, Haase A, Luch A, Dawson KA, Lynch I (2013) Mechanisms of silver nanoparticles release transformation and toxicity: a critical review of current knowledge and recommendations for future studies and applications. Materials 6:2295–2350
História do zinco (2018). https://www.rheinzink.pt/qualidade/historia-do-zinco/. Accessed on 7 June 18)
Rodrígues-Sánchez L, Blanco MC, López-Quintela MA (2000) Eletrochemical synthesis of silver nanoparticles. J Phys Chem B 104:9683–9688
Roe D, Karandikar B, Bonn-Savage Gibbins B, Roullet J-B (2008) Antimicrobial surface funcionalization of plastic catheters by silver nanoparticles. J Antimicrob Chemother 61:869–876
Russel A, Hugo WB (1994) Antimicrobial activity and action of silver. Prog Med Chem 31:351–370
Schultz C, Powell K, Crossley A, Jurkschat K, Kille P, Morgan AJ, Read D, Tyne W, Lahive E, Svendsen C, Spurgeon DJ (2015) Analytical approaches to support current understanding of exposure, uptake and distributions of engineered nanoparticles by aquatic and terrestrial organisms. Ecotoxicology 24:239–261
Sendra M, Yeste MP, Gatica JM, Moreno-Garrido I, Blasco J (2017) Homoagglomeration and heteroagglomeration of TiO_2, in nanoparticle and bulk form, onto freshwater and marine microalgae. Sci Total Environ 592:403–411
Shah M, Fawcett D, Sharma S, Tripathy SK, Poinern GEJ (2015) Green synthesis of metallic nanoparticles via biological entities. Materials 8:7278–7308
Shriver DF, Atkins PW (2008) Química Inorgânica. 4th edn. Bookman
Souza GD, Rodrigues MA, Silva PP, Guerra W (2013) Prata: Breve hitórico, propriedades e aplicações. Educación Química 24:14–16
Suman TY, RadhikaRajasree SR, Kirubagaran R (2015) Evaluation of zinc oxide nanoparticles toxicity on marine algae chlorella vulgaris through flow cytometric, cytotoxicity and oxidative stress analysis. Ecotoxicol Environ Saf 113:23–30
Tapiero H, Tew KD (2003) Trace elements in human physiology and pathology: zinc and metallothioneins. Biomed Pharmacother 57:399–411
Tappin A, Barriada JL, Braungardt CB, Evans EH, Patey MD, Achterberg EP (2010) Dissolved silver in European estuarine and coastal Waters. Water Res 44:4204–4216
Tartaj P, Morales MP, González-Carreni T, Veintemillas-Verdaguer S, Serna CJ (2005) Advances in magnetic nanoparticles for biotechnology applications. J Magn Magn Mater 290:28–34
Thakur RS, Agrawal R (2015) Application of nanotechnology in pharmaceutical formulation design and development. Current Drug Therapy 15:20–34
The Nanodatabase [WWW document] (2018). http://nanodb.dk/en/. Accessed on 12 May 18)
Troester M, Brauch HJ, Hofmann T (2016) Vulnerability of drinking water supplies to engineered nanoparticles. Water Res 96:255–279

Vale G, Mehennaoui K, Cambier S, Libralato G, Jomini S, Domingos RF (2016) Manufactured nanoparticles in the aquatic environment-biochemical responses on freshwater organisms: a critical overview. Aquat Toxicol 170:162–174

Valee BL, Falchuk KH (1993) The biochemical basis of zinc physiology. Physiol Rev 73:79–111

Windler L, Height M, Nowack B (1998) Comparative evaluation of antimicrobials for textile applications. Environ Int 53:62–73

World Silver Surver, Silver Institute (2018). http://www.silverinstitute.org. Accessed on 10 May 18

Yang K, Lin D, Xing B (2009) Interactions of humic acid with nanosized inorganic oxides. Langmuir 25:3571–3576

Yeasmin S, Datta HK, Chaudhuri S, Malik D, Bandyopadhyay A (2017) In vitro anticancer activity of shape controlled silver nanoparticles (AgNPs) in various organ specific cell lines. J Mol Liq 242:757–766

Yin B, Ma H, Wang S, Chen S (2003) Electrochemical synthesis of silver nanoparticles under protection of poly(N-vinylpyrrolidone). J Phys Chem B 107:8898–8904

Young SY, Lee HT, Jeong SH (2003) Preparation of nanocomposite fibers for permanent antibacterial effect. J Mater Sci 38:2143–2147

Yu SJ, Yin YG, Chao JB, Shen MH, Liu JF (2014) Highly dynamics PVP-coated silver nanoparticles in aquatic environments: Chemical and morphology change induced by oxidation of Ag(0) and reduction of Ag(+). Environ Sci Technol 48:403–411

Yung MM, Wong SW, Kwok KW, Liu FZ, Leung YH, Chan WT, Li XY, Djurisic AB, Leung KM (2015) Salinity-dependent toxicities of zinc oxide nanoparticles to the marine diatom *Thalassiosira pseudonana*. Aquat Toxicol 165:31–40

Yung MMN, Fougères PA, Leung YH (2017) Physicochemical characteristics and toxicity of surface-modified zinc oxide nanoparticles to freshwater and marine microalgae. Sci Rep 7:15909

Zhang C, Hu Z, Deng B (2016) Silver nanoparticles in aquatic environments: physiochemical behavior and antimicrobial mechanisms. Water Res 88:403–427

Zhao X, Ren X, Zhu R, Luo Z, Ren B (2016) Zinc oxide nanoparticles induce oxidative DNA damage and ros-triggered mitochondria-mediated apoptosis in zebrafish embryos. Aquat Toxicol 180:56–70

Zhou D, Keller AA (2010) Role of morphology in the aggregation kinetics of ZnO nanoparticles. Water Res 44:2948–2956

Top Countries for Zinc Production (2018). https://investingnews.com/daily/resource-investing/base-metals-investing/zinc-investing/top-zinc-producing-countries/. Accessed on 7 June 18

Chapter 10
A Review on Ecotoxicity of Zinc Oxide Nanoparticles on Freshwater Algae

M. Bhuvaneshwari, V. Iswarya, N. Chandrasekaran and Amitava Mukherjee

Abstract The steady increase in use of ZnO nanoparticles (NPs) based consumer products possess high risk towards aquatic flora and fauna. Understanding the toxicity mechanism of ZnO NPs at an early stage is essential to develop a proper scientific design in innovation process to confine the exposure of nanoparticles at safer levels. Being the primary producers of aquatic food chain, various algal species were used as model organisms in several toxicological studies. This chapter summarizes the aspects that include environmental release, behaviour, dissolution, particle size, morphology, irradiation, reactive oxygen species generation, penetration/internalization of ZnO NPs on various algal species. Dissolved Zn ion toxicity was the major toxicity mechanism exhibited by ZnO NPs towards most of the algal species tested. Additionally, photocatalytic activity of ZnO NPs under UV irradiation also significantly increased the toxicity. To conclude, knowledge gaps and future research needs on the usage, discharge and its eco-toxicological effects of ZnO NPs were addressed.

Keywords ZnO NPs · Algae · Toxicity · Dissolution · Reactive oxygen species

Nomenclature

NPs Nanoparticles
ZnO Zinc Oxide
SEM Scanning Electron Microscopy

M. Bhuvaneshwari · V. Iswarya · N. Chandrasekaran · A. Mukherjee (✉)
Centre for Nanobiotechnology, Vellore Institute of Technology, Vellore 632014, India
e-mail: amit.mookerjea@gmail.com; amitav@vit.ac.in

M. Bhuvaneshwari
e-mail: bhuvimani021@gmail.com

V. Iswarya
e-mail: v.iswaryavelu@gmail.com

N. Chandrasekaran
e-mail: nchandrasekaran@vit.ac.in

M. Rai and J. K. Biswas (eds.), *Nanomaterials: Ecotoxicity, Safety, and Public Perception*, https://doi.org/10.1007/978-3-030-05144-0_10

TEM Transmission Electron Microscopy
ROS Reactive Oxygen Species
LDH Lactate Dehydrogenase
SOD Superoxide Dismutase
GSH Reduced Glutathione
UV Ultra Violet
EC_{50} Effective Concentration

10.1 Introduction

The field of nanotechnology has progressively expanded in the past few decades particularly in the development of nanoparticle (NPs) based consumer products as it exhibits distinct size related physical and chemical properties (Murray et al. 2000). The properties of nano-sized particles are entirely different from its bulk material forms owing to the increased surface area to volume ratio, and hence these NPs are now widely been accepted in numerous commercial usage (Khan et al. 2017). Production of engineered nanoparticles (ENPs) were observed to be from few hundreds to 1500 different types during the year 2005–2015 (Walters et al. 2016). It is estimated that 30-fold increase in NPs incorporated products were available between 2011 and 2015, and its global marketplace value was over 1 trillion dollars in year 2015 (Toensmeier 2004). Among the most commonly exploited NPs, metal and metal oxide ENPs, such as zinc oxide (ZnO NPs), silver (Ag NPs), titanium dioxide (TiO_2 NPs), copper oxide (CuO NPs), and cerium oxide (CeO_2 NPs) nanoparticles are being largely amalgamated in consumer products, where its sole optoelectronic, photocatalytic activity, antimicrobial potential and other distinct properties accomplish them as an expansive scope of utilizations.

Despite the manifold applications and usage of metal and metal oxide NPs in various fields, the ecological risk posed by those NPs and its interaction with aquatic organisms is still the prevailing fact (Snyder-Talkington et al. 2012). As stated by RSRAE Nanoscience and Nanotechnologies, the rapid advancement of nanotechnology industries and its broad applications will inevitably increase the exposure of these NPs in aquatic environment (RSRAE 2004). Any contaminant/toxin above its permissible limit can cause detrimental effects to aquatic as well as terrestrial ecosystem. Several studies have explored the eco-toxicity of NPs against numerous aquatic organisms, which mainly include the primary producer algae (Aravantinou et al. 2015, 2017; Aruoja et al. 2015; Bhuvaneshwari et al. 2015; Suman et al. 2015). Considering all these evidences, it is always worthwhile to examine the toxicological implication of environmentally available NPs in an aquatic ecosystem. In this chapter, the authors aim to discuss the environmental release, behaviour, dissolution and toxicity mechanisms of ZnO nanoparticles of particular concern against various algal isolates.

10.2 Zinc Oxide Nanoparticles (ZnO NPs) and Their Utilization

ZnO is the group II–VI semiconductor type material. It exhibits several unique properties such as wide band gap energy, higher conductivity, good transparency, high electron mobility, and strong optical and luminescence properties at room temperature. ZnO particles has wurtzite crystal phase and hexagonal cell structure with *a* and *c* lattice constraints. The production rate of ZnO NPs was reported to be around 550–5500 tones year^{-1} (Piccinno et al. 2012), which is almost 10–100 times greater than any other nanoparticles produced annually. ZnO NPs were broadly used in numerous industrial constituents as a photo-catalyst (Deng et al. 2008), in pharmaceutical and cosmetic industries (Lu et al. 2015), inhibition of corrosion (Sapkal et al. 2012), in UV light-emitting devices (Izu et al. 2014), as semiconductor in electronic devices (Tsukazaki et al. 2005), and in food packing industries (Krishnakumar et al. 2009). In cosmetic industries, ZnO NPs are mainly used for its high photocatalytic (mostly in sunscreens) activity. ZnO NPs capability of mildew resistance, corrosion inhibition, and its unique stain blocking support makes it an ideal particle to be used in paints. Owing to their high antimicrobial activity, they are used in ointments. ZnO NPs are used in rubber industries for tire manufacturing as they promote the process of vulcanization. Apart from this, ZnO NPs are used to detect, prevent and remove environmental pollutants (Lee et al. 2005) by designing and creating environmentally friendly products (Kamat and Meisel 2002, 2003). In addition to the inherent properties and applications of ZnO NPs, other physicochemical assets like particle size, morphology, and surface charges of ZnO NPs are the prominent factors that contribute to toxicity (Wong et al. 2010).

10.3 Environmental Release and Behaviour of ZnO NPs in the Aquatic System

ZnO NPs are reported to be released into the aquatic environment through unintentional disposal from the products incorporated with nanoparticles (Klaine et al. 2008; Peng et al. 2017). However, ZnO NPs reaching the aquatic system through the consumer products used in our day to day life were reported to be below 1 mg L^{-1} (Gottschalk et al. 2009). When the nanoparticles enter into the aquatic system, their stability, mobility, reactivity and the physicochemical characteristics differ dynamically (Fig. 10.1). Released NPs can undergo physical (size and shape) and chemical modifications upon interactions with the natural colloids (natural organic matter), exposure to sunlight, changes in pH, ionic strength, electrolyte concentration, inorganic salts (Jahan et al. 2017) and other pollutants (soluble salts, heavy metals, pesticides, etc.) present in the system. The behaviour of ZnO NPs in aquatic system includes aggregation, flocculation, redox reactions, speciation, dissolution (release of Zn^{2+} ions), surface modifications (interactions with other metal and pollutants),

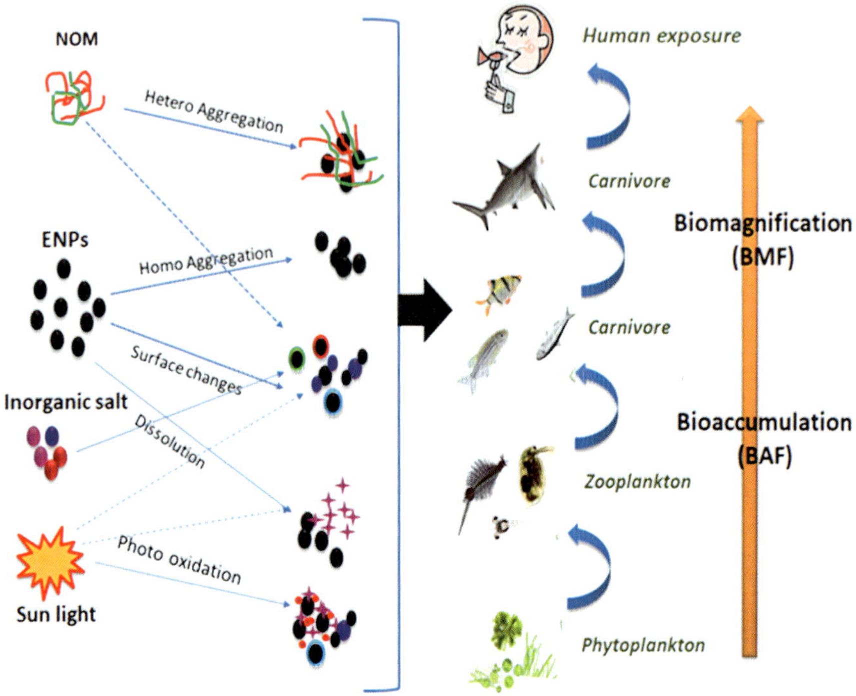

Fig. 10.1 Schematic representation shows the possible routes of exposure pathways of ZnO NPs and its environmental transformation processes in relevant to aquatic organisms in food chain

and complexation with humid acids and natural organic matter (NOM). As a result, these processes end up in sedimentation and deposition onto solid phases of aquatic system. On the other hand, Keller et al. (2010) reported that the presence of NOM (humic acid) stabilizes the nanoparticles in natural system and prevents aggregation. A consequential change in the stability and chemistry of the ZnO NPs will have an impact on their reactivity towards the biological components. Photochemical transformation of the NPs upon exposure to light may become photo reactive and can generate free radicals. Adsorption and surface modification occurs when the nanoparticles interacts with solid surfaces by Van der Waals, electrostatic, or chemical interactions (Dąbrowski 2001; Pan and Xing 2008; Rabe et al. 2011). Also, the nature of experimental medium plays a significant role in the stability of NPs, and thus, the aggregation could likely reduce their bioavailability (Dabrunz et al. 2011).

10.4 Importance of Algae in the Aquatic System

Algae is a diverse group of photosynthetic organisms having the capability to convert carbon dioxide and water into sugar in presence of sunlight, and in turn serve as food for all organisms present in aquatic system, directly or indirectly. Microalgae are one of the primary organisms of aquatic food webs which predominantly involved in the purification of water (Wang et al. 2009; Ji et al. 2011). It has the ability to accumulate pollutants present in aquatic system, and hence they are called bio-indicators of aquatic environment (Barhoumi and Dewez 2013). Water quality parameters and growth of algal cells can be directly quantified by measuring the amount of photosynthetic pigment present. Many studies have been carried out on algal cells to assess the impact of various pollutants present in the aquatic system. Algal cells are well appropriate to study such problems for their small size (10 μm), geometric regularity, ease of use, and easy culturing in laboratory conditions.

10.5 Ecotoxicity of ZnO NPs on Freshwater Algae

Former studies have reported the ecotoxicity of ZnO NPs on various fresh water and marine algal species. Nano-ZnO, bulk ZnO particles and $ZnSO_4$ salts showed an incomparable toxicity to the fresh water algae *P. subcapitata* with 72 h EC_{50} of 0.042, 0.037, and 0.042 mg/L, respectively (Aruoja et al. 2009). Nano ZnO (dispersant) showed higher toxicity on algae with 72 h EC_{50} of 0.049 and 0.044 mgL^{-1} for total (both ZnO particle and dissolved Zn^{2+} ions in suspension) and ionic effect (only dissolved Zn^{2+} ion from ZnO NPs) as compared to its bulk form (Franklin et al. 2007). Also, nano ZnO (powder) showed similar 72 h EC_{50} (0.068 mgL^{-1}) for total and ionic effect as of nano dispersant. Likewise, $ZnCl_2$ and $ZnCl_2$ with surfactant showed equal EC_{50} value of 0.060 mgL^{-1} (total and ionic). Ji et al. (2011) stated the order of Zn^{2+} > nano-ZnO > bulk-ZnO toxicity towards freshwater algae *Chlorella* sp. at the exposure concentration of <50 mgL^{-1}. Once the NPs concentration increased beyond 50 mgL^{-1}, the nanoscale ZnO particles showed higher toxicity than Zn^{2+} ions. On other hand, Gunawan et al. (2013) described that *C. reinhardtii* microalgae was found to be more sensitive to ZnO NPs at a very low concentrations (8-day $EC_{50} \geq 0.01$ mgL^{-1}). Dissolution of ZnO effectively inhibited the marine algae *D. tertiolecta* with the EC_{50} of about 1.94, 3.57 and 0.65 mgL^{-1} for nano ZnO, bulk ZnO and $ZnCl_2$ respectively (Manzo et al. 2013). Wong et al. (2010) reported 96 h LC_{50} of 4.56, 6.55 mgL^{-1} for *T. pseudonana* and 2.36, 2.97 mgL^{-1} for *S. costatum* upon exposure to nZnO and bulk ZnO particles. Nano ZnO was reported to be more toxic than bulk ZnO particles, nZnO effect could predominantly accredited by dissolved Zn^{2+} ions. Miller et al. (2010) reported the toxicity of ZnO NPs on *T. pseudonana, D. tertiolecta, S. marinoi,* and *I. galbana*, mentioning that the dissolution, and uptake of zinc ions apparently inhibited the algal growth. Miao et al. (2010) studied the Fv/Fm, growth rate and Chlorophyll a content upon exposure to ZnO NPs on *T. pseudonana*. The release of Zn^{2+} ions effectively caused growth inhibition and decrease in chlorophyll a content. These studies clearly reveals that the dissolution of Zn^{2+} from ZnO NPs could be the major cause for its toxicity.

A few researchers (Chen et al. 2012; Lee and An 2013) have observed destabilization of the cell membrane caused growth inhibition of algae cells linear to the concentration of ZnO NPs and dissolved Zn^{2+} ions used. The aggregation and adsorption of ZnO NPs compromised the growth, morphology, and membrane integrity of *Chlorella* sp. (Chang et al. 2012). They suggested that the dissolution of ZnO NPs is the possible reason behind the mechanical cell damage induced by the NPs.

ZnO nanoparticles toxicity was observed to be strongly dependent on their particle morphology (Gurr et al. 2005) and stability in the experimental medium (Panessa-Warren et al. 2009). Peng et al. (2011) studied effect of ZnO spheres of 6.3 nm (small spherical) and 15.7 nm (large spherical) against diatoms, and determined that the small sized spherical particles exhibited higher toxicity. Merdzan et al. (2014) reported different coatings such as bare ZnO NPs (20 nm), ZnO PAA (<20 nm) and nZnO HMP (40 nm) on *C. reinhardtii*. They observed enhanced dissolution, toxicity, and accumulation for bare ZnO NPs compared to that of surface modified ZnO NPs. Additional studies reported that the other mechanistic factors like reactive oxygen species generation, imbalance in antioxidant enzymes (SOD, GSH, GST and CAT) were also involved in the toxicity of ZnO NPs on algal cells. Suman et al. (2015) reported that the antioxidant enzyme, SOD (superoxide dismutase) increased, and other antioxidants decreased upon exposure of ZnO NPs to algae *C. vulgaris* with concentration-dependent increase in toxicity. Aravantinou et al. (2015, 2017) reported that the ZnO NPs toxicity strongly depended on type of algal species used, concentration of NPs, the growth media composition, and interaction period.

10.6 Mechanism of ZnO NPs Toxicity on Algal Cells

There are various possible routes of ZnO NPs interaction with algal cells, which includes penetration, endocytosis, membrane rupture, adsorption and internalization (Fig. 10.2). Further, the Zn ions permeate the membrane through ion channels.

(I) ***Penetration***: Algal cell walls are semipermeable with small pores of size less than 5–20 nm which allows the passage of particles in size less than 20 nm into cell wall. Furthermore, exposure of ZnO NPs damages the cell wall thereby it can internalize the particles by increasing its pore size. Once NPs get percolated into cell wall, they can easily cross the cell membrane and make them available to interact with different organelles.

(II) ***Endocytosis***: Once the NPs enter into cell wall, it can easily reach the plasma membrane by endocytosis progressions. The lipid bilayer of the plasma membrane produce cavity-like assembly that engulfs the NPs into cytoplasm. There are various endocytic pathways that were anticipated for NPs assimila-

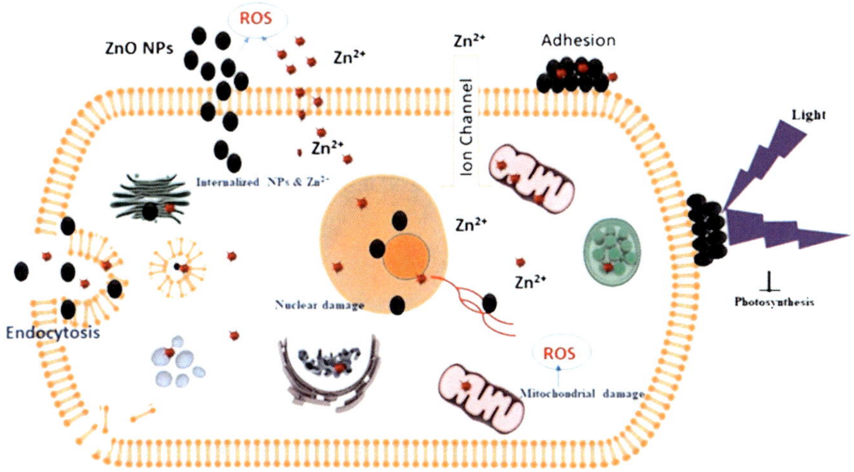

Fig. 10.2 Possible routes of ZnO NPs exposure on algae

tion into cells: clathrin-mediated endocytosis, caveolae-mediated endocytosis, macropinocytosis and phagocytosis (Scown et al. 2010; Behra et al. 2013).

(III) ***Membrane rupture***: Interaction of NPs can damage the algal cell wall, thereby it could easily enter/internalize into the cell membrane and cytoplasm. Once the cell membrane is damaged, internalization of NPs occurs by increasing the permeability of membrane through plasmolysis. Song et al. (2010) reported that the elevated levels of lactate dehydrogenase (LDH) enzyme release reflects the compromised algal cell exposed to ZnO NPs. Localization and distortion of outer cell wall of fresh water algae *P. subcapitata*, indicates that ZnO NPs decline the stability of algal cell membrane (Lee and An 2013). These particles are also available to daughter cells when the NPs internalized parent algal cells undergoes cell division.

(IV) ***Adsorption***: Aggregates of ZnO NPs adhere to algal surface and decrease the light availability to undergo normal photosynthesis process, thereby limiting the biomass and inhibiting the algal growth. Entrapment of NPs on algal cells causes cell membrane damage, as a result of which leakage of chlorophyll content from cells occurs. The adsorption of agglomerates of CuO particles on *Chlorella vulgaris* was observed by Wu et al. (2017). Surface charge based interaction between cell-NPs is also a key point to be addressed in any toxicity assessment. Negatively-charged algal cells could electrostatically interact positively-charged ZnO NPs, and facilitate NPs absorption (Bhuvaneshwari et al. 2015). A recent study by Chen et al. (2012) and Ji et al. (2011) confirmed the adsorption of NPs on algal surface (*Chlorella* sp.) using SEM micrographs. Few active binding sites (polysaccharides, cellulose and glycoproteins) present on cell wall facilitates the adsorption of NPs (Chen et al. 2012), and thereby could disrupt the cellular metabolic pathways of algae. Similarly, Hazeem et al.

(2016) reported that ZnO NPs are found to have adverse impact on algal growth (*Picochlorum* sp.) and chlorophyll concentration. Reduction in expression of genes that are responsible for chlorophyll synthesis was observed in plants exposed to ZnO NPs (Wang et al. 2016). Bhuvaneshwari et al. (2015) reported the internalization and ultra-structural damages caused by ZnO NPs on algal cells under different irradiation condition (Fig. 10.3).

10.7 Factors Influencing the Toxicity of ZnO NPs

10.7.1 Dissolved Zn^{2+} Ion Effect on Algae

ZnO NPs suspension consists both the ZnO particles and Zn^{2+} ions, and the toxicity exerted by that suspension includes ionic, particles or the total effect of both. Available studies on ecotoxicity of ZnO NPs suggest several mechanisms or modes of action on algal cells. Among those, majority of the reports have established that the release of Zn^{2+} mainly contributes on toxicity towards algae (Gunawan et al. 2013; Lee and An 2013; Manzo et al. 2013). It is well known that the ZnO NPs have a tendency to leach Zn^{2+} ions in the experimental medium (Blinova et al. 2010; Bacchetta et al. 2016; Wang et al. 2016), and several prior studies confirmed the toxicity of free Zn^{2+} ions. However, few researchers highlighted that the toxic effects are not specially ascribed to release of Zn ions (Chen et al. 2012; Suman et al. 2015). In contrast to Zhou et al. (2014) and Hazeem et al. (2016) stating the particle effect, few reports emphasized that both the particles and Zn^{2+} ion are probable to produce toxicity of ZnO NPs (Miller et al. 2010; Ji et al. 2011). It is still quiet controversial whether the $NP_{(ion)}$ or $NP_{(particle)}$ of the ZnO NPs would be the key basis of toxicity.

In most of the toxicity studies discussed, the effects of Zn ions were measured using metal salts such as $ZnCl_2$ and $ZnSO_4$ which could result in Zn ions. The toxicity utilised by $ZnCl_2$ and $ZnSO_4$ metal salts is relatively diverse from the toxicity induced by dissolved Zn^{2+} ion and ZnO NPs, due to the release of several anionic metal ions such as Cl^- and SO_4^{2-} along with the Zn ions. To avoid this misconception, in our recent reports (Bhuvaneshwari et al. 2015, 2016) we have separated the dissolved Zn^{2+} ions from ZnO nanoparticle suspension by ultracentrifugation (13,000 rpm) and filtration (0.1 μm followed by 3 kDa filter) method. Figure 10.4 represents Zn ions obtained from different ZnO NPs (50 nm, 100 nm and bulk particles) in natural lake water medium. Those separated dissolved Zn ions were further exposed to fresh water algal cells *S. obliquus* (Bhuvaneshwari et al. 2015) in natural experimental medium, and found that Zn ions can induce significant level of toxicity towards algae.

Released Zn ion can easily penetrate into the algal cells through small pores present in algal cell wall. Also, Zn ions can be carried into the cell through ion channels by several ZIP proteins (Zinc transporter proteins). Released Zn ions could disrupt the zinc homeostasis process which causes lysosomal and mitochondria damage and

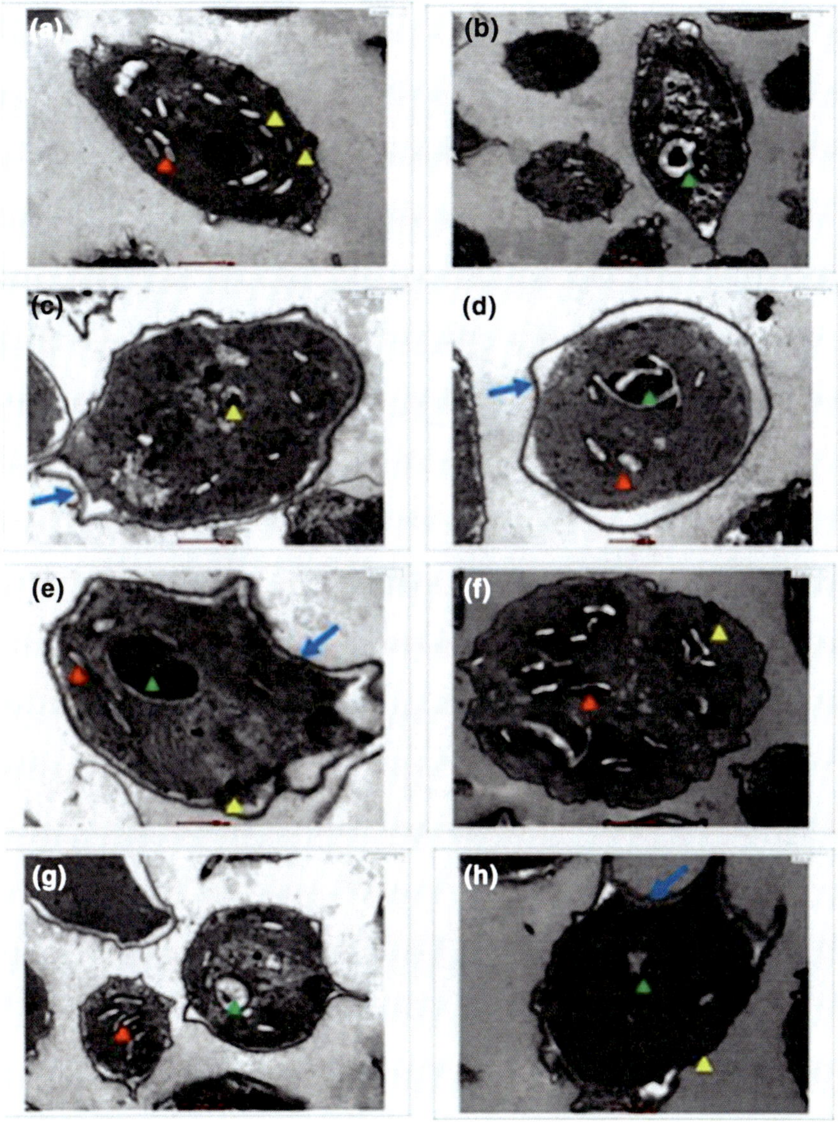

Fig. 10.3 Transmission electron micrographs of algal cells **a** ZnO NPs exposed cells to UV irradiation shows internalization of NPs into algal cell wall, excess of phosphate granule deposition; **b** and **c** lipid vesicle formation due to nanoparticle stress; **d** and **e** shrinkage, NPs aggregates, degradation of nucleus and chlorophyll (**f** and **g**); starch–pyrenoid complex formation and damaged cell membrane (**a**–**h**). *Note* Yellow arrow—NPs internalization; green arrow—phosphate granule; blue arrow—lipid vesicles; blue arrow—membrane damages

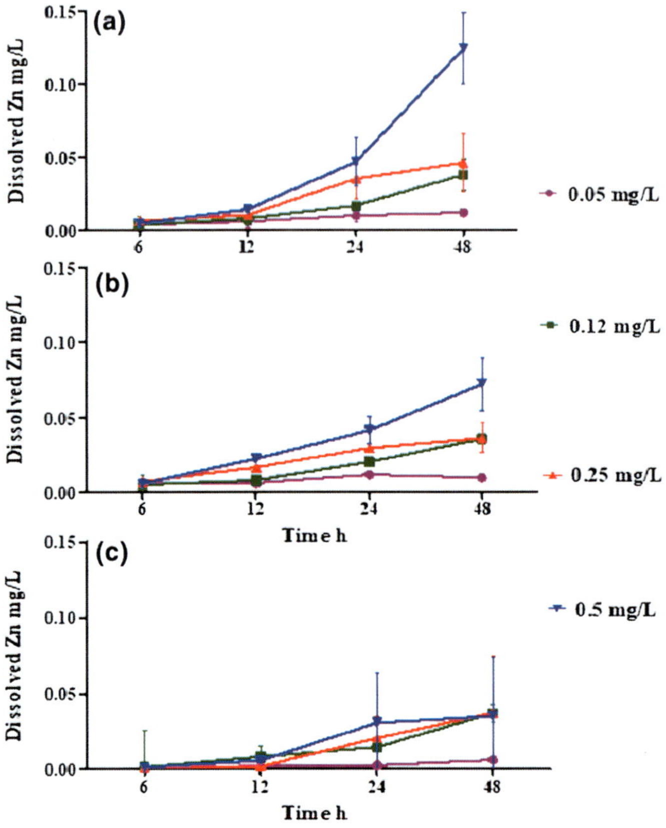

Fig. 10.4 Dissolution of different sizes of ZnO NPs **a** 50 nm, **b** 100 nm and, **c** bulk particles

finally necrosis of algal cells (Xia et al. 2008). But, still the established toxicity pathways are not quite enough to decipher the dissolved Zn ions toxicity to freshwater species (Miller et al. 2010). There are always various physicochemical factors like pH, stability, and the aggregation state of nanoparticles that would also greatly affect the dissolution of particles which may exacerbate its toxicity.

10.7.2 Effect of Particle Size and Morphology

Nanoparticle size is one of the major factors involved in toxicity. Lesser the particle size higher will be the surface area to volume ratio and consequent reactivity of nanoparticles. Even the particles with the same composition can exhibit diverse cytotoxicity index respect to their particle size and surface area. Smaller sized nanoparticles with the enhanced surface area could easily pass through membrane, and get accumulated into algal cells (Bystrzejewska-Piotrowska et al. 2009), whereas aggregated particles are less available to the cells. Particle size dependent toxicity (nano ZnO > bulk ZnO) on algal cells was reported by several authors namely Wong et al. (2010), Manzo et al. (2013), and Bhuvaneshwari et al. (2015) and found that the particle size plays a significant role in the induction of cytotoxicity by ZnO NPs. In contrast, few studies (Franklin et al. 2007; Aruoja et al. 2009; Ji et al. 2011) have reported that both nano and bulk form of ZnO NPs exhibited similar response upon exposure to algal cells. Merdzan et al. (2014) reported that the particle dissolution, metal bio-uptake and toxicity were influenced by the presence of different stabilizers. Peng et al. (2011) observed that rod-shaped ZnO NPs are more toxic to marine diatom as compared to rods and sphere-shaped particles. Similarly, recent study by Zhang et al. (2018) stated that the sensitivity of algal cells *(S. costatum > D. salina > P. tricomutum > P. subcordiforus)* exposed to different morphology and surface area of ZnO NPs. Different morphology of particles such as spheres, rods, cubes, triangles, wires, and coating affects the nanoparticle reactivity and their means of transport in the environment.

10.7.3 Effect of Irradiation

Few studies have reported that the photocatalytic activity of ZnO NPs under UV-C, visible light (Bhuvaneshwari et al. 2015), UV-A, UV-B and visible light (Lee and An 2013) as well as UV-A and UV-B irradiation (Kim and An 2012) and its toxicity towards algal cells. Enhanced toxicity of UV irradiated ZnO NPs was compared to natural sunlight and other laboratory light condition, which suggests the photo toxic nature of ZnO NPs. Photo induced toxicity of ZnO NPs exposed under UV-C and visible light condition was observed against *S. obliquus* at environmentally available exposure concentrations of $<1\ mgL^{-1}$ (Bhuvaneshwari et al. 2015). In our study, we found that the photocatalytic ROS was the sole reason for enhanced toxicity under UV irradiation. Photodecomposition of ZnO NPs under UV-C irradiation compared to visible light and dark condition resulted in additional dissolution. The enhanced dissolution of ZnO NPs under UV-C condition is known to be a photo-dissolution effect. Similar to these findings, Lee and An (2013) as well as Kim and An (2012) reported a substantial ionic dissolution from ZnO NPs at different exposure conditions (UVA, UVB and visible light). The photo enhanced dissolution reactions of ZnO NPs are explained in the equations below (Han et al. 2010).

$$\text{Photocatalysis: } ZnO(s) + h\nu \rightarrow ZnO(s) + h^{+} + e^{-} \ldots . \quad (10.1)$$

$$\text{Re-combination: } h^{+} + e^{-} \rightarrow h\nu \ldots . \quad (10.2)$$

$$ZnO(s) + 2h^{+} \rightarrow Zn^{2+} + 1/2\, O_2\ (\text{ZnO photo-dissolution}) \ldots . \quad (10.3)$$

Where, ZnO(s) is NPs, hν is a photon, h^+ and e^- are photo-generated hole and electron on the ZnO NPs surface, and Zn^{2+} (aq.) is a Zn^{2+} ions in aqueous solution.

10.7.4 Reactive Oxygen Species Generation

Exposure of ZnO NPs under UV radiation increases phototoxic effect by producing reactive oxygen species (ROS) such as hydrogen peroxide, singlet oxygen, hydroxyl radicals and superoxide ions (Gunawan et al. 2013; Zhou et al. 2014). ZnO nanoparticles (NPs) are a semiconductor-type particle with a band gap energy of 3.37 eV, at wavelength of 368 nm. Incident photons of energy higher than 3.37 eV are immediately absorbed by ZnO particles, and electrons states move from the valence band to conduction band. Such electron transfer can initiate a sequence of photoreactions. Positive holes (h^+) are direct oxidant, and form reactive hydroxyl radicals ($OH^•$) which are a foremost oxidant in the photocatalytic system (Ahmed et al. 2010). The nano sized particles with enhanced specific quantum yield generate higher ROS (Tseng et al. 2006). Internalized nanoparticles perhaps produce ROS through mitochondrial damage, as a consequence arrested growth, delayed metamorphosis and DNA damage occurs. The generation of ROS results in the degradation of cell membrane by LDH release (Pathakoti et al. 2013; Rajavel et al. 2014) and leakage of cellular contents (Nel et al. 2006; Xia et al. 2008). A direct relationship between NPs-induced ROS production and growth inhibition is further confirmed by the significant increase in intracellular ROS levels within algae upon exposure to ZnO NPs under solar UV radiation, and it suggests that their phototoxicity is associated with oxidative stress.

10.8 Conclusion and Research Needs

Release of NPs into aquatic system creates a disputation whether it could pose impairment to ecological food web. This literature search provides details on environmental release, behaviour, dissolution and toxicity mechanisms of ZnO nanoparticles against various algal isolates. The intrinsic toxicity mechanisms of ZnO NPs were observed to be dissolution (release of Zn ions), ROS generation, and internalization of NPs. Instead of using metal salts to assess the toxicty of Zn ion effects, Zn ions released from ZnO NPs can be utilized as a direct source of Zn ion in any toxicity assessment. Consequently, these studies demonstrate algae as an imperative route of ZnO uptake in food chain transfer. The discharge of nano-metals to an aquatic system could be taken up by organisms from both water (waterborne exposure) and dietary (algae to daphnia) sources, which is one of the primary concerns in conserving well-proportioned ecological niche. Future research should mainly focus on trophic toxicity of various NPs on ecosystem. Hence, it is, therefore, essential to improve the appropriate systematic design which will restrict the human exposure to lethal engineered nanoparticles in the environment to safe levels. Risk assessment is the fundamental origin to safeguard human health and the environment concerning nanotechnology.

References

Ahmed S, Rasul MG, Martens WN, Brown R, Hashib MA (2010) Heterogeneous photocatalytic degradation of phenols in wastewater: a review on current status and developments. Desalination 261:3–18

Aravantinou AF, Andreou F, Manariotis ID (2017) Long-term toxicity of ZnO nanoparticles to *Scenedesmus rubescens* cultivated in different media. Sci Rep 7:13454

Aravantinou AF, Tsarpali V, Dailianis S, Manariotis ID (2015) Effect of cultivation media on the toxicity of ZnO nanoparticles to freshwater and marine microalgae. Ecotoxicol Environ Saf 114:109–116

Aruoja V, Dubourguier HC, Kasemets K, Kahru A (2009) Toxicity of nanoparticles of CuO, ZnO and TiO_2 to microalgae *Pseudokirchneriella subcapitata*. Sci Total Environ 407:1461–1468

Aruoja V, Pokhrel S, Sihtmäe M, Mortimer M, Mädler L, Kahru A (2015) Toxicity of 12 metal-based nanoparticles to algae, bacteria and protozoa. Environ Sci Nano 2:630–644

Bacchetta R, Maran B, Marelli M, Santo N, Tremolada P (2016) Role of soluble zinc in ZnO nanoparticle cytotoxicity in *Daphnia magna*: a morphological approach. Environ Res 148:376–385

Barhoumi L, Dewez D (2013) Toxicity of superparamagnetic iron oxide nanoparticles on green alga *Chlorella vulgaris*. Biomed Res Int 2013:647974

Behra R et al (2013) Bioavailability of silver nanoparticles and ions: from a chemical and biochemical perspective. J R Soc Interf 10:20130396

Bhuvaneshwari M et al (2015) Cytotoxicity of ZnO NPs towards fresh water algae *Scenedesmus obliquus* at low exposure concentrations in UV-C, visible and dark conditions. Aquat Toxicol 162:29–38

Bhuvaneshwari M, Iswarya V, Nagarajan R, Chandrasekaran N, Mukherjee A (2016) Acute toxicity and accumulation of ZnO NPs in *Ceriodaphnia dubia*: relative contributions of dissolved ions and particles. Aquat Toxicol 177:494–502

Blinova I, Ivask A, Heinlaan M, Mortimer M, Kahru A (2010) Ecotoxicity of nanoparticles of CuO and ZnO in natural water. Environ Pollut 158:41–47

Bystrzejewska-Piotrowska G, Golimowski J, Urban PL (2009) Nanoparticles: their potential toxicity, waste and environmental management. Waste Manag 29:2587–2595

Chang Y-N, Zhang M, Xia L, Zhang J, Xing G (2012) The toxic effects and mechanisms of CuO and ZnO nanoparticles. Materials 5:2850–2871

Chen P, Powell BA, Mortimer M, Ke PC (2012) Adaptive interactions between zinc oxide nanoparticles and *Chlorella* sp. Environ Sci Technol 46:12178–12185

Dąbrowski A (2001) Adsorption—from theory to practice. Adv Colloid Interf Sci 93:135–224

Dabrunz A et al (2011) Biological surface coating and molting inhibition as mechanisms of TiO_2 nanoparticle toxicity in *Daphnia magna*. PLoS ONE 6:e20112

Deng Z, Chen M, Gu G, Wu L (2008) A facile method to fabricate ZnO hollow spheres and their photocatalytic property. J Phys Chem B 112:16–22

Franklin NM, Rogers NJ, Apte SC, Batley GE, Gadd GE, Casey PS (2007) Comparative toxicity of nanoparticulate ZnO, bulk ZnO, and $ZnCl_2$ to a freshwater microalga (*Pseudokirchneriella subcapitata*): the importance of particle solubility. Environ Sci Technol 41:8484–8490

Gottschalk F, Sonderer T, Scholz RW, Nowack B (2009) Modeled environmental concentrations of engineered nanomaterials (TiO2, ZnO, Ag, CNT, Fullerenes) for different regions. Environ Sci Technol 43:9216–9222

Gunawan C, Sirimanoonphan A, Teoh WY, Marquis CP, Amal R (2013) Submicron and nano formulations of titanium dioxide and zinc oxide stimulate unique cellular toxicological responses in the green microalga *Chlamydomonas reinhardtii*. J Hazard Mater 260:984–992

Gurr JR, Wang AS, Chen CH, Jan KY (2005) Ultrafine titanium dioxide particles in the absence of photoactivation can induce oxidative damage to human bronchial epithelial cells. Toxicology 213:66–73

Han J, Qiu W, Gao W (2010) Potential dissolution and photo-dissolution of ZnO thin films. J Hazard Mater 178:115–122

Hazeem LJ, Bououdina M, Rashdan S, Brunet L, Slomianny C, Boukherroub R (2016) Cumulative effect of zinc oxide and titanium oxide nanoparticles on growth and chlorophyll a content of *Picochlorum* sp. Environ Sci Pollut Res Int 23:2821–2830

Izu N, Shimada K, Akamatsu T, Itoh T, Shin W, Shiraishi K, Usui T (2014) Polyol synthesis of Al-doped ZnO spherical nanoparticles and their UV–vis–NIR absorption properties. Ceram Int 40:8775–8781

Jahan S, Yusoff IB, Alias YB, Bakar A (2017) Reviews of the toxicity behavior of five potential engineered nanomaterials (ENMs) into the aquatic ecosystem. Toxicol Rep 4:211–220

Ji J, Long Z, Lin D (2011) Toxicity of oxide nanoparticles to the green algae *Chlorella* sp. Chem Eng J 170:525–530

Kamat PV, Meisel D (2002) Nanoparticles in advanced oxidation processes. Curr Opin Colloid Interf Sci 7:282–287

Kamat PV, Meisel D (2003) Nanoscience opportunities in environmental remediation. C R Chim 6:999–1007

Keller AA et al (2010) Stability and aggregation of metal oxide nanoparticles in natural aqueous matrices. Environ Sci Technol 44:1962–1967

Khan I, Saeed K, Khan I (2017) Nanoparticles: Properties, applications and toxicities. Arab J Chem

Kim SW, An YJ (2012) Effect of ZnO and TiO_2 nanoparticles preilluminated with UVA and UVB light on *Escherichia coli* and *Bacillus subtilis*. Appl Microbiol Biotechnol 95:243–253

Klaine SJ et al (2008) Nanomaterials in the environment: behavior, fate, bioavailability, and effects. Environ Toxicol Chem 27:1825–1851

Krishnakumar T, Jayaprakash R, Pinna N, Singh V, Mehta B, Phani A (2009) Microwave-assisted synthesis and characterization of flower shaped zinc oxide nanostructures. Mater Lett 63:242–245

Lee WM, An YJ (2013) Effects of zinc oxide and titanium dioxide nanoparticles on green algae under visible, UVA, and UVB irradiations: no evidence of enhanced algal toxicity under UV pre-irradiation. Chemosphere 91:536–544

Lee Y, Lee J, Bae CJ, Park JG, Noh HJ, Park JH, Hyeon T (2005) Large-scale synthesis of uniform and crystalline magnetite nanoparticles using reverse micelles as nanoreactors under reflux conditions. Adv Funct Mater 15:503–509

Lu J, Xu C, Dai J, Li J, Wang Y, Lin Y, Li P (2015) Improved UV photoresponse of ZnO nanorod arrays by resonant coupling with surface plasmons of Al nanoparticles. Nanoscale 7:3396–3403

Manzo S, Miglietta ML, Rametta G, Buono S, Di Francia G (2013) Toxic effects of ZnO nanoparticles towards marine algae *Dunaliella tertiolecta*. Sci Total Environ 445–446:371–376

Merdzan V, Domingos RF, Monteiro CE, Hadioui M, Wilkinson KJ (2014) The effects of different coatings on zinc oxide nanoparticles and their influence on dissolution and bioaccumulation by the green alga. C reinhardtii Sci Total Environ 488–489:316–324

Miao AJ, Zhang XY, Luo Z, Chen CS, Chin WC, Santschi PH, Quigg A (2010) Zinc oxide-engineered nanoparticles: dissolution and toxicity to marine phytoplankton. Environ Toxicol Chem 29:2814–2822

Miller RJ, Lenihan HS, Muller EB, Tseng N, Hanna SK, Keller AA (2010) Impacts of metal oxide nanoparticles on marine phytoplankton. Environ Sci Technol 44:7329–7334

Murray CB, Kagan CR, Bawendi MG (2000) Synthesis and characterization of monodisperse nanocrystals and close-packed nanocrystal assemblies. Annu Rev Mater Sci 30:545–610

Nel A, Xia T, Mädler L, Li N (2006) Toxic potential of materials at the nanolevel. Science 311:622–627

Pan B, Xing B (2008) Adsorption mechanisms of organic chemicals on carbon nanotubes. Environ Sci Technol 42:9005–9013

Panessa-Warren BJ, Maye MM, Warren JB, Crosson KM (2009) Single walled carbon nanotube reactivity and cytotoxicity following extended aqueous exposure. Environ Pollut 157:1140–1151

Pathakoti K, Morrow S, Han C, Pelaez M, He X, Dionysiou DD, Hwang HM (2013) Photoinactivation of *Escherichia coli* by sulfur-doped and nitrogen-fluorine-codoped TiO_2 nanoparticles under solar simulated light and visible light irradiation. Environ Sci Technol 47:9988–9996

Peng C et al (2017) Behavior and potential impacts of metal-based engineered nanoparticles in aquatic environments. Nanomaterials (Basel) 7:21

Peng X, Palma S, Fisher NS, Wong SS (2011) Effect of morphology of ZnO nanostructures on their toxicity to marine algae. Aquat Toxicol 102:186–196

Piccinno F, Gottschalk F, Seeger S, Nowack B (2012) Industrial production quantities and uses of ten engineered nanomaterials in Europe and the world. J Nanopart Res 14:1109–1120

Rabe M, Verdes D, Seeger S (2011) Understanding protein adsorption phenomena at solid surfaces. Adv Colloid Interf Sci 162:87–106

Rajavel K, Gomathi R, Manian S, Rajendra Kumar RT (2014) In vitro bacterial cytotoxicity of CNTs: reactive oxygen species mediate cell damage edges over direct physical puncturing. Langmuir 30:592–601

Rsrae N (2004) Nanotechnologies: opportunities and uncertainties RS policy document 19/04'. The Royal Society and Royal Academy of Engineering, London

Sapkal RT, Shinde SS, Waghmode TR, Govindwar SP, Rajpure KY, Bhosale CH (2012) Photocorrosion inhibition and photoactivity enhancement with tailored zinc oxide thin films. J Photochem Photobiol B 110:15–21

Scown TM et al (2010) Effects of aqueous exposure to silver nanoparticles of different sizes in rainbow trout. Toxicol Sci 115:521–534

Snyder-Talkington BN, Qian Y, Castranova V, Guo NL (2012) New perspectives for in vitro risk assessment of multiwalled carbon nanotubes: application of coculture and bioinformatics. J Toxicol Environ Health B Crit Rev 15:468–492

Song W, Zhang J, Guo J, Zhang J, Ding F, Li L, Sun Z (2010) Role of the dissolved zinc ion and reactive oxygen species in cytotoxicity of ZnO nanoparticles. Toxicol Lett 199:389–397

Suman TY, Radhika Rajasree SR, Kirubagaran R (2015) Evaluation of zinc oxide nanoparticles toxicity on marine algae *Chlorella vulgaris* through flow cytometric, cytotoxicity and oxidative stress analysis. Ecotoxicol Environ Saf 113:23–30

Toensmeier PA (2004) Nanotechnology faces scrutiny over environment and toxicity. Plast Eng 60:14–14

Tseng YH, Lin HY, Kuo CS, Li YY, Huang CP (2006) Thermostability of Nano-TiO_2 and its photocatalytic activity. React Kinet Catal Lett 89:63–69

Tsukazaki A et al (2005) Repeated temperature modulation epitaxy for p-type doping and light-emitting diode based on ZnO. Nat Mater 4:42–46

Walters CR, Pool E, Somerset V (2016) Nanotoxicology: a review. In: Toxicology—new aspects to this scientific conundrum. InTech, pp 45–63

Wang H, Wick RL, Xing B (2009) Toxicity of nanoparticulate and bulk ZnO, Al_2O_3 and TiO_2 to the nematode *Caenorhabditis elegans*. Environ Pollut 157:1171–1177

Wang X et al (2016) Zinc oxide nanoparticles affect biomass accumulation and photosynthesis in Arabidopsis. Front Plant Sci 6:1243

Wong SW, Leung PT, Djurisic AB, Leung KM (2010) Toxicities of nano zinc oxide to five marine organisms: influences of aggregate size and ion solubility. Anal Bioanal Chem 396:609–618

Wu F, Bortvedt A, Harper BJ, Crandon LE, Harper SL (2017) Uptake and toxicity of CuO nanoparticles to *Daphnia magna* varies between indirect dietary and direct waterborne exposures. Aquat Toxicol 190:78–86

Xia T et al (2008) Comparison of the mechanism of toxicity of zinc oxide and cerium oxide nanoparticles based on dissolution and oxidative stress properties. ACS Nano 2:2121–2134

Zhang C, Chen X, Wang J, Tan L (2018) Toxicity of zinc oxide nanoparticles on marine microalgae possessing different shapes and surface structures. Environ Eng Ecol Sci 35–8:785–790

Zhou H, Wang X, Zhou Y, Yao H, Ahmad F (2014) Evaluation of the toxicity of ZnO nanoparticles to *Chlorella vulgaris* by use of the chiral perturbation approach. Anal Bioanal Chem 406:3689–3695

Chapter 11
Toxicity of Nanomaterials in Agriculture and Food

Ryan Rienzie and Nadeesh M. Adassooriya

Abstract Nanotechnology is revolutionizing the world with promising breakthroughs that make the human lives more comfortable, and no doubt that this would be the same in the future of agriculture and food industries as well. Besides, the world is still not fully confident about how this technology would work in long run, particularly in relation to the concerns regarding the possible toxic effects that can threaten the living either by direct or indirect entry into living cells. Together with agricultural activities, food production is one of the leading aspects. Lack of proper understanding on mechanisms that nanomaterial behave inside the living cells has directed us to re-think on incorporating them with agricultural purposes and then consumption of such nanomaterial involved products. The kinetics and toxicity of nanomaterials are unique and depending upon the extrinsic and intrinsic factors which hinder predicting their behaviour under various environments, therefore has become more complex to understand the possible impacts. In this context, this chapter aims at broadly discussing on the toxicological aspects of nanomaterials in agriculture and food systems while pointing out the future prospects of applications of nanotechnology to the same.

Keywords Toxicity · Nanomaterials · Nanoparticles · Ecosystem
Microbial community

R. Rienzie
Faculty of Agriculture, University of Peradeniya, Peradeniya 20400, Sri Lanka
e-mail: ryanrienzie@gmail.com

N. M. Adassooriya (✉)
Faculty of Applied Science, University of Sri Jayewardenepura, Nugegoda 10250, Sri Lanka
e-mail: nadeesh@sjp.ac.lk

M. Rai and J. K. Biswas (eds.), *Nanomaterials: Ecotoxicity, Safety, and Public Perception*, https://doi.org/10.1007/978-3-030-05144-0_11

11.1 Introduction

Nanotechnology is an emerging interdisciplinary technology that provides greater opportunities for new innovations with the aim of boosting crop and livestock production and protection throughout the world. Presently there is a number of innovations introduced to the world aiming at bridging the various technological gaps of past era. For instance there is a greater interest on moving towards nanoformulations of agrochemicals such as fertilizers and pesticides which are more efficient than the respective conventional formulations. A number of diagnostic tools have been developed to detect agriculturally important pathogens affecting crop and livestock systems. Despite the array of benefits of nanomaterials in agricultural and food technology, their toxic effects have gained a greater attention within the scientific world.

Living components in an agricultural or food system can be affected by nanomaterials in a different manner compared to their bulk counterparts as the properties of materials at nanoscale are different. Entry of nanomaterials into agricultural and food systems is of greater concern, and they can enter into the agricultural and food systems either by point or non-point sources as deliberate or unintentional means. Effects of various nanomaterials on growth and development of crops and livestock by means of morphology, physiology, biochemistry and molecular biology have been intensively studied and evidences are still compiling. In addition to the impacts on crops and livestock upon direct contact, secondary effects might be extended up to any other living component in the system. For instance, nanomaterials can make significant changes to soil microbes by altering their cellular functions, diversity and community composition, which will ultimately make negative impacts to the nutrient cycling processes. However, information is still not adequate to draw straightforward conclusions on their long term fate to the living. Studies claim that entry of nanomaterials into crop and livestock systems and then to the food systems pose threats due to their potential of bioaccumulation through the food chains that finally affecting the consumers.

Identifying this nature of nanomaterials, many of the countries such as USA, UK, Australia, New Zealand, Japan and some developing countries in the world have set up safety assessment and regulatory systems related to the use of nanotechnologically derived innovations by stressing up their health and environmental safety concerns. In such a background, the toxicological aspects of nanotechnology-based food categories such as nanofoods, nutraceuticals and functional foods and food packaging materials are of greater concern. This chapter tries to broadly discuss the toxicological aspects of nanomaterials in agriculture and food systems with special reference to crop systems while pointing out the future prospective of applications of nanotechnology to the same.

11.2 Factors Contributing to Toxicity of Nanoparticles

11.2.1 Properties of Nanoparticles that Contribute to Toxicity

Attributes of the nanomaterials themselves such as the physicochemical properties affect exerting toxicity to plants when entered. Shape, size and composition of the nanoparticles (NPs) regulate the toxicity (Zhang et al. 2015; Khare et al. 2015). NPs inherently possess higher surface area to volume ratio compared to bulk metals due to which they can be highly reactive (Rico et al. 2011) and they are estimated as absorbed 15–20 times more compared to the respective bulk particles (Zhao et al. 2012b). Moreover, the exposure concentration or load and the type has an impact on level of toxicity (Rizwan et al. 2017). For instance very low surface friction of multiwall carbon nanotubes facilitates the flow of organic substances into the plant body (Rico et al. 2011).

Deliberate attempts to enhance the properties of nanoparticles have favoured in changing the reactivity, while enhancing or adding new properties in addition to the ones they are inherently possessed. One property is the surface charge, which can be changed via various coatings/functionalizing material such as acids 2,3-mesodimercaptosuccinic acid, citric acid, humic acid, 8-mercaptooctanoic acid, hyaluronic acid and polymers or their derivatives such as polyvinyl pyrrolidone, polyethylene glycol, chitosan, gum Arabic, polyvinyl alcohol, sodium bis (2-ethylhexyl) (Lopez-Moreno et al. 2018). This surface charge can either be negative or positive and affects the dispersing ability, thereby interacting more with the environment (Lopez-Moreno et al. 2018). One way of altering the charge is by altering the surface charge of NPs. Zhai et al. (2015) reported that the surface charge affects the location and distribution throughout the plant body. Accordingly, surface coating has enhanced the potential of translocation by means of greater dispersion, solubility and prevented agglomeration (Bombin et al. 2015).

Despite the greater ability of coated/functionalized nanoparticles to enter and translocate throughout the plant body, having a coating does not necessarily indicate that they can absorb via roots. There are reports that indicate the coating has reduced the uptake compared to intact nanoparticles via roots. Stability of the nanoparticles also affects the level of toxicity. For instance, due to the properties of the building components of nanoparticles they can be kinetically unstable thus tend to aggregate while reducing the surface area and thereby the reactivity. Ultimately the toxicity will also be altered (Lopez-Moreno et al. 2018).

11.2.2 Properties of Plants and Growth Medium that Aid Toxicity

From the view of plants, uptake and translocation of nanoparticles may vary with plant species, cultivars growth conditions, growth stage and chemical composition

(Rizwan et al. 2017). Even though it cannot come into a clear cut conclusion in general monocots possess a low uptake capacity compared to dicots (Siddiqi and Husen 2017).

11.2.3 Growth Medium

Nanoparticles have been reported to be affecting the physical, chemical and biological properties of the growth media in which the plants are grown. Likewise, chemical parameters such as the humic substances and, ion content, physical parameters include porosity, hydraulic conductivity (Ben-Moshe et al. 2013), texture, structure, compactness are causal factors that alter the behaviour of NPs (Fierer and Jackson 2006). When present in soil medium, nanoparticles can strongly attach to soil colloids, exhibit low mobility at different ionic concentrations and enhance sorption capacity (Zhao et al. 2012a). In the presence of nanoparticles, soil biological parameters including the soil microbial community can be affected. Alternatively soil microbes can influence the nanoparticles by changing the reactivity (Fierer and Jackson 2006; Gajjar et al. 2009).

11.3 Modes of Action, Entry and Translocation of Nanomaterials

Three major modes of action of nanomaterials can be identified namely dissolution, direct contact and co-transport of other contaminants (Mukherjee et al. 2016). Moreover, activity of most of the nanoparticle species is broad spectrum which exerts both positive and negative impacts to the ecosystems upon release. Figure 11.1 represents the possible modes of entry of nanomaterials into agricultural and food systems and followed by their fate. Many in depth research evidences are existing in terms of plant mediated systems, however rare in relation to human and livestock. Nanomaterials can enter into agricultural and food systems via aerial, growth medium mediated pathways and food/feed either via deliberate or accidental (unintentional) means through point sources and/or non-point sources. After entry, interactions occur directly with both living beings and non-living matter. Subsequently, there is a tendency of transferring between trophic levels.

Entry and translocation of nanoparticles are of major concern due to their ability to cause toxic effects. Entry of nanoparticles into a plant body can occur via roots from the growth medium or through aerial parts with water and air. Followed by upward or downward movement happen (Rizwan et al. 2017). Entry and translocation of nanoparticles is a process involved with a series of active and passive phenomena. Amongst, uptake from the medium which the plant is grown is considered as an active phenomenon (Etxeberria et al. 2009). Entry and translocation can be considered as

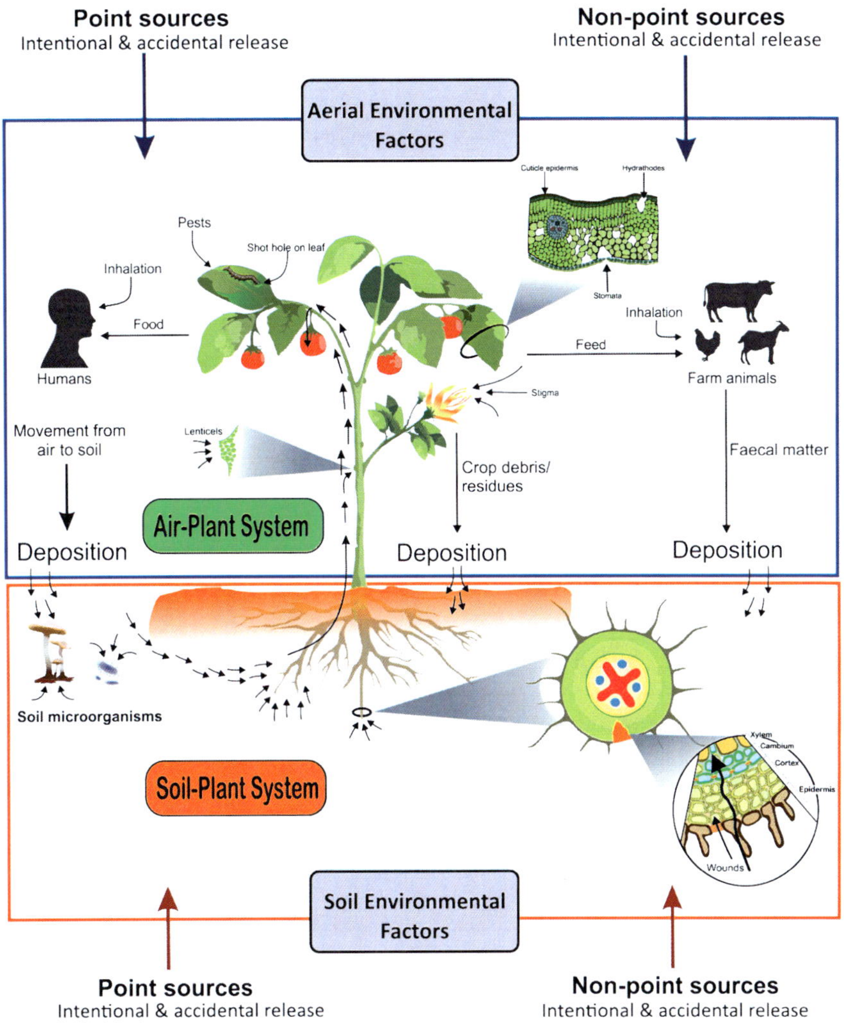

Fig. 11.1 Schematic diagram showing the entry and movement of nanomaterials and in agricultural and food systems (information obtained from Dietz and Herth 2011; Rico et al. 2011; Anjum et al. 2016; Cox et al. 2017; Ruttkay-Nedecky et al. 2017)

more or less independent phenomena, and the ability to entry of nanoparticles does not essentially imply that they can translocate throughout the entire plant. At the same time, these two phenomena depend upon a combination of factors associated with plants, nanoparticles and the growth medium. Accumulation or agglomeration of nanoparticles is another phenomenon which can take place in between entry and translocation.

11.3.1 Root-Mediated Entry and Bottom to Top Movement

Upon get in contact with a plant root, nanoparticles penetrate into the cell wall, followed by to the cellular membrane of the root epidermal cells. Transfer through the cell membrane occurs using the pores located on the cell membrane (Rico et al. 2011). Followed by a series of events, the nanoparticles move into the vascular tissues and then to the stele. Subsequent transfer unto the aerial parts occur via xylem of the plant which is the most vital path for translocation of nanoparticles (Aslani et al. 2014).

Following absorption of intact nanoparticles into the cytoplasm of the root, the release of the elements from the nanoparticles takes place as ions. Ions after releasing, might change their oxidation status and form different other complexes or compounds. Moreover, the nanoparticles can be transported throughout the plant body while accumulating with varying amounts in different organs.

Nanoparticles when present in soil undergo a series of steps and finally subjected to decomposition. Larger nanoparticles and nanoparticle agglomerates might be separated into smaller nanoparticles (Horst et al. 2010; Capaldi-Arruda et al. 2015). Smaller nanoparticles form different ionic forms and complexes that enable the entry into the plant body (Capaldi-Arruda et al. 2015).

11.3.2 Entry via Aerial Parts and Top-Down Movement

Apart from the root-mediated entry pathway, entry can occur via aerial openings of the plants such as cuticles, trichomes, stomata, stigma, and hydathodes, nectarithodes, lenticels, etc. Moreover, through damaged places (wounds) or root junctions (Ruttkay-Nedecky et al. 2017). In either cases, once inside the cells the nanoparticles are transported apoplastically, symplastically or by accomplishing cell to cell movement using plasmodemata. Entry to the stele from the xylem occurs symplastically (Thripathi et al. 2017).

11.4 Toxic Effects of Nanomaterials on Soil Ecosystems

11.4.1 Effects on Soil Microbial Communities

Significance of microbial communities in agricultural environments is high as a major component in the food web and soil microbes is a good indicator of soil productivity and fertility. The higher sorption capacity of bacterial cells encompasses accumulating nanomaterials inside the cells; while at the same time owing to mobilize through subsequent trophic levels of the food chains. Impact of various nanoparticles on the soil microbial communities has been investigated through many studies that revealed the negative or toxic nature of the nanoparticles. Metal and metal oxide nanoparticles have been reported to be exerting higher toxic effects on soil microorganisms; however varied effects have also been shown in terms of carbon based nanomaterials. Effects have been shown to be exerting against almost all microbial categories in soil ecosystems in terms of microbial activity, abundance and diversity. However, with varied degrees of impact depending on the bacterial species subjected (Pallavi et al. 2016). Therefore, in a soil-plant ecosystem the behaviour of nanomaterials is complexer than expected (Feng et al. 2013).

11.4.2 Effects on Nutrient Cycling (Nitrogen and Carbon Cycling)

Nitrogen cycle is one of the major natural phenomena, which enhance the agricultural productivity up to a greater extent. Affecting the microorganisms associated with nitrogen cycle will negatively affect the productivity of the crops. Aerobic denitrifying bacterium, *Pseudomonas putida* has been found to be affected by single wall carbon nanotubes (SWCNTs) due to alteration of the phase transition temperature and elevated levels of unsaturated fatty acids (Rodrigues et al. 2013). *P. putida* has been reported to be affected by ZnO NPs as well. Bandyopadhyay et al. (2015) reported that ZnO NPs exerting bactericidal effects against the nitrogen fixing bacterium, *Sinorhizobium meliloti*. The ill effects of TiO_2 and ZnO NPs on the order Rhizobiales are by declining the bacterial community and reducing the diversity (Ge et al. 2011). Members of the Rhizobiales form symbiotic associations with plants having N fixing ability. ZnO NPs affected thermogenic metabolism of *Azotobacter* species, which are involved in the process of fixing nitrogen (Chai et al. 2015). Furthermore, the same authors reported of affecting the activity of urease, catalase and fluorescein diacetate hydrolysis enzymes in phosphorus and potassium solubilising bacteria. ZnO NPs have also been reported as N mineralization retardants due to their negative impacts on heterotrophic bacterial and fungal communities (Rashid et al. 2017). Long-term effects of Ag NPs on *Bradyrhizobium* spp. associated with arctic soils were reported by Kumar et al. (2011). Effects of SWCNTs have been observed in terms of decreasing biomass of soil fungi that are acting as primary decomposers of soil organic matter, thus are important in nutrient recycling (Rodrigues et al. 2013; Jin et al. 2013). Furthermore, soil fungi has also been reported to be negatively affected after adding ZnO NPs (Rashid et al. 2017).

Soil microbial biomass carbon (SMBC) is a vital determinant of the balance soil carbon release and sequestration in the plant-soilmicrobial environment when considered as a single system (Thakur et al. 2015). It is the most active portion of the soil carbon pool (Balota et al. 2003). Significant decrease of the SMBC has been reported upon exposure to nanoparticles such as Ag NPs/ Ni NPs and Co NPs (Antisari et al. 2015), CuO NPs (Xu et al. 2015) and SWCNT (Jin et al. 2013).

11.4.3 Soil Microbial Community Composition and Diversity

Toxic effects of MWCNTs have been shown to be affecting the soil microbial community composition and diversity (Chung et al. 2011; Kerfahi et al. 2015). Moreover, effect of graphene on soil microbes has been investigated by Ren et al. (2015). They revealed that, upon addition of graphene into soil, it affected the bacterial population at genus level. Furthermore, fullerenes have also been reported to be distressing the soil organisms, but they were later regulated by the soil organic matter present in soil, eliminating the negative effect (Tong et al. 2016). Johansen et al. (2008) reported the ill effects of fullerenes (C_{60}) against soil microbes and protozoan communities. Harmful effects of CuO and ZnO nanoparticles on microbial activity, abundance and diversity have also been studied (Dinesh et al. 2012). Pallavi et al. (2016) demonstrated the concentration dependent impact of Ag NPs on phosphate solubilisers, N fixers and siderophore producers in altering their diversity. They showed that the decline of colony count and the diversity of the above categories in soil with cowpea, wheat and lettuce when treated with Ag NPs at 50 and 75 ppm concentrations. Reduced microbial biomass due to ZnO NPs was also reported by Ge et al. (2011). pH dependent effects on the interactions between bacterial communities and ZnO NPs have been demonstrated by Read et al. (2016). They found that soils with higher pH showing lower susceptibility to the effects due to contact with NPs. The impact of Ag NPs on soil bacterial diversity is dependent on the both concentration and plant species grown, which may be due to the patterns and composition of the root exudates secreted by various plant species. The organic acids released by plants play a vigorous role in influencing the rhizosphere's diversity of microorganisms (Pallavi et al. 2016). The plant growth promoting rhizobacteria (PGPR) has also was reported to be negatively affected by metal oxide nanoparticles. For instance, *P. chlororaphis* which is involved in production of indol acitic acid (IAA), a growth promoter of plants is affected by ZnO nanoparticles (Dimkpa et al. 2011).

All above studies are comparatively short-term (up to 2 months). However, the impacts that depict the long-term impacts are rare. Kumar et al. (2011) studied the long run impact on microbial biomass of arctic soils in presence of Ag, Cu, and Si NPs. They observed a significant decrease of the microbial biomass after 6 months of incubation. In general carbon based nanomaterials are to be less toxic compared to metal based ones (Simonin and Richaume 2015).

11.4.4 Soil Enzyme Activity

Impairments to the soil enzyme activities due to ZnO NPs have been reported by Chai et al. (2015). ZnO and TiO_2 NPs significantly reduced activity for several soil enzymes: soil protease, catalase, and peroxidase (Du et al. 2011). Presence of MWCNTs was reported to be affecting the soil enzymes such as soil phosphatase, 1,4-

ß-N-acetylglucosaminidase, 1,4-ß-glucosidase, cellobiohydrolase and xylosidase (Chung et al. 2011).

11.5 Toxicity to Crop Plants

Under this section, toxicity effects of various nanoparticles on common crops plants are discussed. A number of research evidences can be found in relation to toxic effects of nanomaterial on plants belong to the plant families incuding Poaceae (Gramineae), Leguminosae (Fabaceae), Solanaceae, Amaryllidaceae, Cucurbitaceae and Brassicaceae.

(I) **Wheat (*Triticum aestivum*)**

Wheat is among the most studied crops in terms of toxicity exerted by nanomaterials specifically metal based nanoparticles. Toxicity of Cu, CuO, TiO_2, Zn, ZnO, Ag and Fe_2O_3 nanoparticles on wheat has been widely studied in different growth media.

(a) ***Cu based NPs: CuO NPs***

CuO NPs have been found to be impeding the growth of wheat. For instance Zhou et al. (2011) reported that the uptake by plant and adsorption was increased with the increased CuO NPs concentrations employed in an agar medium, while Lee et al. (2008) reported that seedling growth has significantly reduced in a concentration dependent manner in agar medium and reduction of the root growth of sand grown wheat (Dimkpa et al. 2012). CuO NPs not only reduce the root and shoot growth, but also affect the biomass ultimately. Moreover, they affect negatively to some steps of photosynthesis, such as by reducing the chlorophyll contents of tissues involved, while increasing the peroxydase and catalase activities in roots of treated plants (Dimkpa et al. 2012).

(b) ***Ti based NPs: TiO_2 NPs***

TiO_2 NPs is also important as one of the mostly studied NPs and reports are existing on decreasing the overall growth and yield of wheat (Du et al. 2011; Rafique et al. 2018). At cell level, the toxic potential of TiO_2 has been evident through formation of micronuclei upon exposure (Rafique et al. 2018).

(c) ***Ag based NPs: Ag NPs***

Ag NPs also have been studied for their impact on the growth parameters of wheat. At higher concentrations (10 mg/L), Ag NPs have altered the expression of certain genes that encode proteins related to primary metabolism and cell defense (Vannini et al. 2014).

(d) ***Zn based NPs: ZnO NPs***

Ceasing the growth of wheat crop due to Zn and ZnO NPs has been revealed by a number of researches. Declining of chlorophyll content of photosynthetic tissues and increasing of peroxydase and catalase activities in roots have been reported by Dimkpa et al. (2012). In addition to the vastly studied metal based nanoparticles discussed above, Fe_2O_3 has also been studied for its effects on seed germination, which showed its ability to reduce seed germination at higher concentrations (Feizi et al. 2013).

(II) **Rice (*Oryza sativa*)**

(a) ***Ti based NPs: TiO_2 NPs***

Hydroponically grown rice has shown reduced biomass contents and disturbance to the antioxidant defense system upon exposure to TiO_2 NPs. Furthermore, change of certain metabolite concentrations inside the cells, suppression of carbohydrate synthesis mechanisms and elevation of respiration pathways have been demonstrated with regard to TiO_2 NPs (Wu et al. 2017).

(b) ***Ce based NPs: CeO_2 NPs***

Higher concentrations of CeO_2 NPs have increased the H_2O_2 generation, while enhancing the lipid peroxidation and electrolyte leakage in rice plants (Rico et al. 2013a). Such toxic effect could spread unto the seeds, ultimately reducing the quality of the harvest (Rico et al. 2013b).

(c) ***Cu based NPs: CuO NPs***

Seeds treated with CuO NPs can exert decreased seed germination and growth while causing severe oxidative damages to cells (Shaw et al. 2014).

(d) ***ZnO NPs***

ZnO NPs can negatively affect the growth of shoot and root growth of rice seedlings (Boonyanitipong et al. 2011).

(e) ***Ag based NP: Ag NPs***

Morphological variations due to damage in root cell walls and vacuoles of rice plants can occur due to Ag NPs (Mazumdar and Ahmed 2011) while reducing the root growth by means of elongation and weight that eventually reducing the above ground biomass (Mirzajani et al. 2013; Nair and Chung 2014b). These effects can be exerted with higher concentrations of Ag NPs. Ag NPs have also reported to be affecting the total chlorophyll content while reducing the carotenoid content of rice plants. Reactive oxygen species (ROS) production has also been observed to be increased in a concentration dependent manner (Nair and Chung 2014b).

(f) ***Carbon-based nanomaterials: SWCNT and MWCNT***

Rice has been reported to be sensitive to both SWCNT and MWCNT. SWCNT has been studied for its effects on DNA damage by affecting the viability of cells (Shen et al. 2010). MWCNT is also known in terms of reducing the weight and length of roots and shoots, leaf wilting and reduced size at higher concentrations while with increased chromatin, condensation led to damage the DNA by affecting viability of cells causing genotoxicity (Begum et al. 2012).

(III) Maize (*Zea mays*)

(a) *TiO_2 NPs*

TiO_2 NPs have been reported to be accumulated in maize roots upon entering via root pathway (Larue et al. 2012) which can lead to induce oxidative stress and inhibiting root growth (Castiglione et al. 2011; Laware and Raskar 2014). Furthermore, TiO_2 NPs have inhibited the root hydraulic conductivity and transpiration while reducing the pore diameter of cell walls (Asli and Neumann 2009). TiO_2 NPs could induce chrmosomal anomalies and reduced the mitotic index (MI) in maize by providing evidences on genotoxicity (Castiglione et al. 2011).

(b) *Zn NPs*

Zn NPs shown to be accumulated inside the roots, which was evidenced through the analysis of Zn in the roots. Reducing the root growth parameters such as length has also been revealed (Zhang et al. 2015).

(c) *CuO NPs*

According to the findings of Wang et al. (2012), seed germination of maize does not affect by CuO, however CuO affected the subsequent growth of seedlings resulted.

(d) *Ag and Al NPs*

Another type of nanoparticles is Ag NPs that retard the germination of the maize seeds (Pokhrel and Dubey 2013). In addition to that Al NPs have been also reported to be reducing root length and germination of maize seeds (Lin and Xing 2008).

(IV) Other crops of family Poaceae

In addition to the widely studied cereal crops, perennial rye grass (*Lolium perenne*) is reported to be affected by CuO NPs by inhibiting the root development (Atha et al. 2012), while ZnO NPs are reported to be reducing biomass and affecting roots of the same (Lin and Xing 2008). In addition, Al NPs have been reported as growth retardants of rye (*Secale cereale*) (Lin and Xing 2007).

(V) Legumes

(a) *Cu-based NPs*

Among the legumes studied for the toxicity of CuO NPs, soybean is important. Exposure of soybean plants to CuO can cause reduction of shoot and root growth in terms of their weight and length, while increasing the H_2O_2 activity (Nair and Chung 2014a). Cu NPs has also been reported to be affecting the overall seedling growth of mung bean. Chickpea also affects by CuO NPs by resulting in decreased shoot and root growth. CuO NPs can induce ROS generation (Adhikari et al. 2012; Nair and Chung 2015a). Furthermore, CuO NPs toxicity against roots can be exhibited through root necrosis (Adhikari et al. 2012).

Alfalfa is another crop species that can be affected due to toxicity of CuO due to which decreased the root length while altering the Cu, P, and S concentrations in shoots (Hong et al. 2015). This example is vital to indicate that CuO NPs can alter the nutrient concentrations inside the cells.

(b) ***ZnO NPs***

Upon exposure to ZnO NPs, lengths of roots and shoots of soybean can be affected (Lopez-Moreno et al. 2010; Bandyopadhyay et al. 2015), and the surface area to volume can also be affected while, the plants exposed to higher concentrations can affect the formation of seeds (Yoon et al. 2014). ZnO NPs can also decrease the shoot and root length and biomass, while with retarded leaf area. Furthermore, ZnO NPs can increase the catalase activity inside the roots of soybean plants (Bandyopadhyay et al. 2015).

(c) ***Fe_2O_3 NPs***

Fe_2O_3 NPs can reduce biomass of mycorrhized clover. Furthermore, significant reduction in biomass with reduced glomalin content and root nutrient acquisition was observed in mycorrhized clover (Feng et al. 2013).

(d) ***TiO_2* NPs**

Effects of TiO_2 NPs have been observed on Narbon vetch that exhibited its toxicity by forming micronuclei, chromosome lagging, chromosome bridging in a dose dependent way (Castiglione et al. 2011).

(VI) **Tobacco (*Nicotiana tabacum*)**

(a) ***TiO_2 NPs***

TiO_2 NPs affect the tobacco plants by damaging the DNA (Ghosh et al. 2010), while inhibiting the root growth and inducing multiple root emergence (Frazier et al. 2014).

(VII) **Tomato (*Lycopersicon esculentum*)**

(a) ***TiO_2 NPs***

Tomato has been shown to be affected by TiO_2 NPs. Soil and aerosolized foliar entering of of TiO_2 NPs has been experimentally studied. Accordingly, spray treated plants has shown reduction in root length, while accumulation of the same also shown (Raliya et al. 2015). Both above ground parts (shoots and fruits), and roots have been shown to be affected by TiO_2 NPs.

(b) ***Graphene***

Reduction of number of roots, shoots and leaves of tomato has been shown under high concentrations of graphene. Decrease of active photosynthetic tissue area (leaves) while causing cell death under higher concentrations has also been resulted due to graphene (Begum et al. 2011).

(c) ***Ag NPs***

Upon exposure of tomato roots to Ag NPs, the mature plants have shown reduced root elongation, while reducing the chlorophyll contents and number of fruits yielded. Furthermore, higher superoxide dismutase concentrations were also prominent as reported by Song et al. (2013a). In addition to that the effects on nutrient contents of the plant tissues have been less reported. For instance, entering of Ag NPs via roots has resulted in reducing the amount of K and Fe contents in the leaves, while exhibiting the deficiencies of the same.

(VIII) **Onion (*Allium cepa*)**

(a) ***Zn and Cu based NPs: ZnO and CuO NPs***

ZnO NPs have been studied for their effects on various parameters. Seed germination of onion is affected by ZnO NPs both by means of positively and negatively. For instance onion seed germination has been enhanced by ZnO NPs at minute concentrations while reducing the rate with elevated concentrations. Furthermore, it can increase overall shoot and root length and biomass due to exposure to ZnO NPs (Raskar and Laware 2014). Except the physiological and morphological effects, molecular level impacts are evident with respect to various nanoparticles in onions. Genotoxicity is one such aspect and several indicators such as mitotic index (MI) and related chromosomal level observations are vital to understand the extent of genotoxicity. MI refers to the ration between the numbers of cells in a population undergoing mitosis to the total number of cells in a population.

MI of root tip cells of onions is a good indicator related to cytotoxicity and this has been studied with reference to various nanoparticles. For instance, root tip cells exposed to higher ZnO NPs has shown lower MI along with chromosomal abnormalities at elevated concentrations (Raskar and Laware 2014). CuO NPs also shown to be reducing the MI. In addition to that, chromosomal level abnormalities due to CuO NPs occur at higher concentrations (Nagaonkar et al. 2015).

(b) ***Ti based NPs: TiO_2 NPs***

Damaging DNA of onion root meristem cells can be resulted due to TiO_2 NPs and this phenomenon has been observed in dose dependent (Demir et al. 2014) and irrespectively to concentrations (Ghosh et al. 2010). Onion root exposure to TiO_2-NP can result in chromosomal breakages and clumping, improper polarization and anaphase while chromosomes becoming sticky during metaphase (Pakrashi et al. 2014). Formation of micronuclei in onion due to TiO_2 NPs has been observed at various studies (Ghosh et al. 2010; Pakrashi et al. 2014). As evidenced by various means, root exposure to TiO_2 NPs at higher concentrations ultimately impedes the growth (Pittol et al. 2017).

(c) ***Carbon-based nanomaterials: Fullerols ($C_{60}OH_2$)***

Except metal based nanoparticles carbo related nanoparticles can cause toxic effects to onions and have reveled in certain studies. Fullerols ($C_{60}OH_2$) can cause toxicity i.e. cell necrosis, while generating reactive oxygen species (ROS) and reducing cell integrity (Chen et al. 2010).

(d) ***Ag NPs***

Ag NPs have also been studied for their toxicity effects to onions. Ag NPs mixed agar grown onions have exhibited growth suppression, but similar suppression couldn't be observed in soil while it has changed the properties of Ag NPs in soil (Lee et al. 2012). Coated Ag NPs with fumed silica can reduce the root growth (Pittol et al. 2017).

(IX) **Cucumber (*Cucumis sativus*)**

(a) ***Zn based nanoparticles: Zn NPs and ZnO NPs***

Zn NPs have been reported to affect the root growth of cucumber by reducing the length of the roots (Lin and Xing 2008). Similarly, ZnO NPs also reported to retard the growth of roots (Kim et al. 2011) and reducing shoot growth and biomass (Kim et al. 2012). However, the Zn concentration in the plant tissues has been increased upon exposure of ZnO NPs (Kim et al. 2011).

(b) ***CuO NPs***

CuO NPs are also reported to be decrease the plant biomass while increasing the antioxidant activity inside the tissues (Kim et al. 2012).

(c) ***MWCNTs***

Among the C based nanoparticles, MWCNT has been reported affecting shoot and root weight while causing the cell death in a dose dependent manner (Stampoulis et al. 2009).

(X) **Pumpkin (*Cucurbita maxima*)**

(a) ***Ag NPs***

Among the various nanoparticles tested on pumpkin, Ag NPs are reported reduce the biomass and transpiration of pumpkin upon exposure.

(b) ***Si NPs***

Si NPs are another type of nanoparticles that reduce the plant growth, transpiration (Hawthorne et al. 2012) and inhibition of germination of pumpkin seeds (Stampoulis et al. 2009).

(c) ***MWCNTs***

MWCNTs are also reported to reduce the biomass and transpiration of pumpkin (Stampoulis et al. 2009).

(XI) ***Brassica* spp.**

B. napus (rapeseed), *B. juncea* (mustard) and *B. oleracea* (cabbage) have been reported to be affected by nanoparticles.

(a) ***ZnO NPs***

Kauhi et al. (2014) reported that *B. napus* is affected by ZnO due to decreasing the shoot length, root length and dry mass. Similarly, Kauhi et al. (2015) found that ZnO NPs reduce the overall plant growth and biomass, while causing the anatomical changes in roots and ultrastructure of leaves of *B. napus*. *B. juncea* also has been reported to be affected by ZnO NPs causing shoot and root length and biomass, while increasing the reactive oxygen species generation by antioxidant enzymes (Rao and Shekhawat 2014). *B. napus* is also affected by Zn NPs due to retardation of root length (Lin and Xing 2007).

(b) ***CuO NPs***

CuO NPs are also reported as affecting shoot and root length and changing the leaf size and root architecture of *Brassica spp*. Furthermore, it has led to decrease the photosynthetic pigments, while with increased oxidative stress (Nair and Chung 2015b).

(c) ***Al_2O_3 NPs***

It has been reported that *B. oleracea* is affected by Al_2O_3 NPs by negatively affecting seedling growth and biochemical parameters (Amist et al. 2017).

(XII) **Red Spinach (*Amaranthus tricolor*)**

(a) ***Carbon-based nanomaterials: MWCNT* and graphene**

Notable reduction of weight and length of roots and shoots with the increase of the concentrations has been reported (Begum et al. 2012). Begum and Fugetso (2012) reported the impact of MWCNT on inhibition of stomatal opening and inducing ability of stomatal closing and generation of ROS. Furthermore, Begum et al. (2011) reported of notable reduction of shoot, leaf number and leaf area and cell death with increased concentrations of graphene.

(XIII) **Cotton (*Gossypium* spp.)**

(a) ***SiO_2 NPs***

Toxic effects of SiO_2 NPs have been reported on cotton. For instance, decreased the plant height and shoot and root biomass in a dose dependent manner (Le et al. 2014).

(b) ***CeO_2 NPs***

CeO_2 is one of the least studied NPs. CeO_2 can retard the plant growth i.e. plant height, overall affecting the total biomass of the plants. Another important phenomenon is that CeO_2 can affect both essential and trace elemental concentrations in plants (Li et al. 2014).

(XIV) **Radish (*Raphanus sativus*)**

It has been reported that TiO_2 NPs are toxic to roots of radish (Pittol et al. 2017). CuO NPs also reported to be decreasing the root growth of radish (Atha et al. 2012). Radish is also affected by Zn NPs by reducing the root length (Lin and Xing 2007).

(XV) **Lettuce (*Lactuca sativa*)**

Lettuce is also affected by various nanoparticles including TiO_2, Al_2O_3, ZnO NPs and MWCNTs that in all cased resulted in reduction of root length (Lin and Xing 2007, 2008; Begum et al. 2012; Song et al. 2013b).

11.6 Generation of Reactive Oxygen Species as a Mechanism of Toxicity

Nanoparticles can produce reactive oxygen species (ROS), which negatively affect the biological membranes of any organism living in an ecosystem. With the presence of ions released from the nanoparticles i.e. metal nanoparticles, ground state oxygen (O_2) into O singlets, superoxide anions (O_2^-), hydrogen peroxide (H_2O_2), and hydroxyl radicals ($OH^•$). The effects to the organisms due to ROS depend upon the ROS species generated. The impact of ROS is by induced oxidization of double bonds on fatty acid molecules (phospholipids) present in the biological membranes by affecting the fluidity and permeability, ultimately affecting the substance transfer processes through the membrane while subjecting to osmotic stress. Subsequent damages by ROS can be occurred by damaging genetic material.

Detoxification of ROS mediated toxins occurs inside plant and are mediated by an enzyme system. Super oxide dismutase (SOD), catalase (CAT) and peroxidases (POD) provide the protection to plant cells against ROS, more specifically POD scavenges ROS while SOD and CAT convert $O_2^-$and H_2O_2 into water and O_2 which are non-toxic (Rajput et al. 2017). Under increased concentrations of these enzymes, it indirectly produces phenolic compounds, which have a defense effect over plant pathogens (Cox et al. 2017). However, the mechanism of producing such defense compounds upon exposure of NPs is yet to be revealed. There might be a simulating action by the nanoparticles as pathogens such as viruses and viroids of which the individual units possess dimensions in the nanorange. This in turn might lead the cells to identify the nanoparticles as plant pathogens. In most instances these effects have been reported in a dose dependent manner. Under lower concentrations of nanoparticles, defense systems are effective to scavenge ROS by enzymes. However, at higher concentrations, the scavenging is hindered due to oxidative burst (Rizwan et al. 2017). Levels of generation of these enzymes can vary with the nature of the nanoparticles i.e. bare or coated. For instance, upon exposure to coated nanoparticles, it has been shown of increasing SOD and CAT compared to when exposed to same bare NP species (Lopez-Moreno et al. 2018).

11.7 Nutrient Dynamics and Kinetics Inside Plants due to Nanoparticles

Number of research studies has been demonstrated on curtailing mineral nutrient uptake or reduction of their levels inside certain tissues in the plants treated with nanoparticles (Table 11.1). But it is notable that, the kinetics differ with the factors related to plants and nanoparticles. Interestingly, similar set of factors affect the uptake of nanoparticles themselves. Even though the mechanisms of suppression due to nanoparticles contact is not revealed, might follow the similar mechanisms that involve in exerting toxicity due to exposure to soluble toxic metal ions. Rizwan et al. (2017) and Lopez-Moreno et al. (2018) in their reviews extensively discussed about the impact of nanomaterials on nutrient uptake.

11.8 Toxicity of Nanoparticles in Food Systems

11.8.1 Horizontal and Vertical Transfer of Nanoparticles

Besides the contamination of the primary producers of a food chain as discussed previously, transfer through subsequent trophic levels is also of utmost consideration. Accordingly both vertical and horizontal transfer and trophic accumulation are of utmost importance. Conclusive evidence to prove the potential for biomagnification through the trophic levels of food chains is presently lacking. Several indicative reports provide some supporting information to prove the possibility of nanoparticles through food chains; however, these results yet envisage the probable accumulations at higher trophic levels. Judy et al. (2011) in their study revealed supporting evidence to prove that nanoparticles have potential to biomagnify through food chains. They fed the tobacco hornworms (*Manduca sexta*) on Au NPs surface treated tomato plants, upon which the hornworms were found to contain very low amounts of Au NPs. Contrary to that, the worms fed on plants grown on soil contaminated with Au NPs had significantly higher amounts of accumulated Au NPs, without showing the toxic effects to the hornworms. Moreover, transmission via progeny is a major concern in relation to nanomaterials. Lin et al (2009) proved the ability of natural organic matter modified fullerenes (C_{70}) to present mainly in seeds and in the leaves of the consecutive generation of a rice variety.

11.8.2 Nanoparticles in Food: Toxicological Concerns

The number of research studies on food with respect to human and animal health related consequences upon exposure to nanomaterials is low comparative to the ones based on plants. Therefore, it is worthwhile to note that it needs to have a better

Table 11.1 Effect of different nanoparticles on nutrient dynamics in various plant parts

Plant species	Plant part	Nano particles	Nutrient elements affected													Reference
			N	P	K	Mg	S	Ca	Fe	B	Zn	Cu	Mn	Na	Mo	
Alfalfa	S	CuO		+			+					+				Hong et al. (2015)
Barley	L	Citrate-Au			+				+		+		–			Feichtmeier et al. (2015)
Basil	R	TiO_2							+			+				Tan et al. (2017)
Basil	R	TiO_2-Al_2O_3-dimethicone		–				–				–	+			Tan et al. (2017)
Basil	R	TiO_2-Al_2O_3-glycerol							+							Tan et al. (2017)
Bean	S	CuO						–	–		–			+		Dimkpa et al. (2015)
Bean	R	ZnO				+				+					+	Medina-Velo et al. (2017)
Bean	L	Z-COTE-HP1(ZnO)								+					–	Medina-Velo et al. (2017)
Bean	St	Z-COTE-HP1 (ZnO)								+					+	Medina-Velo et al. (2017)
Bean	R	Z-COTE (ZnO)								+						Medina-Velo et al. (2017)
Bean	R	Z-COTE-HP1 (ZnO)				+		+								Medina-Velo et al. (2017)

(continued)

Table 11.1 (continued)

Plant species	Plant part	Nano particles	Nutrient elements affected													
			N	P	K	Mg	S	Ca	Fe	B	Zn	Cu	Mn	Na	Mo	Reference
Bell pepper	L, F	CuO									–					Rawat et al. (2018)
Cucumber	R	CeO_2									+	+	+		–	Zhao et al. (2014)
Cucumber	R	ZnO													–	Zhao et al. (2014)
Lettuce		CuO		–		–		–					–			Trujillo-Reyes et al. (2014)
Lettuce	S	CuO		–					–							Hong et al. (2015)
Tomato	L, St	SnO_2			–	–	–	–								Antisari et al. (2014)
Transgenic cotton	S, R	CeO_2		–	–	–		–	–	–	–	–	–	–		Li et al. (2014)
Transgenic cotton	S, R	SiO_2				–			–		–	–		–		Le et al. (2014)
Wheat	S	Biochar-Fe_3O_4 Nano Fe complexes	+	+												Joseph et al. (2015)

S shoots, *R* roots, *L* leaves, *St* stems, *F* fruits, + increase in uptake, – decrease in uptake research based on major food related aspects of nanotechnology such as bioavailability, fate, bioactivity and toxicity when exposed in long run. Indeed the importance of such information exists because of the differences in the possibly alike behavior of the materials when at nanoscale, and macro or microscale and due to the toxicity of nanoparticles and the mechanisms might be different to the levels expressed by the respective bulk counterparts (Boumeester et al. 2009). The exposure to nanomaterials could be either direct or indirect as they can be either intentional or unintentional. The direct entry of nanomaterials into the bodies of the higher order organisms occurs via orally, dermally or inhalation pathways.

Inside the body, their behavior can be entirely different to the respective bulk materials with the complex chemical environment, which changes region wise in the gastrointestinal tract (GIT) plus the unique attributes of the nanomaterials themselves. Ultimately, the properties related to adsorption, distribution, metabolism and excretion that govern their toxicity can be changed (Hu et al. 2009). It can expect the nanomaterials to be absorbed at the GIT into the circulation system (Acosta 2009), however they may change the absorption pathway (McClements 2011).

Uptake of nanoparticles into the cells depends upon their size, shape, charge and interfacial chemistry (des Rieux et al. 2006). Upon absorbing into the circulatory system, they can either be metabolized, excrete or accumulated in bodily tissues (Boumeester et al. 2009). Furthermore, NPs with positive charge and hydrophobic nature increase the circulation time. Cationic NPs are likely to be absorbed quickly compared to the anionic ones (Hagens et al. 2007). Accumulation kinetics are also dependent upon their size and tissue type (organ) and the accumulation potential of smaller nanoparticles is higher compared to larger ones (De Jong et al. 2008). Interestingly, the studies have demonstrated the movement of the nanoparticles into the fetus.

11.8.3 Effects of Nanomaterial to the Higher Order Animals

The higher antibacterial activity and low mammalian toxicity of Ag NPs have favored food industry to incorporate them into food to disrupt the microbial activities of food. However, reports indicate that Ag NPs can pass through the GIT into the blood stream (Kim et al. 2008). TiO_2 is also used in the fields which have direct contacts with humans/animals such as cosmetics, food and food packaging (Fujishima et al. 2000). TiO_2 was reported to be non-toxic to humans (Yemmireddy and Hung 2015), but it has now been classified as a group 2B carcinogen by the International Agency for Research on Cancer (IARC) upon inhalation study results. Even though the exposure limits for food grade TiO_2 nanoparticles have been predicted in terms of various age groups, it has been yet failed to establish a clear cut health-based guidance value. Many in vitro studies have shown increased DNA breakage due to exposure to TiO_2 NPs (Magdolenova et al. 2014). ZnO NPs have also been observed to be nontoxic to humans (Gunalan et al. 2012).

11.8.4 Concerns Regarding Delivery Systems

During transportation inside the GIT, the physical state of the nanoparticles can get changed (McClements et al. 2007) due to certain physical phenomena such as flocculation, coalescence, digestion, solubilisation and Ostwald ripening may be encountered by the nanoparticles (Li et al. 2012). Safety concerns on the nanolipids have been gained attention because there is a potential of lipofilic substances to getting changed when encapsulated in nanosized lipid droplets than they are present in simple lipid matrix and leading to change their general absorption pathway (McClements 2011). Generation of free nanoparticles is a major concern with respect to delivery systems based on nanoemulsions (Boumeester et al. 2009; Ahlbom et al. 2012). High bioavailability enhanced by the nanodelivery system sometimes may bring harmful effects to the body (Jiang et al. 2008), thus the effects might only be apparent upon long-term contact with higher frequency (McClements and Xiao 2012).

11.9 Concluding Remarks and Future Implications

The interactions between nanomaterials and living beings are rather complex and evident through the large number of studies, specifically in terms of plant-growth media systems. Due to this complex behaviour of nanomaterials, the room to make firm conclusions has been hindered. The major factors involved in causing toxicity are nanomaterials per se such as, physicochemical parameters, concentration employed, particle size, specific surface area, functionality and stability while at the same time depending upon with other extrinsic factors such as, organismal characteristics including age, species, life cycle stage, pH of the environment that nanoparticles are exposed, characteristics of the growth media (in the case of plants). In such a background, the release of nanomaterials either by deliberate or unintentional attempts to the environment would cause detrimental effects. Analysing the present status of nanomaterials usage and generating a large number of new nanomaterial-based agrochemicals, it can foresee that the humans would be ably replace or reduce the use of agrochemicals presently been used and further with new nano based modes of actions. More attention needs to be paid on studying the contaminations at trophic levels. Addressing the research gaps in the field of nanomaterial research is a timely need. Since two decades of intensive nanomaterial research has been started, a large number of nanomaterial related studies has been conducted, however almost all are confined to the laboratory scale or mini-scale studies and rather short-term. Therefore, it is needed to extend the present research system up to long term field scale studies, under representative environments. Investigating the fate and behaviour of different species of nanomaterials, nanomaterial mixtures, different forms of nanomaterials (pristine and functionalized/coated) and associated risk analysis is a major requirement before upscaling the products. Furthermore long run impacts to the organisms which are directly exposed (crops, farm animals and humans) are to be revealed.

Accordingly, in view of environmental sustainability, interaction of nanomaterials with both biotic and abiotic components in various ecosystems with more emphasis on adaptive modelling is also a need to be addressed. Furthermore, it is needed to be emphasized that more studies on the propensities and impacts of nanomaterials due to trophic transfer and accumulation should be conducted. Finally, stepping ahead, other dimensions such as social acceptance and economic feasibility are to be studied in addition to the associated technological aspects.

References

Acosta E (2009) Bioavailability of nanoparticles in nutrient and nutraceutical delivery. Curr Opin Colloid Interf Sci 14(3):15

Adhikari T, Kundu S, Biswas AK, Tarafdar JC, Rao AC (2012) Effect of copper oxide nanoparticle on seed germination of selected crops. J Agric Sci Technol 2:815–823

Ahlbom A, Bridges J, De Jong W (2012) Scientific committee on emerging and newly identified health risks (SCENIHR). The safety of dental amalgam and alternative dental restoration materials for patients and users, 6 May 2008

Amist N, Singh NB, Yadav K, Singh SC, Pandey JK (2017) Comparative studies of Al^{3+} ions and Al_2O_3 nanoparticles on growth and metabolism of cabbage seedlings. J Biotechnol 254(2017):1–8

Anjum NA, Rodrigo MAM, Moulick A, Heger Z, Kopel P, Zítka O, Adam V, Lukatkin AS, Duarte AC, Pereira E, Kizek R (2016) Transport phenomena of nanoparticles in plants and animals/humans. Environ Res 151:233–243

Antisari LV, Carbone S, Gatti A, Vianello G, Nannipieri P (2014) Uptake and translocation of metals and nutrients in tomato grown in soil polluted with metal oxide (CeO_2, Fe_3O_4, SnO_2, TiO_2) or metallic (Ag, Co Ni) engineered nanoparticles, Environ Sci Pollut Res. http://dx.doi.org/10.1007/s11356-014-3509-0

Antisari LV, Laudicina VA, Gatti A, Carbone S, Badalucco L, Vianello G (2015) Soil microbial biomass carbon and fatty acid composition of earthworm *Lumbricus rubellus* after exposure to engineered nanoparticles. Biol Fertil Soils 51:261–269

Aslani F, Bagheri S, Muhd Julkapli N, Juraimi AS, Hashemi FSG, Baghdadi A (2014) Effects of engineered nanomaterials on plants growth: an overview, Sci World J 1–28

Asli S, Neumann M (2009) Colloidal suspensions of clay or titanium dioxide nanoparticles can inhibit leaf growth and transpiration via physical effects on root water transport. Plant Cell Environ 32:577–584

Atha DH, Wang H, Petersen EJ, Cleveland D, Holbrook RD, Jaruga P et al (2012) Copper oxide nanoparticle mediated DNA damage in terrestrial plant models. Environ Sci Technol 46:1819–1827. https://doi.org/10.1021/es202660k

Balota EL, Colozzi A, Andrade DS, Dick RP (2003) Microbial biomass in soils under different tillage and crop rotation systems. Biol Fertil Soils 38:15–20

Bandyopadhyay S, Plascencia-Villa G, Mukherjee A, Rico CM, Jose-Yacaman M, Peralta-Videa JR, Gardea-Torresdey JL (2015) Comparative phytotoxicity of ZnO NPs bulk ZnO, and ionic zinc onto the alfalfa plants symbiotically associated with *Sinorhizobium meliloti* in soil. Sci Total Environ 515:60–69

Begum P, Fugetsu B (2012) Phytotoxicity of multi-walled carbon nanotubes on red spinach (*Amaranthus tricolor* L) and the role of ascorbic acid as an antioxidant. J Hazard Mater 243:212–222

Begum P, Ikhtiari R, Fugetsu B (2011) Graphene phytotoxicity in the seedling stage of cabbage, tomato, red spinach, and lettuce. Carbon 49:3907–3919. https://doi.org/10.1016/j.carbon.2011.05.029

Begum P, Ikhtiari R, Fugetsu B, Matsuoka M, Akasaka T, Watari F (2012) Phytotoxicity of multi-walled carbon nanotubes assessed by selected plant species in the seedling stage. Appl Surf Sci 262(2012):120–124
Ben-Moshe T, Frenk S, Dror I, Minz D, Berkowitz B (2013) Effects of metal oxide nanoparticles on soil properties. Chemosphere 90(2):640–646. https://doi.org/10.1016/j.chemosphere.2012.09.018
Bombin S, Le Febvre M, Sherwood J, Xu Y, Bao Y, Ramonell KM (2015) Developmental and reproductive effects of iron oxide nanoparticles in *Arabidopsis thaliana*. Int J Mol Sci 16:24174–24193
Boonyanitipong P, Kositsup B, Kumar P, Baruah S, Dutta J (2011) Toxicity of ZnO and TiO_2 nanoparticles on germinating rice seeds. Int J Biosci Biochem Bioinform 1:282–285
Bouwmeester H, Dekkers S, Noordam MY, Hagens WI, Bulder AS, de Heer, ten Voorde SECG, Wijnhoven SWP, Marvin HJP, Sips AJAM (2009) Review of health safety aspects of nanotechnologies in food production, Regul Toxicol Pharmacol 53:52–62
Capaldi-Arruda SC, Silva ALD, Galazzi RM, Azevedo RA, Arruda MAZ (2015) Nanoparticles applied to plant science: a review. Talanta 131:693–705. https://doi.org/10.1016/j.talanta.2014.08.050
Castiglione MR, Giorgetti L, Geri C, Cremonini R (2011) The effects of nano-TiO_2 on seed germination, development and mitosis of root tip cells of *Vicia narbonensis* L. and *Zea mays* L. J Nanopart Res 13:2443–2449
Chai H, Yao J, Sun J, Zhang C, Liu W, Zhu M, Ceccanti B (2015) The effect of metal oxide nanoparticles on functional bacteria and metabolic profiles in agricultural soil. Bull Environ Contam Toxicol 94(4):490–505
Chen R, Ratnikova TA, Stone MB, Lin S, Lard M, Huang G, Hudson JS, Ke PC (2010) Differential uptake of carbon nanoparticles by plant and mammalian cells. Small 6:612–617
Chung H, Son Y, Yoon TK, Kim S, Kim W (2011) The effect of multi-walled carbon nanotubes on soil microbial activity. Ecotoxicol Environ Saf 74:569–575
Cox A, Venkatachalam P, Sahi S, Sharma N (2017) Reprint of: silver and titanium dioxide nanoparticle toxicity in plants: a review of current research. Plant Physiol Biochem 110:33–49
De Jong WH, Hagens WI, Krystek P, Burger MC, Sips A, Geertsma RE (2008) Particle size-dependent organ distribution of gold nanoparticles after intravenous administration. Biomaterials 29:1912–1919
Demir E, Kaya N, Kaya B (2014) Genotoxic effects of zinc oxide and titanium dioxide nanoparticles on root meristem cells of *Allium cepa* by comet assay. Turk J Biol 38:31–39
des Rieux A, Fievez V, Garinot M, Schneider YJ, Preat V (2006) Nanoparticles as potential oral delivery systems of proteins and vaccines: a mechanistic approach. J Control Release 116:1–27
Dietz KJ, Herth S (2011) Plant nanotoxicology. Trends Plant Sci 16:582–589
Dimkpa CO, Calder A, Britt DW, McLean JE, Anderson AJ (2011) Responses of a soil bacterium, *Pseudomonas chlororaphis* O6 to commercial metal oxide nanoparticles compared with responses to metal ions. Environ Pollut 159(7):1749–1756
Dimkpa CO, McLean JE, Britt DW, Anderson AJ (2015) Nano-CuO and interaction with nano-ZnO or soil bacterium provide evidence for the interference of nanoparticles in metal nutrition of plants. Ecotoxicology 24:119–129
Dimkpa CO, McLean JE, Latta DE, Manangon E, Britt DW, Johnson WP, Boyanov MI, Anderson AJ (2012) CuO and ZnO nanoparticles: phytotoxicity, metal speciation, and induction of oxidative stress in sand-grown wheat. J Nanopart Res 14(2012):1–15. https://doi.org/10.1007/s11051-012-1125-9
Dinesh R, Anandaraj M, Srinivasan V, Hamza S (2012) Engineered nanoparticles in the soil and their potential implications to microbial activity. Geoderma 173–174:19–27. https://doi.org/10.1016/j.geoderma.2011.12.018
Du WC, Sun YY, Ji R, Zhu JG, Wu JC, Guo HY (2011) TiO_2 and ZnO Nanoparticles negatively affect wheat growth and soil enzyme activities in agricultural soil. J Environ Monit 13:822–828
Etxeberria E, Gonzalez P, Pozueta J (2009) Evidence for two endocytic transport pathways in plant cells. Plant Sci 177:341–348

Feichtmeier NS, Walther P, Leopold K (2015) Uptake, effects, and regeneration of barley plants exposed to gold nanoparticles. Environ Sci Pollut Res 22:8549–8558

Feizi H, Moghaddam PR, Shahtahmassebi N, Fotovat A (2013) Assessment of concentrations of nano and bulk iron oxide particles on early growth of wheat (*Triticum aestivum* L.). Ann Rev Res Biol 3:752–761

Feng Y, Cui X, He X, Dong G, Chen M, Wang J, Lin X (2013) The role of metal nanoparticles in influencing arbuscular mycorrhizal fungi effects on plant growth. Environ Sci Technol 47:9496–9504. https://doi.org/10.1021/es402109n

Fierer N, Jackson RB (2006) The diversity and biogeography of soil bacterial communities. PNAS 103(3):626–631

Frazier TP, Burklew CE, Zhang BH (2014) Titanium dioxide nanoparticles affect the growth and micro RNA expression of tobacco (*Nicotiana tabacum*). Funct Integr Genom 14(1):75–83. https://doi.org/10.1007/s10142-013-0341-4

Fujishima A, Rao TN, Truk DA (2000) Titanium dioxide photocatalysis. J Photochem Photobiol C Photo Chem Rev 1(1):1–21

Gajjar P, Pettee B, Britt DW, Huang W, Johnson WP, Anderson AJ (2009) Antimicrobial activities of commercial nanoparticles against an environmental soil microbe, *Pseudomonas putida* KT2440, J Biol Eng 3(1):9. http://dx.doi.org/10.1186/1754-1611-3-9

Ge Y, Schimel JP, Holden PA (2011) Evidence for negative effects of TiO_2 and ZnO nanoparticles on soil bacterial communities. Environ Sci Technol 45:1659–1664

Ghosh M, Bandyopadhyay M, Mukherjee A (2010) Genotoxicity of titanium dioxide (TiO_2) nanoparticles at two trophic levels: plant and human lymphocytes. Chemosphere 81:1253–1262

Gunalan S, Sivaraj R, Rajendran V (2012) Green synthesized ZnO nanoparticles against bacterial and fungal pathogens. Prog Nat Sci Mater Int 22(6):693–700

Hagens WI, Oomen AG, de Jong WH, Cassee FR, Sips A (2007) What do we (need to) know about the kinetic properties of nanoparticles in the body? Regul Toxicol Pharmacol 49(217):229

Hawthorne J, Musante C, Sinha SK, White JC (2012) Accumulation and Phytotoxicity of Engineered Nanoparticles to. Int J Phytorem 14(4):429–442

Hong J, Rico CM, Zhao L, Adeleye AS, Keller AA, Peralta-Videa JR, Gardea-Torresdey GL (2015) Toxic effects of copper-based nanoparticles or compounds to lettuce (*Lactuca sativa*) and alfalfa (*Medicago sativa*). Environ Sci Process Impacts 17:177–185

Horst AM, Neal AC, Mielke RE, Sislian PR, Suh WH, Mädler L, Stucky GD, Holden PA (2010) Dispersion of TiO_2 nanoparticle agglomerates by *Pseudomonas aeruginosa*. Appl Environ Microbiol 76:7292–7298

Hu L, Mao ZW, Gao CY (2009) Colloidal particles for cellular uptake and delivery. J Mater Chem 19:3108–3115

Jiang J, Oberdorster G, Biswas P (2008) Characterization of size, surface charge, and agglomeration state of nanoparticle dispersions for toxicological studies. J Nanopart Res 11:77–89

Jin L, Son Y, Yoon TK, Kang YJ, Kim W, Chung H (2013) High concentrations of single-walled carbon nanotubes lower soil enzyme activity and microbial biomass. Ecotoxicol Environ Saf 88:9–15

Johansen A, Pedersen AL, Jensen KA, Karlson U, Hansen BM, Scott-Fordsmand JJ, Winding A (2008) Effects of C_{60} fullerene nanoparticles on soil bacteria and protozoans. Environ Toxicol Chem 27:1895–1903. https://doi.org/10.1897/07-375.1

Joseph S, Anawar HM, Storer P, Blackwell P, Chia C, Lin Y, Munroe P, Donne S, Horvat J, Wang J, Solaiman ZM (2015) Effects of enriched biochars containing magnetic iron nanoparticles on mycorrhizal colonisation, plant growth, nutrient uptake and soil quality improvement. Pedosphere 25(5):749–760

Judy JD, Unrine JM, Bertsch M (2011) Evidence for biomagnification of gold nanoparticles within a terrestrial food chain. Environ Sci Technol 45:776–781

Kerfahi D, Tripathi BM, Singh D, Kim H, Lee S, Lee J, Adams JM (2015) Effects of functionalized and raw multi-walled carbon nanotubes on soil bacterial community composition. PLoS ONE 10:e123042

Khare P, Sonane M, Nagar Y, Moin N, Ali S, Gupta KC, Satish A (2015) Size dependent toxicity of zinc oxide nanoparticles in soil nematode *Caenorhabditis elegans*. Nanotoxicology 9:423–432

Kim S, Kim J, Lee I (2011) Effects of Zn and ZnO nanoparticles and Zn^{2+} on soil enzyme activity and bioaccumulation of Zn in *Cucumis sativus*. Chem Ecol 27:49–55

Kim S, Lee S, Lee I (2012) Alteration of phytotoxicity and oxidant stress potential by metal oxide nanoparticles in *Cucumis sativus*. Wat Air Soil Pollut 223:2799–2806

Kim YS, Kim JS, Cho HS, Rha DS, Kim JM, Park JD, Choi BS, Lim R, Chang HK, Chung YH, Kwon IH, Jeong J, Han BS, Yu IJ (2008) Twenty eight-day oral toxicity, genotoxicity, and gender-related tissue distribution of silver nanoparticles in Spraguee Dawley rats. Inhal Toxicol 20(6):575–583

Kouhi SMM, Lahouti M, Ganjeali A, Entezari MH (2014) Comparative phytotoxicity of ZnO nanoparticles ZnO microparticles, and Zn^{2+} on rapeseed (*Brassica napus* L.): investigating a wide range of concentrations. Toxicol Environ Chem 96:861–868

Kouhi SMM, Lahouti M, Ganjeali A, Entezari MH (2015) Long-term exposure of rapeseed (*Brassica napus* L.) to ZnO nanoparticles: anatomical and ultrastructural responses. Environ Sci Pollut Res 22(14):10733–10743. https://doi.org/10.1007/s11356-015-4306-0

Kumar N, Shah V, Walker VK (2011) Perturbation of an arctic soil microbial community by metal nanoparticles. J Hazard Mater 190(1–3):816–822

Larue C, Laurette J, Herlin-Boime N (2012) Accumulation, translocation and impact of TiO_2 nanoparticles in wheat (*Triticum aestivum* spp.): influence of diameter and crystal phase. Sci Total Environ 431:197–208

Laware SL, Raskar S (2014) Effect of titanium dioxide nanoparticles on hydrolytic and antioxidant enzymes during seed germination in onion. Int J Curr Microbiol Appl Sci 3(7):749–760

Le VN, Rui Y, Gui X, Li X, Liu S, Han Y (2014) Uptake, transport, distribution and Bio-effects of SiO_2 nanoparticles in Bt-transgenic cotton. J Nanobiotechnol 12:50

Lee WM, An YJ, Yoon H, Kweon HS (2008) Toxicity and bioavailability of copper nanoparticles to the terrestrial plants mungbean (*Phaseolus radiatus*) and wheat (*Triticum aestivum*): plant agar test for water insoluble nanoparticles. Environ Toxicol Chem 27:1915–1921

Lee WM, Kwak JI, An YJ (2012) Effect of silver nanoparticles in crop plants *Phaseolus radiatus* and *Sorghum bicolor*: media effect on phytotoxicity. Chemosphere 86:491–499

Li X, Gui X, Rui Y, Ji W, Yu Z, Peng S (2014) Bt-transgenic cotton is more sensitive to CeO_2 nanoparticles than its parental non-transgenic cotton. J Hazard Mater 274:173–180

Li Y, Zheng J, Xiao H, McClements DJ (2012) Nanoemulsion-based delivery systems for poorly water-soluble bioactive compounds: influence of formulation parameters on polymethoxyflavone crystallization. Food Hydrocoll 27:517–528

Lin D, Xing B (2007) Phytotoxicity of nanoparticles: inhibition of seed germination and root growth. Environ Pollut 150:243–250

Lin D, Xing B (2008) Root uptake and phytotoxicity of ZnO nanoparticles. Environ Sci Technol 42:5580–5585

Lin S, Reppert J, Hu Q, Hudson JS, Reid ML, Ratnikova TA, Rao AM, Luo H, Ke PC (2009) Uptake, translocation, and transmission of carbon nanomaterials in rice plants. Small 5(10):1128–1132

Lopez-Moreno ML, Cedeno-Mattei Y, Bailon-Ruizf SJ, Vazquez-Nunez E, Hernandez-Viezcas JA, Perales-Perez OJ, De La Rosa G, Peralta-Videa JR, Gardea-Torresdey JL (2018) Environmental behavior of coated NMs: physicochemical aspects and plant interactions. J Hazard Mater 347:196–217

Lopez-Moreno ML, De La Rosa G, Hernandez-Viezcas JA, Peralta-Videa JR, Gardea-Torresdey JL (2010) X-ray absorption spectroscopy (XAS) corroboration of the uptake and storage of CeO_2 nanoparticles and assessment of their differential toxicity in four edible plant species. J Agric Food Chem 58:3689–3693

Magdolenova Z, Collins AR, Kumar A, Dhawan A, Stone V, Dusinska M (2014) Mechanisms of genotoxicity. Rev Recent In vitro In vivo Stud Eng Nanopart Nanotoxicol 8:233–278

Mazumdar H, Ahmed GU (2011) Phytotoxicity effect of silver nanoparticles on *Oryza sativa*. Int J Chem Technol Res 3:1494–1500

McClements DJ (2011) Edible nanoemulsions: fabrication, properties, and functional performance. Soft Matter 7:2297–2316

McClements DJ, Decker EA, Weiss J (2007) Emulsion-based delivery systems for lipophilic bioactive components. J Food Sci 72:109–124

McClements DJ, Xiao H (2012) Potential biological fate of ingested nanoemulsions: influence of particle characteristics. Food Funct 3:202–220

Medina-Velo IA, Barrios AC, Zuverza-Mena N, Hernandez-Viezcas JA, Chang CH, Ji X, Zink JI, Peralta-Videa JR, Gardea-Torresdey JL (2017) Comparison of the effects of commercial coated and uncoated ZnO NMs and Zn compounds in kidney bean (*Phaseolus vulgaris*) plants. J Hazard Mater 332:214–222

Mirzajani F, Askari H, Hamzelou S, Farzaneh M, Ghassempour (2013) Effect of silver nanoparticles on *Oryza sativa* L. and its rhizosphere bacteria. Ecotoxicol Environ Saf 88:48–54

Mukherjee A, Majumdar S, Servin AD, Pagano L, Dhankher OP, White JC (2016) Carbon nanomaterials in agriculture: a critical review. Front Plant Sci 7:172. https://doi.org/10.3389/fpls.2016.00172

Nagaonkar D, Shende S, Rai M (2015) Biosynthesis of copper nanoparticles and its effect on actively dividing cells of mitosis in *Allium cepa*. Biotechnol Prog 31(2):557–565. https://doi.org/10.1002/btpr.2040

Nair PMG, Chung IM (2014a) A mechanistic study on the toxic effect of copper oxide nanoparticles in soybean (*Glycine max* L.) root development and lignification of root cells. Biol Trace Elem Res 162:342–352

Nair PMG, Chung IM (2014b) Physiological and molecular level effects of silver nanoparticles exposure in rice (*Oryza sativa* L.) seedlings. Chemosphere 112:105–113

Nair PMG, Chung IM (2015a) Changes in the growth, redox status and expression of oxidative stress related genes in chickpea (*Cicer arietinum* L.) in response to copper oxide nanoparticle exposure. J Plant Growth Regul 34(2):350–361. http://dx.doi.org/10.1007/s00344-014-9468-3

Nair PMG, Chung IM (2015b) Study on the correlation between copper oxide nanoparticles induced growth suppression and enhanced lignification in Indian mustard (*Brassica juncea* L.). Ecotoxicol Environ Saf 113:302–313

Pakrashi S, Jain N, Dalai S, Jayakumar J, Chandrasekaran PT, Raichur AM, Chandrasekaran N, Mukherjee A (2014) *In vivo* genotoxicity assessment of titanium dioxide nanoparticles by *Allium cepa* root tip assay at high exposure concentrations. PLoS One e87789

Pallavi CM, Mehta Srivastava R, Arora S, Sharma AK (2016) Impact assessment of silver nanoparticles on plant growth and soil bacterial diversity. Biotech 6:254

Pittol M, Tomacheski D, Simoes DN, Ribeiro VF, Santana RMC (2017) Macroscopic effects of silver nanoparticles and titanium dioxide on edible plant growth. Environ Nanotechnol Monitor Manage 8:127–133

Pokhrel LR, Dubey B (2013) Evaluation of developmental responses of two crop plants exposed to silver and zinc oxide nanoparticles. Sci Total Environ 452:321–332

Rafique R, Zahra Z, Virk N, Shahid M, Pinelli E, Park TJ, Kallerhoff J, Muhammad A (2018) Dose-dependent physiological responses of *Triticum aestivum* L. to soil applied TiO_2 nanoparticles: alterations in chlorophyll content, H_2O_2 production, and genotoxicity. Agric Ecosyst Environ 255:95–101. https://doi.org/10.1016/j.agee.2017.12.010

Rajput VD, Minkina TM, Behal A, Sushkova SN, Mandzhieva S, Singh R, Gorovtsov A, Tsitsuashvili VS, Purvis WO, Ghazaryan KA, Movsesyan HS (2017) Effects of zinc-oxide nanoparticles on soil, plants, animals and soil organisms: a review. Environ Nanotechnol Monitor Manage 9:76–84

Raliya R, Nair R, Chavalmane S, Wangab W, Biswas P (2015) Mechanistic evaluation of translocation and physiological impact of titanium dioxide and zinc oxide nanoparticles on the tomato (*Solanum lycopersicum* L.) plant. Metallomics 7(12):1584–1594

Rao S, Shekhawat GS (2014) Toxicity of ZnO engineered nanoparticles and evaluation of their effect on growth, metabolism and tissue specific accumulation in *Brassica juncea*. J Environ Chem Eng 2:105–114

Rashid MI, Shahzad T, Shahid M, Ismail IM, Shah GM, Almeelbi T (2017) Zinc oxide nanoparticles affect carbon and nitrogen mineralization of *Phoenix dactylifera* leaf litter in a sandy soil. J Hazard Mater 324:298–305

Raskar SV, Laware SL (2014) Effect of zinc oxide nanoparticles on cytology and seed germination in onion. Int J Curr Microbiol App Sci 3:467–473

Rawat S, Pullagurala VLR, Hernandez-Molina M, Sun Y, Niu G, Hernandez-Viezcas JA, Peralta-Videa JR, Gardea-Torresdey JL (2018) Impacts of copper oxide nanoparticles on bell pepper (*Capsicum annum* L.) plants: a full life cycle study. Environ Sci Nano 5:83–95. https://doi.org/10.1039/C7EN00697G

Read DS, Matzke M, Gweon HS, Newbold LK, Heggelund L, Ortiz MD, Lahive E, Spurgeon D, Svendsen C (2016) Environ Sci Pollut Res Int 23(5):4120–4128

Ren W, Ren G, Teng Y, Li Z, Li L (2015) Time-dependent effect of graphene on the structure, abundance, and function of the soil bacterial community. J Hazard Mater 297:286–294

Rico CM, Majumdar S, Duarte-Gardea M, Peralta-Videa JR, Gardea-Torresdey JL (2011) Interaction of nanoparticles with edible plants and their possible implications in the food chain. J Agric Food Chem 59:3485–3498

Rico CM, Morales MI, Barrios AC, McCreary R, Hong J, Lee WY, Nunez J, Peralta-Videa JR, Gardea-Torresdey JL (2013b) Effect of cerium oxide nanoparticles on the quality of rice (*Oryza sativa* L.) grains. J Agric Food Chem 61:11278–11285. https://doi.org/10.1021/jf404046v

Rico CM, Morales MI, McCreary R, Castillo-Michel H, Barrios AC, Hong J, Tafoya A, Lee W-Y, Varela-Ramirez A, Peralta-Videa JR, Gardea-Torresdey JL (2013a) Cerium oxide nanoparticles modify the antioxidative stress enzyme activities and macromolecule composition in rice seedlings. Environ Sci Technol 47:14110–14118. https://doi.org/10.1021/es4033887

Rizwan M, Ali S, Qayyum MF, Ok YS, Adrees M, Ibrahim M, Zia-ur-Rehman M, Farid M, Abbas F (2017) Effect of metal and metal oxide nanoparticles on growth and physiology of globally important food crops: a critical review. J Hazard Mater 322:2–16

Rodrigues DF, Jaisi DP, Elimelech M (2013) Toxicity of functionalized single-walled carbon nanotubes on soil microbial communities: implications for nutrient cycling in soil. Environ Sci Technol 47:625–633. https://doi.org/10.1021/es304002q

Ruttkay-Nedecky B, Krystofova O, Nejdl L, Adam V (2017) Nanoparticles based on essential metals and their phytotoxicity. J Nanobiotechnol 26(15):33. https://doi.org/10.1186/s12951-017-0268-3

Shaw AK, Ghosh S, Kalaji HM, Bosa K, Brestic M, Zivcak M, Hossain Z (2014) Nano-CuO stress induced modulation of antioxidative defense and photosynthetic performance of Syrian barley (*Hordeum vulgare* L.). Environ Exp Bot 102:37–47

Shen CX, Zhang QF, Li J, Bi FC, Yao N (2010) Induction of programmed cell death in *Arabidopsis* and rice by single-wall carbon nanotubes. Am J Bot 97(10):1602–1609

Siddiqi KS, Husen A (2017) Plant response to engineered metal oxide nanoparticles. Nanoscale Res Lett 12:92. https://doi.org/10.1186/s11671-017-1861-y

Simonin M, Richaume A (2015) Impact of engineered nanoparticles on the activity abundance, and diversity of soil microbial communities: a review. Environ Sci Pollut Res 22:13710–13723

Song U, Jun H, Waldman B, Roh J, Kim Y, Yi J, Lee EJ (2013a) Functional analyses of nanoparticle toxicity: a comparative study of the effects of TiO_2 and Ag on tomatoes (*Lycopersicon esculentum*). Ecotoxicol Environ Saf 93:60–67

Song U, Shin M, Lee G, Roh J, Kim Y, Lee EJ (2013b) Functional analysis of TiO_2 nanoparticle toxicity in three plant species. Biol Trace Elem Res 155(1):93–103. https://doi.org/10.1007/s12011-013-9765-x

Stampoulis D, Sinha SK, White JC (2009) Assay-dependent phytotoxicity of nanoparticles to plants. Environ Sci Technol 43:9473–9479

Tan WJ, Du WC, Barrios AC, Jr Armendariz, Zuverza-Mena R, Ji N, Chang ZX, Zink CH, Hernandez-Viezcas JA, Peralta-Videa JR, GardeaTorresdey JL (2017) Surface coating changes the physiological and biochemical impacts of nano-TiO_2 in basil (*Ocimum basilicum*) plants. Environ Pollut 222:64–72

Thakur MP, Milcu A, Manning P, Niklaus PA, Roscher C, Power S, Reich PB, Scheu S, Tilman D, Ai F, Guo H, Ji R, Pierce S, Ramirez NG, Richter AN, Steinauer K, Strecker T, Vogel A, Eisenhauer N (2015) Plant diversity drives soil microbial biomass carbon in grasslands irrespective of global environmental change factors. Glob Change Biol 21:4076–4085

Tong Z, Bischoff M, Nies LF, Carroll NJ, Applegate B, Turco RF (2016) Influence of fullerene (C-60) on soil bacterial communities: aqueous aggregate size and solvent co-introduction effects. Sci Rep 6

Tripathi DK, Singh S, Singh S, Srivastava PK, Singh VP, Singh S, Prasad SM, Singh PK, Dubey NK, Pandey AC, Chauhan DK (2017) Nitric oxide alleviates silver nanoparticles (AgNps)-induced phytotoxicity in *Pisum sativum* seedlings. Plant Physiol Biochemi 110:167–177

Trujillo-Reyes J, Majumdar S, Botez CE, Peralta-Videa JR, Gardea-Torresdey JL (2014) Exposure studies of core-shell Fe/$Fe_3O_4$and Cu/CuO NPs to lettuce (*Lactuca sativa*) plants: are they a potential physiological and nutritional hazard. J Hazard Mater 267:255–263

Vannini C, Domingo G, Onelli E, De Mattia F, Bruni I, Marsoni M, Bracale M (2014) Phytotoxic and genotoxic effects of silver nanoparticles exposure on germinating wheat seedlings. J Plant Physiol 171:1142–1148

Wang Z, Xie X, Zhao J, Liu X, Feng W, White JC, Xing B (2012) Xylem-and phloem-based transport of CuO nanoparticles in maize (*Zea mays* L.). Environ Sci Technol 46:4434–4441

Wu B, Zhu L, Le XC (2017) Metabolomics analysis of TiO_2 nanoparticles induced toxicological effects on rice (*Oryza sativa* L.). Environ Pollut 230:302–310. https://doi.org/10.1016/j.envpol.2017.06.062

Xu C, Peng C, Sun L, Zhang S, Huang H, Chen Y, Shi J (2015) Distinctive effects of TiO_2 and CuO nanoparticles on soil microbes and their community structures in flooded paddy soil. Soil Biol Biochem 86:24–33

Yemmireddy VK, Hung YC (2015) Effect of binder on the physical stability and bactericidal property of titanium dioxide (TiO_2) nanocoatings on food contact surfaces. Food Control 57:82–88

Yoon SJ, Kwak JI, Lee WM, Holden PA, An YJ (2014) Zinc oxide nanoparticles delay soybean development: a standard soil microcosm study. Ecotoxicol Environ Saf 100:131–137

Zhai G, Gutowski SM, Walters KS, Yan B, Schnoor JL (2015) Charge, size, and cellular selectivity for multiwall carbon nanotubes by maize and soybean, Environ Sci Technol 49:7380–7390. http://dx.doi.org/10.1021/acsest.5b01145

Zhang R, Zhang H, Tu C, Hu X, Li L, Luo Y, Christie P (2015) Phytotoxicity of ZnO nanoparticles and the released Zn (II) ion to corn (*Zea mays* L.) and cucumber (*Cucumis sativus* L.) during germination. Environ Sci Pollut Res 22(14):11109–11117. http://dx.doi.org/10.1007/s11356-015-4325-x

Zhao L, Peralta-Videa JR, Ren M, Varela-Ramirez A, Li C, Hernandez-Viezcas JA, Aguilera RJ, Gardea-Torresdey JL (2012a) Transport of Zn in a sandy loam soil treated with ZnO NPs and uptake by corn plants: electron microprobe and confocal microscopy studies. Chem Eng J 184:1–8

Zhao L, Peralta-Videa JR, Varela-Ramirez A, Castillo-Michel H, Li C, Zhang J, Aguilera RJ, Keller AA, Gardea-Torresdey JL (2012b) Effect of surface coating and organic matter on the uptake of CeO_2 NPs by corn plants grown in soil: Insight into the uptake mechanism. J Hazard Mater 225–226:131–138

Zhao L, Peralta-Videa JR, Rico CM, Hernandez-Viezcas JA, Sun Y, Niu G, Servin A, Nunez JE, Duarte-Gardea M, Gardea-Torresdey JL (2014) CeO_2 and ZnO nanoparticles change the nutritional qualities of cucumber (*Cucumis sativus*). J Agric Food Chem 62(13):2752–2759

Zhou D, Jin S, Li L, Wang Y, Weng N (2011) Quantifying the adsorption and uptake of CuO nanoparticles by wheat root based on chemical extractions. J Environ Sci 23:1852–1857

Part II
Risks

Chapter 12
Carbon Nanomaterials: Potential Risks to Human Health and the Environment

O. V. Sinitsyna, G. B. Meshkov and I. V. Yaminsky

Abstract Carbon nanomaterials, such as carbon nanotubes, graphene, and graphene oxide, are perspective for application in electronics, energy storage, aerospace and defense industries, biomedicine, etc. In the chapter, we briefly overview the structure, unique properties, current and future applications of carbon nanomaterials. Their effects on living systems are considered in details. We identify the potential health and environmental risks related to the release of the carbon nanoparticles to the environment. The chapter addresses the issues of bioaccumulation and biodegradation of the carbon nanomaterials.

Keywords Graphene · Graphene oxide · Carbon nanotubes · Toxicity
Biodegradation

Nomenclature

CNT	Carbon Nanotube
GO	Graphene Oxide
SWCNT	Single-Walled Carbon Nanotube
MWCNT	Multi-Walled Carbon Nanotube
TEM	Transmission Electron Microscopy
AFM	Atomic Force Microscope
ROS	Reactive Oxygen Species

O. V. Sinitsyna (✉) · I. V. Yaminsky
Laboratory for Physical Chemistry of Polymers, A. N. Nesmeyanov Institute of Organoelement Compounds of Russian Academy of Sciences, Moscow, Russia
e-mail: sinitsyna@gmail.com

I. V. Yaminsky
e-mail: yaminsky@nanoscopy.org

G. B. Meshkov · I. V. Yaminsky
Faculty of Physics, Lomonosov Moscow State University, Moscow, Russia
e-mail: meshkov@polly.phys.msu.ru

M. Rai and J. K. Biswas (eds.), *Nanomaterials: Ecotoxicity, Safety, and Public Perception*, https://doi.org/10.1007/978-3-030-05144-0_12

NIOSH National Institute for Occupational Safety and Health
CDC Centers for Disease Control and Prevention

12.1 Introduction

In 1991, a Japanese researcher Sumio Iijima of NEC Corporation attracted the attention of the scientific community to the unusual carbon nanostructures, consisting of needle-like tubes (Iijima 1991). From this time, a boom in research on carbon nanotubes (CNTs) has begun. Their outstanding mechanical, electrical, optical and thermal properties promised great prospects for application in aerospace, microelectronics, energy storage, and nanomedicine. In 2004, the attention of researchers working in the field of carbon nanomaterials was switched to another new carbon allotrope, graphene (Novoselov et al. 2004). Konstantin Novoselov and Andre Geim were awarded the Nobel Prize in 2010 "for groundbreaking experiments regarding the two-dimensional material graphene" (Web Site of the Nobel Prize 2018). Graphene is a very strong, transparent, and flexible conductor. The potential of its application in electronics, energy storage, composites industries is enormous. One of the attractive ways to synthesize graphene for the industry is the reduction of graphene oxide (GO), which is obtained by graphite oxide dispersing. An intermediate water-soluble compound, GO deserves special consideration. In near future, it is expected that the GO will be used in supercapacitors, lithium-ion batteries, systems for water treatment, nonlinear optical devices, etc.

But the wider usage of CNTs, graphene, and GO in industry, the more concerns arise about the risks to health and the environment. Can the carbon nanoparticles be harmful? What is their impact on the environment? What are the safe ways to use the unique properties of these nanomaterials? We try to address these questions in this chapter.

12.2 Structure of Carbon Nanoparticles and Instrumentation for Their Visualization and Detection

Graphene is a hexagonal lattice of carbon atoms. There are single-layer graphene with monoatomic thickness, and few layer graphene. Cylinders from graphene are called carbon nanotubes. There are also single-walled and multi-walled carbon nanotubes (SWCNTs and MWCNTs). The more layers in graphene and carbon nanotubes, the more similar to graphite are their properties. Graphene sheets and carbon nanotubes tend to aggregate, so special efforts are required to prepare their fine dispersions. SWCNT paper is shown in Fig. 12.1, where individual nanotubes are assembled into bundles.

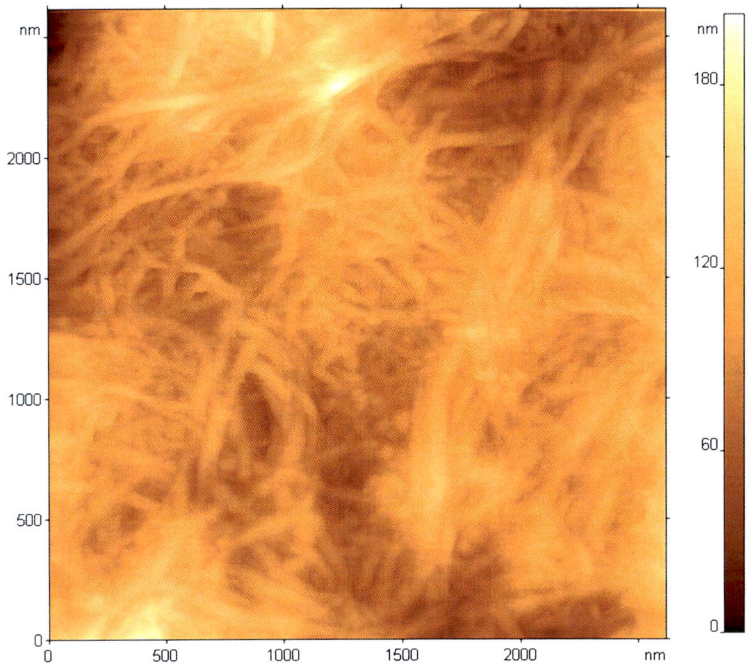

Fig. 12.1 SWCNT paper. The image was obtained with FemtoScan atomic force microscope (AFM)

Raman spectroscopy is very useful technique to confirm the presence of SWCNTs and graphene and their characterization (Liu et al. 2014). It is a quick and non-destructive method. The most intense G band around 1590 cm^{-1} is a feature of sp^2 bonded carbon materials (Dresselhaus et al. 2005). The D band around 1350 cm^{-1} is related to disorder in sp^2 bonded carbons (Pimenta et al. 2001). The ratio of the intensity of the D band to the G band is related to the quality of CNTs and graphene (Qian et al. 2003). The peaks found at low wave number, <350 cm^{-1}, in spectra of SWCNTs are assigned to radial breathing modes (Dresselhaus et al. 2005). These modes are unique to carbon nanotubes. Their origin is related to the radial symmetrical expansion and contraction of the tubes. However, these peaks are absent in the spectra of MWCNTs due to their very low intensity. Knowing the frequency of the radial breathing modes, the diameter of the SWCNTs can be calculated (Dresselhaus et al. 2005). The shape and intensity of 2D band of graphene at 2700 cm^{-1} depend on the number of the layers, which helps to differentiate graphene samples with less than 5 layers (Ferrari et al. 2006).

Significant importance for identification and determination of the structure of carbon nanotubes and graphene belongs to transmission electron microscopy (TEM) and methods of scanning probe microscopy. TEM allows one to determine the number of walls in CNTs, their diameter (Liu et al. 2014) and thickness of graphene

(Hernandez et al. 2008). Using scanning tunneling microscopy and spectroscopy the atomic structure of SWCNTs was resolved and the dependence of their electronic properties (semiconducting or metallic) on wrapping angle and diameter was studied (Wildöer et al. 1998).

It is possible to check the uniformity of the carbon nanomaterial distribution in composites by the evaluation of thin sections. A TEM image of a composite with MWCNTs is shown in Fig. 12.2. It was synthesized by anionic polymerization of ε-caprolactam in the presence of MWCNTs. The addition of nanotubes increases the rigidity and improves the friction characteristics of polycaproamide. Insignificant quantities of the carbon nanotubes (about of 0.1% by weight) reduce the wear and friction coefficient of the composite. One of the roles of MWCNTs is the nucleation of the fine spherulitic polycaproamide structure (Krasnov et al. 2009; Vygodskii et al. 2009).

The most popular chemical methods of graphite oxide preparation are Hummers' method (Hummers and Offeman 1958) and its modifications (Marcano et al. 2010). In Hummers' method, graphite is oxidized by a mixture of concentrated sulfuric acid, potassium permanganate, and sodium nitrate. Graphite oxide is dispersed on single layer graphene oxide in diluted alkaline solutions. Graphite intercalated compounds with sulfuric acid are formed at the first stage of Hummers' method followed by graphite layers oxidation. If the graphite precursor has a well ordered microstructure, its oxidation proceeds very slowly and can take several hours (Dimiev and Tour 2014). The quenching of insufficiently oxidized graphite with water may result in blister formation (Sinitsyna et al. 2018) (Fig. 12.3). Sinitsyna et al. (2018) suggest that the blisters are filled with water. Thus, the samples of few layer graphene oxide may contain the products of incomplete decomposition of the graphite intercalation

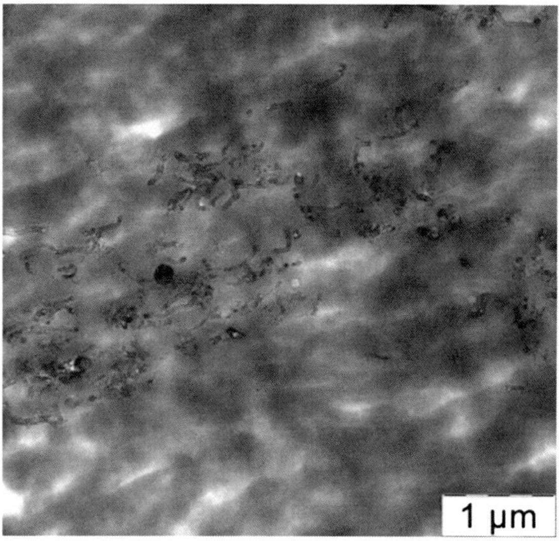

Fig. 12.2 MWCNTs in a polymer composite. TEM image of a thin section

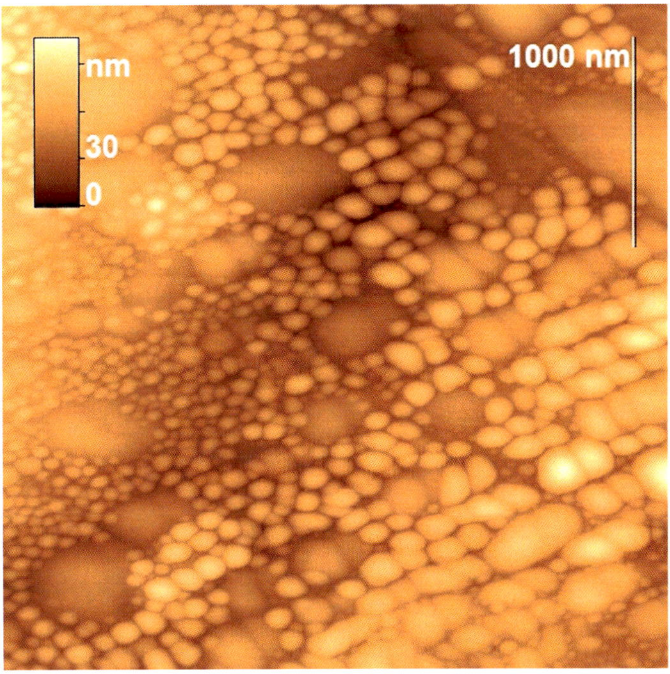

Fig. 12.3 Blisters on graphite surface. The image was obtained with FemtoScan AFM

compound with sulfuric acid and inclusions of water drops with the reactants and reaction intermediates residues.

The oxidized graphite can be created electrochemically with the aid of an AFM probe with a conductive coating (Sinitsyna et al. 2010) (Fig. 12.4). The process is called local anodic oxidation. The probe is a cathode, the graphite surface is an anode, and the water film adsorbed from the atmosphere acts as an electrolyte.

At the present moment, the Lerf-Klinowski model of graphite oxide (Lerf et al. 1998) is the most accepted. It is assumed that the layers of graphite oxide contain domains of two types: the regions of pure graphene with sp^2-hybridization of carbon atoms and the oxidized regions with sp^3-hybridization of carbon atoms. In the latter regions, epoxy and hydroxyl groups are attached to the graphite layers. Carbonyl and carboxyl groups are bonded to the edges of the layers. High-resolution transmission electron microscopy data confirm the presence of two types of domains, as well as the formation of nanometer holes in the layers (Erickson et al. 2010).

Since the carbon nanomaterials lack the defined chemical structure, and have a broad range of diameters and lengths, their detection in carbon rich biological matrices is a difficult task (Petersen et al. 2011).

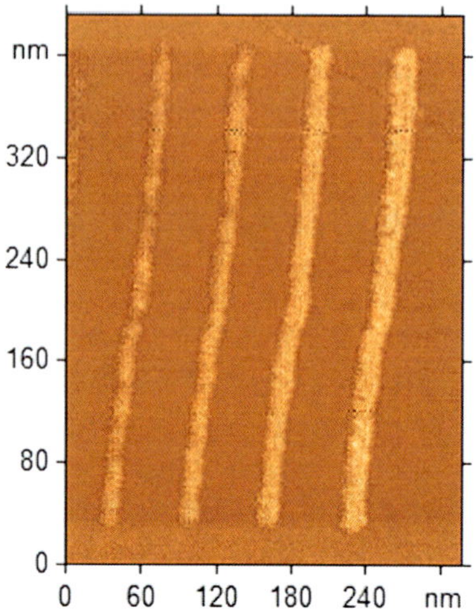

Fig. 12.4 Lines, consisting of graphite oxide, were obtained by local anodic oxidation of graphite surface. The image was obtained with FemtoScan AFM

12.3 Properties and Applications of Carbon Nanomaterials

Graphene and CNTs have remarkable mechanical, electrical and optical properties. Some experimental data are summarized in Table 12.1.

Products loaded with carbon nanoparticles have been already on the market according to the Nanodatabase (Nanodatabase 2018) and the nanomaterial consumer products inventory (Vance et al. 2015). Polymer composites with CNTs are strong, light-weighted and non-corrosive, so they are used in cars and aircrafts. The composites with CNTs become conductive, this is necessary for the electrostatic spray painting of the automotive body parts.

CNT incorporation improves the strength and torsional stability of composite materials utilized in sporting goods (hockey sticks, badminton strings, badminton

Table 12.1 Physico-chemical properties of the carbon nanomaterials

	Graphene	CNTs
Young's modulus	1.0 TPa (Lee et al. 2008)	0.410–4.150 TPa—MWCNT (Treacy et al. 1996, Wong et al. 1997)
Specific surface area	2630 $m^2\ g^{-1}$ (Bonaccorso et al. 2015)	144–662 $m^2\ g^{-1}$—SWCNTs, 22–180 $m^2\ g^{-1}$—MWCNTs (Birch et al. 2013)
Intrinsic mobility	200 000 $cm^2\ v^{-1}\ s^{-1}$ (Bolotin et al. 2008)	>100 000 $cm^2\ v^{-1}\ s^{-1}$—Semiconducting CNT (Dürkop et al. 2004)
Thermal conductivity	5000 $Wm^{-1}\ K^{-1}$ (Balandin et al. 2008)	1500 $Wm^{-1}\ K^{-1}$ (Koziol et al. 2017)

and tennis rackets, golf stick, baseball bats, bicycles, spikes). CNTs are already used in industry for the manufacture of transparent conductive films for flexible electronics, diaphragms in headphones, heated clothing, primer designed to protect metal surfaces in corrosive environments. It is believed that a high density nonvolatile random access memory chip with suspended nanotube junctions as memory bits will be released to the market in the near future. The emergence of products with working CNT tips, such as field emitters for screens (Harris 1999), and very sharp cantilevers for probe microscopy (Harris 1999; Demicheva et al. 2008), is expected.

The number of products containing graphene in the market is constantly growing. These include tennis rackets, battery strap, oil-drilling mud, phone touch screen (Zurutuza and Marinelli 2014).

The carbon nanomaterials are also perspective for biological and biomedical applications. Graphene can be an excellent carrier for water insoluble anti-cancer drugs such as hydroxycamptothecin and paclitaxel (Liu et al. 2008). Small biologically important molecules can be intercalated between graphene sheets (Lu et al. 2010). It was suggested to exploit CNTs for the delivery of the therapeutics across the skin (Kuche et al. 2018).

Graphene substrates can be used as a growth surface for neurons (Li et al. 2011). The proliferation of human mesenchymal stem cells is accelerated on graphene in comparison to glass or Si wafers (Nayak et al. 2011). It was shown that bone tissue could be regenerated without growth factors in the presence of graphene (Nayak et al. 2011).

GO-based materials are perspective for applications in wastewater treatment (Hashim et al. 2016). GO has outstanding photocatalytic and adsorption properties due to its large specific surface area and a variety of functional groups. GO-based nanocomposites can be used for photocatalytic degradation of heavy metals, organic dyes, carcinogenic chemicals, and insecticides. In commonly used photocatalysts (TiO_2, ZnO, ZnS, and WO_3) the photogenerated electrons and holes recombine rapidly (Azarang et al. 2014). The addition of GO in photocatalytic composites as an electron acceptor can slow down the recombination rate (Hu et al. 2013; Gupta et al. 2015). SWCNTs can be useful for remediation since they effectively adsorb polychlorinated biphenyl (Parks et al. 2014).

12.4 Potential Risks to Human Health

CNTs are the particles with high aspect ratio (length to diameter ratio). It is assumed that the CNTs can pose the same risks to human health as biopersistent industrial fibers, including asbestos (Tran et al. 2008). The inhalation of these fibers leads to lung damage related to frustrated phagocytosis and prolonged inflammation (Tran et al. 2008). Inhalation of the carbon materials can lead to harmful effects on the lungs, causing granulomas formation and lung fibrosis (Schinwald et al. 2012). The pulmonary toxicity of CNTs was investigated on rodents by Shvedova et al. (2005). It was found that the exposure to CNTs results in acute inflammatory response, granuloma formation, diffuse fibrosis, and the thickening of alveolar wall (Shvedova et al. 2005). The toxicity of CNTs depends on many factors. Brown et al. (2007) observed frustrated phagocytosis for rigid, but no for entangled MWCNTs with the length of 50 μm. The similar inflammation reactions and fibrosis were reported for pristine and ground (shortened) MWCNTs (Muller et al. 2005). According to Murphy et al. (2011), macrophages produced TGF-β1 and platelet-derived growth factor under the influence of MWCNTs, which promotes the transformation of lung fibroblasts to myofibroblasts, a major factor in fibrosis development.

The toxicity of the CNTs can be related both to their intrinsic properties and to the presence of impurities. For example iron in SWCNTs can induce oxidative stress and lipid peroxidation in the RAW 264.7 macrophage cell line (Kagan et al. 2006). CNTs can induce the reactive oxygen species (ROS) generation and the activation of ROS-associated intracellular signaling pathways in mesothelial cells (Johnston et al. 2010; Shvedova et al. 2012; Campagnolo et al. 2013; Clift et al. 2014).

Current Intelligence Bulletin of NIOSH (CDC 2013) summarized the adverse respiratory effects (pulmonary fibrosis, granulomatous inflammation) and systemic responses (an increase in inflammatory mediators in the blood, oxidative stress in aortic tissue and increase plaque formation) found in animals, and drew a conclusion about the need to protect workers from the exposure to CNTs. After inhalation, MWCNT can reach the alveoli, and finally translocate to the pleural cavity (Mercer et al. 2013). The effectiveness of the translocation depends on the MWCNT shape and size (Xu et al. 2014). The migration of MWCNT into the intrapleural space could potentially cause mesothelioma (CDC 2013).

CNTs cause cytotoxicity in human keratinocytes (Shvedova et al. 2003; Monteiro-Riviere et al. 2005). The activation of antigen/antibody response systems and granuloma formation were found in mice after the subcutaneous implantation of CNTs for 3 months (Koyama et al. 2006).

At present, the effects of CNTs on human health have been evaluated only in a small number of epidemiological studies. Biomarkers of inflammation and fibrosis were found in the biofluids of workers at the plant for the MWCNT production in Russia (Fatkhutdinova et al. 2016). In another study (Lee et al. 2015), the exposure to MWCNTs results in increase of malondialdehyde, 4-hydroxy-2-hexenal, and n-hexanal levels in exhaled breath condensates of workers.

Despite the fact that in some works graphene and GO were found to be biocompatible (Chang et al. 2011; Yue et al. 2012; Conroy et al. 2014), there are many studies showing the cytotoxicity and in vivo toxicity of these nanomaterials. Graphene can induce dose and shape dependent cytotoxicity (Zhang et al. 2010; Sasidharan et al. 2011; Li et al. 2012). Strong hydrophobic interaction with cell membranes and the depletion of the potential of the mitochondrial membranes can be related to graphene cytotoxicity (Sasidharan et al. 2011; Li et al. 2012). The intravenous administration of graphene led to a reduction of the body weight in mice (Sasidharan et al. 2015). Lung inflammation was observed in the case of the graphene pulmonary administration (Duch et al. 2011).

It was suggested that GO uptake occurs by phagocytosis because non-phagocytic cells did not engulf the GO flakes (Yue et al. 2012). It was shown that GO accumulated at the surface of monkey kidney cells (Sasidharan et al. 2011). But the functionalization of GO can change the interaction mechanisms with the cells. So carboxylated GO flakes were internalized by the cells (Sasidharan et al. 2011).

Initial GO accumulation in the reticuloendothelial system after intravenous administration was reported (Yang et al. 2010a). A dose of 20 mg kg^{-1} was not toxic with follow-up period of 3 months. The accumulation of GO was predominantly observed in the lungs (Yang et al. 2010b; Zhang et al. 2011a). Unlike other carbon nanomaterials, GO circulated several hours. Pathological changes in lungs were observed at a dose of 10 mg kg^{-1}. In another study (Wang et al. 2011) granulomas in lungs, spleen and liver were induced by GO injection in mice. The direct injection of GO in lungs also led to severe pulmonary injury (Duch et al. 2011). Graphene and GO dispersion in surfactants and their functionalization can reduce the toxicity (Duch et al. 2011; Sasidharan et al. 2011; Zhang et al. 2011b). The GO cytotoxicity can be related to its contamination by manganese during the chemical synthesis (Yue et al. 2012).

The low haemolytic activity of GO was demonstrated (Sasidharan et al. 2012), which can further reduced by chitosan coating (Liao et al. 2011). The high risk of thrombosis was reported (Singh et al. 2011, 2012) due to intravenous administration of GO in mice. The elicited blood platelets aggregation was associated with the presence of charged functional groups in GO (Singh et al. 2011).

12.5 Potential Ecological Impacts

There are several ways in which carbon nanoparticles can be released to the environment. The exposure can be during the production of carbon nanoparticles, the production of goods containing nanoparticles, and their transportation and disposal. Laboratory experiments show that the nanotubes can be released into the environment, for example, during in vitro degradation of poly(DL-Lactide-co-glycolide) (PLGA) matrix containing carboxylated SWCNTs (Armentano et al. 2008) or during the burning of a nanocomposite waste material with MWCNTs as the modeling of end-of-life combustion (Bouillard et al. 2010). It is interesting to note that in vitro degradation of the PLGA nanocomposite with unfunctionalized SWCNTs results in accumulation of the nanotubes on the surface without their release to the surrounding media (Armentano et al. 2008). The same results are obtained in the case of UV-photodegradation. A dense entangled MWCNT layer is observed on the surface of MWCNT-epoxy composite (Nguyen et al. 2009). A large amount of graphene is found on the surface of polyurethane-graphene oxide composite (Bernard et al. 2011).

The potential for biodegradation of the carbon nanoparticles depends strongly on their chemical functionalization. For example, a soil enzyme, horseradish peroxidase, does not transform pristine SWCNTs but reduces the length of carboxylated SWCNTs (Allen et al. 2009). Similarly to the nanotubes, GO degrades in the presence of horseradish peroxidase and hydrogen peroxide, and the reduced GO is much more stable under the same conditions (Kotchey et al. 2011). SWCNTs are not easily biodegraded by fungal cultures or environmental microbial communities (Parks et al. 2015). The reduction of GO by *E. coli* and *S. aureus* was reported (Gurunathan et al. 2013).

The study of toxicity of the carbon nanoparticles to biological processes in wastewater treatment is important to predict their influence on the operation of wastewater treatment facilities. SWCNTs toxicity for *Escherichia coli*, *Pseudomonas aeruginosa*, *Bacillus subtilis*, and *Staphylococcus epidermis* was shown by Kang et al. (2009). The dose-dependent respiration inhibition of the microbial communities in activated sludge was found in the presence of MWCNTs (Luongo and Zhang 2010). Changes in microbial community structure were found (Goyal et al. 2010) due to exposure to SWCNTs. The internalization of SWCNTs by the ciliated protozoan *Tetrahymena thermophila* was reported (Ghafari et al. 2008). *T. thermophila* regulates microbial populations and it is important for wastewater treatment. The exposure to SWCNTs results in the aggregation of *T. thermophila,* which alters its ability to ingest and digest bacteria (Ghafari et al. 2008).

The uptake of CNTs was studied in several organisms e.g. *Daphnia magna* (Petersen et al. 2009; Edgington et al. 2010), fruit flies (Leeuw et al. 2007). In a review (Petersen et al. 2011) it was concluded that CNTs are localized in the intestine, and their absorption across epithelial membranes is negligible. CNTs can be eliminated by soil and sediment organisms, and their ecotoxicological effects on soils and sediments is minimal (Petersen et al. 2011).

The MWCNTs can pierce root cell walls of wheat plants (Wild and Jones 2009). The data on MWCNTs absorption by tomato plant tissues are contradictory (Khodokovskaya et al. 2009, 2011). The toxicity of GO on cabbage, tomato, and red spinach growth were found (Begum et al. 2011). The main effects, the inhibition of growth and biomass, visible symptoms of necrosis, and increased ROS, are dose and time dependent. Lettuce seedlings were resistant to the GO exposure (Begum et al. 2011). GO inhibited the germination of rice seeds, which may be caused by the blockage of the pores (Nair et al. 2012).

The fate of carbon nanoparticles in aquatic systems, as discussed by Zhang et al. (2011c), depends on the water composition. So, in seawater, most likely MWCNTs will be sorbed by sediments, and in water with the high content of dissolved organic matter, they can stay in the water column for a long time. The aggregation of CNTs was related to an increase in their toxicity in *Ceriodaphnia dubia* (Kennedy et al. 2009). Any adverse effects of functionalized GO in zebrafish was not detected (Nair et al. 2012). It was shown with an example of a model Fenton reaction, that few-layer graphene is significantly transformed by the Fenton's reagents, Iron(II) and hydrogen peroxide (Feng et al. 2015). The accumulation of graphene by *Daphnia magna* is reduced by its chemical transformation.

The toxicity of dispersed SWCNTs to fish can be a result of the disruption of gill function, which leads to oxygen deprivation followed by changes in oxidative stress indicators in tissues, and vascular changes in the brain (Smith et al. 2007). The dietary exposure to high level of SWCNTs was not harmful for fish (Fraser et al. 2011). Freixa et al. (2018) concluded that sufficiently small concentrations of the carbon nanomaterials in aquatic environments (in the order of ng L^{-1}) are not toxic to aquatic organisms. But the carbon nanomaterials can change the toxicity of other micropolutants (Freixa et al. 2018).

12.6 Conclusions

The question of the interaction between living systems and the carbon nanomaterials becomes particularly relevant with an increase in their consumption. The toxicity of graphene, GO and CNTs was demonstrated in many studies in vitro and in vivo. There is a consensus that CNTs with their fibrous structure pose a danger to human health. However, it is difficult to assess the role of the particle size, physico-chemical properties, chemical functionalization and purity in the toxicity of the carbon nanomaterials because of their high heterogeneity.

Acknowledgements This work was supported by the Russian Foundation for Basic Research (grant nos. 16-29-06290 ofi_m).

References

Allen BL, Kotchey GP, Chen YN, Yanamala NVK, Klein-Seetharaman J, Kagan VE, Star A (2009) Mechanistic investigations of horseradish peroxidase-catalyzed degradation of single-walled carbon nanotubes. J Am Chem Soc 131(47):17194–17205

Armentano I, Dottori M, Puglia D, Kenny JM (2008) Effects of carbon nanotubes (CNTs) on the processing and in-vitro degradation of poly(DL-lactide-co-glycolide)/CNT films. J Mater Sci Mater Med 19(6):2377–2387

Azarang M, Shuhaimi A, Yousefi R, Golsheikh AM, Sookhakian M (2014) Synthesis and characterization of ZnO NPs/reduced graphene oxide nanocomposite prepared in gelatin medium as highly efficient photo-degradation of MB. Ceram Int Part B 40(7):10217–10221

Balandin AA, Ghosh S, Bao W, Calizo I, Teweldebrhan D, Miao F, Lau CN (2008) Superior thermal conductivity of single-layer graphene. Nano Lett 8(3):902–907

Begum P, Ikhtiari R, Fugetsu B (2011) Graphene phytotoxicity in the seedling stage of cabbage, tomato, red spinach, and lettuce. Carbon 49:3907–3919

Bernard C, Nguyen T, Pelligrin B, Holbrook RD, Zhao M, Chin J (2011) Fate of graphene in polymer nanocomposite exposed to UV radiation. J Phys Conf Ser 304:012063

Birch ME, Ruda-Eberenz TA, Chai M, Andrews R, Hatfield RL (2013) Properties that influence the specific surface areas of carbon nanotubes and nanofibers. Ann Occup Hyg 57(9):1148–1166

Bolotin KI, Sikes KJ, Jiang Z, Klima M, Fudenberg G, Hone J, Kim P, Stormer HL (2008) Ultrahigh electron mobility in suspended graphene. Solid State Commun 146(9–10):351–355

Bonaccorso F, Colombo LYG, Stoller M, Tozzini V, Ferrari AC, Ruoff RS, Pellegrini V (2015) Graphene, related two-dimensional crystals, and hybrid systems for energy conversion and storage. Science 347(6217):1246501

Bouillard JX, Vignes A, Rmili B, Moranviller D, Fleury D, Le Bihan O, Frejafon E, Ustache A (2010) Nanowastes: risk assessment from the end-of-life combustion of nanomaterials. In: International conference on safe production and use of nanomaterials, NANOSAFE 2010, Grenoble, France, 16–18 November 2010

Brown DM, Kinloch IA, Bangert U, Windle AH, Walter DM, Walker GS, Scotchford CA, Donaldson K, Stone V (2007) An in vitro study of the potential of carbon nanotubes and nanofibres to induce inflammation mediators and frustrated phagocytosis. Carbon 45:1743–1756

Campagnolo L, Massimiani M, Palmieri G, Bernardini R, Sacchetti C, Bergamaschi A, Vecchione L, Magrini A, Bottini M, Pietroiusti A (2013) Biodistribution and toxicity of pegylated single wall carbon nanotubes in pregnant mice. Part Fibre Toxicol 10(1):21

CDC (2013) NIOSH Publications and Products, Current Intelligence Bulletin 65: Occupational Exposure to Carbon Nanotubes and Nanofibers (2013–145)

Chang Y, Yang S-T, Liu JH, Dong E, Wang Y, Cao A, Liu Y, Wang H (2011) In vitro toxicity evaluation of graphene oxide on A549 cells. Toxicol Lett 200:201–210

Clift MJ, Endes C, Vanhecke D, Wick P, Gehr P, Schins RP, Petri-Fink A, Rothen-Rutishauser B (2014) A comparative study of different in vitro lung cell culture systems to assess the most beneficial tool for screening the potential adverse effects of carbon nanotubes. Toxicol Sci 137(1):55–64

Conroy J, Verma NK, Smith RJ, Rezvani E, Duesberg GS, Coleman JN, Volkov Y (2014) Biocompatibility of pristine graphene monolayers, nanosheets and thin films. arXiv:1406:2497

Demicheva OV, Meshkov GB, Sinitsyna OV, Tomishko AG, Yaminsky IV (2008) Multiwall carbon nanotube tips for scanning probe microscopy. Nanotechnol Russia 3(11–12):704–709

Dimiev AM, Tour JM (2014) Mechanism of graphene oxide formation. ACS Nano 8:3060–3068

Dresselhaus MS, Dresselhaus G, Saito R, Jorio A (2005) Raman spectroscopy of carbon nanotubes. Phys Rep 409:47–99

Duch MC, Budinger GS, Liang YT, Soberanes S, Urich D, Chiarella SE, Campochiaro LA, Gonzalez A, Chandel NS, Hersam MC (2011) Minimizing oxidation and stable nanoscale dispersion improves the biocompatibility of graphene in the lung. Nano Lett 11(12):5201–5207

Dürkop T, Getty SA, Cobas E, Fuhrer MS (2004) Extraordinary mobility in semiconducting carbon nanotubes. Nano Lett 4(1):35–39

Edgington AJ, Roberts AP, Taylor LM, Alloy MM, Reppert J, Rao AM, Ma JD, Klaine SJ (2010) The influence of natural organic matter on the toxicity of multiwalled carbon nanotubes. Environ Toxicol Chem 29(11):2511–2518

Erickson K, Erni R, Zonghoon Lee Z, Alem N, Will Gannett A, Zettl A (2010) Determination of the local chemical structure of graphene oxide and reduced graphene. Oxide Adv Mater 22:4467–4472

Fatkhutdinova LM, Khaliullin TO, Vasil'yeva OL, Zalyalov RR, Mustafin IG, Kisin ER, Birch ME, Yanamala N, Shvedova AA (2016) Fibrosis biomarkers in workers exposed to MWCNTs. Toxicol Appl Pharmacol 299:125–131

Feng Y, Lu K, Mao L, Guo X, Gao S, Petersen EJ (2015) Degradation of ^{14}C-labeled few layer graphene via Fenton reaction: reaction rates, characterization of reaction products, and potential ecological effects. Water Res 84:49–57

Ferrari AC, Meyer JC, Scardaci V, Casiraghi C, Lazzeri M, Mauri F, Piscanec S, Jiang D, Novoselov KS, Roth S, Geim AK (2006) Raman spectrum of graphene and graphene layers. Phys Rev Lett 97:187401–187404

Fraser TWK, Reinardy HC, Shaw BJ, Henry TB, Handy RD (2011) Dietary toxicity of single walled carbon nanotubes and fullerenes (C60) in rainbow trout (Oncorhynchus mykiss). Nanotoxicology 5:98–108

Freixa A, Acuña V, Sanchís J, Farré M, Barceló D, Sabater S (2018) Ecotoxicological effects of carbon based nanomaterials in aquatic organisms. Sci Total Environ 619–620:328–337

Ghafari P, St-Denis CH, Power ME, Jin X, Tsou V, Mandal HS, Bols NC, Tang XW (2008) Impact of carbon nanotubes on the ingestion and digestion of bacteria by ciliated protozoa. Nat Nanotechnol 3(6):347–351

Goyal D, Zhang XJ, Rooney-Varga JN (2010) Impacts of singlewalled carbon nanotubes on microbial community structure in activated sludge. Lett Appl Microbiol 51(4):428–435

Gupta VK, Eren T, Atar N, Yola ML, Parlak C, Karimi-Maleh H (2015) $CoFe2O_4@TiO_2$ decorated reduced graphene oxide nanocomposite for photocatalytic degradation of chlorpyrifos. J Mol Liq 208:122–129

Gurunathan S, Han JW, Eppakayala V, Kim JH (2013) Microbial reduction of graphene oxide by *Escherichia coli*: a green chemistry approach. Colloids Surf B 102:772–777

Harris PJF (1999) Carbon nanotubes and related strutures. Cambridge University Press

Hashim N, Muda Z, Hussein MZ, Isa IM, Mohamed A, Kamari A, Bakar SA, Mamat M, Mohamad A (2016) A brief review on recent graphene oxide-based material nanocomposites: synthesis and applications. J Mater Environ Sci 7(9):3225–3243

Hernandez Y, Nicolosi V, Lotya M, Blighe FM, Sun Z, De S, McGovern IT, Holland B, Byrne M, Gun'Ko YK, Boland JJ, Niraj P, Duesberg G, Krishnamurthy S, Goodhue R, Hutchison J, Scardaci V, Ferrari AC, Coleman JN (2008) High-yield production of graphene by liquid-phase exfoliation of graphite. Nat Nanotechnol 3:563–568

Hu C, Lu T, Chen F, Zhang R (2013) A brief review of graphene–metal oxide composites synthesis and applications in photocatalysis. J Chin Adv Mater Soc 1(1):21–39

Hummers WS Jr, Offeman RE (1958) Preparation of graphitic oxide. J Am Chem Soc 80:1339–1339

Iijima S (1991) Helical microtubules of graphitic carbon. Nature 354:56–58

Johnston HJ, Hutchison GR, Christensen FM, Peters S, Hankin S, Aschberger K, Stone V (2010) A critical review of the biological mechanisms underlying the in vivo and in vitro toxicity of carbon nanotubes: the contribution of physico-chemical characteristics. Nanotoxicology 4(2):207–246

Kagan VE, Tyurina YY, Tyurin VA, Konduru NV, Potapovich AI, Osipov AN, Kisin ER, Schwegler-Berry D, Mercer R, Castranova V, Shvedova AA (2006) Direct and indirect effects of single walled carbon nanotubes on RAW 264.7 macrophages: role of iron. Toxicol Lett 165(1):88–100

Kang S, Mauter MS, Elimelech M (2009) Microbial cytotoxicity of carbon-based nanomaterials: implications for river water and wastewater effluent. Environ Sci Technol 43(7):2648–2653

Kennedy AJ, Gunter JC, Chappell MA, Goss JD, Hull MS, Kirgan RA, Steevens JA (2009) Influence of nanotube preparation in aquatic bioassays. Environ Toxicol Chem 28(9):1930–1938

Khodakovskaya M, Dervishi E, Mahmood M, Xu Y, Li ZR, Watanabe F, Biris AS (2009) Carbon nanotubes are able to penetrate plant seed coat and dramatically affect seed germination and plant growth. ACS Nano 3(10):3221–3227

Khodakovskaya MV, de Silva K, Nedosekin DA, Dervishi E, Biris AS, Shashkov EV, Galanzha EI, Zharov VP (2011) Complex genetic, photothermal, and photoacoustic analysis of nanoparticle-plant interactions. Proc Natl Acad Sci USA 108(3):1028–1033

Kotchey GP, Allen BL, Vedala H, Yanamala N, Kapralov AA, Tyurina YY, Klein-Seetharaman J, Kagan VE, Star A (2011) The enzymatic oxidation of graphene oxide. ACS Nano 5:2098–2108

Koyama S, Endo M, Kim Y-A, Hayashi T, Yanagisawa T, Osaka K, Koyama H, Hania H, Kuroiwa N (2006) Role of systemic T-cells and histopathological aspects after subcutaneous implantation of various carbon nanotubes in mice. Carbon 44:1079–1092

Koziol KK, Janas D, Brown E, Hao L (2017) Thermal properties of continuously spun carbon nanotube fibres. Physica E Low Dimens Syst Nanostruct 88:104–108

Krasnov AP, Afronicheva OV, Mit' VA, Bazhenova VB, Volkova TV, Sinitsyna OV, Vygodskii YS, Rashkovan IA, Kazakov ME (2009) Role of nanofiller in friction of polymers based on polycaproamide: indirect effect. J Fric Wear 30(5):350–355

Kuche K, Maheshwari R, Tambe V, Mak KK, Jogi H, Raval N, Pichika MR, Tekade RK (2018) Carbon nanotubes (CNTs) based advanced dermal therapeutics: current trends and future potential. Nanoscale 10:8911–8937

Lee C, Wei X, Kysar JW, Hone J (2008) Measurement of the elastic properties and intrinsic strength of monolayer graphene. Science 321(5887):385–388

Lee JS, Choi YC, Shin JH, Lee JH, Lee Y, Park SY, Baek JE, Park JD, Ahn K (2015) Health surveillance study of workers who manufacture multi-walled carbon nanotubes. Nanotoxicology 9(6):802–811

Leeuw TK, Reith RM, Simonette RA, Harden ME, Cherukuri P, Tsyboulski DA, Beckingham KM, Weisman RB (2007) Single-walled carbon nanotubes in the intact organism: near-IR imaging and biocompatibility studies in Drosophila. Nano Lett 7 (9):2650–2654

Lerf A, He H, Forster M, Klinowski J (1998) Structure of graphite oxide revisited. J Phys Chem 102:4477–4482

Li N, Zhang X, Song Q, Su R, Zhang Q, Kong T, Liu L, Jin G, Tang M, Cheng G (2011) The promotion of neurite sprouting and outgrowth of mouse hippocampal cells in culture by graphene substrates. Biomaterials 32(35):9374–9382

Li Y, Liu Y, Fu Y, Wei T, Le Guyader L, Gao G, Liu RS, Chang YZ, Chen C (2012) The triggering of apoptosis in macrophages by pristine graphene through the MAPK and TGF-beta signaling pathways. Biomaterials 33(2):402–411

Liao KH, Lin Y-S, Macosko CW, Haynes CL (2011) Cytotoxicity of graphene oxide and graphene in human erythrocytes and skin fibroblasts. ACS Appl Mater Interfaces 3(7):2607–2615

Liu Z, Robinson JT, Sun XM, Dai HJ (2008) PEGylated nanographene oxide for delivery of water-insoluble cancer drugs. J Am Chem Soc 130:10876e7

Liu W-W, Chai S-P, Mohamed AR, Hashim U (2014) Synthesis and characterization of graphene and carbon nanotubes: a review on the past and recent developments. J Indus Eng Chem 20:1171–1185

Lu CH, Zhu CL, Li J, Liu JJ, Chen X, Yang HH (2010) Using graphene to protect DNA from cleavage during cellular delivery. Chem Commun 46:3116e8

Luongo LA, Zhang XQ (2010) Toxicity of carbon nanotubes to the activated sludge process. J Hazard Mater 178(13):356–362

Marcano DC, Kosynkin DV, Berlin JM, Sinitskii A, Sun Z, Slesarev A, Alemany LB, Lu W, Tour JM (2010) Improved synthesis of graphene oxide. ACS Nano 4(8):4806–4814

Mercer RR, Scabilloni JF, Hubbs AF, Wang L, Battelli LA, McKinney W, Castranova V, Porter DW (2013) Extrapulmonary transport of MWCNT following inhalation exposure. Part Fibre Toxicol 10:38

Monteiro-Riviere NA, Nemanich RJ, Inman AO, Wang YY, Riviere JE (2005) Multi-walled carbon nanotube interactions with human epidermal keratinocytes. Toxicol Lett 155:377–384

Muller J, Huaux F, Moreau N, Misson P, Heilier J-F, Delos M, Arras M, Fonseca A, Nagy JB, Lison D (2005) Respiratory toxicity of multi-wall carbon nanotubes. Toxicol Appl Pharmacol 207(3):221–231

Murphy FA, Poland CA, Duffin R, Al-Jamal KT, Ali-Boucetta H, Nunes A, Byrne F, Prina-Mello A, Volkov Y, Li S, Mather SJ, Bianco A, Prato M, MacNee W, Wallace WA, Kostarelos K, Donaldson K (2011) Length-dependent retention of carbon nanotubes in the pleural space of mice initiates sustained inflammation and progressive fibrosis on the parietal pleura. Am J Pathol 178(6):2587–2600

Nair R, Mohamed MS, Gao W, Maekawa T, Yoshida Y, Ajayan PM, Kumar DS (2012) Effect of carbon nanomaterials on the germination and growth of rice plants. J Nanosci Nanotechnol 12:2212–2220

Nanodatabase (2018) The nanodatabase. http://nanodb.dk

Nayak TR, Andersen H, Makam VS, Khaw C, Bae S, Xu X, Ee PLR, Ahn J-H, Hong BH, Pastorin G, Özyilmaz B (2011) Graphene for controlled and accelerated osteogenic differentiation of human mesenchymal stem cells. ACS Nano 5(6):4670–4678

Nguyen T, Pelligrin B, Mermet L, Shapiro A, Gu X, Chin J (2009) Network aggregation of CNTs at the surface of epoxy/MWCNT composite exposed to UV radiation. Nanotechnology 2009

Novoselov KS, Geim AK, Morozov SV, Jiang D, Zhang Y, Dubonos SV, Grigorieva IV, Firsov AA (2004) Electric field effect in atomically thin carbon films. Science 306(5696):666–669

Parks AN, Chandler GT, Portis LM, Sullivan JC, Perron MM, Cantwell MG, Burgess RM, Ho KT, Ferguson PL (2014) Effects of single-walled carbon nanotubes on the bioavailability of PCBs in field-contaminated sediments. Nanotoxicology 8:111–117

Parks AN, Chandler GT, Ho KT, Burgess RM, Ferguson PL (2015) Environmental biodegradability of [^{14}C] single-walled carbon nanotubes by *Trametes versicolor* and natural microbial cultures found in New Bedford Harbor sediment and aerated wastewater treatment plant sludge. Env Toxicol Chem 34(2):247–251

Petersen EJ, Akkanen J, Kukkonen JVK, Weber WJ Jr (2009) Biological uptake and depuration of carbon nanotubes by *Daphnia magna*. Environ Sci Technol 43(8):2969–2975

Petersen EJ, Zhang L, Mattison NT, O'Carroll DM, Whelton AJ, Uddin N, Nguyen T, Huang Q, Henry TB, Holbrook RD, Chen KL (2011) Potential release pathways, environmental fate, and ecological risks of carbon nanotubes. Environ Sci Technol 45:9837–9856

Pimenta MA, Jorio A, Brown SDM, Souza Filho AG, Dresselhaus G, Hafner JH, Lieber CM, Saito R, Dresselhaus MS (2001) Diameter dependence of the Raman-band in isolated single-wall carbon nanotubes. Phys Rev B 64:041401–041404

Qian W, Liu T, Wei F, Yuan H (2003) Quantitative Raman characterization of the mixed samples of the single and multi-wall carbon nanotubes. Carbon 41:1851–1854

Sasidharan A, Panchakarla LS, Chandran P, Menon D, Nair S, Rao CN, Koyakutty M (2011) Differential nano-bio interactions and toxicity effects of pristine versus functionalized graphene. Nanoscale 3(6):2461–2464

Sasidharan A, Panchakarla LS, Sadanandan AR, Ashokan A, Chandran P, Girish CM, Menon D, Nair SV, Rao CN, Koyakutty M (2012) Hemocompatibility and macrophage response of pristine and functionalized graphene. Small 8:1251–1263

Sasidharan A, Swaroop S, Koduri CK, Girish CM, Chandran P, Panchakarla L, Somasundaram VH, Gowd GS, Nair S, Koyakutty M (2015) Comparative *in vivo* toxicity, organ biodistribution and immune response of pristine, carboxylated and PEGylated few-layer graphene sheets in Swiss albino mice: a three month study. Carbon 95:511–524

Schinwald A, Murphy FA, Jones A, MacNee W, Donaldson K (2012) Graphene-based nanoplatelets: a new risk to the respiratory system as a consequence of their unusual aerodynamic properties. ACS Nano 6:736–746

Shvedova AA, Castranova V, Kisin ER, Schwegler-Berry D, Murray AR, Gandelsman VZ, Maynard A, Baron P (2003) Exposure to carbon nanotube material: assessment of nanotube cytotoxicity using human keratinocyte cells. J Toxicol Environ Heal A 66:1909–1926

Shvedova AA, Kisin ER, Mercer R, Murray AR, Johnson VJ, Potapovich AI, Tyurina YY, Gorelik O, Arepalli S, Schwegler-Berry D, Hubbs AF, Antonini J, Evans DE, Ku BK, Ramsey D, Maynard A, Kagan VE, Castranova V, Baron P (2005) Unusual inflammatory and fibrogenic pulmonary responses to single-walled carbon nanotubes in mice. Am J Physiol Lung Cell Mol Physiol 289:L698–L708

Shvedova AA, Pietroiusti A, Fadeel B, Kagan VE (2012) Mechanisms of carbon nanotube-induced toxicity: focus on oxidative stress. Toxicol Appl Pharmacol 261(2):121–133

Singh SK, Singh MK, Nayak MK, Kumari S, Shrivastava S, Grácio JJA, Dash D (2011) Thrombus inducing property of atomically thin graphene oxide sheets. ACS Nano 5(6):4987–4996

Singh SK, Singh MK, Kulkarni PP, Sonkar VK, Grácio JJA, Dash D (2012) Amine-modified graphene: thrombo-protective safer alternative to graphene oxide for biomedical applications. ACS Nano 6(3):2731–2740

Sinitsyna OV, Meshkov GB, Yaminsky IV (2010) A novel tool for the local anodic oxidation of graphite. Proc Inst Mech Eng Part N: J Nanoeng Nanosyst 223:133–138

Sinitsyna OV, Meshkov GB, Grigorieva AV, Antonov AA, Grigorieva IG, Yaminsky IV (2018) Blister formation during graphite surface oxidation by hummers' method. Beilstein J Nanotechnol 9:407–414

Smith CJ, Shaw BJ, Handy RD (2007) Toxicity of single walled carbon nanotubes to rainbow trout, (*Oncorhynchus mykiss*): Respiratory toxicity, organ pathologies, and other physiological effects. Aquat Toxicol 82(2):94–109

Tran CL, Hankin SM, Ross B, Aitken RJ, Jones AD, Donaldson K, Stone V, Tantra R (2008) An outline scoping study to determine whether high aspect ratio nanoparticles (HARN) should raise the same concerns as do asbestos fibres. Report on Project CB0406 (An outline scoping study to determine whether high aspect ratio nanoparticles (HARN) should raise the same concerns as do asbestos fibres)

Treacy MMJ, Ebbesen TW, Gibson JM (1996) Exceptionally high Young's modulus observed for individual carbon nanotubes. Nature 381:678–680

Vance ME, Kuiken T, Vejerano EP, McGinnis SP, Hochella MF Jr, Rejeski D, Hull MS (2015) Nanotechnology in the real world: redeveloping the nanomaterial consumer products inventory. Beilstein J Nanotechnol 6:1769–1780

Vygodskii YS, Volkova TV, Zabegaeva ON, Chistyakova ZY, Shanditsev VA, Buzin MI, Zubavichus YV, Sinitsyna OV, Nikiforova GG, Krasnov AP, Garbuzova IA, Belavtseva EM (2009) Synthesis and characteristics of the composites based on poly(caproamide) and multiwalled carbon nanotubes. Polym Sci Ser C 51(1):63–73

Wang Y, Liu J, Liu L, Sun DD (2011) High-quality reduced graphene oxide-nanocrystalline platinum hybrid materials prepared by simultaneous co-reduction of graphene oxide and chloroplatinic acid. Nanoscale Res Lett 6(1):241

Web Site of the Nobel Prize (2018) www.nobelprize.org

Wild E, Jones KC (2009) Novel method for the direct visualization of in vivo nanomaterials and chemical interactions in plants. Environ Sci Technol 43(14):5290–5294

Wildöer JWG, Venema LC, Rinzler AG, Smalley RE, Dekker C (1998) Electronic structure of atomically resolved carbon nanotubes. Nature 391:59–62

Wong EW, Sheehan PE, Lieber CM (1997) Nanobeam mechanics: elasticity, strength and toughness of nanorods and nanotubes. Science 277:1971–1975

Xu J, Alexander DB, Futakuchi M, Numano T, Fukamachi K, Suzui M, Omori T, Kanno J, Hirose A, Tsuda H (2014) Size-and shape-dependent pleural translocation, deposition, fibrogenesis, and mesothelial proliferation by multiwalled carbon nanotubes. Cancer Sci 105:763–769

Yang K, Wan J, Zhang S, Zhang Y, Lee S-T, Liu Z (2010a) *In Vivo* pharmacokinetics, long-term biodistribution, and toxicology of PEGylated graphene in mice. ACS Nano 5:516–522

Yang K, Zhang S, Zhang G, Sun X, Lee ST, Liu Z (2010b) Graphene in mice: ultrahigh in vivo tumor uptake and efficient photothermal therapy. Nano Lett 10:3318–3323

Yue H, Wei W, Yue Z, Wang B, Luo N, Gao Y, Ma D, Ma G, Su Z (2012) The role of the lateral dimension of graphene oxide in the regulation of cellular responses. Biomaterials 33:4013–4021

Zhang Y, Ali SF, Dervishi E, Xu Y, Li Z, Casciano D, Biris AS (2010) Cytotoxicity effects of graphene and single-wall carbon nanotubes in neural phaeochromocytoma-derived PC12 cells. ACS Nano 4(6):3181–3186

Zhang X, Yin J, Peng C, Hu W, Zhu Z, Li W, Fan C (2011a) Huang Q (2011a) Distribution and biocompatibility studies of graphene oxide in mice after intravenous administration. Carbon 49:986–995

Zhang S, Yang K, Feng L, Liu Z (2011b) In vitro and in vivo behaviors of dextran functionalized graphene. Carbon 49(12):4040–4049

Zhang L, Petersen EJ, Huang QG (2011c) Phase distribution of 14C-labeled multiwalled carbon nanotubes in aqueous systems containing model solids: Peat. Environ Sci Technol 45(4):1356–1362

Zurutuza A, Marinelli C (2014) Challenges and opportunities in graphene commercialization. Nat Nanotechnol 9:730–734

Chapter 13
Nanoparticles Emitted by Biomass Burning: Characterization and Monitoring of Risks

Maria Angélica M. Costa, Henrique M. Fogarin, Ana F. M. Costa, Lorena O. Pires, Débora D. V. Silva, Michele Lima-Souza and Kelly J. Dussán

Abstract Using biomass (e.g. sugarcane bagasse) as biofuel in industries and distilleries is already well known and has recently been increasingly more utilized to produce heat and energy, thus emissions from burning this fuel have reached local and global populations. Biomass burning in general is the major anthropogenic source of

M. A. M. Costa · H. M. Fogarin · L. O. Pires · D. D. V. Silva · K. J. Dussán (✉)
Department of Biochemistry and Chemical Technology, Institute of Chemistry, São Paulo State University-UNESP, Av. Prof. Francisco Degni, 55 - Jardim Quitandinha, 14800-900 Araraquara, São Paulo, Brazil
e-mail: kelly.medina@unesp.br

M. A. M. Costa
e-mail: maria.am.costa@unesp.br

H. M. Fogarin
e-mail: henriquemfogarin@gmail.com

L. O. Pires
e-mail: lorena.pires@unesp.br

D. D. V. Silva
e-mail: debora.virginio@unesp.br

A. F. M. Costa
Department of Bioprocess and Biotechnology, School of Pharmaceutical Sciences of São Paulo State University-UNESP, Rodovia Araraquara - Jáu Km 1, S/N – Campos Ville, 14800-903 Araraquara, São Paulo, Brazil
e-mail: anaflavia.mc@outlook.com

M. Lima-Souza
Department of Analytical Chemistry, Institute of Chemistry, São Paulo State University-UNESP, Av. Prof. Francisco Degni, 55 - Jardim Quitandinha, 14800-900 Araraquara, São Paulo, Brazil
e-mail: michelelsouza@gmail.com

M. Rai and J. K. Biswas (eds.), *Nanomaterials: Ecotoxicity, Safety, and Public Perception*, https://doi.org/10.1007/978-3-030-05144-0_13

many air pollutants. Environmental assessments must be carried out frequently and should consider not only burning fields, which are still recurrent, but also the use of bagasse as fuel, its emissions and the control in industry, taking into account air quality and harmful effects on health. In order to evaluate the atmospheric changes and the effect of the compounds emitted during biomass combustion on human health, particulate matter (PM) emission control and characterization must be carried out. Particles emitted by biomass burning have varied compositions, including organic, inorganic as well as elemental carbon. These analyses can identify different emission sources, such as fixed sources, air mass trajectories and quantification of open burning. However, ambient air quality standards should be reassessed, including the effects of ultrafine particulate matter (nanoparticles) and solutions should be sought to reduce the concentration of these materials in the atmosphere and hence improve public health.

Keywords Biomass · Burning · Emission · Characterization · Deposition
Air quality

13.1 Introduction

Historically, biomass burning has influenced the global climate, and therefore there is a vast body of research related to Earth Systems Science. Over the last decades, air pollution has gained more attention and consequently more strict air quality regulations have been applied. As a result, some topics which were considered secondary beforehand, such as the intercontinental transport of pollutants, have now become more significant (Chen et al. 2017).

The activity of burning biomass generates air pollutants such as carbon monoxide (CO), carbon dioxide (CO_2), sulfur dioxide (SO_2), nitrogen oxides (NOx), methane and particulate matter (PM) which may contain nanoparticles and contributes to changes in atmospheric chemistry and negative human health effects (Levine 1996; Buseck and Adachi 2008).

The atmospheric composition is the result of complex interaction factors, such as the magnitude of emissions, transport of pollutants, chemical and physical changes, deposition (dry and humid), topographical and climatic conditions, which may or may not be conducive to the dispersal of pollutants (Andreae 2007; Calvo et al. 2013). Gas composition and PM in the atmosphere are constantly modified by natural activities on the planet surface. Nevertheless, anthropogenic sources have caused abrupt changes in pollutant concentration and great impacts on their composition.

Atmospheric nanoparticles are spatially and temporally variable in concentration and in composition. These nano-compounds are distributed as liquid droplets, semi-volatile materials, solid or low-volatility nanoparticles (Buseck and Adachi 2008). Regarding their concentration, the number of nanoparticles is affected by the emission intensities, environmental conditions and proximity to sources (Buseck and Adachi 2008).

The size distribution of PM varies from small aggregates of molecules to dust. They can be classified based on their aerodynamic diameter which is smaller than 10 μm (PM10), respirable fractions as fine particles smaller than 2.5 μm (PM2.5), and PM defined as those between PM10 and PM2.5. PM1 is identified as ultrafine particles, while particles smaller than 0.1 μm are referred to as nanoparticles (Hinds 1999). Knowledge of particle compositions in relation to size can provide relevant information about possible sources, behavior and impact on atmospheric processes and human health, since coarse particles are usually formed by mechanical processes such as particle resuspension (a matter of concern due to their influence on climate and health) resulting largely from combustion (fossil and biomass).

All these emissions produced by biomass burning affect both local and global air quality, also affecting public health and the environment (Seiler and Crutzen 1980; Phuleria et al. 2005; Sapkota et al. 2005). Thus, it is evident that methods which are able to monitor the burning of biomass and its pollutants emitted into the ambient are of the utmost importance.

The effects of pollution are mainly associated with the concentration and composition in the emission source. Legislation governing air quality is specific to each country, therefore the compounds monitored and the limits established for each compound differ in terms of the bodies responsible for monitoring and control, depending on the region (Harrison 1999). Regulatory bodies set standards in terms of the main compounds emitted and concentration limits considered safe for the environment and for health (US EPA 2018).

In terms of industries, these standards should be evaluated routinely and should fit within the limits established by the bodies responsible for that emission source (Vallero 2014). However, when it comes to biomass burning, it is difficult to establish the significant impacts from this type of burning because they are open environments that depend on the magnitude of the emissions, topography and meteorological conditions that contribute to the dispersion and transport of these compounds (Harrison and Hester 2009). Thus, biomass burning is considered to be a significant source of pollution to the atmosphere, including a wide variety of gases and particles, with local, regional and even global impact (Finlayson-Pitts and Pitts Jr 2000b).

The chapter focuses mainly on the description of emission processes of particulate matter (PM), including emission sources, environmental control, chemical and physical characterization of the pollutants. An approach regarding the dry deposition of pollutants produced by simple combustion and other industrialization processes is also shown. However, regulations on the effect of biomass combustion on ambient air should be more stringent due to the strong impact that this process has on the formation of pollutants.

13.2 Sources: Open Burning and Industrial Burning

13.2.1 Nanoparticle Sources

Solid biomass fuels mainly consist mainly of carbon, hydrogen and oxygen, which are converted to about 90% in CO_2, CO and water. However, incomplete combustion of biomass produces both gaseous and particulate air pollution, such as particles in the nanosize range, fine and coarse particles, CO, volatile organic compounds (VOCs), NO_2 and formaldehyde. In particulate matter from biomass burning, many trace species such as metals (Na, K, S, Zn, Ti, Fe etc.) and inorganic ions (nitrate, sulphate etc.) have been reported, as well as polycyclic aromatic hydrocarbons (PAH) both in the particle and gas phase (Kumar et al. 2013).

Nanoparticles are formed through the natural or human mediated disintegration of larger structures or by controlled assembly processes, such as material fabrication and transportation utilizing internal combustion and jet engines. These examples unintentionally release nanoparticles into the atmosphere. However, in recent years a new type of nanoparticle source has been introduced, within the sphere of intentionally engineered nanoscale components of consumer products and advanced technologies (SCENIHR 2005).

The number of nanoparticles in the air can be surprisingly similar in urban and rural areas, with as much as 10^6–10^8 nanoparticles per liter of air depending on conditions. In rural areas, nanoparticles mostly originate from the oxidation of volatile compounds of biogenic or anthropogenic origin, including secondary organic aerosols. In urban areas, the primary sources of these particles are diesel engines or cars with defective or cold catalytic converters. Photo-oxidation processes also lead to significant numbers of nanoparticles in urban areas (SCENIHR 2005).

13.2.2 Open Burning

Open burning sources are mainly deforestation, shifting cultivation, savanna fires, fuel wood and burning agricultural residues (Koppmann and von Czapiewski 2005). Thus, biomass burning includes source such as wildland fires and agricultural residue or waste burning, power generation, prescribed burning and residential wood combustion (Kumar et al. 2013). Biomass incineration is a major domestic source of energy in developing countries. Sometimes, deliberate or accidental burning of vegetation becomes uncontrolled, and reaches large forest areas, savannahs or other less dense vegetation (Arbex et al. 2004).

Power plants burn a blend of fuels, including coal and biomass (e.g. wood, waste and agricultural residue), and therefore differ in the formation mechanism and size distribution of nanoparticles (Kumar et al. 2013).

Burning agriculture residues and forest fires emit notable emissions of particulate materials (including nanoparticles) to the atmospheric environment. In flaming combustion (higher temperatures), particle formation begins with the creation of condensation nuclei such as polycyclic aromatic hydrocarbons from ejected fuel gases, as well as from a variety of soot-like species onto which condensable vapors can condensate. In smouldering combustion (lower temperature), the mass fraction of soot in smoke particles is very low and particles are largely formed by the condensation of volatilized organics on any available particles or surfaces (Koppmann and von Czapiewski 2005; Kumar et al. 2013).

Savanna fires are the single largest category of biomass burning emissions worldwide. The reason for the high incidence of savanna fires is the seasonal cycle of the wet season, during which biomass is produced, and the dry season, during which the biomass is turned into highly flammable material. Although in most cases fires are human induced, also natural induced savanna fires are common, when lightning strikes the dry vegetation (Koppmann and von Czapiewski 2005).

Pollution episodes associated with wildfire emissions have been recognized worldwide, and are ever increasing. The particle number emission factor (PNEF) estimated from burning biomass during forest fires can vary in the range of (1–30) $\times 10^{15}$ kg^{-1}, the same order as traffic nanoparticle emission factors (2.2–11) $\times 10^{15}$ kg^{-1} (Kittelson et al. 2004), but 2–3 orders of magnitude higher than PNEFs form domestic biomass burning. However, freshly emitted forest fire particles are typically larger than traffic particles; the size distribution has reached approximately 120 nm (Kumar et al. 2013).

13.2.3 *Industrial Burning*

Freshly generated smoke particles from biomass combustion consist mostly of organic carbon with approximately 15% carbon black and inorganic species and are found predominantly in the accumulation mode. The particle size distributions vary depending on combustion appliance, fuel type, combustion phase and the sampling method (Vu et al. 2015).

Carbon black (CB) is a low solubility particle produced industrially from incomplete thermal decomposition of hydrocarbons in which the process is controlled to achieve pre-defined and reproducible particle sizes and properties suitable for a wide range of industrial applications. In thermal-oxidative processes, such as the furnace black process, various types of hydrocarbon are sprayed into a natural gas-fired furnace and quenched with water to prevent complete burning (Donaldson et al. 2005).

As an example of biomass, bagasse can be mentioned, which is the main residue from alcohol production and sugar from sugarcane. Bagasse has attracted interest due to its low cost and calorific potential, about 4000 kcal/kg, in contrast to the calorific potential of coal (6000 kcal/kg). These values make bagasse an interesting waste. For this reason, its main use is as fuel in furnaces. Using of sugarcane bagasse as fuel generates other sub-residues: soot (very fine particles, resulting from combustion and deposited in the boiler), ash generated in boiler and fly ash. Fly ash can be defined as the fine particles that rise together with the gases generated by combustion. It is captured in the chimneys, usually by hydrostatic filters, before being released into the environment. The composition of the ash varies according to the material used and according to the incineration process parameters, such as temperature, incineration time and percentage of humidity of the incinerated material. Thus, ash is a material of heterogeneous composition and morphology. The average ash generated by the combustion is 6.2 kg of ash per 260 kg of incinerated dry cane bagasse (Cacuro and Waldman 2015).

The burning process forms metallic compounds due to the degradation of the organic fraction to a degree that depends on the combustion efficiency. At the same time, pyrolysis chemistry generates other complex organic molecules, some of which may persist along with elemental carbon nanoparticles. The combustion materials and mode of combustion will ultimately determine the characteristics of the nanoparticles, including the chemical composition, particle size and particle solubility. The large surface area of these nanoparticles present a maximal opportunity for dissolution of soluble species from the surface of the insoluble core. For insoluble nanoparticles, the large surface area provides a surface on which catalytic chemistry can occur that favors the formation of free radicals (Donaldson et al. 2005).

Pulverized coal combustion is a commonly-used and efficient method of coal burning in power stations. It occurs in air-blown, oxygen enriched furnaces, which produce nanoparticles and particulate material in abundance. In pulverized coal power stations, pulverized coal is blown into the furnace and burned off producing a fly ash emission. This particulate emission is controlled by the number or methodologies including electrostatic precipitators, filter scrubbers and mechanical collectors. However, these control measures are not 100% effective and some particles are released into the environment (Donaldson et al. 2005).

13.3 Monitoring: Environmental and in Chimneys

There are basically two groups of sampling required within an environmental control program: sampling of ambient air for monitoring air quality; sampling at the source, or in pipelines (De Nevers 1995). The first group aims to determine the characteristics of public air, usually to verify that the concentration of the pollutants is within the standards, while the second group aims to determine the emission rates of a polluting source (De Nevers 1995).

13.3.1 Field Observations in Outdoor Studies

Field fires can be defined as burning matter under conditions where products of combustion generated by burning are emitted into the environment without going through any duct or chimney. Open fires or field fires includes agricultural open burning, forest and basin fires, scales and fuel issues such as residue field burning and biofuel waste open burning (Li et al. 2003; Rocha et al. 2005; Li et al. 2007; Martins et al. 2009; Huang et al. 2012). Regarding biomass burning, there are some chemical markers (e.g. potassium, mannosan, galactosan), diagnostic ratios and specific target particles that are commonly used to trace biomass burning in the field and help make source apportionment of biomass burning emissions (Guo et al. 2004; Bo et al. 2008; Han et al. 2009; Li and Shao 2009; Wang et al. 2011; Fu et al. 2012; Saffari et al. 2013; Li et al. 2015, 2017). It is reported that biomass burning is a potential source of ambient primary and secondary carbonaceous aerosols and is related to pollution issues such as haze, fog and photochemical smog (Kaul et al. 2011; Li et al. 2011; Wang et al. 2013; Yang et al. 2015).

Field observations are described as a practical way to evaluate and characterize the properties and dynamic changes regarding the pollutants of biomass burning (Reid et al. 2005). Studies in outdoor environments concerning biomass burning aim to show how smoke plumes contribute to regional or global reactive trace gases, cloud condensation nuclei concentrations and PM2.5 (Du et al. 2011; Ding et al. 2015; Bougiatioti et al. 2016). External observations are also used to understand the role of aerosols in the formation of haze and formation of new particles (Wang et al. 2013; Yang et al. 2015; Sun et al. 2016). The third function of this kind of monitoring is to discover the relation between smoke aerosol emissions and loss in air quality and public health hazards (Lai and Sequeira 2001; Fullerton et al. 2008; Saffari et al. 2013). Furthermore, this approach attempts to elucidate the mechanism of aging of smoke particles and their internal mixing with aerosols, which may include changes in chemical compositions, redox activity and hygroscopicity, as well as influence the atmospheric chemical process, for example diurnal O_3 formation (Cape et al. 2011; Jaffe and Wigder 2012; Nie et al. 2014; Hu et al. 2016).

Field observations have advantages compared to studies carried out in the laboratory as they observe the dynamic of burning, as the investigations are conducted close to the real biomass burning (Ding et al. 2013; Brito et al. 2014). On the other hand, many factors make it difficult to obtain quality results, such as variable environmental conditions, chemical contamination, low concentration of target compounds and random burning process. Other disadvantages are related to the challenges in performing the practical work and the lack of the relationship between field research and laboratory studies (Dhammapala et al. 2007; Aurell et al. 2015; Carrico et al. 2016). Thus, it is reported that more accurate and quality results are obtained as a set of results derived from field research, along with laboratory studies and numerical simulations (Calfapietra et al. 2016; Huo et al. 2016; Pokhrel et al. 2016).

13.3.2 Satellite Remote Sensing

Continuous measurements of atmospheric aerosols in a location or a region can be carried out by using ground-based instruments. These instruments are important to estimate the aerosol impact of biomass burning on the local and regional climate (Badarinath et al. 2009; Mielonen et al. 2013). For global monitoring of biomass-burning aerosols, various satellite sensors can be used, such as MODIS (Moderate Resolution Imaging Spectroradiometer), MISR (Multi-Angle Imaging Spectroradiometer), CALIPSO (Cloud-Aerosol Lidar Infrared Pathfinder Satellite Observation), OMI (Ozone Monitoring Instrument), MOPITT (Measurements of Pollution in the Troposphere) and AIRS (Atmospheric Infrared Sounder) among others. Biomass burning aerosols are analyzed by these sensors, collecting products of physical and optical properties of aerosols, precursor gases, characteristics of smoke-plume and mapping the burned areas (Guan et al. 2010; Giglio et al. 2013; Kumar et al. 2015; Vadrevu et al. 2015; Zhu et al. 2016).

Significant applications of satellite sensing include monitoring vulnerable regions (e.g. tropical wildfires, agricultural burning, biofuel burning, etc.) resulting in important information concerning the number of fires, fire location, smoke plume distribution, which are key variables for local and regional climates, issues related to ecosystems and human health (Barnaba et al. 2011; Kaskaoutis et al. 2012; Chakrabarty et al. 2016).

13.3.3 Laboratory Studies and Observations

Investigations of biomass burning inside laboratories are normally conducted with the objective of characterizing the performance of domestic heaters and specially to quantify emission factors of heaters when operated in different conditions, regarding burning rates, type of wood fuel and wood conditions (dry or wet). Thus, laboratory studies do not establish burning conditions equivalent to those in open fires. As a result, emission factors obtained from biomass burning in the laboratory cannot be used to qualify emissions from wild fires because small changes in the experiment conditions may result in large variations in the emission factor (Chen et al. 2017). Research has been conducted to study the emission of pollutants in different burners, such as the characterization of nanoparticle emissions from wood burning in a fireplace (McDonald et al. 2000) and burning wood logs in various combustion systems (Wieser and Gaegauf 2000). Generally, the literature reports that most particles from biomass burning had diameters of less than 2.5 μm (Hays et al. 2002; Hedberg et al. 2002; Ferge et al. 2005).

13.3.4 Monitoring Biomass Burning in Ducts

In combustion processes, the emission rate of PM into the atmosphere can be quantified with chimney measurements, determining the concentration of particles within the effluent gas (Costa 1998). Therefore, sampling of particulate matter in the case of biomass combustion consists of collecting a sample of the gas that is moving through the duct or chimney of the biomass burner in order to obtain a representative fraction of the total flow (Carvalho Jr. and Lacava 2003). The concentration within

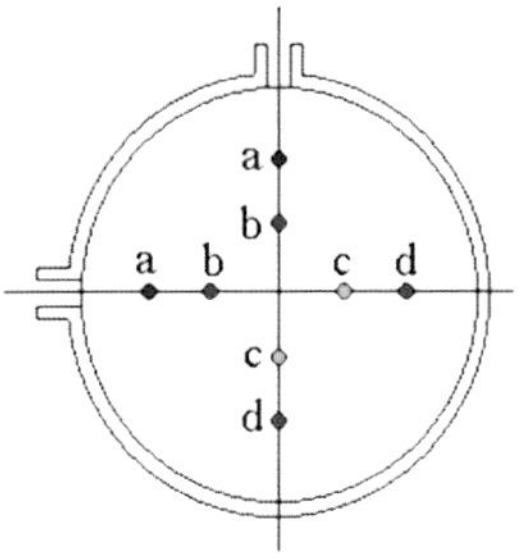

Sampling location points	Distance from duct wall
a	0.065 D
b	0.250 D
center	center
c	0.750 D
d	0.935 D

Fig. 13.1 Location of sample probes inside a duct (D: duct diameter)

the instrument used to carry out the measurement must necessarily be the same as that in the flow from which the sample is drawn. The sampling rate must be identical to the flow velocity, which characterizes an isokinetic sampling (Carvalho Jr. and Lacava 2003).

Monitoring particles emitted by biomass burning inside ducts or chimneys requires the acquisition of representative samples of these particles, since variations in duct properties, as in velocity and temperature, are common (Cieslinski 2014). The fine and ultrafine particles follow the flow lines of the gaseous stream and can easily deviate from the collectors, and therefore this is the reason why samplings must occur in an isokinetic manner, following a standardization (Cieslinski 2014). An example of specifications to determine where the sample probes should be located inside a duct is shown in Fig. 13.1. The number of points inside the duct were made according to national regulations (CETESB 1990) in which the sampling probes are to be positioned as shown in Fig. 13.1, where the distance between the points was calculated as a function of the duct diameter.

13.3.5 Campaigns Organized to Monitor Biomass Burning

Various campaigns (such as PRIDE-PRD "Program of Regional Integrated Experiments of Air Quality over the Pearl River Delta", CARE-Beijing "Campaign of Air Quality Research in Beijing and Surrounding Region") have been conducted to investigate air quality both on regional and large scales, providing details about pollution conditions, physicochemical profiles, sources and atmospheric chemistry. These campaigns usually support the regulation and enforcement of some environmental policies. Furthermore, the database obtained from these movements are helpful for numerical model simulations and further studies (Ho et al. 2010; Wu et al. 2011; Liu et al. 2012; George et al. 2016).

13.4 Concentration and Size Distribution of Particles

The most used concepts to determine the particle concentration measures are: mass concentration, number concentration and size distribution (Obaidullah et al. 2013).

Mass concentration is one of the most commonly used properties in sampling analyses. It is defined as the ratio between the particulate matter mass and the air volume unit (Vincent 2007) and is commonly expressed in $\mu g\ m^{-3}$ or $mg\ m^{-3}$. Another parameter that is usually used to evaluate the particulate matter emissions is the Emission Factor (EF). The EF is defined as the ratio between the amount of mass emitted from a compound and the amount of dry fuel consumed and it is usually expressed in $g\ kg^{-1}$ (Alves et al. 2010). The number of particles concentration is determined as being the ratio of the number of particles to the unit volume of aerosol, and generally has units of particles per cubic meter or particles per cubic centimeter (Kulkarni et al. 2011).

Size distribution is one of the most important properties in PM sampling because it is related to harmful health effects, selecting industrial emission control equipment and particle behavior. It is also very important for determining possible emission sources and particulate trajectories. Kantová et al. (2017) reported the particulate matter size distribution as the most important property for determining particle behavior disperse in a gas. The varied sizes of particles are related to their behavior and these are governed by different physical laws (Kulkarni et al. 2011).

Likewise, the particles found in the atmosphere are not spherical and the determination of the diameter and consequently size are more difficult and, therefore it is necessary to create a definition to determine these properties. Vincent (2007) describes some of the most used definitions to calculate particle diameter: the equivalent volume diameter, the equivalent projected area diameter, the equivalent surface area diameter and the aerodynamic diameter.

The aerodynamic diameter is one of the most used definitions, it is defined as the diameter of a 1 $g\ cm^{-3}$ density sphere (Souza 2016). This definition has a great relevance from a physical point of view, geometric size, density, shape and behavior (Vincent 2007).

The particulate matter can be classified according to the particle size. According to the US EPA (2016), particulate matter has a large size range. It contains particles large enough to be seen with the naked eye and such small particles that an electron microscope is needed to observe them.

According to the US EPA (2016), particulate matter is divided into PM10 (inhalable particles with diameters between 2.5 and 10 μm) and PM2.5 (inhalable particle with diameters smaller than 2.5 μm). Figure 13.2 shows a size comparison for particulate matter diameters.

Whitby and Sverdrup (1980) suggested 3 different groups of particles: the nucleation or Aitken mode (0.01–0.08 μm), the accumulation mode (0.08 e 1–2 μm) and the coarse particle mode (>2.5 μm). The fine particle mode is represented by the combination of the Aitken and accumulation modes. Figure 13.3 shows the particulate matter distribution according to the size range, where the mode introduced

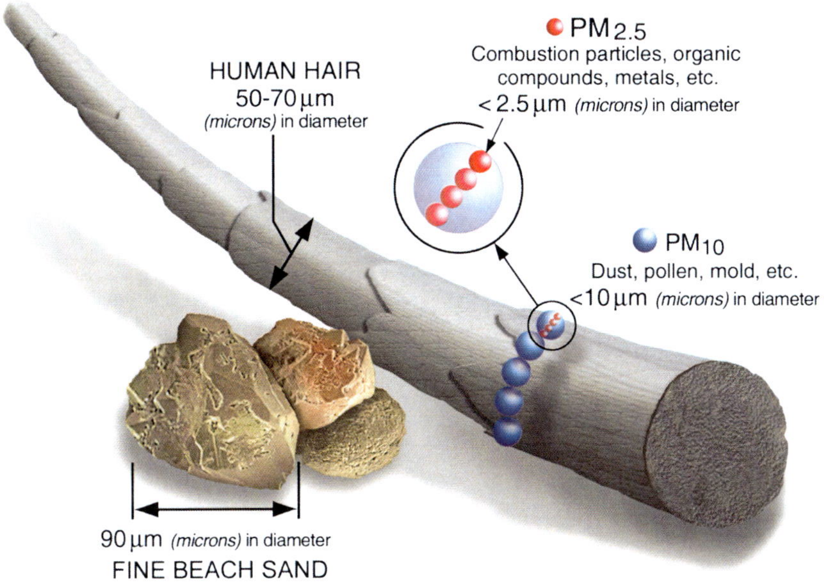

Fig. 13.2 Size comparison for particulate matter diameters. Image courtesy of US EPA (2016)

by Whitby and Sverdrup (1980) are represented by the solid curve. In addition to having different size ranges, the particulates have different chemical characteristics, formation mechanisms and behaviors (Finlayson-Pitts and Pitts Jr 2000a).

The ultrafine particulate matter formation can be a heterogeneous or homogeneous mechanism. The homogeneous mechanism is characterized by the low pressure gas phase molecules condensation into very small particles (Souza et al. 2017). The heterogeneous mechanism is different as it occurs with molecule condensation onto existing small particles (Souza et al. 2017). The Aitken particles are formed by homogeneous processes (conversion gas-particle). The accumulation particles are formed by heterogeneous processes (condensation and coagulation onto existing particles) and the coarse particles are formed by mechanical processes (Alves 2005).

Studies carried out in rural areas in eastern central India throughout the rice straw burning campaign determined the particulate matter mass and size distribution (Nirmalkar and Deb 2016). Researchers sampled size distribution in the coarse particle fraction during pre and post burning periods, and during the burning period most of the particles were present in the fine particulate matter fraction (Nirmalkar and Deb 2016).

A large body of research has evaluated emission factors such as pollutant emission effects, mass concentration, emission factor, number concentration, size distribution and chemical composition in different countries. For example, a study conducted in the South Atlantic coastal area at Baía de Todos os Santos bay, Brazil, evaluated the effects of pollutant emission on the atmosphere, resulting in a mean concentration

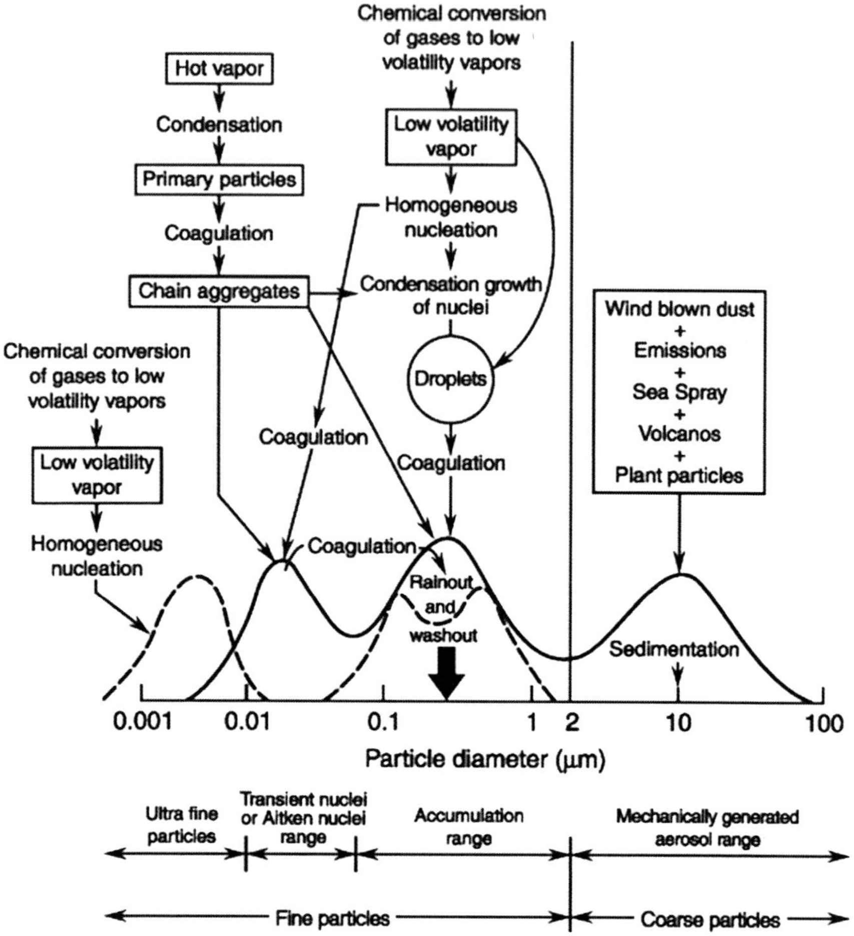

Fig. 13.3 Schematic of an atmospheric aerosol size distribution. Reprinted from Chemistry of the Upper and Lower Atmosphere: Theory, Experiments, and Applications, Finlayson-Pitts, Barbara J. and Pitts Jr, James N., Chapter 9—Particles in the Troposphere, 349–435, Copyright (2000), with permission from Elsevier

of total carboxylates of 78 ng m^{-3} in fine fraction and 81 ng m^{-3} in coarse fraction (Mkoma et al. 2014). The inorganic ions presented concentrations of 2143 ng m^{-3} in fine fraction and 4880 ng m^{-3} in coarse fraction (Mkoma et al. 2014).

The sampling campaign done during July, 2001 and June, 2002 in Pittsburgh, USA reported differences between the particulate matter sampled in the winter campaign and those sampled in the summer campaign (Cabada et al. 2004). The concentration number was higher in the winter if compared with the summer data, and in the summer samples the carbonaceous material, sulfate and ammonium contributed respectively in 50%, 28% and 12% of the ultrafine particulate matter mass (Cabada et al. 2004).

Studies carried out in Guangzhou, China showed the total amine concentration in the range of 79.6–140.9 ng m^{-3} for PM10 and the most abundant compound was methylamine with a concentration in the range of 29.2–70.1 ng m^{-3} (Liu et al. 2017).

A recent study conducted in China reported higher mass concentration in the range of 1.0–1.6 μm particles on cloudy days, but the total number of particles was lower on cloudy days when compared to those on clear days (Yu et al. 2018).

Alves et al. (2010) calculated the EF for substances such as methanol, propene, ethane, acetylene, NH_3, N_2O, NOx, SO_2 and PM2.5 and obtained EF of 1.8 for methanol, 1.7 for propene, 1.3 for ethane, 3.4 for acetylene, 1.3 for NH_3, 0.122 for N_2O, lower than 0.3 for NOx, 0.16 for SO_2 and 3.4 for PM2.5.

Taiwo et al. (2014) analyzed the particle number concentration, water soluble ions, metals adsorbed on the particulate matter and evaluated the size distribution of particulate material. A sampling campaign was carried out in two different sites in the United Kingdom. The authors obtained concentrations of 100 particles per cubic centimeter in the range of 0.3–0.4 μm. The mass distribution at the Elms Road Observatory was higher in the fine fraction and in Port Talbot the concentrations were higher in the coarse fraction of the particulate matter.

13.5 Chemical Characterization

Chemical characterization, performed by instrumental techniques, is important in terms of identifying and quantifying the emitted compounds. These techniques have enabled the monitoring of gases and particles, thus contributing to control more accurately.

Air quality standards are based on monitoring of some compounds called "Criteria pollutants", which are: Ground-level ozone, Carbon monoxide, Sulphur oxides, Nitrogen oxides, Lead (Pb), Particulate matter (PTS, PM10, PM2.5), whose regulation is based on environmental and human health criteria. Reference methods and equivalence were established by the US Environmental Protection Agency to measure these compounds, as shown in Table 13.1 (US EPA 2018).

However, monitoring and control methods need to be developed regarding compounds that are not regulated. There is a multitude of compounds that deserve attention regarding their harmful potential, whether for health, for example, Polycyclic Aromatic Hydrocarbons (PAH), which have a carcinogenic potential, or in relation to the climate, as is the case of Black Carbon (BC), which contributes positively to climate forcing. However, systematic monitoring of air quality is restricted to a particular group of pollutants which are defined in terms of their importance and the resources available for their monitoring.

The first step of chemical characterization is sampling, which should be representative for the range of particle sizes of interest. Filters, cyclones, and impactors are used to achieve particle size separation. This size distribution is important since particles with smaller diameters (PM2.5) have negative effects on human health (Finlayson-Pitts and Pitts Jr 2000b).

Samplers are based on filtration, centrifugal or gravitational sedimentation, inertial impaction, diffusion, interception, thermal or electrostatic precipitation. Selecting the sampler will depend on the purpose of the analysis, as well as the choice of the filter, which should take into account the efficiency of collection, adsorption of artifacts or water vapor.

The particles have a complex composition, including inorganic material, ions, organic and water and, for each class of compounds, different techniques are used which characterize each compound.

13.5.1 Ion Analysis

The ions can be determined by colorimetric methods, ion selective electrodes, Raman spectroscopy, mass spectrometry and also via ion chromatography. Among these techniques, ion chromatography can include a higher number of ions during the same analysis. In this technique, the ions are separated through a column and are detected by electrical conductivity. Nitrate and sulfate, for example, pass through a column of basic resin containing carbonate solution as the eluent. As the eluent has high conductivity and this could mask the nitrate and sulfate signal, the eluent passes through a suppressor containing a strongly acidic resin, converting the carbonate to CO_2 and H_2O, which has limited conductivity, and nitrate and sulfate in their respective acids, which have high conductivity and can be easily detected with the suppressed eluent. Thus, a variety of cations and anions can be detected by ion chromatography (Finlayson-Pitts and Pitts Jr 2000b; Skoog et al. 2013).

Ions may be indicative of sources present in the environment. Potassium ion (K^+) can be used as a biomass burning tracer, as in the case of Araraquara, São Paulo state, Brazil in 2010. Biomass burning of sugarcane straw was associated with high concentrations of potassium and also the presence of secondary aerosols SO_4^{2-},

Table 13.1 Reference or equivalent methods for Criteria pollutants by the U.S. Environmental Protection Agency (US EPA 2018)

Gas	Reference or equivalent method
SO_2	UV fluorescence Differential optical absorption spectrometry Pararosaniline
O_3	UV absorption Chemiluminescence Differential optical absorption spectrometry
NO_2	Ozone chemiluminescence Differential optical absorption spectrometry Sodium arsenite
CO	Nondispersive infrared

NO^{3-} e NH_4^+, produced by reactions involving gaseous precursors (including NOx, SO_2, and NH_3) (Caetano-Silva et al. 2013).

13.5.2 Analyzing Inorganic Elements

Inorganic elements can be analyzed adopting several techniques, including: colorimetry, atomic absorption spectrometry (AA), X-ray fluorescence analysis (XRF), mass spectrometry (MS), inductively coupled plasma spectroscopy (ICP), particle-induced X-ray emission (PIXE) and X-ray photoelectron spectroscopy (XPS).

Inductively coupled plasma spectroscopy has the advantage of being less susceptible to chemical interferences; good spectral emission for most elements under a single set of excitation conditions, allowing the spectrum of dozens of elements simultaneously; plasma sources also allow the determination in low concentrations of elements and also the determination of nonmetals (Skoog et al. 1998).

The sample is introduced into a plasma formed by the radiofrequency discharge of a highly energetic ionized gas such as argon. The generated ions and electrons are induced to traverse annular paths by interaction with a floating magnetic field generated by induction of tesla coil. Elements in the plasma are excited and emit characteristic wavelengths (Skoog et al. 1998; Finlayson-Pitts and Pitts Jr 2000b).

13.5.3 Analysis of Organic Elements

The most common identification technique is mass spectrometry coupled to a separation technique such as gas chromatography (GC) or liquid (HPLC). To analyze organic compounds in general, a pre-separation in the column containing silica gel, with different solvents of increasing polarity, is carried out to separate the following fractions: n-alkanes, polycyclic aromatic hydrocarbons, carbonyl compounds, hydroxyl containing compounds, acids and sugars. Each group of compounds is then subjected to chromatography coupled to masses for separation and identification. If the organic compounds exhibit high volatility, GC is used. If they are of low volatility, HPLC is used.

The characterization of each compound is of significant importance, whether due to its effects on the environment or health. The results obtained in Araraquara showed that emissions from sugarcane burning can cause unsafe concentrations of carcinogenic compounds and mutagenic potentials such as benzo(a)pyrene of the class of Polycyclic aromatic hydrocarbons (PAHs) (Urban et al. 2016). In Limeira, 130 km from Araraquara, it was observed that nitroaromatics are the major cause of mutagenicity, which may occur as a result of the mixture of emissions from biomass burning with anthropogenic emissions (Alves et al. 2016).

13.5.4 Analysis of Organic Carbon (OC) and Elemental Carbon (EC)

The separation of organic carbon from elemental carbon is mainly done by thermal methods, digestion, extraction and optical techniques (Finlayson-Pitts and Pitts Jr 2000b). The thermal-optical analysis is currently considered as the reference methodology to quantify atmospheric particulate EC and OC deposited on filters. The sample is subjected to temperature scales in the presence of O_2. The method is based on the principle that the volatile organic compounds will vaporize first and then other organic compounds will be oxidized. EC does not volatilize until very high temperatures are reached, and their release occurs only when oxygen is present. Carbonaceous vapors are converted after volatilization in CH_4 and quantified with a flame ionization detector, FID (Karanasiou et al. 2015).

13.5.5 Real-Time Measurements

Progress in technology has allowed the development of increasingly sensitive instruments that allow the monitoring of chemical composition in real time. An example is the Aerosol Chemical Speciation Monitor (ACSM) that allows real-time measurement, non-refractory aerosol particle mass and chemical composition: Organic, Sulfate, Nitrate, Ammonium and Chloride. In Beijing, China, the measures using ACSM demonstrated a high potential impact of regional transport of neighboring regions on pollution (Sun et al. 2015).

Although these techniques allow continuous monitoring of chemical compositions, traditional techniques have not been excluded for organic compounds due to their complexity.

13.6 Dry Deposition of Fine and Manometric Particles

The particles present in the atmosphere undergo different processes, such as atmospheric dispersion, chemical reactions, dry deposition or wet deposition. Dry deposition is the process of removal through which the aerosol particles are absorbed by the surface of the Earth, while in the wet deposition process the particles need to be first incorporated into the droplets and then reach the surface by precipitation (Pellerin et al. 2017). In contrast to the nature of wet deposition, dry deposition is a continuous and effective process that is part of the cleaning procedure of the atmosphere (Wu et al. 2018). The dry deposition process is the transport of gaseous and particulate species from the atmosphere to the different surfaces in the absence of precipitation, and may cause strong impacts including in biogeochemical cycles. Dry deposition removes gases and particles at the interfaces of the air surface and is responsible in a manner comparable to wet deposition (Wu et al. 2018). This dry phenomenon involves different sources, such as industrial emissions, natural dust, trace metals, chemicals and even nuclear emissions from serious accidents. Studies relate dry deposition in terms of deposition velocity. One of the current challenges is to quantify the impact of pollutant deposition by correlating the deterioration of environmental conditions, as well as the effects on health (Pellerin et al. 2017). The dry deposition process is associated with all meteorological, chemical and biological phenomena that influence a gas flow and particulate pollutants, interacting with the soil surface without involving water in the atmosphere (Giardina and Buffa 2018). Understanding dry particle deposition is far from complete due to the deposition dependence in relation to particle size, density, soil, vegetation, deposition site characteristics and chemical compounds. The phenomenon of dry deposition is governed by the complex interaction between turbulent transport processes and deposition mechanisms, such as gravitational and Brownian sedimentation (Huang et al. 2014).

The main phenomena that are considered and affect dry deposition can be described as follows: transport due to turbulence in the low atmospheric layer (Planetary Boundary Layer PBL), which depends only on the level of atmospheric turbulence and is independent of nature physics and chemistry of the pollutant; the diffusion related to air and land interface called the Quasi-Laminar Sublayer (QLS), in which the molecular diffusion to gas, Brownian motion, turbulent diffusion to particles and gravity to heavier particles become dominant; and transfer to the soil (e.g., interception, impaction and rebound), which shows a significant dependence on the type of surface with which the pollutant interacts (Giardina and Buffa 2018). Ultrafine particles have a similar behavior to gaseous molecules and suffer deposition through diffusion processes while the deposition of thicker particles occurs by sedimentation, because they are more subject to gravitational action (Janhäll 2015).

These mechanisms of collection, inertial impaction, interception and Brownian diffusion are imperative in determining the rate of deposition and the concentration of the deposition fluxes of the species (Zhu et al. 2013; Qiao et al. 2015).

According to Giardina and Buffa (2018), the effects of Brownian diffusion and turbulence are significant and contribute to the total velocity of dry deposition of particles ranging in size from 0.01 μm to about a few micrometers. These effects dominate the quasi-laminar sublayer diffusion processes (QLS) of the surface. For particles of intermediate diameter in the range of about 0.1–1 μm, the process depends heavily on atmospheric conditions, surface characteristics and particle size. For particles with diameters above this range, deposition is dominated by other phenomena, such as inertial and turbulent impaction.

In the inertial impaction mechanism, the inertia of particles is very large, and therefore it is carried by the flow towards an obstacle; the particles are not able to follow the gaseous flow due to the inertia and consequently, collide with the obstacle and remain on the surface. In turbulent impaction, particles have a high velocity, swirling turbulent flows occur and the particles present sufficient momentum to reach the surface (Giardina and Buffa 2018).

The most commonly used method to evaluate dry deposition aerosol flows is the velocity of deposition of particles from an altitude z to the receiving surface. Dry deposition velocity varies over several orders of magnitude, depending on the nature of the receiving surfaces (forests, snow, cities, meadows, bare ground, water, rivers, ocean), flow conditions and characteristics of the aerosol (Calec et al. 2017).

Several authors have evaluated dry deposition fluxes on different surfaces using mathematical models that predict this phenomenon. Dry deposition of particles on solid surfaces such as soil and vegetation has also been studied (Slinn 1982; Zhang et al. 2001; Zhang et al. 2003; Pryor et al. 2008; Pryor et al. 2009; Zhang et al. 2014; Wasiuta et al. 2015). Studies of the dry aerosol deposition on the water surface evaluated two different situations: results determined in wind tunnels and those measured in situ in lakes or oceans. Wu et al. (2018) evaluated the fluxes of dry deposition of nutrients in oceans in china and Pellerin et al. (2017) evaluated and quantified the dry deposition velocity of the particles as a function of their size (from 2.5 nm to 1.2 μm), relating the micro meteorological parameters and surface characteristics. Calec et al. (2017) reported that the first wind tunnel study was conducted in the 1970s by Möller and Schumann (1970), followed by Schumann (1975), for particle sizes ranging from 0.010 to 1 μm. This study showed that the dry deposition on the water surface depends on the size of the particle. They also cited the work of Sehmel and Sutter (1974) who measured the rate of dry deposition of particles with several micrometers in diameter in a wind tunnel on a water surface. They observed that the contribution of phoretic effects was comparable to those of Brownian and turbulent diffusions.

Considering the deposition rates on different surfaces, various authors recognize the complexity and limited theoretical understanding of the mechanism for 100 nm particle sizes. For this reason, models to predict particle deposition at different locations with velocities and deposition fluxes for particles from 100 nm to 2.5 μm were proposed (Wasiuta et al. 2015). The deposition flux of particles is usually defined as a product of the local atmospheric concentration at an altitude relative to the velocity of deposition, responsible for an appreciable fraction of the total deposition flux (Nho-Kim et al. 2004). These authors proposed deposition velocity for the model in the range of 0.04–0.14 cm s^{-1} for particles with diameters of 0.1–0.5 μm, according to the values measured 0.01 ± 0.03 cm^{-1}. A minimum value was obtained for the deposition rate for particles between 0.5 and 1.0 μm in diameter. The mean values obtained for 0.1, 1 and 0.5 μm particles were respectively 0.09, 0.03 and 0.52 cm s^{-1}. Parameterization of dry particle deposition was developed for the Canadian Aerosol Module (CAM) and the Slinn (1982) model (Mammarella et al. 2011). Knowledge about dry particle deposition is far from complete due to the complex dependence of deposition on various properties. Many other authors use this model in predicting velocity and deposition fluxes (Mammarella et al. 2011). Important studies have evaluated dry deposition flows with results relevant to different areas. According to Wu et al. (2018) estimates of atmospheric deposition of nitrogen, phosphorus, and silicon in coastal waters suggest that atmospheric deposition may be a significant source of these nutrients. Nitrogen deposition monitoring showed very high nitrogen deposition rates on dozens of observation sites in different regions of China. The authors concluded that dry nutrient deposition might increase the risk of imbalanced nutrient status in the seawater column of Daya Bay. The ratio of NO_3/NH_4^+ in the dry deposition fluxes was 9.0, while this ratio was 2.5% in the concentration. Ammonia in the aerosol particles was mainly from anthropogenic sources and was considered as being volatilized from agricultural areas, while nitrate was mainly from the combustion of fossil fuels from vehicles and industries. This scale of nutrient deposition may contribute to algae bloom and partially induces eutrophication in the Daya Bay.

The influence of sugarcane biomass burning on the particle size distribution, deposition rate according to the Zhang model and the determination of flows in the city of Araraquara during the open burning period was evaluated in the work of sampling and chemical composition. The deposition fluxes were calculated for all species and were larger during the season of the year when the fires occur. Although the highest deposition rates were obtained for calcium, sulfate and nitrate, they were triggered by processes only partially affected by burning and associated activities. The increase in the annual nitrogen deposition flow was approximately 9% due directly to the burning of sugarcane. These observations may be relevant to any region where dry season biomass burning is used in land management. The replacement production model requires combustion of green waste under controlled conditions to produce electricity, with expected gains from both economic and environmental perspectives. In terms of emissions, the net effect must be to replace a large, spatially diffuse source by a number of point sources of pollutants, the magnitudes of which will largely depend on the efficiency of stack scrubber systems (Rocha et al. 2005).

13.7 Conclusion and Future Perspectives

Nanotechnology is important in developing sustainable technologies for the future, for humanity and the environment. However, care needs to be taken to ensure that nanotechnology is not creating future problems. It is advisable to examine and address environmental and human health concerns before the adopting specific nanotechnologies on a large scale, and incorporate green chemistry metrics into the production and distribution of nanoparticles. The negative effects of biomass burning on the environment and on human health are evident and regulations in the world are still limited. Therefore, further studies are needed to more accurately quantify the adverse effects of biomass combustion on health. This should include comparative studies between the effects of biomass combustion products and fossil fuels. Consequently, studies on emissions from biomass combustion can reveal when this process is performed inappropriately, having an impact on local air quality. On the other hand, properly operated, state-of-the-art combustion systems exhibit low environmental impact and effectively replace fossil fuels and mitigate the increase in atmospheric CO_2. Beside the use of modern technologies, a proper plant operation and a stringent enforcement of emission regulations need to be ascertained.

Acknowledgements The authors are grateful to FAPESP—São Paulo Research Foundation (grant numbers #2016/23209-0 and #2018/00697-5)

References

Alves C (2005) Aerossóis atmosféricos: perspectiva histórica, fontes, processos químicos de formação e composição orgânica. Quim Nova 28:859–870

Alves CA, Gonçalves C, Pio CA, Mirante F, Caseiro A, Tarelho L, Freitas MC, Viegas DX (2010) Smoke emissions from biomass burning in a Mediterranean shrubland. Atmos Environ 44(25):3024–3033. https://doi.org/10.1016/j.atmosenv.2010.05.010

Alves DKM, Kummrow F, Cardoso AA, Morales DA, Umbuzeiro GA (2016) Mutagenicity profile of atmospheric particulate matter in a small urban center subjected to airborne emission from vehicle traffic and sugar cane burning. Environ Mol Mutagen 57(1):41–50. https://doi.org/10.1002/em.21970

Andreae MO (2007) Aerosols before pollution. Science 315(5808):50–51. https://doi.org/10.1126/science.1136529

Arbex MA, Cançado JED, Pereira LAA, Braga ALF, Saldiva PHdN (2004) Queima de biomassa e efeitos sobre a saúde. J Bras Pneumol 30:158–175

Aurell J, Gullett BK, Tabor D (2015) Emissions from southeastern U.S. Grasslands and pine savannas: comparison of aerial and ground field measurements with laboratory burns. Atmos Environ 111:170–178. https://doi.org/10.1016/j.atmosenv.2015.03.001

Badarinath KVS, Kumar Kharol S, Rani Sharma A (2009) Long-range transport of aerosols from agriculture crop residue burning in Indo-Gangetic Plains—a study using LIDAR, ground measurements and satellite data. J Atmos Solar Terr Phys 71(1):112–120. https://doi.org/10.1016/j.jastp.2008.09.035

Barnaba F, Angelini F, Curci G, Gobbi GP (2011) An important fingerprint of wildfires on the European aerosol load. Atmos Chem Phys 11(20):10487–10501. https://doi.org/10.5194/acp-11-10487-2011

Bo Y, Cai H, Xie SD (2008) Spatial and temporal variation of historical anthropogenic NMVOCs emission inventories in China. Atmos Chem Phys 8:7297–7316. https://doi.org/10.5194/acp-8-7297-2008

Bougiatioti A, Bezantakos S, Stavroulas I, Kalivitis N, Kokkalis P, Biskos G, Mihalopoulos N, Papayannis A, Nenes A (2016) Biomass-burning impact on CCN number, hygroscopicity and cloud formation during summertime in the eastern Mediterranean. Atmos Chem Phys 16(11):7389–7409. https://doi.org/10.5194/acp-16-7389-2016

Brito J, Rizzo LV, Morgan WT, Coe H, Johnson B, Haywood J, Longo K, Freitas S, Andreae MO, Artaxo P (2014) Ground-based aerosol characterization during the South American biomass burning analysis (SAMBBA) field experiment. Atmos Chem Phys 14:12069–12083. https://doi.org/10.5194/acp-14-12069

Buseck PR, Adachi K (2008) Nanoparticles in the atmosphere. Elements 4(6):389–394. https://doi.org/10.2113/gselements.4.6.389

Cabada JC, Rees S, Takahama S, Khlystov A, Pandis SN, Davidson CI, Robinson AL (2004) Mass size distributions and size resolved chemical composition of fine particulate matter at the Pittsburgh supersite. Atmos Environ 38(20):3127–3141. https://doi.org/10.1016/j.atmosenv.2004.03.004

Cacuro TA, Waldman WR (2015) Cinzas da Queima de Biomassa: Aplicações e Potencialidades. Revista Virtual Quimica 7(6):2154–2165

Caetano-Silva L, Allen AG, Lima-Souza M, Cardoso AA, Campos MLAM, Nogueira RFP (2013) An analysis of diurnal cycles in the mass of ambient aerosols derived from biomass burning and agro-industry. J Geophys Res Atmos 118(15):8675–8687. https://doi.org/10.1002/jgrd.50630

Calec N, Boyer P, Anselmet F, Amielh M, Branger H, Mailliat A (2017) Dry deposition velocities of submicron aerosols on water surfaces: laboratory experimental data and modelling approach. J Aerosol Sci 105:179–192. https://doi.org/10.1016/j.jaerosci.2016.11.012

Calfapietra C, Morani A, Sgrigna G, Di Giovanni S, Muzzini V, Pallozzi E, Guidolotti G, Nowak D, Fares S (2016) Removal of ozone by urban and peri-urban forests: evidence from laboratory, field, and modeling approaches. J Environ Qual 45(1):224–233. https://doi.org/10.2134/jeq2015.01.0061

Calvo AI, Alves C, Castro A, Pont V, Vicente AM, Fraile R (2013) Research on aerosol sources and chemical composition: past, current and emerging issues. Atmos Res 120–121:1–28. https://doi.org/10.1016/j.atmosres.2012.09.021

Cape JN, Cornell SE, Jickells TD, Nemitz E (2011) Organic nitrogen in the atmosphere—Where does it come from? A review of sources and methods. Atmos Res 102(1):30–48. https://doi.org/10.1016/j.atmosres.2011.07.009

Carrico CM, Prenni AJ, Kreidenweis SM, Levin EJT, McCluskey CS, DeMott PJ, McMeeking GR, Nakao S, Stockwell C, Yokelson RJ (2016) Rapidly evolving ultrafine and fine mode biomass smoke physical properties: comparing laboratory and field results. J Geophys Res Atmos 121(10):5750–5768. https://doi.org/10.1002/2015JD024389

Carvalho JA Jr, Lacava PT (2003) Emissões em processos de combustão. Editora Unesp, Sao Paulo

CETESB (1990) Dutos e Chaminés de Fontes Estacionárias - Determinação dos Pontos de Amostragem - L9.221. Companhia Ambiental do Estado de São Paulo http://www.esaat.com.br/docs/met_cetesb/CETESB-L9.221.pdf. Accessed 21 March 2018

Chakrabarty RK, Gyawali M, Yatavelli RLN, Pandey A, Watts AC, Knue J, Chen L-WA, Pattison RR, Tsibart A, Samburova V, Moosmüller H (2016) Brown carbon aerosols from burning of boreal peatlands: microphysical properties, emission factors, and implications for direct radiative forcing. Atmos Chem Phys 16:3033–3040. https://doi.org/10.5194/acp-16-3033-2016

Chen J, Li C, Ristovski Z, Milic A, Gu Y, Islam MS, Wang S, Hao J, Zhang H, He C, Guo H, Fu H, Miljevic B, Morawska L, Thai P, Lam YF, Pereira G, Ding A, Huang X, Dumka UC (2017) A review of biomass burning: emissions and impacts on air quality, health and climate in China. Sci Total Environ 579:1000–1034. https://doi.org/10.1016/j.scitotenv.2016.11.025

Cieslinski JEF (2014) Estudo da emissão e do controle dos gases e particulados provenientes da queima de biomassa. Doctorate degree, Universidade Estadual Paulista "Júlio de Mesquita Filho", Guaratingueta

Costa MAM (1998) Amostragem de partículas dispersas em correntes gasosas confinadas. Universidade Estadual de Maringá, Maringá, Master degree

De Nevers N (1995) Air pollution control engineering. McGraw-Hill, New York; Singapore

Dhammapala R, Claiborn C, Simpson C, Jimenez J (2007) Emission factors from wheat and Kentucky bluegrass stubble burning: comparison of field and simulated burn experiments. Atmos Environ 41(7):1512–1520. https://doi.org/10.1016/j.atmosenv.2006.10.008

Ding AJ, Fu CB, Yang XQ, Sun JN, Petäjä T, Kerminen V-M, Wang T, Xie Y, Herrmann E, Zheng LF, Nie W, Liu Q, Wei XL, Kulmala M (2013) Intense atmospheric pollution modifies weather: a case of mixed biomass burning with fossil fuel combustion pollution in eastern China. Atmos Chem Phys 13:10545–10554. https://doi.org/10.5194/acp-13-10545-2013

Ding K, Liu J, Ding A, Liu Q, Zhao TL, Shi J, Han Y, Wang H, Jiang F (2015) Uplifting of carbon monoxide from biomass burning and anthropogenic sources to the free troposphere in East Asia. Atmos Chem Phys 15(5):2843–2866. https://doi.org/10.5194/acp-15-2843-2015

Donaldson K, Tran L, Jimenez LA, Duffin R, Newby DE, Mills N, MacNee W, Stone V (2005) Combustion-derived nanoparticles: a review of their toxicology following inhalation exposure. Part Fibre Toxicol 2(1):10. https://doi.org/10.1186/1743-8977-2-10

Du H, Kong L, Cheng T, Chen J, Du J, Li L, Xia X, Leng C, Huang G (2011) Insights into summertime haze pollution events over Shanghai based on online water-soluble ionic composition of aerosols. Atmos Environ 45(29):5131–5137. https://doi.org/10.1016/j.atmosenv.2011.06.027

Ferge T, Maguhn J, Hafner K, Mühlberger F, Davidovic M, Warnecke R, Zimmermann R (2005) On-line analysis of gas-phase composition in the combustion chamber and particle emission characteristics during combustion of wood and waste in a small batch reactor. Environ Sci Technol 39(6):1393–1402. https://doi.org/10.1021/es049493o

Finlayson-Pitts BJ, Pitts Jr JN (2000a) Particles in the troposphere. In: Chemistry of the upper and lower atmosphere. Academic Press, San Diego, pp 349–435. https://doi.org/10.1016/B978-012257060-5/50011-3

Finlayson-Pitts BJ, Pitts Jr JN (2000b) Chemistry of the upper and lower atmosphere: theory, experiments, and applications. Academic Press, San Diego

Fu H, Zhang M, Li W, Chen J, Wang L, Quan X, Wang W (2012) Morphology, composition and mixing state of individual carbonaceous aerosol in urban Shanghai, vol 12. https://doi.org/10.5194/acp-12-693-2012

Fullerton D, Bruce N, Gordon S (2008) Indoor air pollution from biomass fuel is a major health concern in the developing world, vol 102. https://doi.org/10.1016/j.trstmh.2008.05.028

George M, Suhail S, Chandran S, Chen J, Lu K, Ruth A, Venables D, Varma R (2016) Open-path in situ measurement of the nitrate radical concentrations during the CAREBeijing-NCP 2014 summer campaign. In: EGU general assembly conference abstracts. http://adsabs.harvard.edu/abs/2016EGUGA.1812189G. Accessed 05 April 2018

Giardina M, Buffa P (2018) A new approach for modeling dry deposition velocity of particles. Atmos Environ 180:11–22. https://doi.org/10.1016/j.atmosenv.2018.02.038

Giglio L, Randerson JT, Werf GR (2013) Analysis of daily, monthly, and annual burned area using the fourth-generation global fire emissions database (GFED4). J Geophys Res Biogeosci 118(1):317–328. https://doi.org/10.1002/jgrg.20042

Guan H, Esswein R, Lopez J, Bergstrom R, Warnock A, Follette-Cook M, Fromm M, Iraci LT (2010) A multi-decadal history of biomass burning plume heights identified using aerosol index measurements. Atmos Chem Phys 10(14):6461–6469. https://doi.org/10.5194/acp-10-6461-2010

Guo H, Wang T, Simpson I, Blake DR, Yu XM, Kwok YH, Li YS (2004) Source contributions to ambient VOCs and CO at a rural site in Eastern China, vol 38. https://doi.org/10.1016/j.atmosenv.2004.05.004

Han YM, Lee SC, Cao JJ, Ho KF, An ZS (2009) Spatial distribution and seasonal variation of char-EC and soot-EC in the atmosphere over China. Atmos Environ 43(38):6066–6073. https://doi.org/10.1016/j.atmosenv.2009.08.018

Harrison RM (1999) Introduction. In: Harrison RM (ed) Understanding our environment: an introduction to environmental chemistry and pollution, 3rd edn. The royal society of chemistry. Cambridge. http://dx.doi.org/10.1039/9781847552235-00001

Harrison RM, Hester RE (2009) Air quality in urban environments. Royal society of chemistry, London, UK. https://doi.org/10.1039/9781847559654

Hays MD, Geron CD, Linna KJ, Smith ND, Schauer JJ (2002) Speciation of gas-phase and fine particle emissions from burning of foliar fuels. Environ Sci Technol 36(11):2281–2295. https://doi.org/10.1021/es0111683

Hedberg E, Kristensson A, Ohlsson M, Johansson C, Johansson P-Å, Swietlicki E, Vesely V, Wideqvist U, Westerholm R (2002) Chemical and physical characterization of emissions from birch wood combustion in a wood stove. Atmos Environ 36(30):4823–4837. https://doi.org/10.1016/S1352-2310(02)00417-X

Hinds WC (1999) Aerosol technology: properties, behavior, and measurement of airborne particles. Wiley, Michigan University

Ho KF, Lee SC, Ho SSH, Kawamura K, Tachibana E, Cheng Y, Zhu T (2010) Dicarboxylic acids, ketocarboxylic acids, α-dicarbonyls, fatty acids, and benzoic acid in urban aerosols collected during the 2006 campaign of air quality research in Beijing (CAREBeijing-2006). J Geophys Res Atmos 115:(D19). https://doi.org/10.1029/2009JD013304

Hu W, Hu M, Hu WW, Niu H, Zheng J, Wu Y, Chen W, Chen C, Li L, Shao M, Xie S, Zhang Y (2016) Characterization of submicron aerosols influenced by biomass burning at a site in the Sichuan Basin, southwestern China. Atmos Chem Phys 16(20):13213–13230. https://doi.org/10.5194/acp-16-13213-2016

Huang C-W, Launiainen S, Grönholm T, Katul GG (2014) Particle deposition to forests: an alternative to K-theory. Atmos Environ 94:593–605. https://doi.org/10.1016/j.atmosenv.2014.05.072

Huang X-F, He L-Y, Xue L, Sun T-L, Zeng L-W, Gong Z-H, Hu M, Zhu T (2012) Highly time-resolved chemical characterization of atmospheric fine particles during 2010 Shanghai World Expo. Atmos Chem Phys 12:4897–4907. https://doi.org/10.5194/acp-12-4897-2012

Huo J, Lu X, Wang X, Chen H, Ye X, Gao S, Gross DS, Chen J, Yang X (2016) Online single particle analysis of chemical composition and mixing state of crop straw burning particles: from laboratory study to field measurement. Front Environ Sci Eng 10(2):244–252. https://doi.org/10.1007/s11783-015-0768-z

Jaffe DA, Wigder NL (2012) Ozone production from wildfires: a critical review. Atmos Environ 51:1–10. https://doi.org/10.1016/j.atmosenv.2011.11.063

Janhäll S (2015) Review on urban vegetation and particle air pollution—deposition and dispersion. Atmos Environ 105:130–137. https://doi.org/10.1016/j.atmosenv.2015.01.052

Kantová N, Holubčík M, Jandačka J, Čaja A (2017) Comparison of particulate matters properties from combustion of wood biomass and brown coal. Procedia Eng 192:416–420. https://doi.org/10.1016/j.proeng.2017.06.072

Karanasiou A, Minguillón MC, Viana M, Alastuey A, Putaud JP, Maenhaut W, Panteliadis P, Močnik G, Favez O, Kuhlbusch TAJ (2015) Thermal-optical analysis for the measurement of elemental carbon (EC) and organic carbon (OC) in ambient air a literature review. Atmos Meas Tech Discuss 2015:9649–9712. https://doi.org/10.5194/amtd-8-9649-2015

Kaskaoutis DG, Gautam R, Singh RP, Houssos EE, Goto D, Singh S, Bartzokas A, Kosmopoulos PG, Sharma M, Hsu NC, Holben BN, Takemura T (2012) Influence of anomalous dry conditions on aerosols over India: transport, distribution and properties. J Geophys Res Atmos 117:(D9). https://doi.org/10.1029/2011JD017314

Kaul DS, Gupta T, Tripathi SN, Tare V, Collett JL (2011) Secondary organic aerosol: a comparison between foggy and nonfoggy days. Environ Sci Technol 45(17):7307–7313. https://doi.org/10.1021/es201081d

Kittelson DB, Watts WF, Johnson JP (2004) Nanoparticle emissions on Minnesota highways. Atmos Environ 38(1):9–19. https://doi.org/10.1016/j.atmosenv.2003.09.037

Koppmann R, von Czapiewski K (2005) A review of biomass burning emissions, part I: Gaseous emissions of carbon monoxide, methane, volatile organic compounds, and nitrogen containing compounds. Atmos Chem Phys Discuss 5:10455–10516. https://doi.org/10.5194/acpd-5-10455-2005

Kulkarni P, Baron PA, Willeke K (2011) Aerosol measurement: principles, techniques, and applications, 3rd edn. Wiley, Hoboken. https://doi.org/10.1002/9781118001684

Kumar KR, Yin Y, Sivakumar V, Kang N, Yu X, Diao Y, Adesina AJ, Reddy RR (2015) Aerosol climatology and discrimination of aerosol types retrieved from MODIS, MISR and OMI over Durban (29.88°S, 31.02°E), South Africa. Atmos Environ 117:9–18. https://doi.org/10.1016/j.atmosenv.2015.06.058

Kumar P, Pirjola L, Ketzel M, Harrison RM (2013) Nanoparticle emissions from 11 non-vehicle exhaust sources: a review. Atmos Environ 67:252–277. https://doi.org/10.1016/j.atmosenv.2012.11.011

Lai LY, Sequeira R (2001) Visibility degradation across Hong Kong: its components and their relative contributions. Atmos Environ 35(34):5861–5872. https://doi.org/10.1016/S1352-2310(01)00395-8

Levine JS (1996) Biomass burning and global change: remote sensing, modeling and inventory development, and biomass burning in Africa, vol 1. MIT Press, London

Li C, Hu Y, Zhang F, Chen J, Ma Z, Ye X, Yang X, Wang L, Tang X, Zhang R, Mu M, Wang G, Kan H, Wang X, Mellouki A (2017) Multi-pollutant emissions from the burning of major agricultural residues in China and the related health-economic effects. Atmos Chem Phys 17:4957–4988. https://doi.org/10.5194/acp-17-4957-2017

Li C, Ma Z, Chen J, Wang X, Ye X, Wang L, Yang X, Kan H, Donaldson DJ, Mellouki A (2015) Evolution of biomass burning smoke particles in the dark. Atmos Environ 120:244–252. https://doi.org/10.1016/j.atmosenv.2015.09.003

Li J, Pósfai M, Hobbs PV, Buseck PR (2003) Individual aerosol particles from biomass burning in southern Africa: 2, Compositions and aging of inorganic particles. J Geophys Res Atmos 108:(D13). https://doi.org/10.1029/2002JD002310
Li W, Shao L (2009) Transmission electron microscopy study of aerosol particles from the brown hazes in Northern China. J Geophys Res Atmos 114:(D9). https://doi.org/10.1029/2008JD011285
Li X, Li P, Yan L, Chen J-M, Cheng T, Xu S (2011) Characterization of polycyclic aromatic hydrocarbons in fog-rain events, vol 13. https://doi.org/10.1039/c1em10543d
Li X, Wang S, Duan L, Hao J, Li C, Chen Y, Yang L (2007) Particulate and trace gas emissions from open burning of wheat straw and corn stover in China. Environ Sci Technol 41(17):6052–6058. https://doi.org/10.1021/es0705137
Liu F, Bi X, Zhang G, Peng L, Lian X, Lu H, Fu Y, Wang X, Pa Peng, Sheng G (2017) Concentration, size distribution and dry deposition of amines in atmospheric particles of urban Guangzhou, China. Atmos Environ 171:279–288. https://doi.org/10.1016/j.atmosenv.2017.10.016
Liu Z, Wang Y, Gu D, Zhao C, Huey LG, Stickel R, Liao J, Shao M, Zhu T, Zeng L, Amoroso A, Costabile F, Chang C-C, Liu S-C (2012) Summertime photochemistry during CAREBeijing-2007: ROx budgets and O3 formation. Atmos Chem Phys 12:7737–7752. https://doi.org/10.5194/acp-12-7737-2012
Mammarella I, Rannik Ü, Aalto P, Keronen P, Vesala T, Kulmala M (2011) Long-term aerosol particle flux observations. Part II: Particle size statistics and deposition velocities. Atmos Environ 45(23):3794–3805. https://doi.org/10.1016/j.atmosenv.2011.04.022
Martins JA, Gonçalves FLT, Morales CA, Fisch GF, Pinheiro FGM, Leal Júnior JBV, Oliveira CJ, Silva EM, Oliveira JCP, Costa AA, Silva Dias MAF (2009) Cloud condensation nuclei from biomass burning during the Amazonian dry-to-wet transition season. Meteorol Atmos Phys 104(1):83–93. https://doi.org/10.1007/s00703-009-0019-6
McDonald JD, Zielinska B, Fujita EM, Sagebiel JC, Chow JC, Watson JG (2000) Fine particle and gaseous emission rates from residential wood combustion. Environ Sci Technol 34(11):2080–2091. https://doi.org/10.1021/es9909632
Mielonen T, Aaltonen V, Lihavainen H, Hyvärinen A-P, Arola A, Komppula M, Kivi R (2013) Biomass burning aerosols observed in Northern Finland during the 2010 wildfires in Russia. Atmosphere 2013 4:17–34. https://doi.org/10.3390/atmos4010017
Mkoma SL, Rocha GO, Domingos JSS, Santos JVS, Cardoso MP, Silva RL, Andrade JB (2014) Atmospheric particle dry deposition of major ions to the South Atlantic coastal area observed at Baía de Todos os Santos, Brazil. An Acad Bras Ciênc 86:37–55
Möller U, Schumann G (1970) Mechanisms of transport from the atmosphere to the earth's surface. Oceans and atmospheres. J Geophys Res 75:3013–3019
Nho-Kim EY, Michou M, Peuch VH (2004) Parameterization of size-dependent particle dry deposition velocities for global modeling. Atmos Environ 38(13):1933–1942. https://doi.org/10.1016/j.atmosenv.2004.01.002
Nie WJ, Ding A, Xie Y, Xu Z, Mao H, Kerminen V-M, F. Zheng L, Qi X, Yang X-Q, Sun J, Herrmann E, Petäjä T, Kulmala M, Fu C (2014) Influence of biomass burning plumes on HONO chemistry in eastern China, vol 14. https://doi.org/10.5194/acpd-14-7859-2014
Nirmalkar J, Deb MK (2016) Impact of intense field burning episode on aerosol mass loading and its possible health implications in rural area of eastern central India. Air Qual Atmos Health 9(3):241–249. https://doi.org/10.1007/s11869-015-0330-y
Obaidullah M, Bram S, Verma VK, De Ruyck J (2013) A review on particle emissions from small scale biomass combustion, vol 2
Pellerin G, Maro D, Damay P, Gehin E, Connan O, Laguionie P, Hébert D, Solier L, Boulaud D, Lamaud E, Charrier X (2017) Aerosol particle dry deposition velocity above natural surfaces: quantification according to the particles diameter. J Aerosol Sci 114:107–117. https://doi.org/10.1016/j.jaerosci.2017.09.004
Phuleria HC, Fine PM, Zhu Y, Sioutas C (2005) Air quality impacts of the October 2003 Southern California wildfires. J Geophys Res Atmos 110(D7). https://doi.org/10.1029/2004jd004626

Pokhrel RP, Wagner NL, Langridge JM, Lack DA, Jayarathne T, Stone EA, Stockwell CE, Yokelson RJ, Murphy SM (2016) Parameterization of single-scattering albedo (SSA) and absorption Ångström exponent (AAE) with EC/OC for aerosol emissions from biomass burning. Atmos Chem Phys 16:9549–9561. https://doi.org/10.5194/acp-16-9549-2016

Pryor SC, Barthelmie RJ, Spaulding AM, Larsen SE, Petroff A (2009) Size-resolved fluxes of sub-100-nm particles over forests. J Geophys Res Atmos 114(D18). https://doi.org/10.1029/2009jd012248

Pryor SC, Gallagher M, Sievering H, Larsen SE, Barthelmie RJ, Birsan F, Nemitz E, Rinne J, Kulmala M, Grönholm T, Taipale R, Vesala T (2008) A review of measurement and modelling results of particle atmosphere–surface exchange. Tellus B: Chem Phys Meteorol 60(1):42–75. https://doi.org/10.1111/j.1600-0889.2007.00298.x

Qiao X, Xiao W, Jaffe D, Kota SH, Ying Q, Tang Y (2015) Atmospheric wet deposition of sulfur and nitrogen in Jiuzhaigou National Nature Reserve, Sichuan Province, China. Sci Total Environ 511:28–36. https://doi.org/10.1016/j.scitotenv.2014.12.028

Reid JS, Eck TF, Christopher SA, Koppmann R, Dubovik O, Eleuterio DP, Holben BN, Reid EA, Zhang J (2005) A review of biomass burning emissions part III: Intensive optical properties of biomass burning particles. Atmos Chem Phys 5:827–849. https://doi.org/10.5194/acp-5-827-2005

Rocha GO, Allen AG, Cardoso AA (2005) Influence of agricultural biomass burning on aerosol size distribution and dry deposition in Southeastern Brazil. Environ Sci Technol 39(14):5293–5301. https://doi.org/10.1021/es048007u

Saffari A, Daher N, Samara C, Voutsa D, Kouras A, Manoli E, Karagkiozidou O, Vlachokostas C, Moussiopoulos N, Shafer MM, Schauer JJ, Sioutas C (2013) Increased biomass burning due to the economic crisis in Greece and its adverse impact on wintertime air quality in Thessaloniki. Environ Sci Technol 47(23):13313–13320. https://doi.org/10.1021/es403847h

Sapkota A, Symons JM, Kleissl J, Wang L, Parlange MB, Ondov J, Breysse PN, Diette GB, Eggleston PA, Buckley TJ (2005) Impact of the 2002 canadian forest fires on particulate matter air quality in Baltimore City. Environ Sci Technol 39(1):24–32. https://doi.org/10.1021/es035311z

SCENIHR (2005) The appropriateness of existing methodologies to assess the potential risks associated with engineered and adventitious products of nanotechnologies. European Commission Health & Consumer Protection Directorate-General and Scientific Committee on Emerging and Newly Identified Health Risks https://ec.europa.eu/health/ph_risk/committees/04_scenihr/docs/scenihr_o_003b.pdf. Accessed 20 April 2018

Schumann G (1975) The process of direct deposition of aerosols at the sea surface. Paper presented at the XVI general assembly of the international union of geodesy and geophysics, University of Heidelberg, August–September 1975

Sehmel G, Sutter S (1974) Particle deposition rates on a water surface as a function of particle diameter and air velocity. J Rech Atmos 912–920

Seiler W, Crutzen PJ (1980) Estimates of gross and net fluxes of carbon between the biosphere and the atmosphere from biomass burning. Clim Change 2(3):207–247. https://doi.org/10.1007/bf00137988

Skoog DA, James Holler F, Nieman TA (1998) Principles of instrumental analysis, 5th edn. Harcourt Brace & Company, Philadelphia

Skoog DA, West DM, James Holler F, Crouch SR (2013) Fundamentals of analytical chemistry, 96th edn. Brooks/Cole, Belmont, CA

Slinn WGN (1982) Predictions for particle deposition to vegetative canopies. Atmos Environ (1967) 16(7):1785–1794. https://doi.org/10.1016/0004-6981(82)90271-2

Souza ML (2016) Estudo da distribuição de tamanho e composição iônica de aerossóis e seus efeitos na capacidade de nuclear gotas de nuvens. Universidade Estadual Paulista, Araraquara, Doctorate degree

Souza ML, Allen AG, Cardoso AA (2017) Understanding aerosol formation mechanisms in a subtropical atmosphere impacted by biomass burning and agroindustry. Atmos Res 183:94–103. https://doi.org/10.1016/j.atmosres.2016.08.019

Sun LY, Wang ZB, Du W, Zhang Q, Wang Q, Fu P, Pan X, Li J, Jayne J, Worsnop DR (2015) Long-term real-time measurements of aerosol particle composition in Beijing, China: seasonal variations, meteorological effects, and source analysis. Atmos Chem Phys 15:10149–10165. https://doi.org/10.5194/acpd-15-14549-2015

Sun Y, Jiang Q, Xu Y, Ma Y, Zhang Y, Liu X, Li W, Wang F, Li J, Wang P, Li Z (2016) Aerosol characterization over the North China Plain: Haze life cycle and biomass burning impacts in summer. J Geophys Res Atmos 121(5):2508–2521. https://doi.org/10.1002/2015JD024261

Taiwo AM, Beddows DCS, Shi Z, Harrison RM (2014) Mass and number size distributions of particulate matter components: comparison of an industrial site and an urban background site. Sci Total Environ 475:29–38. https://doi.org/10.1016/j.scitotenv.2013.12.076

Urban RC, Alves CA, Allen AG, Cardoso AA, Campos MLAM (2016) Organic aerosols in a Brazilian agro-industrial area: speciation and impact of biomass burning. Atmos Res 169:271–279. https://doi.org/10.1016/j.atmosres.2015.10.008

US EPA (2016) Particulate Matter (PM) Pollution. United States Environmental Protection Agency. https://www.epa.gov/pm-pollution/particulate-matter-pm-basics#PM. Accessed 20 April 2018

US EPA (2018) Criteria Air Pollutants. United States Environmental Protection Agency. https://www.epa.gov/criteria-air-pollutants. Accessed 19 April 2018

Vadrevu KP, Lasko K, Giglio L, Justice C (2015) Vegetation fires, absorbing aerosols and smoke plume characteristics in diverse biomass burning regions of Asia. Environ Res Lett 10(10):105003

Vallero D (2014) Fundamentals of air pollution, 5th edn. Academic Press, USA

Vincent JH (2007) Aerosol sampling: science, standards, instrumentation and applications. Wiley, Hoboken. https://doi.org/10.1002/9780470060230

Vu TV, Delgado-Saborit JM, Harrison RM (2015) Review: particle number size distributions from seven major sources and implications for source apportionment studies. Atmos Environ 122:114–132. https://doi.org/10.1016/j.atmosenv.2015.09.027

Wang W, Jariyasopit N, Schrlau J, Jia Y, Tao S, Yu T-W, Dashwood RH, Zhang W, Wang X, Simonich SLM (2011) Concentration and photochemistry of PAHs, NPAHs, and OPAHs and toxicity of PM2.5 during the Beijing Olympic Games. Environ Sci Technol 45(16):6887–6895. https://doi.org/10.1021/es201443z

Wang ZB, Hu M, Yue DL, He LY, Huang XF, Yang Q, Zheng J, Zhang RY, Zhang YH (2013) New particle formation in the presence of a strong biomass burning episode at a downwind rural site in PRD, China. Tellus B: Chem Phys Meteorol 65(1):19965. https://doi.org/10.3402/tellusb.v65i0.19965

Wasiuta V, Lafrenière MJ, Norman A-L (2015) Atmospheric deposition of sulfur and inorganic nitrogen in the Southern Canadian Rocky Mountains from seasonal snowpacks and bulk summer precipitation. J Hydrol 523:563–573. https://doi.org/10.1016/j.jhydrol.2015.01.073

Whitby KT, Sverdrup GM (1980) California aerosols: their physical and chemical characteristics. Adv Environ Sci Technol 8:477–525

Wieser U, Gaegauf CK (2000) Nanoparticle emissions of wood combustion processes. Paper presented at the 1st world conference and exhibition on biomass for energy and industry, Sevilla, Spain, 5–9 June

Wu QZ, Wang ZF, Gbaguidi A, Gao C, Li LN, Wang W (2011) A numerical study of contributions to air pollution in Beijing during CAREBeijing-2006. Atmos Chem Phys 11(12):5997–6011. https://doi.org/10.5194/acp-11-5997-2011

Wu Y, Zhang J, Liu S, Jiang Z, Huang X (2018) Aerosol concentrations and atmospheric dry deposition fluxes of nutrients over Daya Bay, South China Sea, vol 128. https://doi.org/10.1016/j.marpolbul.2018.01.019

Yang YR, Liu XG, Qu Y, An JL, Jiang R, Zhang YH, Sun YL, Wu ZJ, Zhang F, Xu WQ, Ma QX (2015) Characteristics and formation mechanism of continuous hazes in China: a case study during the autumn of 2014 in the North China plain. Atmos Chem Phys 15:8165–8178. https://doi.org/10.5194/acp-15-8165-2015

Yu F, Wang Q, Yan Q, Jiang N, Wei J, Wei Z, Yin S (2018) Particle size distribution, chemical composition and meteorological factor analysis: a case study during wintertime snow cover in Zhengzhou, China. Atmos Res 202:140–147. https://doi.org/10.1016/j.atmosres.2017.11.016

Zhang J, Shao Y, Huang N (2014) Measurements of dust deposition velocity in a wind-tunnel experiment. Atmos Chem Phys 14(17):8869–8882. https://doi.org/10.5194/acp-14-8869-2014

Zhang L, Brook JR, Vet R (2003) A revised parameterization for gaseous dry deposition in air-quality models. Atmos Chem Phys 3(6):2067–2082. https://doi.org/10.5194/acp-3-2067-2003

Zhang L, Gong S, Padro J, Barrie L (2001) A size-segregated particle dry deposition scheme for an atmospheric aerosol module. Atmos Environ 35(3):549–560. https://doi.org/10.1016/S1352-2310(00)00326-5

Zhu J, Xia X, Che H, Wang J, Zhang J, Duan Y (2016) Study of aerosol optical properties at Kunming in southwest China and long-range transport of biomass burning aerosols from North Burma. Atmos Res 169:237–247. https://doi.org/10.1016/j.atmosres.2015.10.012

Zhu L, Chen Y, Guo L, Wang F (2013) Estimate of dry deposition fluxes of nutrients over the East China Sea: the implication of aerosol ammonium to non-sea-salt sulfate ratio to nutrient deposition of coastal oceans. Atmos Environ 69:131–138. https://doi.org/10.1016/j.atmosenv.2012.12.028

Part III
Safety Issues, Regulations and Public Perception

Chapter 14
Nanomaterials: Toxicity, Risk Managment and Public Perception

Bushra Jamil, Rabia Javed, Asma Saleem Qazi and Muhammad Ali Syed

Abstract It is believed that the future of medicine lies in nanotechnology where the medicines would be tailor-made according to the requirement of the individual. With the amassed recognition of the potential of nanomaterials they are not only used in medicines but in many other consumer products. Nanomaterials exists in nature everywhere and many others are engineered. Because of the omnipresence of these nanoparticles human beings are blatantly exposed to them. Nevertheless, our knowledge about health and environmental risks of these engineered nanomaterials still remains limited and largely incomplete. Many studies confer their toxicity in animal models where they can lead to teratogenicity, genotoxicity and also defects in organs like liver, lungs, kidneys and immune system. However, the toxic responses depends upon many variables including size, shape, charge, composition and surface area. Nanoparticles (NPs) of same compositions can diverge in term of toxicity by changing the size, shape and charge of NPs. Likewise the time of exposure also plays very vita role in toxicity. Many available studies provide incomplete information by conducting few in vitro analyses. However, the process is so intricate that the in vitro studies are not sufficient to provide enough information to elucidate the full toxic potential. Therefore, in vivo studies are pre-requisite. This chapter provides an insight

B. Jamil (✉) · R. Javed · A. S. Qazi
Department of Bio-Sciences, National University of Medical Sciences, The Mall, Rawalpindi, Pakistan
e-mail: Bushra.jamil@numspak.edu.pk; bushrajamil2000@gmail.com

R. Javed
e-mail: rabia.javed@ymail.com; rabia.javed@numspak.edu.pk

A. S. Qazi
e-mail: samaaqazi_ksa@hotmailmail.com; asma.qazi@numspak.edu.pk

M. A. Syed
Department of Microbiology, University of Haripur, Haripur, Pakistan
e-mail: mirwah2000@yahoo.de

M. Rai and J. K. Biswas (eds.), *Nanomaterials: Ecotoxicity, Safety, and Public Perception*, https://doi.org/10.1007/978-3-030-05144-0_14

from toxicologist's perspective on the matter, outlining possible routes of uptake by humans, known or suspected toxic effects, and the possible practical implication for human health risk assessment and public perception.

Keywords Nano-toxicity · Risk assessment · Public perception · Nano-particles

Nomenclature

NP	Nano-particles
UFP	Ultra Fine Particles
QSAR	Quantitative Structure Activity Relationship
LDH	Lactate Dehydrogenase
nm	Nanometer
TiO_2	Titanium Oxide
TNF	Tumor Necrosis Factor
SiO_2	Silicon Dioxide
Ni NPs	Nickel Nano-particles
Co NPs	Cobalt Nano-particles
IL	Interleukin
Au	Gold
Ag	Silver
Si NPs	Silicon Nano-particles
PBS	Phosphate-buffered Saline
BBB	Blood Brain Barrier
CNS	Central Nervous System
FDA	Food and Drug Administration

14.1 Introduction

It is generally assumed that the nanoparticles (NPs) include all those particles that have a size less than 100 nm (Bouallegui 2017). When particles are in this size range then the properties exhibited by them are different from that of bulk materials. For instance, gold is inert in its bulk form but is highly reactive and exhibits different properties when it is in the form of NPs (Subbenaik 2016). The NPs of same composition can vary in term of toxicity by changing physicochemical properties. That's why there are many variants of nanoforms as compare to non nanoforms (Gatoo 2014).

Since, NPs are omnipresent and can be generated as a result of anthropogenic (as a result of human activity) activities or may produce naturally (Lu et al. 2014). Utilization of NPs has also been increased and they are used in many (if not every) industries like medical (medical imaging, disease diagnoses), pharmaceutical (drug

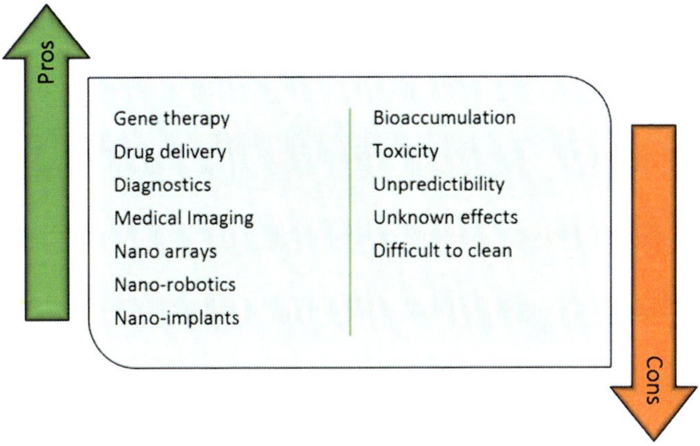

Fig. 14.1 Pros and cons of nanpmaterials

delivery, cancer treatment, and gene therapy), agriculture, cosmetology, electronics, computer science, material engineering etc. (Fig. 14.1). Though the utilization of NPs has been increased, but their toxicity is unknown in many cases and despite their well-known toxic potential on bacteria, algae, fish, vertebrates and invertebrates, still the extensive research in this domain is still lacking (Barnard et al. 2006; Kahru and Dubourguier 2010; Lu et al. 2014) (Fig. 14.2). The number of nano products is rising steadily in the market and a huge number of the products are in the queue (Vance 2015). Yet, finding basic information on manufactured nanomaterials and their potential risks to human health as well as environment is not an easy task. The absence of regulations and oversight of the nanotech industry may add to the general lack of knowledge of the potential risks to human and environmental health. This dearth of information has engaged the nano community debating whether to mandate widespread testing of the materials (Chatterjee 2009).

In this book chapter, we highlight the possible toxic human health effects of nanomaterials that can result from exposure to these ultrafine particles (UFPs) and their health outcomes. The comparability of engineered NPs to UFPs suggests that the human health effects are likely to be similar. Therefore, it is prudent to elucidate their toxicological effect to minimize occupational and environmental exposure. Highlighting the human health outcomes caused by UFPs is not intended to give a lesser importance to either the unprecedented technological and industrial rewards of the nanotechnology or their beneficial human uses (Gwinn and Vallyathan 2006)

The toxicity of metal based nanomaterials recently emerged as a worldwide concern due to their nano-specific properties and transformation in the environment, beside their ability to release toxic metals has resulted in interest to nano-biological interactions. NPs are omnipresent, but mostly concentrated in aquatic and terrestrial environments and are consumed by all living organisms. As these NPs are foreign substances to the living body so they interfere with normal body's physiological functions and accumulate in it. They not only interfere with body's normal function

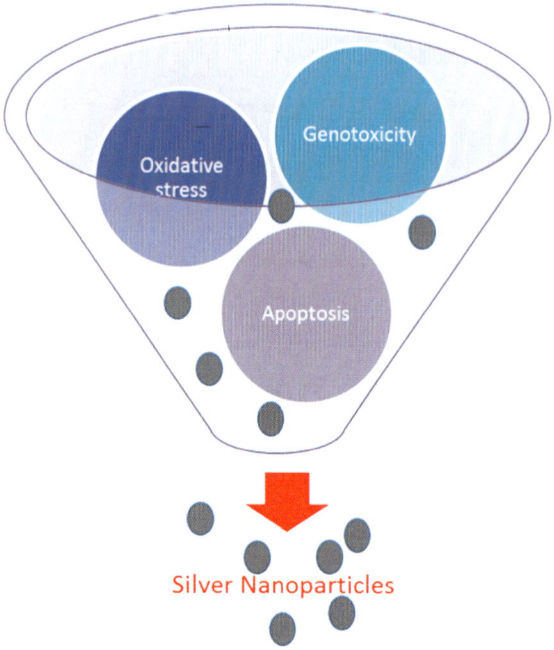

Fig. 14.2 Possible toxic effects of silver nanoparticles in humans. In order to find their optimum therapeutic effects, these effects have to be eliminated

but also interfere with the development of embryo and may cause certain lethal malformations. Because of their extremely minute size, they may not easily be detected by body's immune system. However, their small size helps them to penetrate the individual cells by disrupting the cell membrane and interfering with normal functioning of cells leading to cell death. In addition to positive charge on NPs, they also affect the normal composition and function of cell membrane such as ion transport or signal transduction (Exbrayat et al. 2015).

A report based on studies, surveys, inquiries and articles highlighting possible hazards associated with nanomaterials was prepared by The Royal Society and The Royal Academy of Engineering. This report though indicates the valid potential risks associated with the NPs, but is not providing any frame work to contain them or dealing with them (Seaton 2009). Based upon this knowledge neither we predict the interaction of NPs with body and body organs nor the extent of damage. Nevertheless, the interaction is so complex and having so many variables (physico chemical properties, biological effect testing, risk analysis etc.) that it is very hard to predict the toxicity. In addition to that, the minute size of NPs, as compared to biomolecules, makes this interaction more profound and far less predictive. We exactly don't know where they will go and bind. So there is a dire need to be more focused and more critical to develop strategies to protect the public health and the environment (Nel et al. 2006).

Once a risk assessment is complete, then risk management decisions can be made about the details of the products' life cycle (manufacture, use, and disposal). For any new technology, it is vital that this process goes ahead without stifling the innovation itself, or hampering the development of business. Clearly, no regulatory agency wants to curtail economic development, but at the same time there is a need for product and public safety

14.2 Current Knowledge About Toxicity of Nanomaterials

As the living organisms are not generally encountered by metallic NPs during their course of evolution, so there are no defensive mechanisms evolved in them against these NPs (Moore 2006). Therefore, the toxicity of metallic NPs has been emerging as a worldwide concern. There are many reasons for toxicity of nanomaterials, such as:

- their extremely small size and large surface area
- their non-specific binding
- accumulation in vital organs
- inability to be released by body excretory system

Ability to constantly release toxic free radicals (Dhawan and Sharma 2010). The toxicity of nanomaterials can differ according to the physicochemical properties of NPs, and therefore, it is essential to determine the risk by varying each parameter, which requires a lot of experience, money, time and laboratory animals. Nonetheless, many approaches have been practiced for a more sustainable solution. One such approach is utilizing tools such as Quantitative structure activity relationships (QSAR), grouping and high throughput screening. The objectives of all these tools are to priorities only hazardous materials. However, current scientific knowledge is not sufficient to define cutoff values. Therefore, currently, regulatory frame work of chemicals i.e., REACH (Registration, Evaluation, Authorization and Restriction of Chemicals) is being applied to nano materials as well. Furthermore, those aspects of exposure, kinetics and hazard assessment that are most likely to be influenced by the nano specific properties of the material under assessment are required to be identified. These aspects are summarized in six elements, which play a key role in the strategy: exposure potential, dissolution, nanomaterial transformation, accumulation, genotoxicity and immunotoxicity (Dekkers et al. 2016).

14.3 Factors Contributing to Toxicity of Nanomaterials

The important aspect of their toxicity is their potential ability to cross tissue or cellular membranes to enter individual cell and interact with their subcellular compartments. The mediators of toxicity of nanomaterials are mentioned in detail in the following.

14.3.1 Size

Size is one of the major indicator that contributes in nano-toxicity. Reduced size means more surface area and more molecules at the surface of any type of nanoparticles (Donaldson et al. 2002; Wang et al. 2013). This factor makes them more reactive and more miscible as compared to bulk material. Particles of same composition can vary in term of size dependent toxicity. Size effect not only penetration but the distributions of particles in body compartment. Braakhuis et al. (2016) prepared silver NPs of 18, 34, 60, and 160 nm size and which were administered by inhalation in rats. They measured not only the amount of silver in lungs but also the extent of damage caused by them by measuring lactate dehydrogenase (LDH) expression (a marker of cell damage). It was found that the particles of 18 and 34 nm are because more damage as compared to the large particles though the quantity of large particles was more as compared to the smaller ones.

Size of NPs also play a vital role in the bio-distribution. Particles of less than 6 nm cannot be excreted by the kidney and retain in the body. Most common site of retention of NPs is liver and spleen where they cause severe side effects. However, the distribution pattern vary according to the size of NPs. De Jong and Borm (2008), studied the bio-distribution of gold NPs from a size range of 10–250 nm. They concluded that the small size NPs were distributed widely in blood, liver, spleen, kidneys, testes, thymus, heart, lungs and brain whereas the NPs larger than 50 nm in size were more concentrated in blood, liver and spleen only.

14.3.2 Chemical Composition

Toxicity caused by the NPs also depend upon their chemical composition (Donaldson et al. 2005). For example, effects caused by carbon black have been proved to be more toxic than those of titanium dioxide (Tinkle et al. 2003). Reactivity of NPs may also be affected by the chemical soak up on the cell surface (Xia et al. 2004; Sharma 2010). It has also been observed that cytotoxicity can be reduced by mere surface modification of carbon black NPs. For example, by coating the carbon black NPs with pullulan, cytotoxicity of superparamagnetic iron oxide NPs can be reduced significantly. Another example is of albumin and dextran derivatives of iron oxide NPs where also reduced cytotoxicity was noted (Poland et al. 2008).

A study conducted by Nel et al. (2006) revealed that out of polyvinyl chloride, TiO_2, SiO_2, Co and Ni, only Co generated toxicity in endothelial cells along with pro-inflammatory cytokine IL8. Leakage of metal ions from NP core is another reason of toxic effect of NPs (Hoet et al. 2004). Some metal ions like silver and cadmium are toxic enough to cause damage to the cells. Iron and Zinc are usually useful but at higher concentrations they can damage cellular pathways. However, coating these ions with layers of silica (Si) or gold (Au) can result in enhanced chemical stability against NP degradation and ion leakage, thus decreasing toxic effect (Schuppan and Schattenberg 2013).

14.3.3 Shape

NPs are available in different shapes like tubes, crystals, sphere, planes, rings and fibers (Fig. 14.3). The toxic effect due to shape has been reported for myriads of carbon nanotubes, silica, nickel, gold, titanium NPs (Allegri et al. 2016). Shape related toxicity influences the process of membrane wrapping during endocytosis and phagocytosis. For example, endocytosis of spherical NPs is faster and easier as compared to rod shaped or fiber shaped NPs and they are less toxic whether they are homogenous or heterogeneous in shape (Geiser et al. 2008).

14.3.4 Surface Area

Surface area of NPs plays a vital role in interaction of material with biological systems. Smaller sized particles tend to have exponential increase in the surface area relative to volume that tend to increase the reactivity of NPs (Schmid and Stoeger 2016). Surface area is responsible not only for the reactivity but for the distribution and elimination of the material. For these reasons, surface area is one of the significant factors in determining the toxicity of NPs. Increased surface area causes a dose dependent increase in the oxidation and DNA damaging abilities of NPs, which is quite higher than the larger particles of same mass dose (Maynard 2006).

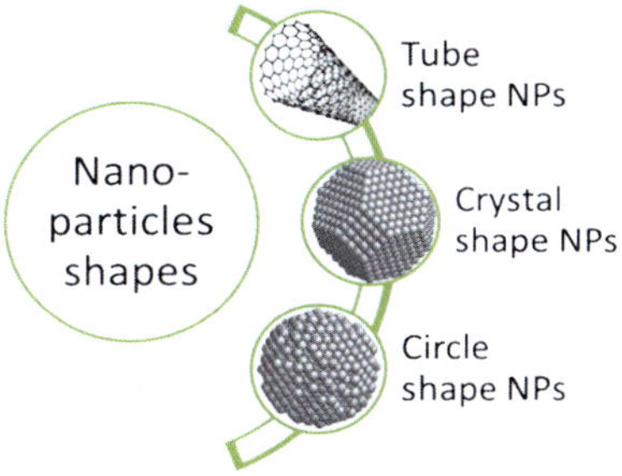

Fig. 14.3 Few shapes of nanoparticles

14.3.5 Surface Charge

Surface charge defines the interactions of NPs, so it plays a major role in their toxicity. NPs selective adsorption, colloidal behavior, protein binding, blood brain integrity and transmembrane permeability are all regulated by their surface charge (Oberdörster et al. 2005). The positively charged NPs show significant cellular uptake compared to negatively charge or neutral particles owing to increased opsonization by plasma proteins along with induced hemolysis and platelet aggregation owing to severe toxicity to the system (Dobrovolskaia and McNeil 2007). Surface charge can also alter blood brain barrier (BBB) integrity and transmembrane permeability. It is also the key detriment of colloidal behavior, and specifically influences the organism‘s response upon exposure to nanoparticles by altering their shape and size through aggregate formation (Bowden et al. 2011).

14.3.6 Aggregation and Concentration

The aggregation and concentration of NPs also affect their toxicity. Although it depends upon their size, surface charge and composition among others (Schmid and Stoeger 2016). For example, it is observed that carbon nanotubes get accumulated mainly in liver, spleen, and lungs and exert cytotoxic effects because of their accumulation for longer duration (Nowack and Bucheli 2007).

14.3.7 Solvent/Media

Through agglomeration, the NPs get affected in their dispersion, thus, affecting their size and in turn toxicity of NPs. For example, particles of titanium oxide, zinc oxide and carbon black have greater size in PBS than in water (Hsiao and Huang 2011). Similarly, the toxic effect of NPs show variation depending upon solvent they are in. Due to this the NPs exhibit different toxic manifestations when dissolved or suspended in different medium/solvents (Georgieva et al. 2011).

14.4 Adverse Effects of Nanomaterials on Human Health

Contrary to benefits and applications of NPs in different areas of science and technology, there are certain associated toxicities of NPs. Basic knowledge and understanding of NPs related toxicities and their effects on human organs are essential in order to use NPs in the modern medicine or elsewhere. Even the environmental toxicities caused by these nanoparticles are important to understand (Boverhof et al.

2015). NPs can enter the environment through soil, water and air and ultimately access human through different routes like respiratory (inhalation) or digestive tract (indigestion) and can cause damage to human body depending upon the size, type and physiochemical properties (Oberdörster et al. 2005). They can be translocated to different organs and tissues of human body and can cause toxic effects (Table 14.1).

Furthermore, they can also affect other animals and plants. As stated before, the toxic effect of NPs depends upon their various factors (Boogaard et al. 2011). For example, magnetic NPs, due to their high reactivity can cause more severe toxic and harmful effects in the host cells, which is unusual for the same sized counterparts (Delfino et al. 2016). Another example is of silver NPs (AgNPs) that are used in various consumer products and their release in the aquatic environment may provide a continuous source of dissolved Ag for aquatic life.

NPs can also cause toxicity due to their ability to organize around the protein concentration. For example, NPs binding with proteins may cause adverse health effects by interfering with protein unfolding, fibrillation, thiol cross linking or loss of enzymatic activity. Similarly, thermodynamic properties of NPs may cause release of toxic ions due to particle dissolution in the biological environment. NPs are also greatly influenced by the presence of organic matter or other natural particles, especially in fresh water (Oberdörster et al. 2005).

Studies that are specifically related to NPs toxicities are also evolving in the literature but their numbers are not very encouraging and still a lot a scientific evidence is required to understand the actual process of NPs toxicity (Kreyling et al. 2004; De Jong and Borm 2008). Nonetheless, harmful effects of NPs results from interaction with living organisms and their environment. In order to study the toxicology of NPs, it is essential that both in vitro and in vivo analyses should be performed. In vitro analysis can guide us about many adverse effects at cellular level. From this knowledge one can get information about apoptosis (programmed cell death), organelle damage, DNA damage, oxidative stress, mutagenesis, and protein up/down regulation. Whereas, the in vivo assessment should be done at whole animal model, as the whole animal is much more complicated than a cell line. In addition to that we can study many other indicators that are representative of general health like behavioral abnormality, weight loss, percent of mortality, and average life span. Likewise, tissue specific studies, for instance effect of NPs on various organs like liver, kidneys, blood, brain, etc. could be done easily. In addition to that drug related parameters like absorption, distribution, metabolism and elimination should also be conducted to obtain the approval for therapeutic purposes.

Since the in vitro analyses are easier to perform, many groups focus on in vitro or cell culture analysis. Nonetheless, they do not give the complete picture.

14.4.1 Gold Nanoparticles

Although gold nanoparticles (AuNP) are very useful in the diagnosis and therapeutics of central nervous system (CNS) and are a valuable imaging modality, yet some

Table 14.1 Physiological effects of NPs on different Cells depending upon their dose, surface area and route of administration in the human body

Type of NPs	Cell	Physiological effects	References
Gold	Renal	Cytotoxicity	L'Azou et al. (2008)
	Myocardium	Cytotoxicity	Liu and Ye (2013)
	Hepatocytes	Atrophy/nacrosis	Hillyer and Albrecht (2001)
	Lungs fibroblast	Genotoxicity/autophagy	Shimida et al. (2006)
	Bone marrow	Cytotoxicity	Qian et al. (2008)
Silver	Lungs fibroblast	Cytotoxicity/DNA damage	Anderson et al. (2015)
	Bone marrow	Genotoxicity/cytotoxicity/carcinogenicity	Chen et al. (2018)
	Hepatocytes	Cytotoxicity/induction of oxidative stress	Bergin and Witzmann (2013)
	Osteoblast	Cytotoxicity/generation of ROS	Park et al. (2010)
Superparamagnetic	Renal	Genotoxicity/cytotoxicity	L'Azou et al. (2008)
	Fibrolast	Cytotoxicity/loss of adhesion skill	Barnes et al. (2007)
Quantum dots	Lymphocyte	Induction of DNA damage/formation of micronuclei/generation of DNA adduct	Zhang et al. (2008)
Polymeric	Lungs fibroblast/epithelial	Cytotoxicity/Inflammatory	Pan et al. (2016)
Carbon nanotubes	Renal	Cytotoxicity	Goulaouic et al. (2008)
	Neurons	Neurotoxic effect	Kreuter et al. (1995)
	Hepatocytes	Inflammatory/genotoxic effect	Elsaesser and Howard (2012)
	Lungs epithelial	Cytotoxicity	Borm et al. (2005)
Silica	Hepatocytes	Hepatotoxicity	Lee et al. (2011) Ahmad et al. (2012)
	Neurons	Dopaminergic neurons damage pathway	Choi et al. (2010)
	Lungs epithelial	Inflammatory response/toxicity	Athinarayanan et al. (2014)
	Lymphocyte	Cytotoxicity/genotoxicity	Demir and Castranova (2016)
	Fibroblast	Cytotoxicity	Dekkers et al. (2013)

(continued)

Table 14.1 (continued)

Type of NPs	Cell	Physiological effects	References
Titanium dioxide	Renal	Induction of oxidative stress/cytotoxicity	Gui et al. (2011)
	Neurons	Neurotoxicity	Wang et al. (2007a)
	Hepatocytes	Genotoxicity/carcinogenicity/inflammation	Jania et al (1994)
	Lungs epithelial	Genotoxicity/mutagenicity/carcinogenicity	Andersson et al. (2011)
	Lymphocyte	Induction of oxidative stress/genotoxicity	Wang et al. (2007a, b)

of the available data reveals their interaction with the cells of CNS. There is an induction in pro-inflammatory cytokines (e.g. TNF-α, IL-β and IL-2) after exposure to AuNPs that leads to rat's brain toxicity. However, moderate toxicity was observed with smaller size (3 nm) AuNP than to the bit larger (5 nm) ones. Smaller sized NPs can easily be distributed to all tissues and organs of body systems than the larger sized particles. One reason could be the hindrance by the blood brain barrier (Alkilany et al. 2008). Smaller AuNPs can penetrate deeper into tissues than larger ones which are largely built up in the superficial epidermis and dermis. Au compounds are generally considered safe and have been in routine clinical use for many years, e.g. in the treatment of rheumatoid arthritis (Liu and Ye 2013). However, many groups have reported their toxicity. Dobrovolskaia et al. (2008) reported the hemolysis and immunotoxic potential of AuNPs. Likewise, Chen et al. (2009) reported the accumulation and subsequent toxicity (inflammation and cellular damage) of AuNPs in the liver and spleen.

14.4.2 Silver Nanoparticles

Silver NPs (AgNPs) exhibit neurodegenerative abilities. Past studies have revealed the disruption of blood brain barrier, pulmonary inhalation and brain edema due to interactions of AgNPs with organ systems along with astrocyte swelling and neuronal degeneration in rat models. AgNPs toxicity is size and concentration dependent (Chen et al. 2018). It is also suggested that mitochondrial dysfunction is one of the neurodegenerative process caused by AgNPs toxicity. The accumulation of AgNPs of smaller size is in liver and that of large sized AgNps is in spleen. In case of repeated accumulation of these NPs, dose dependent accumulation has been observed in the organs and tissues (Liu et al. 2018). A study conducted on zebrafish has shown evidences of development of neurotoxins and changes in synaptic functioning (Truong et al. 2012). AgNPs induced neurotransmitter hyperactivity has shown its effects on behavioral changes with a lower anxiety threshold.

14.4.3 Superparamagnetic Nanoparticles

These NPs consists of an iron oxide core and a variable carbohydrate coating that determine their uptake in the cells and their biological half-life (Yin et al. 2005). In cellular uptake, degree of surface coverage has been considered a major parameter (Lin et al. 2010). Fully coated superparamagnetic nanoparticles can escape opsonization that can also prolong, plasma half-life. This is also one of the few FDA approved NPs that can be used in MRI and imaging of vasculature and lymph nodes. They are also efficient in visualizing the intracerebral malignancies and neurological lesions within the central nervous system (De Domenico et al. 2007). Still their potential nano-toxic effect has to be evaluated extensively to get the complete picture and understanding of mechanism of spread of toxicity. Due to their large surface area to volume ratio, they are considered very important for novel diagnostic and therapeutic applications.

14.4.4 Quantum Dots

Quantum dots are also used for both diagnostic and therapeutic purpose (Aillon et al. 2009). There is not much experimental evidence present to support their applications in visualizing neural stem and progenitor cells but they do act as nano-vehicles for brain surface (Hoshino et al. 2004). For quantum dots, several routes of entry in the brain are suggested. They can be uptaken through phagocytosis or pinocytosis (Derfus et al. 2004). The available data is still scarce on the toxic effects of quantum dots so there is still a need for considerable attention to the quantum dots, especially intended for brain specific barriers (Aillon et al. 2009).

14.4.5 Polymeric Nanoparticles

Polymeric NPs have many excellent properties like biocompatibility, biodegradability and surface modifiability. They also have the capability for sustained drug release (Murthy 2007). Nowadays, Polymeric NPs are being used for the treatment of many patients suffering from pulmonary diseases like TB, coronary obstructive disease, asthma and lungs cancer. Polymeric NPs include collagen, chitosan, gelatin, alginate and bovine serum albumin. It is said that polymeric NPs have improved therapeutic outcome through sustained targeted drug delivery (Fröhlich 2012). In addition to that they have bio-degradative ability, reduced toxic effects and biocompatibility that make them a choice for surface modifications. However, details on enzymatic degradation because of biophysical resemblance and details of toxic effects caused by them needs detailed investigations.

14.4.6 Carbon Nanotubes

These are frequently used in the inhalation models and their toxic effect is dependent on the exposed surface area (Baird et al. 2005). They are more appropriate for the use in drug delivery, tissue-engineering, scaffolds and photodynamic therapy (Goulaouic et al. 2008). Cytotoxicity caused by carbon nanotubes is mediated by the cytokines such as TNF-α upregulation. There are still discrepancies regarding carbon nanotubes composition and mode of administration that requires detailed research in the field (Borm et al. 2005).

14.4.7 Silica Nanoparticles

These NPs are widely used as additives, polish, varnish and cosmetics (usually non-medical uses). They are now introduced as biomarkers in the field of biomedicine, cancer therapeutics and in drug delivery system (Murugadoss et al. 2017). They are usually considered as safe in moderate dosages like less than 20 μg/mL but in higher doses even at 25 μg/mL, the cell show cytotoxic activities and starts agglomeration (Lee et al. 2011). These NPs cause cytotoxic effects by reactive oxygen species (ROS) induction and by lipid peroxidation (Kasaai 2015).

14.4.8 Titanium Dioxide Nanoparticles

These particles are mostly used in sun screens and cosmetics. Their UV light blocking property and esthetic properties have made them advantageous for commercial use (Chulz et al. 2002). Their toxicity cause adverse effects in terms of proliferation, differentiation, mobility and apoptosis. They may get penetrated to open part of hair follicle (Senzui et al. 2010). They can have the different degree of permeation depending upon their concentration, surface coating and dose (Wang et al. 2007a, b). However, like other NPs evidence based study is required for the complete understanding of the process.

14.5 Possible Routes of Entry and Target Sites of Nanomaterials in Human Body

Toxicity of NPs on human organs also depends on the exposure time of NPs to that particular organ. Respiratory system, as mentioned above, present a distinctive target of the toxicity caused by NPs as they are not only the entry point for inhaled particles, but also gate to the cardiac outcome (Donaldson et al. 2002) (Fig. 14.4). Previous

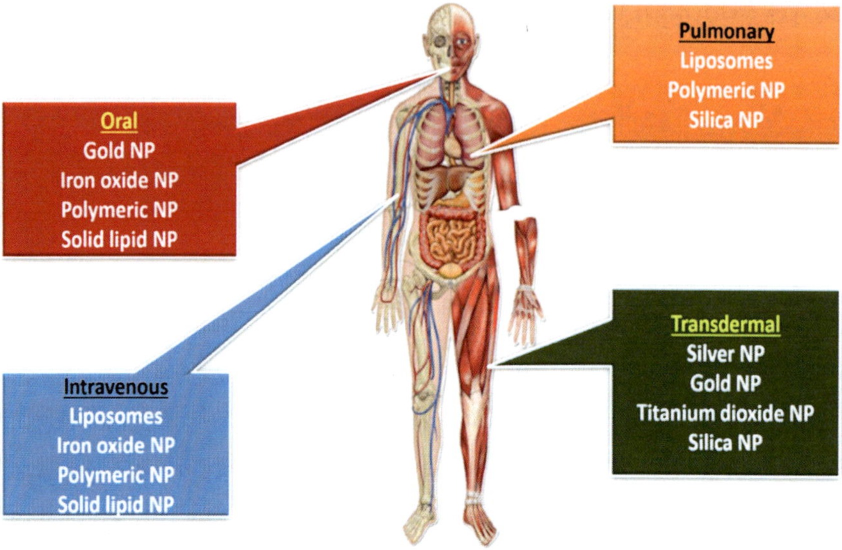

Fig. 14.4 Possible routes of administration of NPs in human organs

studies have revealed that inhaled NPs are removed from the body less efficiently as compared to the larger particles that are removed by macrophages in the lungs. NPs can translocate in the circulatory, nervous and lymphatic system and can cause damage to the associated organs of that particular system (Park and Park 2009).

The basic reason of NP toxicity is their minute size that allows them to penetrate into basic organs and systems of the body, disrupting their normal functioning and damaging the organs. Other toxic effects include tissue inflammation, altered cellular redox balance towards oxidation and cell death (Park and Park 2009).

NPs accumulation in the body, their bio-distribution, absorption, metabolism, toxic effects and excretion also depends on their route of administration (Cho et al. 2009). So far, not much information and reviews are available focusing on association of specific route of administration of NPs and toxic effects.

14.6 Risk Assessment and Evaluation of Nanoparticles

The environmental risk assessment (ERA) is designed to find the harmful effects of engineered nanoparticles. The domains of nanoparticle characterization, release and fate of nanoparticle, and risk/hazard characterization are integrated in ERA strategy. Nanoparticles should be characterized at every step of their life cycle, for example, during production, transport, use and disposal. Data kinetics is involved to understand the problem statement (Scott-Fordsmand et al. 2017). Any novel nanoparticle brought to the market needs evaluation in relation to potential health and environment. Inhalation and dermal risks pose serious threats to public healthcare workers. Hazards are identified on the basis of characteristics of nanoparticles, e.g., size, shape, composition, coating, etc. Exposure and toxicity are assessed on the basis of entry routes and particle behavior of the former and uptake, distribution, metabolism and excretion of the later. The risk assessment involve simple 4-part process; identification of hazard, identification of cause of hazard and hazard affectees, hazard evaluation, and lastly the prioritization of specific control/elimination measures. The first step in this process is to identify which materials are toxic because NPs act differently as compare to the materials we interact daily. Both the desirable and undesirable properties of NPS depends not only on size but on many different variable like composition, impurities, surface chemistry, shape and charge etc. Many of these are intrinsically linked with each other.

The assessment include practical recommendations for the risk control. After implementation of the corresponding recommendations, the re-calculation of risk elucidates whether the level of risk mitigation is up to the acceptable level or not. The contemporary measures to control/eliminate risks conferred by nanoparticulate substances needs to be managed in order to safeguard the employees working in an environment exposed to nanoparticles. After risk assessment, risk characterization is performed on the basis of likelihood and nature of effects as well as effectiveness of controls (Tsuji et al. 2005).

Nanoparticles are thought to cause chemical hazards imposed on the workplace. So, it is dire need of today to assess the risks imposed by nanoparticles on the healthcare workers. A system or regulatory agency should be developed that gathers information about the emerging issues. Then, the emerging issues should be interpreted in the presence of evidences whose appraisal is required for developing guidelines. In fact, the quantity of exposure and likelihood of toxicity of nanoparticles is measured by collecting relevant evidences. Hence, such authorities work in the mitigation of the effects of hazardous chemicals/nanoparticles (Kuempel et al. 2012). Occupational safety and health (OSH), also called Workplace health and safety (WHS) is an organization that works for safe and healthy environment for the welfare of people in workplace. Safety of employees is the duty of employers according to law rules. OSH program has been developed that functions in collecting the basic research data, developing guidelines, and actions of workplace. The biological system responds to nanoparticles with respect to their size, shape, solubility and few other parameters. The response to nanoparticles is variable so the instrument for their detection and measurement should be sensitive and specific. Moreover, instruments for protection and control of nanoparticles' toxicity are needed. The overall process of OSH program involves problem formulation, evaluation of dose-dependency of risks associated with problem, and the filling of data gaps by experimental identification and after critically analyzing the information. Another major concern with regard to nanoformulations is environmental health and safety (EHS). German Federal Environment Agency and the German Federal Institute for Risk Assessment on the carcinogenic potential of nanomaterials presented a critical review on the subject matter and they concluded that the potential carcinogenic potential of NPs can only be assessed on a case-to-case basis. The existing database is not adequate for assessing only one toxic parameter i.e. the carcinogenic potential. There is insufficient characterization techniques, differences in experimental designs and the use of different animal models in different studies.

14.7 Risk Management of Nanoparticles

Societal and workplace risk management/anticipation of engineered nanoparticles is essential to protect the healthcare workers from adverse effects of nanoparticles. Personal protective equipment (PPE) and controls function in minimizing risk exposures. The work-related illness should be assessed by medical screening. The assessment of effectiveness of risk management comes under the broad term "medical surveillance" and is considered a second line of defense to ensure the effectiveness of risk management practices that have been implemented. The workers and employers should work in cooperation for development of strict guidelines for protection of workforce from attack of undesirable engineered nanomaterials. So, nanomaterials should be synthesized responsibly considering their after-effects on human health and safety. Research laboratories and pilot operation systems are the main exposure places for toxicity of nanoparticles (Schulte et al. 2013).

Various management systems have been developed on national and international level for safety and health of workers. For instance, CAN/CSA-Z1000-14 for Canada, AS/NZS 4801-2001 for Australia, ANSI/ASSE Z10-2012 for United States, have been developed. ILO-OSH 2001 published by the International labour organization (ILO) works at international level to help healthcare organizations by developing guidelines for occupational safety and health management systems. All of these standardized systems work for improvement of the healthcare system. ISO 45001 has been developed in 2018 working as an international organization for standardization purpose. Similarly, different legislations have been passed in Denmark, Spain, Sweden, India, Malaysia, China, Singapore, South Africa and Taiwan for safety and protection of individuals from health risks. Although the guidelines of different regions and countries differ, the basic purpose of all occupational management systems is to provide protection to healthcare professionals.

A comprehensive list of all available nanoparticles is maintained and updated by the Project on Emerging Technologies, at the Woodrow Wilson International Center for Scholars (www.wilsoncenter.org/nano) for reference and future research studies.

14.8 Public Perception

The nanomaterials are present ubiquitously and human beings are exposed to them through pollution in the air, polluted and contaminated water, food and medicines. However, unlike the situation with other new technologies (e.g. genetically modified foods), the public has not yet raised a huge concern about nanotechnology (Anderson et al. 2015). It might be because the public do not yet understand the implications of the technical information on nanomaterials (Gaskell et al. 2005). The hype of nanotechnology is created so much amongst public where they are seeing such slogans like "the race of small is huge" so the public is still in awe to see the magic of nanotechnology, where the nanobots would be delivering precise medicines to their targets only and then would automatically be dispersed or eliminated by safe route. Likewise, the concept of wearable devices and the concept of molecular machines where they would be working with precision and elegance. All these concepts of nanotechnology are an attempt to make a better and healthy world. However, these should not be at the cost of balanced environment for one health concept, where we have to care about not only humans, but animals and environment as well.

It is possible that scientists in general, and toxicologists in particular, have a perception and expectation of the risks of nanotechnology that is different from the general public, because they understand better than the masses.

14.9 Conclusion

Smaller size, extraordinary reactivity and enhanced capacity are the major advantages of all the NPs. These NPs can become one of the most important factors in small sized counterparts, despite having severe adverse effects on different human cells. NPs enter in the living body during the process of inhalation/ingestion and can migrate within the body from one organ or tissue to another, where there is possibility for the NPs to show some reactivity and produce toxic effects. A number of studies are being carried out to address the toxic behavior of NPs on animal and plants cell using different types of NPs, yet there is a dire need to extend the circle of research on cytotoxic effects of NPs to address many ongoing unsolved issues created by uncertainty.

The available data on toxicology of NPs is still not sufficient so to use NPs in technologies associated particularly with health problems or drug delivery systems. Adverse health effects need to be evaluated and well understood. This chapter was meant to provide an overview of possible risks of NPs according to their types and routes they follow in human system. The smaller size and larger surface area increases their importance in many applications, along with health hazards, due to their large applications in current era. The hindrance is still present in determining the dose of administration in human body and immunological response from human body.

References

Ahmad J, Ahamed M, Akhtar MJ, Alrokayan SA, Siddiqui MA, Musarrat J, Al-Khedhairy AA (2012) Apoptosis induction by silica nanoparticles mediated through reactive oxygen species in human liver cell line HepG2. Toxicol Appl Pharmacol 259:160–168

Aillon KL, Xie Y, El-Gendy N, Berkland CJ, Forrest ML (2009) nanomaterials physiochemical properties on in vivo toxicity. Adv Drug Deliv Rev 61:457–466

Alkilany AM, Frey RL, Ferry JL, Murphy CJ (2008) Gold nanorods as nanoadmicelles: 1-naphthol partitioning into a nanorod-bound surfactant bilayer. Langmuir 24:10235–10239

Allegri M, Bianchi MG, Chiu M, Varet J, Costa AL, Ortelli S, Blosi M, Bussolati O, Poland CA, Bergamaschi E (2016) Shape-related toxicity of titanium dioxide nanofibres. PLoS ONE 11:e0151365

Anderson DS, Silva RM, Lee D, Edwards PC, Sharmah A, Guo T, Pinkerton KE, Van Winkle LS (2015) Persistence of silver nanoparticles in the rat lung: Influence of dose, size, and chemical composition. Nanotoxicology 9:591–602

Andersson POLC, Ekstrand-Hammarstrom B, Akfur C, Ahlinder L, Bucht A, Osterlund L (2011) Polymorph- and size-dependent uptake and toxicity of TiO_2 nanoparticles in living lung epithelial cells. Small 7:514–523

Athinarayanan J, Periasamy VS, Alsaif MA, Al-Warthan AA, Alshatwi AA (2014) Presence of nanosilica (E551) in commercial food products: TNF-mediated oxidative stress and altered cell cycle progression in human lung fibroblast cells. Cell Biol Toxicol 30:89–100

Baird WM, Hooven LA, Mahadevan B (2005) Carcinogenic polycyclic aromatic hydrocarbon-DNA adducts and mechanism of action. Environ Mol Mutagen 45:106–114

Barnard AS (2006) Nanohazards: knowledge is our first defence. Nat Mater 5:245

Barnes AL, Wassel RA, Mondale F, Chen K, Dormer KL, Kopke RD (2007) Magnetic characterization of superparamagnetic nanoparticles pulled through model membranes. BioMagn Res Technol 5:1

Bergin IL, Witzmann FA (2013) Nanoparticle toxicity by the gastrointestinal route: evidence and knowledge gaps. Int J Biomed Nanosci Nanotechnol 3(1–2)

Boogaard H, Kos GP, Weijers EP, Janssen NA, Fischer PH, van der Zee SC, de Hartog JJ, Hoek G (2011) Contrast in air pollution components between major streets and background locations: particulate matter mass, black carbon, elemental composition, nitrogen oxide and ultrafine particle number. Atmos Environ 45:650–658

Borm PJ, Cakmak G, Jermann E, Weishaupt C, Kempers P, van Schooten FJ, Oberdörster G, Schins RP (2005) Formation of PAH–DNA adducts after in vivo and vitro exposure of rats and lung cells to different commercial carbon blacks. Toxicol Appl Pharmacol 205:157–167

Bouallegui Y, Ben Younes R, Turki F, Oueslati R (2017) Impact of exposure time, particle size and uptake pathway on silver nanoparticle effects on circulating immune cells in mytilus galloprovincialis. J Immunotoxicol 14:116–124

Boverhof DR, Bramante CM, Butala JH, Clancy SF, Lafranconi M, West J, Gordon SC (2015) Comparative assessment of nanomaterial definitions and safety evaluation considerations. Regul Toxicol Pharmacol 73:137–150

Bowden LP, Royer MC, Hallman JR, Lewin-Smith M, Lupton GP (2011) Rapid onset of argyria induced by a silver-containing dietary supplement. J Cutan Pathol 38:832–835

Braakhuis HM, Cassee FR, Fokkens PH, De La Fonteyne LJ, Oomen AG, Krystek P, De Jong WH, Van Loveren H, Park MV (2016) Identification of the appropriate dose metric for pulmonary inflammation of silver nanoparticles in an inhalation toxicity study. Nanotoxicology 10:63–73

Chatterjee R (2009) Calculating the costs of nanohazard testing. ACS Publications, pp 3405–3405

Chen YS, Hung YC, Liau I, Huang GS (2009) Assessment of the in vivo toxicity of gold nanoparticles. Nanoscale Res Lett 4:858

Chen Z, Sheng X, Wang J, Wen Y (2018) Silver nanoparticles or free silver ions work? An enantioselective phytotoxicity study with a chiral tool. Sci Total Environ 610:77–83

Cho M, Cho WS, Choi M, Kim SJ, Han BS, Kim SH, Kim HO, Sheen YY, Jeong J (2009) The impact of size on tissue distribution and elimination by single intravenous injection of silica nanoparticles. Toxicol Lett 189:177–183
Choi J, Zheng Q, Katz HE, Guilarte TR (2010) Silica-based nanoparticle uptake and cellular response by primary microglia. Environ Health Perspect 118:589–595
Chulz J, Hohenberg H, Pflucker F, Gartner E, Will T, Pfeiffer S, Wepf R, Wendel V, Gers-Barlag H, Wittern KP (2002) Distribution of sunscreens on skin. Adv Drug Deliv Rev (Suppl 1):S157–163
De Domenico I, Ward DM, Kaplan J (2007) Hepcidin regulation: ironing out the details. J Clin Investig 117:1755–1758
De Jong WH, Borm PJ (2008) Drug delivery and nanoparticles: applications and hazards. Int J Nanomed 3:133
Dekkers S, Bouwmeester H, Bos PM, Peters RJ, Rietveld AG, Oomen AG (2013) Knowledge gaps in risk assessment of nanosilica in food: evaluation of the dissolution and toxicity of different forms of silica. Nanotoxicology 7:367–377
Dekkers S, Oomen AG, Bleeker EA, Vandebriel RJ, Micheletti C, Cabellos J, Janer G, Fuentes N, Vázquez-Campos S, Borges T, Silva MJ (2016) Towards a nanospecific approach for risk assessment. Regul Toxicol Pharmacol 80:46–59
Delfino RJ, Staimer N, Tjoa T, Arhami M, Polidori A, Gillen DL, George SC, Shafer MM, Schauer JJ, Sioutas C (2016) Associations of primary and secondary organic aerosols with airway and systemic inflammation in an elderly panel cohort. Epidemiology (Cambridge, Mass) 21(6)
Demir E, Castranova V (2016) Genotoxic effects of synthetic amorphous silica nanoparticles in the mouse lymphoma assay. Toxicol Rep 3:807–815
Derfus AM, Chan WC, Bhatia SN (2004) Probing the cytotoxicity of semiconductor quantum dots. Nano Lett 4:11–18
Dhawan A, Sharma V (2010) Toxicity assessment of nanomaterials: methods and challenges. Anal Bioanal Chem 398:589–605
Dobrovolskaia MA, Aggarwal P, Hall JB, McNeil SE (2008) Preclinical studies to understand nanoparticle interaction with the immune system and its potential effects on nanoparticle biodistribution. Mol Pharm 5:487–495
Dobrovolskaia MA, McNeil SE (2007) Immunological properties of engineered nanomaterials. Nat Nanotechnol 2:469
Donaldson K, Brown D, Clouter A, Duffin R, MacNee W, Renwick L, Tran L, Stone V (2002) The pulmonary toxicology of ultrafine particles. J Aerosol Med 15:213–220
Donaldson K, Tran L, Jimenez LA, Duffin R, Newby DE, Mills N, MacNee W, Stone V (2005) Combustion-derived nanoparticles: a review of their toxicology following inhalation exposure. Particle Fibre Toxicol 2:10
Elsaesser A, Howard CV (2012) Toxicology of nanoparticles. Adv Drug Deliv Rev 64:129–137
Exbrayat JM, Moudilou EN and Lapied E (2015) Harmful effects of nanoparticles on animals. J Nanotechnol
Fröhlich E (2012) The role of surface charge in cellular uptake and cytotoxicity of medical nanoparticles. Int J Nanomed 7:5577–5591
Gaskell G, Eyck TT, Jackson J, Veltri G (2005) Imagining nanotechnology: cultural support for technological innovation in Europe and the United States. Pub Underst Sci 14(1):81–90
Gatoo MA, Naseem S, Arfat MY, Mahmood Dar A, Qasim K, Zubair S (2014) Physicochemical properties of nanomaterials: implication in associated toxic manifestations. BioMed Res Int. http://dx.doi.org/10.1155/2014/498420
Geiser M, Casaulta M, Kupferschmid B, Schulz H, Semmler-Behnke M, Kreyling W (2008) The role of macrophages in the clearance of inhaled ultrafine titanium dioxide particles. Am J Respir Cell Mol Biol 38:371–376
Georgieva JV, Kalicharan D, Couraud PO, Romero IA, Weksler B, Hoekstra D, Zuhorn IS (2011) Surface characteristics of nanoparticles determine their intracellular fate in and processing by human blood–brain barrier endothelial cells in vitro. Mol Ther 19:318–325

Goulaouic S, Foucaud L, Bennasroune A, Laval-Gilly P, Falla J (2008) Effect of polycyclic aromatic hydrocarbons and carbon black particles on pro-inflammatory cytokine secretion: impact of PAH coating onto particles. J Immunotoxicol 5:337–345
Gui S, Zhang Z, Zheng L, Cui Y, Liu X, Li N, Sang X, Sun Q, Gao G, Cheng Z (2011) Molecular mechanism of kidney injury of mice caused by exposure to titanium dioxide nanoparticles. J Hazard Mater 195:365–370
Gwinn MR, Vallyathan V (2006) Nanoparticles: health effects—pros and cons. Environ Health Perspect 114:1818
Hillyer JF, Albrecht RM (2001) Gastrointestinal persorption and tissue distribution of differently sized colloidal gold nanoparticles. J Pharm Sci 90:1927–1936
Hoet PH, Brüske-Hohlfeld I, Salata OV (2004) Nanoparticles-known and unknown health risks. J Nanobiotechnol 2:12
Hoshino A, Fujioka K, Oku T, Nakamura S, Suga M, Yamaguchi Y, Suzuki K, Yasuhara M, Yamamoto K (2004) Quantum dots targeted to the assigned organelle in living cells. Microbiol Immunol 48:985–994
Hsiao I, Huang Y (2011) Effects of various physicochemical characteristics on the toxicities of ZnO and TiO_2 nanoparticles toward human lung epithelial cells. Sci Total Environ 409:1219–1228
Jania P, McCarthya D, Florence AT (1994) Titanium dioxide (rutile) particle uptake from the rat GI tract and translocation to systemic organs after oral administration. Int J Pharm 105:157–168
Kahru A, Dubourguier HC (2010) From ecotoxicology to nanoecotoxicology. Toxicology 269:105–119
Kasaai MR (2015) Nanosized particles of silica and its derivatives for applications in various branches of food and nutrition sectors. J Nanotechnol
Kreuter J, Alyautdin RN, Kharkevich DA, Ivanov AA (1995) Passage of peptides through the blood–brain barrier with colloidal polymer particles (nanoparticles). Brain Res 674:171–174
Kreyling WG, Semmler M, Möller W (2004) Dosimetry and toxicology of ultrafine particles. J Aerosol Med 17:140–152
Kuempel ED, Geraci CL, Schulte PA (2012) Risk assessment and risk management of nanomaterials in the workplace: translating research to practice. Annals Occup Hyg 56:491–505
L'Azou B, Jorly J, On D, Sellier E, Moisan F, Fleury-Feith J, Cambar J, Brochard P, Ohayon-Courtès C (2008) In vitro effects of nanoparticles on renal cells. Particle Fibre Toxicol 5:22
Lee S, Yun HS, Kim SH (2011) The comparative effects of mesoporous silica nanoparticles and colloidal silica on inflammation and apoptosis. Biomaterials 32:9434–9443
Lin MM, Kim HH, Kim H, Dobson J, Kim DK (2010) Surface activation and targeting strategies of superparamagnetic iron oxide nanoparticles in cancer-oriented diagnosis and therapy. Nanomed 5(1):109–133
Liu A, Ye B (2013) Application of gold nanoparticles in biomedical researches and diagnosis. Clin Lab 8:23–36
Liu S, Wang C, Hou J, Wang P, Miao L, Fan X, You G, Xu Y (2018) Effects of Ag and Ag_2S nanoparticles on denitrification in sediments. Water Res 137:28–36
Lu X, Zhu T, Chen C, Liu Y (2014) Right or left: the role of nanoparticles in pulmonary diseases. Int J Mol Sci 15:17577–17600
Maynard A (2006) Nanotechnology: a research strategy for addressing risk. Nanotechnologies, Woodrow Wilson International Center for Scholars Project on Emerging. Washington DC
Moore MN (2006) Do nanoparticles present ecotoxicological risks for the health of the aquatic environment? Environ Int 32:967–976
Murthy SK (2007) Nanoparticles in modern medicine: state of the art and future challenges. Int J Nanomed 2:129
Murugadoss S, Lison D, Godderis L, Van Den Brule S, Mast J, Brassinne F, Sebaihi N, Hoet PH (2017) Toxicology of silica nanoparticles: an update. Arch Toxicol 91:2967–3010
Nel A, Xia T, Mädler L, Li N (2006) Toxic potential of materials at the nanolevel. Science 311:622–627

Nowack B, Bucheli TD (2007) Occurrence, behavior and effects of nanoparticles in the environment. Environ Pollut 150:5–22

Oberdörster G, Oberdörster E, Oberdörster J (2005) Nanotoxicology: an emerging discipline evolving from studies of ultrafine particles. Environ Health Perspect 113:823

Pan D, Vargas-Morales O, Zern B, Anselmo AC, Gupta V, Zakrewsky M, Mitragotri S, Muzykantov V (2016) The effect of polymeric nanoparticles on biocompatibility of carrier red blood cells. PLoS ONE 11(3):e0152074

Park EJ, Park K (2009) Oxidative stress and pro-inflammatory responses induced by silica nanoparticles in vivo and in vitro. Toxicol Lett 184:18–25

Park EJ, Bae E, Yi J, Kim Y, Choi K, Lee SH, Yoon J, Lee BC, Park K (2010) Repeated-dose toxicity and inflammatory responses in mice by oral administration of silver nanoparticles. Environ Toxicol Pharmacol 30(2):162–168

Poland CA, Duffin R, Kinloch I, Maynard A, Wallace WA, Seaton A, Stone V, Brown S, MacNee W, Donaldson K (2008) Carbon nanotubes introduced into the abdominal cavity of mice show asbestos-like pathogenicity in a pilot study. Nat Nanotechnol 3:423

Qian X, Peng XH, Ansari DO, Yin-Goen Q, Chen GZ, Shin DM, Yang L, Young AN, Wang MD, Nie S (2008) In vivo tumor targeting and spectroscopic detection with surface-enhanced Raman nanoparticle tags. Nat Biotechnol 26:83

Schmid O, Stoeger T (2016) Surface area is the biologically most effective dose metric for acute nanoparticle toxicity in the lung. J Aerosol Sci 99:133–143

Schulte PA, Geraci CL, Hodson LL, Zumwalde RD, Kuempel ED, Murashov V, Martinez KF, Heidel DS (2013) Overview of risk management for engineered nanomaterials. J Phys Conf Ser 429

Schuppan D, Schattenberg JM (2013) Non-alcoholic steatohepatitis: pathogenesis and novel therapeutic approaches. J Gastroenterol Hepatol 28:68–76

Scott-Fordsmand JJ, Peignenburg WJGM, Semenzin E, Novack B, Hunt N, Hristozov D, Marcomini A, Irfan M-A, Jimenez AS, Landsiedel R, Tran L, Oomen AG, Bos PMJ, Hund-Rinke K (2017) Environmental risk assessment strategy for nanomaterials. Int J Environ Res Public Health 14:1251

Seaton A, Tran L, Aitken R, Donaldson K (2009) Nanoparticles, human health hazard and regulation. J Roy Soc Interface 7(Suppl 1):S119–S129

Senzui M, Tamura T, Miura K, Ikarashi Y, Watanabe Y, Fujii M (2010) Study on penetration of titanium dioxide (TiO(2)) nanoparticles into intact and damaged skin in vitro. J Toxicol Sci 35:107–113

Sharma M (2010) Understanding the mechanism of toxicity of carbon nanoparticles in humans in the new millennium: a systemic review. Indian J Occup Environ Med 14(1):3–5

Shimida A, Kawamura N, Okajima M, Kaewamatawong T, Inoue H, Morita T (2006) Translocation pathway of the intratracheally instilled ultrafine particles from the lung into blood circulation in the mouse. Toxicol Pathol 34:949–957

Subbenaik SC (2016) Physical and chemical nature of nanoparticles. In: Plant nanotechnology. Springer, Cham, pp 15–27

Tinkle SS, Antonini JM, Rich BA, Robert JR, Salmen R, DePree K, Adkins EJ (2003) Skin as a route of exposure and sensitization in chronic beryllium disease. Environ Health Perspect 111:1202–1208

Truong L, Saili KS, Miller JM, Hutchison JE, Tanguay RL (2012) Persistent adult zebrafish behavioral deficits results from acute embryonic exposure to gold nanoparticles. Comp Biochem Physiol C: Toxicol Pharmacol 8:269–274

Tsuji JS, Maynard AD, Howard PC, James JT, Lam C-W, Warheit DB, Santmaria AB (2005) Research strategies for safety evaluation of nanomaterials, Part IV: Risk assessment of nanoparticles. Toxicol Sci 89:42–50

Vance ME, Kuiken T, Vejerano EP, McGinnis SP, Hochella MF Jr, Rejeski D, Hull MS (2015) Nanotechnology in the real world: redeveloping the nanomaterial consumer products inventory. Beilstein J Nanotechnol 6:1769

Wang J, Zhou G, Chen C, Yu H, Wang T, Ma Y, Jia G, Gao Y, Li B, Sun J (2007a) Acute toxicity and biodistribution of different sized titanium dioxide particles in mice after oral administration. Toxicol Lett 168:176–185

Wang JJ, Sanderson BJ, Wang H (2007b) Cyto- and genotoxicity of ultrafine TiO_2 particles in cultured human lymphoblastoid cells. Mutat Res 628:99–106

Wang X, Ding B, Li B (2013) Biomimetic electrospun nanofibrous structures for tissue engineering. Mater Today 16:229–241

Xia Y, Yu H, Jansen R, Seringhaus M, Baxter S, Greenbaum D, Zhao H, Gerstein M (2004) Analyzing cellular biochemistry in terms of molecular networks. Annu Rev Biochem 73:1051–1087

Yin H, Too HP, Chow GM (2005) The effects of particle size and surface coating on the cytotoxicity of nickel ferrite. Biomaterials 26:5818–5826

Zhang WL, Yu WW, Vicki LC, Monteiro-Riviere NA (2008) Biological interactions of quantum dot NPs in skin and in human epidermal keratinocytes. Toxicol Appl Pharmacol 228:200–211

Chapter 15
Microparticle-Supported Nanocomposites for Safe Environmental Applications

Sanchita Mandal, Binoy Sarkar, Raj Mukhopadhyay, Jayanta Kumar Biswas and K. M. Manjaiah

Abstract Utilization of nanotechnology and nanomaterials is an integral part of modern life. The increased use of nanomaterials in various consumer products, industries, medical instruments, and information technology and energy sectors has created research interest because of their potential toxicity for the environment. Nanomaterials can also be added directly and indirectly to the soil and water treatment plants to reduce pollutant concentrations. Nanoparticles may enter the aquatic system through runoff and industrial effluent discharge, therefore, potentially contaminate both the aquatic and terrestrial systems. Researchers have used laboratory experiments to understand the effect of nanomaterials and their transformation in water and soil environment. The transformation of nanoparticles in the environment involves various physical and chemical processes, and their degradation/transformation products may introduce further toxicity. Various microparticles or host materials can be used to support or coat nanoparticles in order to reduce their toxicity. Microparticles including clay minerals, polymer, carbon-based materials (biochar) are popular to support nanoparticles due to their large surface area and improved functional char-

S. Mandal · B. Sarkar
Future Industries Institute, University of South Australia, Mawson Lakes, SA 5095, Australia
e-mail: sanchita.mandal@mymail.unisa.edu.au

B. Sarkar (✉)
Department of Animal and Plant Sciences, The University of Sheffield, Sheffield S10 2TN, UK
e-mail: b.sarkar@sheffield.ac.uk

R. Mukhopadhyay
Division of Irrigation and Drainage Engineering, ICAR-Central Soil Salinity Research Institute, Karnal 132001, India
e-mail: rajssaciari@gmail.com

J. K. Biswas
Department of Ecological Studies, International Centre for Ecological Engineering, University of Kalyani, Kalyani, Nadia 741235, West Bengal, India
e-mail: biswajoy2008@gmail.com

K. M. Manjaiah
Division of Soil Science and Agricultural Chemistry, ICAR-Indian Agricultural Research Institute, New Delhi 110012, India
e-mail: manjaiah.math@gmail.com

M. Rai and J. K. Biswas (eds.), *Nanomaterials: Ecotoxicity, Safety, and Public Perception*, https://doi.org/10.1007/978-3-030-05144-0_15

acteristics. This chapter aims to give an overview of the risks associated with using nanoparticles, and how to reduce the possible toxicity related to nanoparticles by using microparticle host materials.

Keywords Microparticles · Nanoparticles · Clay mineral · Polymer · Biochar Toxicity

15.1 Introduction

Environmental pollution is a global problem that is considerably increasing day by day due to the rapid industrialisation, rise in global population, pace of urbanization, intensive agriculture and other adverse anthropogenic activities. The emergence of nanotechnology has given potential scope and opportunities to treat pollutants because the nanoparticles are very small in size (<100 nm in one dimension) (British Standards, BSI 2007; Zekic et al. 2018), having high surface-to-volume ratios, specific surface area, and excellent chemical reactivity. Due to the rapid expansion of nanotechnology and the frequent utilization of nanomaterials, it is highly important to understand their fate, stability and possible risk related to the environment and human health (Stone et al. 2010). Because of their small size, altered surface properties and high reactivity, nanomaterials can be used in a variety of consumer products including paints, cosmetics, medicines, food, suntan lotion etc., and their application could release nanomaterials into the environment (Aitken et al. 2006; Stone et al. 2010). The possible deleterious effects upon the release of various nanomaterials into the environment require thorough investigations. Several previous studies have observed the toxic effects of nanomaterials on the environment and humans. However, advanced knowledge is still required to understand how nature and properties of these nanomaterials could influence their fate, behaviour and potential toxicity to different environmental receptors (Colvin 2003; Klaine et al. 2008). For example, Scarano and Morelli (2003) reported that upon exposure to a marine phytoplankton, metal nanoparticles (Cd-nanoparticles) could transform to a stable nano-crystal formulation in the aqueous environment. Similarly, Tong et al. (2007) reported that C_{60} nanoparticles after application to soils posed only a little change of soil bacterial community diversity. However, other studies demonstrated that the application of nanoparticles to soils could be toxic to soil microbes (Klaine et al. 2008). Therefore, toxicity testing strategies and environmental risk assessment of nanoparticles should be developed in order to monitor the environmental safety of these materials (Samrot et al. 2018). To reduce the toxicity of nanomaterials there have been increasing research interests recently in using microparticle-supported nanomaterials because the microparticles could reduce the free-movement of nanoparticles in the environment. The supporting microparticles, such as clay minerals and biochar, often could improve the contaminant remediation performance of the resultant nano-composite materials (Bhat et al. 2018). The selection of host microparticles for nanoparticles greatly depends on the targeted pollutant to be remediated in a given environment. A range of effective host microparticles have been used by scientists, and among them, clay minerals, biochar, polymer (callouses, resin, chitosan, etc.) are the most popular ones because of their unique matrix-nanoparticle interactions, easy scalability of production and effective properties for remediation (Zhao et al. 2011; Zhang et al. 2012a; Sarkar et al. 2015). The current chapter discusses the scope of microparticle-supported nanomaterials, their properties and safe environmental applications that potentially reduce the unwanted risk of nanoparticles to the environmental receptors.

15.2 Environmental Risk of Nanomaterials

Nanotechnology has gained widespread research and public interests due to the application of nanomaterials in various sectors including agriculture, medicine and public health. Besides, there are several well-known environmental applications of nanomaterials including air purification, water filtration, antibacterial application, and food packaging (Shatkin 2017). Researchers have considered nanotechnology and nanomaterials as the next logical step of science because it enables us to create functional materials and devices to study environmental processes at molecular levels (Chen and Mao 2007; Kumar and Jakhmola 2007; Ray et al. 2009). Furthermore, engineered nanomaterials are becoming an integral part of our daily life because of their use in the form of cosmetics, food packages, drugs, and other products (Ray et al. 2009; Shatkin 2017). Therefore, pollution owing to nanotechnologies and nanomaterials are increasing day by day. Despite the beneficial use of nanomaterials, the big question is how the use of nanomaterials may affect our environment. The release of nanomaterials into the environment may happen through various routes such as atmospheric emission, solid and liquid waste streams from the material production industries and discharge of daily household wastes (Fig. 15.1). Additionally, nanomaterials in the fabric, personal care products and health care products are entering into the environment through their daily uses. From these activities, nanoparticles ultimately deposit to water and landfill sites, which can contaminate the soil and water bodies. Particles in solid wastes, industrial effluents, accidental spilling during transportation, wind-assisted dispersion and rainwater runoff could aggravate the problem of environmental pollution due to nanomaterials' exposure. The most significant environmental deposition of nanomaterials comes from spillages during the manufacturing, transportation, and intentional discharge of waste products containing nanomaterials into the soil and agricultural lands. Many studies have been conducted to understand the environmental effects and fates of nanomaterials in soil and water bodies (Brayner et al. 2006; Tong et al. 2007; Kunhikrishnan et al. 2015; Lead et al. 2018). Batley et al. (2013) reported that the direct deposition of nanomaterials through fertilization and runoff into the aquatic system could cause toxicity to algae involving adsorption of the nanomaterials to cell surfaces, and causing a disruption to the membrane transport system. Another important aspect of nanomaterials toxicity is the kinetic dissolution or solubility of the materials after their exposure to the environment. Most metal-based nanomaterials are hydrophilic, and

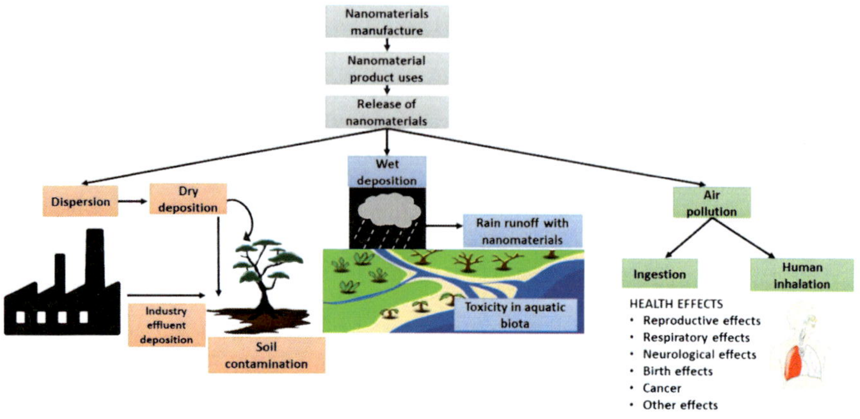

Fig. 15.1 Pathways of nanomaterials entering into the environment and their potential health risks

have low dissolution or solubility. However, despite the common belief that ZnO was "insoluble", the nanoparticle could rapidly dissolve in water and be highly toxic to most of the aquatic biota (Franklin et al. 2007; Batley et al. 2013).

The risks associated with products containing nanoparticles are not only for the environment, but also extend directly or indirectly to the human health. Due to the very small size of these materials, inhalation exposure can occur to airborne particles composed of various nanomaterials. Nanomaterials may attach to larger particles that can change their properties and behaviour in the indoor and outdoor environments so that their entry and potential exposure to the human body can be reduced (Lam et al. 2004; Lee et al. 2007; Stahlhofen et al. 1989). Nanomaterials can be deposited in the human respiratory system, and can cause nanostructure-influenced toxicity due to the high surface area, small diameter, unusual morphology and/or degradation into smaller particles after nanomaterial deposition (Ray et al. 2009). Due to all these toxic effects of nanomaterials, there is a need to develop "smarter" nanomaterials for contaminant remediation purposes. For example, new coating and modification could enhance the effectiveness of nanomaterials, and minimise their potential toxicity.

15.3 Supported Nanomaterials for Environmental Applications

15.3.1 Clay-Supported Nanomaterials and Nanocomposites

Clay minerals, as a naturally abundant resource, can be an appropriate supporting material to attract contaminants and nanoparticles. Clay minerals have a wide range of applications especially for treating environmental contaminants because of their unique properties such as large surface area, high cation exchange capacity (CEC), layered structure and good adsorption capacity. Previous studies showed that clay minerals are a popular choice to remediate organic and inorganic pollutants in the environment (Churchman et al. 2006; Sarkar et al. 2012). Depending on the geographical origin of clay minerals, their properties may vary widely. The classification of clay minerals depends on their layered structure, particle size, exchangeable ions, and the kind of the substituted metals (Shichi and Takagi 2000). The basic structure of clay minerals consists of polymerized silica tetrahedral sheets and aluminium octahedral sheets (Shichi and Takagi 2000). Diversity in the surface structure of clay minerals along with their adsorption capacity and ion exchange properties not only enhance their utilization in treating pollutants from the environment but also allows these minerals to be used as a suitable support of nanoparticles (like nZVI) (Cho et al. 2005; Shi et al. 2011; Ezzatahmadi et al. 2017). Both naturally occurring and modified clay minerals can be used to remove contaminants from the environment. In recent years, a specific focus has been given to clay-supported nanocomposites for contamination remediation due to their high contaminant removal efficiency and fast degradation rate (Xi et al. 2011; Ezzatahmadi et al. 2017). For example, Son et al. (2012) found that the nano zero valent iron (n-ZVI)-montmorillonite composite material showed high magnetization inherited to ZVI, and high catalytic efficiency in decomposing Rhodamine B dye due to the large surface area and reductive activity of ZVI. Zhang et al. (2012c) reported that the application of nZVI-bentonite nanocomposite could reduce chromate (Cr(VI)) concentration from solution. Authors found that the ZVI accelerated the adsorption of Cr(VI) by enhancing the mass transfer of Cr(VI) from aqueous solution to the iron surface. The complete Cr(VI) removal from solution could last longer due to the buffering capacity of the bentonite clay (Zhang et al. 2012c). Ezzatahmadi et al. (2017) suggested that clay mineral-supported nZVI nanocomposites could be an effective remedial material for contaminated water and wastewater treatment. Similarly, superparamagnetic iron oxide nanoparticles supported on palygorskite clay mineral not only enhanced the heavy metal lead removal from aqueous solutions but also facilitated an energy-neutral magnetic separation of the spent adsorbent following the water treatment (Rusmin et al. 2017). Studies have also found that the synthesis of carbon nanoparticles on fibrous palygorskite clay mineral from starch contained oxygen-containing functional groups, and showed variable affinity to cationic and anionic dye compounds (Sarkar et al. 2015). Authors observed that the heat-CO_2-treated nanocomposite adsorbed a large amount of anionic orange II dye compounds (23 mg g^{-1}) (Sarkar et al. 2015). The above-mentioned examples of clay supported nanocomposites and their contaminant remediation efficiency suggested that these materials have enormous potential for in situ immobilization of persistent organic pollutants in soils (Duan et al. 2014). However, most of the previous researches on the use of clay-supported nanocomposites were carried out under laboratory conditions. Therefore, it is important to assess their effectiveness under field situations.

15.3.2 Biochar-Supported Nanomaterials and Nanocomposites

Biochar is a carbon-based material that is produced during thermal decomposition of organic biomasses at a temperature below 800 °C under oxygen-free environment (Joseph and Lehmann 2009). Recently there has been increasing interest in biochar research because of its unique surface properties and ability to act as a potential remediation agent of contaminants in the soil and water bodies (Clough and Condron 2010; Lehmann et al. 2011; Kloss et al. 2012). However, raw biochar has limited effectiveness on contamination remediation, particularly for the high concentration of pollutant in contaminated water. Therefore, recent studies have been focused on using synthesized biochar-supported nanomaterials/nanocomposites to overcome the toxicity of nanomaterials, and to enhance biochar's performance to treat contaminants (Tan et al. 2016; Mian and Liu 2018; Zhang et al. 2018). Previous studies suggested that biochar-supported nanomaterials contained various surface functional groups and high pore volume, which can serve as remediating agents as well as reduce nanomaterial toxicity to environmental receptors. Multiple advantages of using biochar as a supporting material for nanocomposites can be harnessed. Firstly, the feedstock/raw materials of biochar generally come from renewable resources such as agricultural residues and solid wastes (Lehmann and Joseph 2015; Sohi 2012). Secondly, biochar preparation involves low energy requirement, which can be produced under relatively lower temperature than carbon nanotubes (CNTs) (Meyer et al. 2011). Finally, the thermal decomposition processes produce biofuel and syngas accompanied with biochar (Lehmann and Joseph 2015). Therefore, biochar-supported nanomaterials may achieve four integrated goals, including remediation, waste management, carbon sequestration and energy saving.

A range of different nanomaterials (chitosan, graphene, nanocrystal, CNT and nZVI) can be supported on biochar, and the production technology greatly depends on the target element enrichment for remediation. The resulting biochar-nanocomposites show great improvement of some key factors that regulate the contamination remediation processes, including specific surface area, surface charge density, pore-size distribution, pore volume, reactive adsorption sites and effective oxygen-containing surface functional groups (Tan et al. 2016). Previous studies reported the effectiveness of biochar-nanocomposites in pollutants removal from aqueous systems. For example, Zhang et al. (2012a) reported that the MgO-biochar (sugar beet tailing and peanut shells derived biochar) nanocomposite removed phosphate and nitrate (maximum adsorption capacity 835 and 95 mg g^{-1}, respectively) from aqueous solutions with greater capacities than other conventional adsorbents. High adsorption capacity of biochar-nanocomposites made the materials highly suitable for wastewater decontamination. For example, Zhou et al. (2013) found that chitosan-based biochar-nanocomposite and carbon nanotube-coated biochar enhanced lead adsorption due to the presence of amine and oxygen-containing functional groups on the biochar-nanocomposite surfaces (Inyang et al. 2015). Previous studies also suggested that the enhanced heavy metal removal capacity of biochar-nanocomposites could be

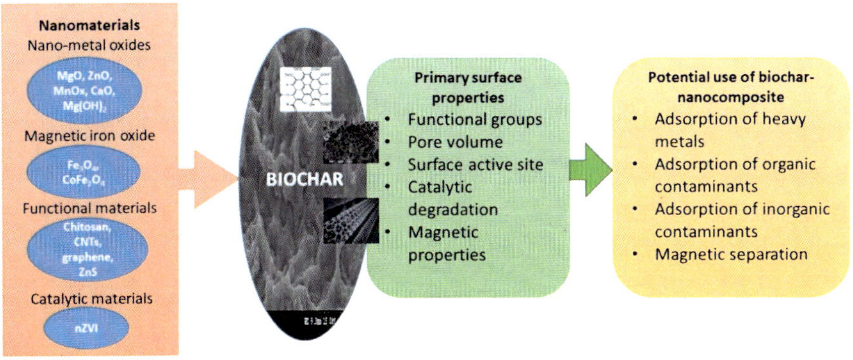

Fig. 15.2 Mechanisms of biochar-supported nanocomposites for the removal of environmental contaminants (adapted from Tan et al. 2016)

attributed to both the surface properties and increased pH of the reaction medium driven by the biochar (Zhang et al. 2012b; Zhou et al. 2013; Wang et al. 2014; Mandal et al. 2016). Figure 15.2 represents plausible mechanisms of biochar-supported nanocomposites for the removal/remediation of environmental contaminants.

Despite the advantages of biochar-nanocomposites as an effective group of pollutant removing agents, the potential environmental and biological risk of using nanomaterials (ZnO, MgO, TiO_2, CuO, MnOx) should be considered in future applications (Kahru et al. 2008; Tan et al. 2016). The toxicity of biochar-nanocomposite materials greatly depends on their nature, stability and medium of use (Tan et al. 2016). Therefore, extensive research is needed to improve the stability of biochar-nanocomposites, and to minimize the release of toxic nanoparticles from the nanocomposites.

15.3.3 Polymer-Supported Nanomaterials and Nanocomposites

Due to the risk associated with the use of nanoparticles, a wide range of host materials have been used to minimize the toxicity and enhance the effectiveness for the removal of pollutants. Among the host materials, granular activated carbon (Vaughan and Reed 2005), silica (Hirano et al. 2004), cellulose (Hubbe et al. 2008), sands (Hansen et al. 2001) and polymeric materials are proven effective. Polymeric hosts for the nanocomposites are comparatively more attractive than other materials because of their surface chemistry (charge, surface area, functional groups), controllable pore size along with their excellent mechanical strength and thermal properties, which make them fit for long-term use (Zhao et al. 2011; Bhat et al. 2018). Some inherited properties of nanoparticles can be retained in the polymer-based nanoparticles, while polymer-supported nanomaterials or polymer-nanocomposites have higher stability and some unique improvement caused by nanoparticle-polymer matrix interactions.

There are various techniques to prepare polymer-supported nanoparticles, and they can be divided into two categories, namely, direct compounding and in situ synthesis. The direct compounding method is the most used procedure, comparatively cost effective and suitable for large-scale production (Zhao et al. 2011). The in situ synthesis method is also widely used. A range of starting materials and fabrication processes can be used to prepare polymer-nanocomposites (Zhao et al. 2011). Numerous studies have applied polymer-nanocomposites for contaminants removal from various environmental media like industrial effluents, gases, groundwater, and soils. The main mechanism of removing pollutants are attributed to the catalytic degradation and adsorption capacity of polymer-nanocomposites. For example, Lin et al. (2005) synthesized nZVI supported on a cation exchange resin to remove carbon tetrachloride (CCl_4) from water, and the removal of CCl_4 in solution was achieved due to the combined effects of degradation by nZVI and adsorption by the host resin (Lin et al. 2005). Studies were also done on polymer-supported magnetic nanocomposites, and the big advantage of using magnetic nanocomposites is that they can be easily separated from water under a magnetic field. Magnetic nanoparticles supported polymer materials were previously produced and applied to remove heavy metals like Cr(VI), Ni(II), Co(II), Pb(II) and organic dyes (Ngomsik et al. 2009; Rocher et al. 2008). Another major application of polymer-nanocomposites is their pollutant sensing and detection ability. Polymer-nanocomposites have created research interests as sensor for air-born pollutants (like NH_3, CO, NO_2) because of their adjustable transport properties, large surface area, chemical specificity, easy processing and scalable production (Jiang et al. 2004; Ram et al. 2005).

The wide range of environmental applications of polymer-nanocomposite is interesting and endless. However, further investigations into the interplay between the host polymer and encapsulated nanoparticles are still required. The successful laboratory-based applications need to be tested in the field scale for commercial uses.

15.4 Mechanism for Nanomaterial Risk Reduction

Nanotechnology has already advanced in all forms including goods, instrument, literature and profit (Peralta-Videa et al. 2011). The boom in 'nanotechnology' includes both natural and man-made nanoparticles. Many articles reviewed the risk assessment, safety survey and toxicity associated with the use of nanomaterials (Conti et al. 2008; Hoyt and Mason 2008; Wiesner et al. 2009). Conti et al. (2008) reviewed the safety procedure of 82 organizations that either manufacture or use nanoparticles in Australia, North America, Asia and Europe, and found around 75% of these organisations do not have a well-established safety program for assessing the risk of using nanoparticles. Therefore, it is important to understand the environmental risks associated with the use of nanoparticles. It was observed from the above discussion (Sect. 15.3) that the use of various host materials could be useful for reducing the toxicity of nanoparticles. For example, studies have found that biochar-supported nanoparticles could detect and remove emerging contaminants (personal care product, pesticides, and heavy metals) in aqueous solution. The mechanisms were attributed to the unique surface chemistry including oxygen-containing functional groups and well-distributed pore space of biochar-supported nanoparticles (Marti et al. 2014; Petrie et al. 2015). Furthermore, polymer-supported nanoparticles can be used as an alternative of harmful materials, reducing the energy requirement and waste generation (Yew et al. 2006). The toxicity of different nanomaterials such as metal oxides, single-enzyme nanoparticles could be reduced by using polymeric supports because of their high stability (Jiang et al. 2011; Katsoyiannis and Zouboulis 2002). Another possible way to reduce nanoparticle toxicity is by using clay minerals as an encapsulating or supporting agent (Xiaoying et al. 2008). The supporting matrix holds the nanoparticles and prevent their unwanted dissipation from the target site.

Some metal oxide nanoparticles like CoO, CuO, Ni_2O_3, and Co_3O_4 can cause severe toxicity in animals. The risk of using these nanoparticles can be reduced through surface passivation. A study by Cai et al. (2017) found that methylene phosphonic acid coating with metal oxide nanoparticles reduced the cytotoxicity and hazardous effects of nanoparticles by reducing the surface activity of metal oxides (85–99%). When nanoparticles are supported on microparticles, the chance of penetration of the nanoparticles into the cell walls of organisms is significantly reduced. As a result, the nanoparticles are prevented from interacting with the cellular materials or adducting with the nucleic acids.

Developing advanced tools to measure and monitor the fate and behaviour of manufactured nanoparticles is highly required these days. The increasing toxicity of nanoparticles needs to be minimised, and the detection of nanoparticles can be achieved by improvising the existing analytical methods. Karn et al. (2009) reported that the organisms are affected by certain nanoparticles under certain environmental conditions, but supported nanoparticles could significantly reduce such toxicity. However, future research needs to develop and understand the mechanisms related to the reduction of toxicity by supported nanoparticles and their ultimate fate in the environment.

15.5 Conclusions

Microparticle-supported nanocomposites have the potential not only to reduce the overall cost of large-scale contamination remediation, but also to reduce the clean-up time and toxicity of nanomaterials. Furthermore, microparticle-supported nanocomposites improve the surface chemical properties and provide excellent host-nanoparticle interaction for pollutant adsorption from soil and water. A range of host materials can be used to coat nanoparticles to reduce their toxicity level. The types of host microparticles greatly depend on their mechanical and thermal behaviours including hydrophobicity/hydrophilicity, chemical stability, biocompatibility and chemical functionalities. Carbon-based materials (e.g., biochar), polymers and clay minerals can be used to support nanoparticles. These host materials qualify for this purpose due to their high surface area, convenient pore size distribution and excellent mechanical strength for long-term use. However, the current information on the production and characteristics of microparticle-supported nanocomposites are inadequate. The potential application of these hybrid materials for environmental remediation also requires further research. The key challenges will be to undertake the analyses under natural field-level applications of microparticle-supported nanocomposites in soil and water.

References

Aitken RJ, Chaudhry MQ, Boxall ABA, Hull M (2006) Manufacture and use of nanomaterials: current status in the UK and global trends. Occup Med 56:300–306

Batley GE, Kirby JK, McLaughlin MJ (2013) Fate and risks of nanomaterials in aquatic and terrestrial environments. Acc Chem Res 46:854–862

Bhat AH, Rehman WU, Khan IU, Khna I, Ahmad S, Ayoub M, Usmani MA (2018) Nanocomposite membrane for environmental remediation. In: Jawaid M, Khan MM (eds) Polymer-based nanocomposites for energy and environmental applications. Woodhead Publishing, Cambridge, MA, pp 407–440

Brayner R, Ferrari-Iliou R, Brivois N, Djediat S, Benedetti MF, Fiévet F (2006) Toxicological impact studies based on *Escherichia coli* bacteria in ultrafine ZnO nanoparticles colloidal medium. Nano Lett 6:866–870

British Standards (BSI) (2007). Terminology for nanomaterials. Public Available Specification No. 1362007. British Standards Institution, London, 16

Cai X, Lee A, Ji Z, Huang C, Chang CH, Wang X, Liao YP, Xia T, Li R (2017) Reduction of pulmonary toxicity of metal oxide nanoparticles by phosphonate-based surface passivation. Part Fibre Toxicol 14:13

Chen X, Mao SS (2007) Titanium dioxide nanomaterials: synthesis, properties, modifications, and applications. Chem Rev 107:2891–2959

Cho HH, Lee T, Hwang SJ, Park JW (2005) Iron and organo-bentonite for the reduction and sorption of trichloroethylene. Chemosphere 58:103–108

Churchman GJ, Gates WP, Theng BKG, Yuan G (2006) Clays and clay minerals for pollution control. In: Bergaya F, Theng BKG, Lagaly G (eds) Developments in clay science. Elsevier, New York, pp 625–675

Clough TJ, Condron LM (2010) Biochar and the nitrogen cycle: introduction. J Environ Qual 39:1218–1223

Colvin VL (2003) The potential environmental impact of engineered nanomaterials. Nat Biotechnol 21:1166

Conti JA, Killpack K, Gerritzen G, Huang L, Mircheva M, Delmas M, Harthorn BH, Appelbaum RP, Holden PA (2008) Health and safety practices in the nanomaterials workplace: results from an international survey. Environ Sci Technol 42:3155–3162

Duan L, Palanisami T, Liu Y, Dong Z, Mallavarapu M, Kuchel T, Semple KT, Naidu R (2014) Effects of ageing and soil properties on the oral bioavailability of benzo[a]pyrene using a swine model. Environ Int 70:192–202

Ezzatahmadi N, Ayoko GA, Millar GJ, Speight R, Yan C, Li J, Li S, Zhu J, Xi Y (2017) Clay-supported nanoscale zero-valent iron composite materials for the remediation of contaminated aqueous solutions: a review. Chem Eng J 312:336–350

Franklin NM, Rogers NJ, Apte SC, Batley GE, Gadd GE, PS Casey PS (2007) Comparative toxicity of nanoparticulate ZnO, bulk ZnO, and ZnCl to a freshwater microalga (Pseudokirchneriella subcapitata): the importance of particle solubility. Environ Sci Technol 41(24):8484–8490

Hansen BØ, Kwan P, Benjamin MM, Li CW, Korshin GV (2001) Use of iron oxide-coated sand to remove strontium from simulated hanford tank wastes. Environ Sci Technol 35:4905–4909
Hirano M, Ota K, Iwata H (2004) Direct formation of anatase (TiO_2)/silica (SiO_2) composite Nanoparticles with high phase stability of 1300 °C from acidic solution by hydrolysis under hydrothermal condition. Chem Mater 16:3725–3732
Hoyt VW, Mason E (2008) Nanotechnology: emerging health issues. J Chem Health Safety 15:10–15
Hubbe MA, Rojas OJ, Lucia LA, Sain M (2008) Cellulosic nanocomposites: a review. Bio Resources 3:929–980
Inyang MI, Gao B, Yao Y, Xue Y, Zimmerman A, Mosa A, Pullammanappallil P, Ok YS, Cao X (2015) A review of biochar as a low-cost adsorbent for aqueous heavy metal removal. Crit Rev Env Sci Technol 46:406–433
Jiang C, Markutsya S, Pikus Y, Tsukruk VV (2004) Freely suspended nanocomposite membranes as highly sensitive sensors. Nat Mater 3:721
Jiang Z, Lv L, Zhang W, Du Q, Pan B, Yang L, Zhang Q (2011) Nitrate reduction using nanosized zero-valent iron supported by polystyrene resins: role of surface functional groups. Water Res 45:2191–2198
Joseph S, Lehmann J (2009) Biochar for environmental management: science and technology. Earthscan Pub., London
Kahru A, Dubourguier HC, Blinova I, Ivask A, Kasemets K (2008) Biotests and biosensors for ecotoxicology of metal oxide nanoparticles: a minireview. Sensors 8:5153
Karn B, Kuiken T, Otto M (2009) Nanotechnology and in situ remediation: a review of the benefits and potential risks. Environ Health Perspect 117:1813–1831
Katsoyiannis IA, Zouboulis AI (2002) Removal of arsenic from contaminated water sources by sorption onto iron-oxide-coated polymeric materials. Water Res 36:5141–5155
Klaine SJ, Alvarez PJJ, Batley GE, Fernandes TF, Handy RD, Lyon DY, Mahendra S, McLaughlin MJ, Lead JR (2008) Nanomaterials in the environment: behavior, fate, bioavailability, and effects. Environ Toxicol Chem 27:1825–1851
Kloss S, Zehetner F, Dellantonio A, Hamid R, Ottner F, Liedtke V, Schwanninger M, Gerzabek MH, Soja G (2012) Characterization of slow pyrolysis biochars: effects of feedstocks and pyrolysis temperature on biochar properties. J Environ Qual 41:990–1000
Kumar A, Jakhmola A (2007) RNA-mediated fluorescent Q-PbS nanoparticles. Langmuir 23:2915–2918
Kunhikrishnan A, Shon HK, Bolan NS, El Saliby I, Vigneswaran S (2015) Sources, distribution, environmental fate, and ecological effects of nanomaterials in wastewater streams. Crit Rev Env Sci Technol 45:277–318
Lam CW, James JT, McCluskey R, Hunter RL (2004) Pulmonary toxicity of single-wall carbon nanotubes in mice 7 and 90 days after intratracheal instillation. Toxicol Sci 77:126–134
Lead JR, Batley GE, Alvarez PJJ, Marie-Noële Croteau, Handy RD, McLaughlin MJ, Judy JD, Schirmer K (2018) Nanomaterials in the environment: behavior, fate, bioavailability, and effects: an updated review. Environ Toxicol Chem 37:2029–2063
Lee KJ, Nallathamby PD, Browning LM, Osgood CJ, Xu XHN (2007) In vivo imaging of transport and biocompatibility of single silver nanoparticles in early development of zebrafish embryos. ACS Nano 1:133–143
Lehmann J, Joseph S (2015) Biochar for environmental management: science, technology and implementation. Routledge
Lehmann J, Rillig MC, Thies J, Masiello CA, Hockaday WC, Crowley D (2011) Biochar effects on soil biota: a review. Soil Biol Biochem 43:1812–1836
Lin CJ, Lo SL, Liou YH (2005) Degradation of aqueous carbon tetrachloride by nanoscale zerovalent copper on a cation resin. Chemosphere 59:1299–1307
Mandal S, Sarkar B, Bolan N, Ok YS, Naidu R (2016) Enhancement of chromate reduction in soils by surface modified biochar. J Environ Manage 186:277–284
Marti E, Variatza E, Balcazar JL (2014) The role of aquatic ecosystems as reservoirs of antibiotic resistance. Trends Microbiol 22:36–41

Meyer S, Glaser B, Quicker P (2011) Technical, economical, and climate-related aspects of biochar production technologies: a literature review. Environ Sci Technol 45:9473–9483

Mian MM, Liu G (2018) Recent progress in biochar-supported photocatalysts: synthesis, role of biochar, and applications. RSC Adv 8:14237–14248

Ngomsik AF, Bee A, Siaugue JM, Talbot D, Cabuil V, Cote G (2009) Co(II) removal by magnetic alginate beads containing Cyanex 272®. J Hazard Mater 166:1043–1049

Peralta-Videa JR, Zhao L, Lopez-Moreno ML, de la Rosa G, Hong J, Gardea-Torresdey JL (2011) Nanomaterials and the environment: a review for the biennium 2008–2010. J Hazard Mater 186:1–15

Petrie B, Barden R, Kasprzyk-Hordern B (2015) A review on emerging contaminants in wastewaters and the environment: current knowledge, understudied areas and recommendations for future monitoring. Water Res 72:3–27

Ram MK, Yavuz Ö, Lahsangah V, Aldissi M (2005) CO gas sensing from ultrathin nano-composite conducting polymer film. Sens Actuators, B 106:750–757

Ray PC, Yu H, Fu PP (2009) Toxicity and environmental risks of nanomaterials: challenges and future needs. J Environ Sci Health Part C 27:1–35

Rocher V, Siaugue JM, Cabuil V, Bee A (2008) Removal of organic dyes by magnetic alginate beads. Water Res 42:1290–1298

Rusmin R, Sarkar B, Tsuzuki T, Kawashima N, Naidu R (2017) Removal of lead from aqueous solution using superparamagnetic palygorskite nanocomposite: material characterization and regeneration studies. Chemosphere 186:1006–1015

Samrot A, Bhavya KS, Sahithya CS, Sowmya N (2018) Evaluation of toxicity of chemically synthesised gold nanoparticles against *Eudrilus eugeniae*. J Clust Sci https://doi.org/10.1007/s10876-018-1440-0

Sarkar B, Liu E, McClure S, Sundaramurthy J, Srinivasan M, Naidu R (2015) Biomass derived palygorskite–carbon nanocomposites: synthesis, characterisation and affinity to dye compounds. Appl Clay Sci 114:617–626

Sarkar B, Xi Y, Megharaj M, Krishnamurti GSR, Bowman M, Rose H, Naidu R (2012) Bioreactive organoclay: a new technology for environmental remediation. Crit Rev Env Sci Technol 42:435–488

Scarano G, Morelli E (2003) Properties of phytochelatin-coated CdS nanocrystallites formed in a marine phytoplanktonic alga (*Phaeodactylum tricornutum*, Bohlin) in response to Cd. Plant Sci 165:803–810

Shatkin JA (2017) Nanotechnology: health and environmental risks, 2nd ed. CRC Press Taylor & Francis Pub., London

Shi L, Lin YM, Zhang X, Chen Z (2011) Synthesis, characterization and kinetics of bentonite supported nZVI for the removal of Cr(VI) from aqueous solution. Chem Eng J 171:612–617

Shichi T, Takagi K (2000) Clay minerals as photochemical reaction fields. J Photoch Photobio C 1:113–130

Sohi SP (2012) Carbon storage with benefits. Science 338:1034–1035

Son YH, Lee JK, Soong Y, Martello D, Chyu MK (2012) Heterostructured zero valent iron–montmorillonite nanohybrid and their catalytic efficacy. Appl Clay Sci 62–63:21–26

Stahlhofen WG, Rudlof G, James AC (1989) Intercomparison of experimental regional aerosol deposition data. J Aerosol Med 2:285–308

Stone V, Nowack B, Baun A, van den Brink N, von der Kammer F, Dusinska M, Handy R, Hankin S, Hassellöv M, Joner E, Fernandes TF (2010) Nanomaterials for environmental studies: classification, reference material issues, and strategies for physico-chemical characterisation. Sci Total Environ 408:1745–1754

Tan X, Liu Y, Gu Y, Xu Y, Zeng G, Hu X, Liu S, Wang X, Liu S, Li J (2016) Biochar-based nanocomposites for the decontamination of wastewater: a review. Bioresour Technol 212:318–333

Tong Z, Bischoff M, Nies L, Applegate B, Turco RF (2007) Impact of fullerene (C_{60}) on a soil microbial community. Environ Sci Technol 41:2985–2991

Vaughan RL, Reed BE (2005) Modeling As(V) removal by a iron oxide impregnated activated carbon using the surface complexation approach. Water Res 39:1005–1014
Wang MC, Sheng GD, Qiu YP (2014) A novel manganese-oxide/biochar composite for efficient removal of lead(II) from aqueous solutions. Int J Environ Sci Technol 12:1719–1726
Wiesner MR, Lowry GV, Jones KL, Hochella JMF, Di Giulio RT, Casman E, Bernhardt ES (2009) Decreasing uncertainties in assessing environmental exposure, risk, and ecological implications of nanomaterials. Environ Sci Technol 43:6458–6462
Xi Y, Megharaj M, Naidu R (2011) Dispersion of zerovalent iron nanoparticles onto bentonites and use of these catalysts for orange II decolourisation. Appl Clay Sci 53:716–722
Xiaoying W, Yumin D, Jiwen L (2008) Biopolymer/montmorillonite nanocomposite: preparation, drug-controlled release property and cytotoxicity. Nanotechnology 19:1–7
Yew SP, Tang HY, Sudesh K (2006) Photocatalytic activity and biodegradation of polyhydroxybutyrate films containing titanium dioxide. Polym Degrad Stab 91:1800–1807
Zekic E, Vukovic Z, Halkijevic I (2018) Application of nanotechnology in wastewater treatment. Gradevinar 70:315–323
Zhang M, Gao B, Yao Y, Xue Y, Inyang M (2012a) Synthesis of porous MgO-biochar nanocomposites for removal of phosphate and nitrate from aqueous solutions. Chem Eng J 210:26–32
Zhang M, Gao B, Yao Y, Xue Y, Inyang M (2012b) Synthesis, characterization, and environmental implications of graphene-coated biochar. Sci Total Environ 435–436:567–572
Zhang Y, Li Y, Li J, Sheng G, Zhang Y, Zheng X (2012c) Enhanced Cr(VI) removal by using the mixture of pillared bentonite and zero-valent iron. Chem Eng J 185–186:243–249
Zhang D, Li Y, Tong S, Jiang X, Wang L, Sun X, Li J, Liu X, Shen J (2018) Biochar supported sulfide-modified nanoscale zero-valent iron for the reduction of nitrobenzene. RSC Adv 8:22161–22168
Zhao X, Lv L, Pan B, Zhang W, Zhang S, Zhang Q (2011) Polymer-supported nanocomposites for environmental application: a review. Chem Eng J 170:381–394
Zhou Y, Gao B, Zimmerman AR, Fang J, Sun Y, Cao X (2013) Sorption of heavy metals on chitosan-modified biochars and its biological effects. Chem Eng J 231:512–518

Chapter 16
Regulatory Framework for Nanomaterials in Agri-Food Systems

Kizhaeral S. Subramanian and S. K. Rajkishore

Abstract The use of nano-materials (NMs) in agri-food system is alarmingly increasing in the past few years, and this warrants to evolve a robust regulatory framework to gain the fruition of the nanotechnology while ensuring environmental safety. This is one of the maiden attempts to bring the information together for the readers to learn the current status of understanding. This chapter outlines the state-of-the-art information on risk assessment strategy for NMs and reviewed the current status of regulatory frameworks in a diverse countries with scientific advancement besides highlighting some key criteria for evolving a foolproof framework. The data have unequivocally demonstrated that the nano-regulatory frameworks available at present are digitally divided, and still lacks clarity. Among them, the European Union is taking a lead step in formulating a nano-regulatory framework for agricultural sector which needs to be further fine-tuned to evolve globally competitive nano-framework ensuring societal acceptance and environmental health. Based on the literature review, a conceptual framework has been suggested that encompasses pre-assessment phase, exposure assessment and risk characterization for regulating the applications of agricultural nanotechnologies.

Keywords Nano-materials · Agriculture · Nano-toxicity · Regulatory framework Agri-food systems

16.1 Introduction

Nanotechnology deals with atomic manipulations with intent to develop processes and products that are precised to deliver inputs in biological systems. In the past 6 years ever since nanotechnology comes to the light of the scientific world, the appli-

K. S. Subramanian (✉) · S. K. Rajkishore
Department of Nano Science & Technology, TNAU, Coimbatore, India
e-mail: kssubra2001@rediffmail.com

S. K. Rajkishore
e-mail: rajkishoresk@gmail.com

M. Rai and J. K. Biswas (eds.), *Nanomaterials: Ecotoxicity, Safety, and Public Perception*, https://doi.org/10.1007/978-3-030-05144-0_16

cations have been far and wide to take the unique advantage of extensive surface-mass ratio and its associated benefits on the ultimate products. Nanotechnology has profound impacts on a wide spectrum of disciplines particularly in energy, environment, electronics and health sciences (Subramanian et al. 2018). Despite fact that it can influence any areas in biological or material sciences, agricultural scientists have begun to employ the emerging technologies as a tool to develop processes and products that enable smart delivery of inputs in agri-food systems without associated environmental hazard. The use of nanotechnology in agricultural sciences in Indian perspective has already been reviewed (Subramanian and Tarafdar 2009; Mishra et al. 2017; Subramanian et al. 2018). The global investment in nanotechnology was just 30 billion USD in 2000 and reached 1000 billion USD in 2015 and expected to get doubled in the year 2020. Such a phenomenal growth in investments to the tune of 2000 times in 20 years that commensurate with the use of nano-materials in various industries. Consequently, the global market of manufactured nano-materials has been valued as $7.3 billion in 2016 and will gallop to $16.8 billion by 2022 (ZMR 2016). About 150–200 billion €worth of classical nano-materials (NMs) such as carbon black, silicon dioxide (SiO_2), titanium dioxide (TiO_2) and silver (Ag) account for more than 90% of the production volume (Haas 2013). Such a skyrocketing growth and applications of nanotechnology has alarming the people across the globe wondering whether this will cause irreparable damage to science and society. Indeed, most of the reported ill-effects of engineered NMs are mostly microcosm studies or paradoxical in many cases. The public perception of nanotechnology is equally divided as 41% of Europeans are positive about nanotechnology, and another 40% are still undecided as indicated by NanoView project (Epp 2015). On the other hand, the people's perception is widely varying with countries. For instance, the acceptance of the technology was the least in EU (46%) and the highest in Switzerland (76%), and others like Germany (65%) and USA (62%) in between (Epp 2015). The overall acceptance of the nanotechnology synchronizes with its uses in consumer products. The surface coating and care products had registered 70% acceptance, textiles 60%, cosmetics 30% and the least with agri-foods (20%). Recent statistics suggests that there is a potential to use nanotechnology to promote productivity of crops while ensuring ecological security.

Within the sphere of agriculture, nanotechnology appears to play a pivotal role in design and fabrication of nano-agri inputs. Nano-fertilizers with nutrient use efficiency, nano-composites for slow and controlled release of agrochemicals, nanomaterials for food packaging to enhance shelf-life and nano-biosensors for plant diagnosis are some of the notable areas where nanotechnology researches are in progress. Despite nanotechnology applications in agriculture begin to expand the surface, ecotoxicological reports suggest that the potential nanomaterials which are likely to hit the farm gate may cause serious ill effects (Rajkishore et al. 2013). Regardless of the promising development of nanotechnology in varied fields, its agricultural applications have not been translated to meet global needs primarily due to shallow awareness and biosafety concerns (Mishra et al. 2017). The foremost reason for scarcity of commercial development of nanotechnology in agriculture is geographically limited existence of legislative framework, regulatory guidelines and negative public opinion (Arts et al. 2014). The growing challenge of global food security and climate change strongly underlines the commercial applications of nanotechnology-based products for the agricultural sector (Rossi et al. 2014). Therefore, there is an urgent need to thoroughly assess the risks and management factors associated with the application of NMs in the agricultural sector, prior to implementation of regulatory guidelines (Amenta et al. 2015).

The number of publications related to biosafety concerns of nanoparticles are increasing exponentially (Shvedova et al. 2018). Nanoparticles have been found to exhibit toxicity in a wide array of biological organisms such as algae, microbes, plants and fishes. On the other hand, there is a phenomenal increase in use of nanotechnology in a wide array of consumer products and food packaging. Such scenarios gained the attention of researchers and policy makers to bring out a nanotechnology regulatory framework for ensuring environmental and health safety while taking the fruition of the fascinating technologies. So far, there is a paucity of work done on regulatory processes and policies. In order to develop a road map on policy framework, significant database and risk assessment strategies are pertinent for evaluating the safety of nano-materials. In general, there has been a little academic analysis of the drivers for this regulatory process and policy. Furthermore, nanotechnology policy requires a regulatory science in the form of a significant database and risk assessment foundation for evaluating the safety of nano-materials. Considering this into account, an attempt was made to gather information of current regulatory frameworks in diverse countries with scientific advancement, and propose a conceptual framework for India to regulate nano-products that are likely to be used from 'farm to fork'. To assess the strengths and limitations of the existing nanotechnology frameworks for agricultural sector, we adopted a multi-perspective methodology that combined a classical review of peer reviewed papers and a collection of information such as guidelines and latest reports that are published in the public domain by leading national regulatory bodies of selected countries (European Union, US, Canada) besides international institutions like Organization of Economic Co-operation and Development (OECD 2016) and International Organization for Standardization. In addition, the existing guidelines for best practices in laboratories and industries put forth by India's nano-regulatory taskforce of Department of Science & Technology, Government of India, New Delhi (DST 2016) is consulted and analyzed to propose a conceptual framework for India.

16.2 Extent of Nano-products from Farm to Fork

Nanotechnology is widely considered as one of the key enabling technologies for sustainable growth, and hence researchers are focusing on this cross-cutting technology for improving the farming system (Subramanian et al. 2018). Nanotechnology has a role to play in the entire agricultural value chain from the farm till it reaches the consumer. Nano-fertilizers for enhanced nutrient use efficiency and farm productivity (Subramanian et al. 2015), nano-particles for seed invigoration (Subramanian et al. 2018), encapsulated form of herbicides for effective weed control (Viji and Chinnamuthu 2015), nano-metal oxides for regulated release of insect pheromone (Dharanivasan et al. 2017) and nano-emulsion to extend self-life of fruits (Anusuya et al. 2016; Jincy et al. 2017) are some of the key inputs that were tested and proved effective in agri-food system. Nanobiotechnology has the capability to resource bioactive constituents in eatables while improvement in foodstuffs at the nanoscale due to antipathogenic ability and enhanced properties of NMs (Martirosyan and Schneider 2014; Eleftheriadou et al. 2017).

Nanoscale biomaterials are used to detect food pathogens and purification besides improving food quality. Some of the prime areas in nano-food systems include nano-additives (Jeon 2016), intelligent packaging, control and nutraceuticals delivery (Berekaa 2015; Ghaani et al. 2016a, b), nano-coding of packaging materials (Bhushani and Anandharamakrishnan 2014), and nano-encapsulation for smart delivery (Katouzian and Jafari 2016). Nanoparticles emulsions are reported to improve the texture and uniformity of the ice cream (Kumar 2015). Complementing with the use of NMs in agri-food systems, the food quality monitoring can be done through "nanosensors" which detects the oxidation process in food vis-à-vis spoliage (Thakur et al. 2018). The integration of nanotechnology in agri-food systems is expected to induce an economic growth to the tune of US$3.4 trillion by 2020 (Sodano and Verneau 2014; Sabourin and Ayande 2015). One of the most abundantly used NMs in food system is AgNPs which accounts for 25% of the nano-enabled products, and its importance is primarily attributed to the antimicrobial property.

16.3 Risk Assessment and Management

Nano-materials are quite enigmatic, and their actions, reactions and associated adverse effects will be varying with size, shape, structure, dose and environmental conditions. This warrants a clearcut risk assessment strategy for developing a robust regulatory framework. Risk assessment is a complex process integrating hazard, exposure, and dose–response data to evolve a comprehensive system for the risk identification and management of nano-materials. In order to acquire better understanding of the subject, it is pertinent to know the terminologies used in the risk assessment. According to the USEPA, the risk is defined as "a measure of the probability that damage to life, health, property, and/or the environment will occur as a result of a given hazard" (USEPA 2007). The term "hazard" refers to the "inherent toxicity of a compound." These substances are chemically toxic and causing adverse health effects as a consequence of exposure. Thus, the risk characterization combines hazard and exposure to estimate risk (Hristozov et al. 2016). The terminology "exposure" is dealing with the combination of the concentration of a substance multiplied by the duration of contact. Finally, the word "dose" is the quantity of substance that enters a biological system or in a specific organ (skin, lung, liver, etc.). Despite the fact that the risk assessment has to be done in a systematic and sequential pattern, in practice, the commercialization of nano-products continues to outpace the development of a prudent risk management mechanisms that are widely accepted by the scientific community and enforced by regulators. The risk assessment management is quite complex due to the fact that the steps or components involved in the processes are varying with literature or individual's perspectives. However, it is universally accepted and adopted from UN reports that the risk assessment comprises four steps namely hazard characterization, dose-response determination, exposure assessment and risk characterization

16.3.1 Hazard

Hazard identification is the key step in risk assessment and management of manufactured nano-materials (NM). There are two approaches, namely weight of evidence (WoE) and Bayesian Networks (BN) that are quite pertinent to quantitatively measure the hazard potential. The WoE is a statistical framework that is based on expert elicitation. On the other hand, BN is a probabilistic, machine learning approach that can predict potential hazard of metal and metal-oxide NMs like TiO_2, Ag, and ZnO.

In both WoE and BN, the NMs are subjected to hazard screening to scout the intrinsic hazard potential of several NMs against one another with a reference to the benchmark. The relative hazard screening helps to read-across experimentally demonstrated adverse effects for a specific NM to one with similar physico-chemical characteristics. Despite the fact that both models (BN and WoE) utilize physico-chemical characteristics and toxicity data to infer the hazard potential, the BN exhibits more stability when the models are fed with new data (Sheehan et al. 2018). Thus, BN approach is gaining support in the context of NM risk assessment (Tsang et al. 2017), while the WoE have been criticized for being overly reliant on expert judgment and qualitative data.

16.3.2 Dose-Response Relationship

The dose-response relationship for NMs is paradoxical and may be totally different from bulk materials. The dose-response assessment quantitatively determines the relationship between adverse effects (i.e., hazards) and a concentration of a substance in a given environment. The strong and robust correlations between dose and biologically relevant endpoints or biomarkers are employed to determine no-observed-adverse-effect levels (NOAELs) and human health exposure thresholds, such as recommended exposure limits (RELs) or occupational exposure limits (OELs) (Sheehan et al. 2018).

16.3.3 Exposure

The exposure assessment formulates exposure scenarios describing how a chemical is used or how it is released into the environment. The exposure will be assessed through one or more routes (i.e. inhalation, ingestion or dermal) or environmental compartments (e.g. water, sediment, soil) under varying exposure scenarios. This may involve monitoring of indoor or outdoor concentrations by means of suitable analytical instruments and/or the estimation of the amount of the substance reaching humans or target environmental species by means of models. Multiple-path particle dosimetry (MPPD) model is commonly used for the simulation of exposure to inhaled NMs in occupational and other scenarios, including animal experimentation (Cassee et al. 2002).

16.3.4 Risk Characterization

The judicious combination of information from the hazard characterization, dose relationship and the exposure assessment is done in order to derive conclusions about the nature and magnitude of risks. Therefore, the risk characterization involves estimation of the potential for adverse health effects to occur, evaluation of uncertainty and summarization of risk information. This is important for implementing additional risk management measures. All the four sequential steps are key elements to develop a tailor-made regulatory framework for risk assessment.

16.4 Tools for Risk Assessment

Risk assessment (RA) is carried out using various parameters that are often referred as "Tools" which are implements used to accomplish specific tasks (e.g. to estimate exposure or to build a dose-response curve). To facilitate the RA of NMs forty eight tools have been identified from the published literature (Hristozov et al. 2016). The tools are categorized into control banding, risk screening, occupational and consumer exposure, environmental fate and exposure, hazard assessment, physicochemical characterization, and decision support tools.

16.4.1 Control Banding

This is one of the tools that involve qualitative and semi-quantitative models to determine the hazard and exposure potentials of chemicals, and to relate the levels of precaution required to control their risks. These are measured on normalized scales called "bands" that are typically plotted on a two-dimensional matrix, which results in establishing a control band associated to a specified level of precaution and a set of risk management measures to be implemented (Hristozov et al. 2016).

16.4.2 Risk Screening Tools

These are qualitative or semi-quantitative besides possessing quantitative risk scoring procedures. They are simple requiring smaller amount of data and the majority of them can cover more than one lifecycle stages (e.g. occupational scenarios in both the production and end-of-life stages).

16.4.3 Occupational and Consumer Exposure Tools

It is a quantitative exposure model for chemicals which is developed as a response to requirements under the European REACH regulation.

16.4.4 Environmental Fate and Exposure Tools

One major drawback in the ecological RA of NMs is the difficulties to detect them in natural environments due to lack of adequate tools to discriminate them from background nanoparticles. In order to enable environmental exposure assessment of NMs in the absence of reliable measurements, several environmental fate models were developed, which can be broadly categorized as "material flow" and "mechanistic" fate models. The majority of these tools is quantitative, and is not easy to apply by non-experts. Most of them released into the environment are evaluated from life cycle perspective. Transformation of NMs and their potential interaction with environmental matrix is an important aspect to be considered in order to avoid misinterpretation of potential effects (Hund-Rinke and Schlich 2014).

16.4.5 Hazard Assessment

The hazard assessment is done based on ecotoxicological effects and dose-response relationships. The different outcomes of in vivo and in vitro toxicity testing achieved so far are considered to be in significant part due to a missing dose correlation between the systems (Demokritou et al. 2013; DeLoid et al. 2015).

16.4.6 Physicochemical Characterization

The physicochemical characterization encompasses size, surface area and surface mass ratio. Doak et al. (2009) have suggested a three-phase approach for nanoparticle characterization. The phase one covers the powder state and includes assessment of chemical composition, size, size distribution, surface area and morphology. The phase two covers characterization of the dispersions, and includes tests on agglomeration or aggregation, as well as on the formation of reactive species. The phase three covers the nano–bio interface, which needs to be assessed in a relevant biological medium. In addition to a comprehensive characterization of the test materials and the employed media, an appropriate dosimetry is a crucial prerequisite for a consistent quantification of the material to which an animal or cellular system is exposed and thus the establishment of realistic exposure scenarios (Oberdörster 2012).

16.4.7 Decision Support Tools

The risks of NMs are assessed on a case-by-case basis for each individual nanoform with its specific size, shape, surface chemistry, etc. (e.g. SCENIHR 2009; EFSA Scientific Committee 2011). These analytical procedures require a lot of experimental animals as well as time, effort, and money to obtain for each case the necessary physicochemical, exposure and hazard data for all relevant exposure scenarios and endpoints (Dekkers et al. 2016). In order to reduce the complicated analytical tools and techniques for the assessment of ecotoxicity and cytotoxicity, several initiatives have been taken to explore ways that enable a risk assessment of NMs using computational modeling. Dekkers et al. (2016) proposed a risk assessment strategy where approaches for grouping, read-across and quantitative structure–activity/toxicity relationships (QSAR/QSTR) modeling are integrated in prioritization of the NMs that may lead to high human health risks. However, for successful applicability of such new approaches it is crucial that sufficient good quality nano-specific information becomes available (OECD 2014; Tantra et al. 2015). The QSTR-perturbation model has been successfully employed to determine the ecotoxicity and cytotoxicity of NMs under different experimental conditions, including diverse measures of toxicities, multiple biological targets, compositions, sizes and conditions to measure those sizes, shapes, times during which the biological targets were exposed to NPs, and coating agents (Kleandrova et al. 2014).

The toxicity-based toxicokinetic/toxicodynamic (TBTK/TD) modeling is a robust mechanistic approach enabling integration of toxic effects on multiple endpoints over time (Chou et al. 2006; Tsai et al. 2009; Ashauer and Escher 2010). Toxicokinetics deal with the time course encompassing absorption, distribution, biotransformation, and elimination of toxicants by linking external exposure concentrations to biologically effective doses, whereas toxicodynamics describe processes leading from toxic actions to subsequent impairments in organisms. Therefore, TBTK/TD modeling provides a rigorously quantitative framework to understand the diverse nanotoxicological issues better (Yang et al. 2017).

The grouping and read-across is yet another approaches that can be utilized to meet these goals. A group represents a number of NMs which share a commonality relevant for risk, which can be one or more common properties in a physical, chemical, exposure, (eco)toxicological, toxicokinetics or fate sense. A NM can belong to more than one group. Read-across refers to predicting hazard information on one or more endpoints relevant for risk assessment of one NM or more NMs (target materials) by using data from another NM or more NMs (source materials).

These approaches are well-established for chemical substances, but a number of challenges prevent their successful implementation for NMs include physicochemical and structural properties, and transformation of NMs in a given environmental or biological media that may affect their interactions with organisms and cells, and thus modify their toxicity. Overcoming these challenges require a profound understanding of which (combinations of) the intrinsic and extrinsic structural and physicochemical characteristics of NMs dispersed in environmental or biological media or embedded in product matrices affect their release potential, uptake, biokinetic behaviour, toxicological modes of action or adverse outcome pathways. In order to develop such understanding, it is useful to generate data on the intrinsic and extrinsic properties derived from existing physicochemical analyses of pristine, released, aged and modified NMs in different biological and environmental media as well as the observed adverse effects based on (eco)toxicological studies.

Large number of modeling tools that is available currently can only provide screening level assessments. However, there is growing evidence that frameworks are being developed that are more quantitative, higher-tier models, capable of incorporating uncertainty into their analyses. Furthermore, there is a sense of developing validated protocols for identification and toxicity testing, reproducible across laboratories. The literature review has clearly indicated that there is no clear cut or well structured regulatory framework encompassing relevant tools for risk assessment for engineered nano-materials (ENMs), and the statement closely coincides with the observations of Hristozov et al. (2016) who have also suggested that no single existing frameworks for ENMs were able to fulfill all evaluation criteria.

16.5 Life Cycle Assessment

The Life Cycle Assessment (LCA) is the best developed and standardized methodology for assessing the environmental aspects and potential impacts throughout a product's life from raw materials and till it gets recycled in the ecosystem (i.e. cradle-to-grave). The environmental and resource impacts include climate change, stratospheric ozone depletion, toxicological stress on human health and ecosystem, the depletion of resources, water use and many others. The ISO-framework for LCA (ISO 14040:2006) has been found to be fully applicable for nanomaterials and nanoproducts, even if data regarding the elementary flows and impacts might be uncertain and scarce (Klöpffer et al. 2007). Indeed, a very limited number of research reports is available on LCA for nano-products. In addition to the qualitative environmental assessments of the different manufacturing methods (Steinfeldt et al. 2007; Sengül et al. 2008), quantified material and energy flow data exist for only a very small number of manufacturing processes and/or for individual nano-materials. It shows that studies have mainly focused on cradle-to-gate assessments. Cradle-to-gate is an assessment of a partial product life cycle; representing the stage from manufacture to its exit from the factory gate. The use phase and the after use phase (recycling, disposal) of the product are usually omitted (Meyer et al. 2009). For both of these phases, there is almost no data investigating their environmental impact.

Gallagher et al. (2017) reported that the upstream and downstream environmental impacts of NMs are increasingly realized, and more challenges arise when it is applied to upstream impacts from nanoparticle synthesis. Thus, it seems crucial to identify the alterations of an NM during its different life cycle stages as these may change the toxicological properties of a material (Laux et al. 2018).

Overall, RA and LCA are the two analytic perspectives on environmental assessment for formulating an effective environmental policy and decision making. RA focuses on management of a toxicological hazard in a specific exposure scenario, while LCA seeks a holistic estimation of impacts of thousands of substances across multiple media, including non-toxicological and non-chemically deleterious effects. Although recommendations to integrate the two approaches have been continuing for the past 15 years there are several practical obstacles. Because the pace of nanomaterials innovation far surpasses the speed of acquisition of environmentally relevant data. Therefore, it is clear that a further integration of RA and LCA based on dataset completion on nano-products will remain ineffectual. Thus, experts suggest to apply both the methods in parallel and integrate only after obtaining separate results for providing clear guidance to decision makers (Linkov et al. 2017).

16.6 Standard Protocols for Assessing Ecotoxicity in Agri-Food Systems

The biosafety concerns of NPs can be effectively addressed by employing appropriate assays to assess nanoparticle toxicity. In general, nanotoxicity studies employ in vitro, in vivo tests, and in silico models to assess toxicity of chemicals and underlying mechanisms (e.g. oxidative stress, immunotoxicity, genotoxicity) (Dusinska et al. 2009). Three major categories of in vitro assays namely cytotoxic, genotoxic and alterations in gene expression assays are widely used to evaluate the toxicity of NMs (Rajkishore et al. 2013). Some of these methods are already developed into OECD guidelines, and can be used for regulatory purposes (Dusinska et al. 2012). In 2006, the OECD Working Party on Manufactured Nanomaterials (WPMN) was established to provide a global forum for nano-safety deliberations to assess the strengths of existing test guidelines (TG) besides evolving or improving the TG

(Rasmussen et al. 2016). The outcome of 10 years of OECD's work is that it covered the testing of eleven nanomaterials with 59 endpoints to address relevant physico-chemical properties, mammalian and environmental toxicity, environmental fate and material safety. In 2013, based on outcomes of the WPMN testing programme in consultation with test results outside the WPMN, the OECD concluded that the approaches for the testing and assessment of traditional chemicals are in general appropriate for assessing the safety of nanomaterials (OECD 2009), but may have to be adapted to the specificities of nanomaterials. It also highlighted that the hazard data collected using such test guidelines will fall under OECD's system of Mutual Acceptance of Data (MAD) which is a legally binding instrument to facilitate the international acceptance of information for the regulatory assessment. But for legal validity, it is mandatory that OECD Test Guidelines and the OECD Principles of Good Laboratory Practice have been applied during the collection of the data.

A total of 59 endpoints were identified by WPMN of OECD for the testing of NMs and grouped into 6 domains namely (1) nanomaterial information/identification, (2) physical-chemical properties and material characterization, (3) environmental fate, (4) environmental toxicology, (5) mammalian toxicology, and (6) material safety (Rasmussen et al. 2016). Among the six, physical-chemical data are very unique for characterizing nanomaterials. In addition, it is generally agreed that some TG needs to be optimized, and few endpoints are needed for new nano-specific test guidelines.

The globally recognized biosafety tests include cytotoxicity, genotoxicity and gene expression assays. These test protocols are used as part of regulatory guidelines. Cell lines of different origins are used to predict acute cytotoxicity. The classic cytotoxic assays are tryphan blue exclusion, lactate dehydrogenase (LDH), tetrazolium assays (MTT, XTT, WST) and alamar blue assay (Dusinska et al. 2009). These are based on colorimetric assays involving optical activity of organic dyes. But they are prone to interference due to interaction between the dye and NPs, and wash-out can be difficult due to interactions. The 2D and 3D systems with clonogenic or plating efficiency assay (adapted from OECD test guideline 476) are considered the most reliable assays for cytotoxicity testing of NMs.

Genotoxicity assays have different endpoints, such as single- and double-strand breaks, point mutations, deletions, chromosomal aberrations, micronuclei formation, DNA repair and cell cycle interactions. The most commonly applied methods for detecting genotoxicity include the bacterial Ames test, DNA strand break measurements in cells (e.g. comet assay, alkaline unwinding and hydroxyapatite chromatography, alkaline elution), and cytogenetic assays (micronucleus and chromosomal aberration assays, including the use of fluorescence in situ hybridization and chromosome painting). According to Dusinska et al. (2012), for regulatory genotoxicity/mutagenicity testing of chemicals, three different in vitro tests with two to three different end points are initially chosen. If two out of three tests are positive, the compound is considered genotoxic, and is also likely to be carcinogenic. If two tests are negative, then the substance is regarded as "non-genotoxic". Demonstrated genotoxic potential in vitro should be confirmed in vivo (often applying the MN assay and DNA repair assays to liver samples).

The Ames test (bacterial reverse mutation assay) is the most commonly used assay which is based on induction of reverse mutation in the histidine gene that enables the bacteria to synthesize histidine and form visible colonies in minimal histidine medium. There are a number of concerns regarding the suitability of the Ames test for NP testing due to the bacterial cell wall composition and the fact that larger NPs cannot cross the cell membrane. It is also unknown how NMs could possibly interfere with histidine synthesis and induce false-negative (downregulation) or positive (upregulation) results.

The comet assay or single cell gel electrophoresis is one of the most common tests for exploring the genotoxic potential of NPs. It detects DNA damage at the level of a single cell. DNA breaks cause relaxation of supercoiled DNA which then extends to form comet like tails under electrophoresis. The relative intensity of DNA in the tail can be quantified, and indicates the DNA break frequency. This method represents a rapid, sensitive, reliable, robust and relatively inexpensive way to study DNA damage (including DNA oxidation), and repair in different cell types both in vitro as well as in vivo. The micronucleus (MN) assay can be applied both in vitro and in vivo, and has been recommended for assessing mutagenic and clastogenic effects of pollutants. The MN assay can easily be applied in different target cells and tissues, and has been used also in nanotoxicology.

In general, for genotoxicity testing, the in vitro and in vivo micronucleus assays, the in vitro mammalian gene mutation assays, and the in vitro and in vivo Comet assays are promising candidates for hazard identification of nano-materials (Gonzalez et al. 2011; Doak et al. 2012; Pfuhler et al. 2013). A review on nanogenotoxicity shows that the most frequently used genotoxicity test is the comet assay, followed by the MN test, chromosome aberrations and the Ames test (Magdolenova et al. 2014). Some of these assays have been applied in both 2D and 3D culture systems. However, the results from the Ames test were almost always negative, indicating that the method may not be suitable for NMs testing. Overall, the most promising assays for nanogenotoxicological testing in vitro seem to be the MN and comet assays, but identifying the most sensitive cell types for these assays remains to be established for most NMs. In addition to validated in vitro test methods, ex vivo and in silico approaches, validated in vivo tests are highly warranted in the field of NMs.

Human toxicity can be assessed in mammalian in vitro assays with primary cell culture or transformed cell lines of different origin (e.g. human, monkey, mouse, hamster or rat). The widespread use of cell lines for toxicity testing reflects the fact that they provide a more standardized system compared to primary cell cultures, as well as higher throughput (Dusinska et al. 2012). Furthermore, an understanding of the transgenerational and trophic chain transfer effects of nanoparticles applications must be included to gain comprehensive knowledge of nanotoxicity. Interestingly, selection of permissible level together with studying transgenerational and trophic chain transfer effects could provide adequate safety assessments (Mishra et al. 2017).

Overall, it is inferred that more in vivo-like cell culture systems combined with validated high throughput screening and in silico data handling will help us to understand the insights of nanotoxicity and ways to combat risks (Dusinska et al. 2012). Hence, it should be kept in mind that regulatory authorities will require reliable routine methods for any monitoring in compliance with the specification of the respective NM. Regarding safety studies with NMs used in food, members of the EFSA Nano Network note that unrealistically high dosing can lead to outcomes that may not be related to the inherent toxicity of the material but to the high amounts administered. Therefore, attention should be paid to the application of physiologically relevant test doses (Laux et al. 2018).

16.7 Current Status of Regulatory Frameworks for Agri-Food Systems

Reviewing the regulatory frameworks across the spectrum of countries suggests that currently the European Union along with Switzerland is the only region in the globe that has specific provisions of "nano" in the legislation (Amenta et al. 2015). The European Commission regulation (No. 1169/201) insists on mandatory labeling for the content of nano-materials. Considering the biosafety concerns of nano-materials, in the year 2004, European Union has taken initiatives to develop an exclusive regulatory policy for nano-materials (Justo-Hanania and Dayan 2014). The FDA in US is yet to adopt a regulatory definition for nano-materials. But, they issued specific guidance for food related applications of nano-materials/nanotechnology. Health Canada and Canadian Food Inspection Agency (CFIA) have jointly stipulated the guidelines for the use of nano-materials while the Public Health Agency of Canada (PHAC) is made responsible for food regulation in Canada. No specific regulation for nano-products is available but such products are regulated under the existing legislative and regulatory frameworks. In the case of India, nano-specific provision is not yet included in the Food Safety and Standards Act (2006).

The literature review unequivocally demonstrated that the nano-regulatory frameworks available at present are digitally divided, and still lacks clarity and inconsistency in defining the term 'size of the nano-materials' (Boverhofet al. 2015). The definition of an NM is based on different criteria, according to its foreseen purpose in industry, research, or regulation. According to the International Organization for Standardization (ISO) "nanotechnology" is the "application of scientific knowledge to manipulate and control matter predominantly in the nanoscale (i.e., approximately from 1 to 100 nm) to make use of size- and structure-dependent properties and phenomena distinct from those associated with individual atoms or molecules, or extrapolation from larger sizes of the same material". The ISO also classifies and categorizes NMs systematically. It defines an NM as material with any external or internal structures or surface structures in the nanoscale (ISO 2015). Furthermore, the European Commission (EC) has recommended a definition of the term "nanomaterial" specifically to be used for regulatory purposes (EC 2011). According to this recommendation, a material, regardless of its origin (i.e., natural, incidental or manufactured), is considered being an NM if it contains 50% or more of unbound, agglomerated or aggregated particles with one or more external dimensions in the size range between 1 and 100 nm. As an exemption from this rule, for fullerenes, graphene flakes and single-wall carbon nanotubes the lower size limit remains ineffectual, while the upper limit of 100 nm still applies. Furthermore, the 50% threshold referring to the number of particles with a size between 1 and 100 nm may be lowered down to 1% in specific cases and whenever warranted due to environmental, health, and safety concerns. Even more so, as NMs suffer from the tendency to agglomerate in the atmosphere or in liquid media, a feature that often hinders their identification by particle analysis techniques. Agglomeration may also lead to the loss of nanoscale-mediated properties such as increased surface area (and reactivity), translucency and specific particle motion behavior while being in dispersion (Laux et al. 2018). Thus, the first step in risk assessment is to determine whether the substance of interest falls under the definition of nano-material. Subsequently, the risk assessment is determined to know whether the existing safety data is sufficient enough to foresee potential health impacts. Our literature survey highlights the need for a common regulatory definition for nano-material that is recognized globally by regulatory agencies/bodies. The size variability in accordance to the country's perspectives and respective regulatory bodies is summarized in Table 16.1. Amenta et al. (2015) and Adholeya et al. (2017) have carried out a comparative analysis of current regulatory frameworks for agri-food systems and some of the highlights are tabulated (Table 16.2) for ready reference.

The "framework" is a conceptual paradigm of how risk assessment should be understood and performed. Our literature survey in concomitant with others' viewpoints, few criteria namely nano-specific requirements, lifecycle thinking, pre-assessment phase, exposure driven approach, iterative and adaptive structure, documented applications and transparency for developing a structure for exercising the process of regulation in a fool proof manner have been identified (Hristozova et al. 2016). Among these criteria, nano-specific requirements are one of the most priority areas for research and development. This is attributed to the fact that novel properties of nano-scale particles demand for a new testing methods and protocols for assessment of hazard. In recent times, this seems to be the most debatable topic in nanoscience for effective risk assessment. One decade of in-depth research by the working group of OECD has led to a preliminary conclusion that current approaches for the testing and assessment of traditional chemicals are appropriate for assessing the safety of nano-materials. However, OECD acknowledged that certain test methods may need to be adapted to allow for appropriate compatibility with nanomaterials. Recent data suggest that the number of publications relating to nanotoxicity has increased exponentially from dozens in 2005 and >3700 in 2017 (Shvedova et al. 2018). This clearly demonstrates the significance of specific set of characterization that is unique for assessing the properties of materials for its toxicity at nano-scale. Hence, our analysis indicates the need for appropriate test methods for nano-materials that are harmonized at the international level by regulatory institutions.

Table 16.1 Comparison of NM definition in current regulatory frameworks in selected countries available for food sector

Country	Regulatory body/responsible organization	Definition of nanomaterial			Relevant legislation
		Size	Distribution threshold	Nanoscale dimension	
European Union	European Commission (European Chemical agency)	1–100 nm	50% by number	At least one	EC No. 1907/2006; Regulation, Evaluation and Authorization of Chemicals (REACH) for plant protection products and food contact materials
	Specific to agri sector European Parliament and the Council of the European Union on the Provision of Food Information to Consumers	1–100 nm and extend[a] up to 1 μm	Not mentioned	At least one	Regulation on the Provision of Food Information to Consumers (1169/2011) EC Regulation No 258/97 concerning "novel foods" and "novel food ingredients"
United States	United States Food and Drug Administration	1–100 nm and extends[a] up to 1 μm	Not mentioned	At least one	Federal Food, Drug, and Cosmetic Act (FFDCA)
Canada	Health Canada	1–100 nm and larger[a]	Not mentioned	At least one	Food and Drugs Act

[a]Larger size particles (up to 1 μm are considered as nanomaterials if it exhibits nanoscale properties)

Table 16.2 Current global status for regulation of nanoproducts in agri-food systems

Country	Regulatory body/responsible organization	Legislation	Provisions in the available legislation	Remarks
USA	Food and Drug Administration (FDA) & US-Environmental Protection Agency (US-EPA)	Federal Food, Drug and Cosmetic Act (FFDCA) Federal Insecticide, Fungicide and Rodenticide Act (FIFRA)	No specifications on nanoproducts in FFDCA or FIFRA	FDA issued a specific guidance that recommends preliminary safety assessments for food/food contact substances produced at nanoscale FDA yet to issue a regulatory definition of NMs
Canada	Canadian Food Inspection Agency (CFIA) and Public Health Agency of Canada (PHAC)	Food and Drugs Act	Nanoproducts are regulated under existing legislative	No nano specific regulatory framework
EU	European commission (EC), European Parliament and Council	1. Regulation on the provision of food information to consumers (1169/2011) 2. Regulation on plastic food contact materials and articles (10/2011) 3. Regulation on active and intelligent materials and articles (450/2009) 4. The biocidal products regulation (528/2012) 5. The cosmetic products regulation (1223/2009) 6. Registration, Evaluation, Authorization and Restriction of Chemicals (REACH) regulation (1907/2006)	These legislations have nano specific provisions. But no exclusive NM legislation	Taking a lead step in regulating nanoproducts

(continued)

Table 16.2 (continued)

Country	Regulatory body/responsible organization	Legislation	Provisions in the available legislation	Remarks
Switzerland	Federal Office for the Environment (FOEN) Swiss Federal office of public health (FOPH)		Existing regulations are ensured for safety of NMs	Information on physico-chemical parameters of NMs is mandatory while application submission
Australia and New Zealand	Food standards Australia New Zealand (FSANZ)	Australian New Zealand Food Standards Code	Amended FSANZ Application Handbook to support new food regulations to manage risks from nanoproducts	First regulatory program for "industrial nanomaterials was introduced by" National Industrial Chemicals Notification and Assessment Scheme (NICNAS) Tasmanian Institute of Agriculture has published a comprehensive review on Nanotechnology in Agricultural Products Logistics Management
China	Ministry of Agriculture Ministry of Health	Food Safety Law of China	National Centre for Nanoscience and Technology (NCNST) and the commission on nanotechnology standardization	Nanoproducts in agri-food systems is not yet approved by the authority
South Korea	Ministry of Food and Drug safety (MFDS) Korean Food and Drug Administration (KFDA) Korean Agency for Technology and Science (KATS)	National Nano-Safety Strategic Plan (2012/2016) and Food Sanitation Act	No NM specifications	Ministry of Knowledge and Economy (MKE) and the Korean Agency for Technology and Standards published Guidance on safety management of nano-based products

(continued)

Table 16.2 (continued)

Country	Regulatory body/responsible organization	Legislation	Provisions in the available legislation	Remarks
Japan	Ministry of Health, Labour and Welfare	Food Sanitation Law	No NM specific legislation	Basic survey on safety of nanoproducts was done
Iran	Iran Nanotechnology Initiative Council (INIC) Nanotechnology Committee of Food and Drug Organization			Food and Drug Organization (FDO) constituted guidelines for nanoproducts in food, beverages, pharmaceutical, medical equipment but agriculture is not yet included
Taiwan	Taiwan Nanotechnology Industry Development Association (TANIDA)		Nanoproducts are certified	TANIDA established nanoMark system to certify the nanoproducts
Thailand	Food & Drug Administration of the Ministry of Public Health			NanoQ label has been introduced for nano products (but not for agri/food) that are certified by the Nanotechnology Association of Thailand
India	Food Safety and Standards Authority of India	Food Safety and Standards Act (2006)		DST Nano mission has come up with safe handling of NMs in workplace and industry. But no specific legislation/regulation for agriculture

16.8 Strategies for Fool-Proof Regulatory Framework

We propose a conceptual nano-regulatory framework for agricultural sector in India which includes the above listed criteria for effective risk assessment and management (Table 16.3). The regulatory framework includes three phases namely pre-assessment, exposure assessment and risk characterization. In the first phase, the endpoints include nano-specific requirement (NM of interest should fulfill the globally recognized regulatory definition of nano-material), physicochemical characterization and risk categorization (low to high). Since risks of NMs varies case-by-case and cost intensive, based on toxicological studies, risk need to be categorized from low to high (e.g., nanobiosensors for plant diagnosis may be included as low risk and agri-inputs like nano-fertilizers and agrochemicals that are added in large quantity in the environment may be included under high risk category). Sub-category includes low risk (for natural materials e.g., clay, zeolite) and high risk (e.g. artificially synthesized e.g., active ingredients). The second phase is exposure assessment that considers possible routes of NM exposure (e.g. soil, water etc.). In the phase II, the endpoints are exposure driven approach and life cycle thinking. This exercise will certainly help to exclude irrelevant exposure routes so as to focus target areas and reduce cost and energy. NMs are likely to change during their lifetime, and hence characterization will take in account of cradle-to-grave approach. The last phase is the risk characterization which includes testing the biosafety of NMs using recognized protocols that are harmonized globally, iterative and adaptive, documented applications and transparency. Iterative and adaptive structure will allow for rapid inclusion of new information on safety in the decision process. Documented applications will understand the strength and limitations of the framework. The most important endpoint is transparency which considers the interests of stakeholders besides gaining public confidence well before commercialization.

With respect to India's progress in nanotechnology regulation, it is a fact that a nano-regulatory task force of DST Nano Mission has come up with guidelines for safe handling of nano-materials in laboratory and industries. This document is

Table 16.3 Proposed conceptual nano-regulatory framework for agricultural sector in India

Phases of framework	Endpoints
Pre-assessment	1. Nano-specific requirement (to confirm with regulatory definition) 2. Physicochemical characterization 3. Risk categorization (low to high)
Exposure assessment	1. Exposure driven approach 2. Life cycle thinking
Risk characterization	1. Testing the biosafety using recognized protocols that are harmonized globally 2. Iterative and adaptive 3. Documented applications 4. Transparency

more of advisory in nature, and has not attempted to provide regulatory definition for the term nano-material itself. However, the taskforce has generously indicated that the EU's comprehensive regulatory framework, including **REACH (Registration, Evaluation, Authorisation and Restriction of Chemicals)**, can be implemented and where necessary adapted to effectively regulate nano-materials. On the other hand, India's rapid development in the sphere of nanotechnology warrants the establishment of a national nanotechnology regulatory authority to address risks associated with this emerging science. Such regulatory authority will focus on testing methods and protocols for ensuring biosafety, workplace and occupational health safety, consumer product safety and end-of-life disposal issues.

16.9 Conclusion and Future Perspectives

A robust risk assessment comprising of hazard and exposure assessment, dose-response relationship and risk characterization is pivotal for a foolproof regulatory framework to manage the risks associated with nanotechnology. Best use of appropriate tools for risk assessment is very much essential to predict the risks with utmost scientific caliber. The other approach namely life cycling thinking for nanoproducts is also needed for better decision making to regulate the nano-products in the market. Though the database is over pouring with more than 3500 publications on nanotoxicity, still scientific uncertainty on the adverse effects of NMs on biological systems prevails in the public domain. Some of the knowledge gaps in nanotoxicity and nanosafety need to addressed at the earliest in coordination with the validated test guidelines of OECD. Such exercises will favour for a strong data set that is required for strengthening a comprehensive regulatory framework. Currently, there is clearly a lack of emphasis on risk related research in nanotechnology. Therefore, it is crucial to examine and estimate the risk for regulating the production, use, consumption and disposal of these materials (TERI 2010). On the other hand, other experts argue that research related to toxicity, and risk assessment must be undertaken once the applicability of specific nano-applications are ascertained, especially when the prototype of the product has been developed and is available for field testing. Furthermore, the main challenges faced by regulatory institutions currently relate to the regulatory capacity, information asymmetry and absence of interagency coordination.

As of now, our literature review suggests that the European Union is taking a lead step in formulating a nano-regulatory framework. Since the use of nano-materials in agriculture is very restricted, it is plausible to adopt the EU framework for the moment, and begin to work and evolve globally competitive nano-framework for agricultural sector for better risk assessment and environmental safety.

References

Adholeya A, Das RK, Dubey MK, Kochar M, Singh R (2017) Zero draft policy on regulation of nanoproducts in agriculture. The Energy and Resources Institute (TERI), New Delhi

Amenta V, Aschberger K, Arenaa M, Bouwmeester H, Moniz FB, Brandhoff P, Gottardo S, Marvin HJP, Mech A, Pesudo Q, Rauscher H, Schoonjans R, Vettori MV, Weigel S, Peters RJ (2015) Regulatory aspects of nanotechnology in the agri feed food sector in EU and non-EU countries. Regul Toxicol Pharmacol 73:463–476

Anusya P, Nagaraj R, Janavi GJ, Subramanian KS, Paliyath G, Subramanian J (2016) Pre-harvest sprays of hexanal formulation for extending retention and shelf-life of mango (*Mangifera indica* L.) fruits. Sci Hort 210:221–230

Arts JHE, Hadi M, Irfan AM, Keene AM, Kreiling R, Lyon D, Maier M, Michel K, Petry T, Sauer UG, Warheit D, Weinch K, Wohlleben W, Landseidel R (2014) A critical appraisal of existing concepts for the grouping of nanomaterials. RegulToxicol Pharmacol 70:492–506

Ashauer R, Escher BI (2010) Advantages of toxicokinetic and toxicodynamic modelling in aquatic ecotoxicology and risk assessment. J Environ Monit 12(11):2056–2061

Berekaa MM (2015) Nanotechnology in food Industry; Advances in food processing, packaging and food safety. Int J Curr Microbiol App Sci 4:345–357

Bhushani JA, Anandharamakrishnan C (2014) Electrospinning and electrospraying techniques: potential food-based applications. Trends Food Sci Technol 38:21–33

Boverhof DR, Bramante CM, Butala JH, Clancy SF, Lafranconi M, West J, Gordon SC (2015) Comparative assessment of nanomaterial definitions and safety evaluation considerations. Regul Toxicol Pharmacol 73:137–150

Cassee FR, Muijser H, Duistermaat E, Freijer JJ, Geerse KB, Marijnissen JC, Arts JH (2002) Particle size-dependent total mass deposition in lungs determines inhalation toxicity of cadmium chloride aerosols in rats. Application of a multiple path dosimetry model. Arch Toxicol 76(5–6):277–286

Chou BYH, Liao CM, Lin MC, Cheng HH (2006) Toxicokinetics/toxicodynamics of arsenic for farmed juvenile milkfish *Chanoschanos* and human consumption risk in BFD-endemic area of Taiwan. Environ Int 32(4):545–553

Dekkers S, Oomen AG, Bleeker EAJ, Vandebriel RJ, Micheletti C, Cabellos J, Janer G, Fuentes N, Azquez-Campos SV, Borges T, Silva MJ, Prina-Mello A, Movia D, Nesslany F, Ribeiro AR, Leite PE, Groenewold M, Cassee FR, Sips AJAM, Dijkzeul A, Teunenbroek TV, Wijnhoven SWP (2016) Towards a nanospecific approach for risk assessment. Regul Toxicol Pharmacol 80:46–59

DeLoid GM, Cohen JM, Pyrgiotakis G (2015) Advanced computational modeling for in vitro nanomaterial dosimetry. Part Fibre Toxicol 12989-015-0109-1

Demokritou P, Gass S, Pyrgiotakis G, Cohen JM, Goldsmith W, McKinney W, Frazer D, Ma J, Schwegler- Berry D, Brain J, Castranova V (2013) An in vivo and in vitro toxicological characterisation of realistic nanoscale CeO_2 inhalation exposures. Nanotoxicology 7(8):1338–1350

Dharanivasan G, Sithanantham S, Kannan M, Chitra S, Kathiravan K, Janarthanan S (2017) Metal Oxide Nanoparticles Assisted Controlled Release of Synthetic Insect Attractant for Effective and Sustainable Trapping of Fruit Flies. J Clust Sci 28(4):2167–2183

Doak SH, Griffiths SM, Manshian B, Singh N, Williams PM, Browns AP, Jenkins GJ (2009) Confounding experimental considerations in nanogenotoxicology. Mutagenesis 24(4):285–293

Doak SH, Manshian B, Jenkins GJS, Singh JN (2012) In vitro genotoxicity testing strategy for nanomaterials and the adaptation of current OECD guidelines. Mutat Res 745(1–2):104–111

DST Dept of Science and Technology (2016) Guidelines and best practices for safe handling of nano-materials in research laboratories and industries. Centre for Knowledge Management of Nanoscience & Technology, Department of Science & Technology, Govt. of India, pp 1–24

Dusinska M, Fjellsbo L, Magdolenova Z (2009) Testing strategies for the safety of nanoparticles used in medical applications. Nanomedicine (Lond) 4:605–607

Dusinska M, Pran ER, Carreira SC, Saunders M (2012) Critical evaluation of toxicity tests. In: Fadeel B, Pietroiusti A, Shvedova AA (eds) Adverse effects of engineered nanomaterials. Engineered

nanomaterials: hazard, exposure and safety assessment. Academic press, Imprint of Elsevier, pp 6–63
EC (2006) Regulation (EC) No 1907/2006 of the European Parliament and the Council of 18 December 2006 concerning the Registration, Evaluation, Authorisation and Restriction of Chemicals (REACH), establishing a European Chemicals Agency, amending Directive 1999/45/EC and repealing Council Regulation (EEC) No 793/93 and Commission Regulation (EC) No 1488/94 as well as Council Directive 76/769/EEC and Commission Directives 91/155/EEC, 93/67/EEC, 93/105/EC and 2000/21/EC. Off J Eur Union L396:1–525
EC (2011) Commission recommendation of 18 October 2011 on the definition of nanomaterial (Text with EEA relevance) (2011/696/EU). Off J Eur Union L275:38–40
EC (2012) Communication from the Commission to the European Parliament, the Council and the European Economic and Social Committee e Second Regulatory
EC (2013) Commission delegated Regulation (EU) No 1363/2013 of 12 December 2013 amending Regulation (EU) No 1169/2011 of the European Parliament and of the Council on the provision of food information to consumers as regards the definition of engineered nanomaterials. Off J Eur Union L343:26–28
ECHA (2012) Guidance on Information Requirements and Chemical Safety Assessment e Appendix R7-1 Recommendations for Nanomaterials Applicable to Chapter R7a Endpoint Specific Guidance. Guidance for the Implementation of REACH. European Chemicals Agency (ECHA), Helsinki, Finland
EFSA Scientific Committee (2011) Scientific opinion on guidance on the risk assessment of the application of nanoscience and nanotechnologies in the food and feed chain. EFSA J. 9, 2140
Eleftheriadou M, Pyrgiotakis G, Demokritou P (2017) Nanotechnology to the rescue using nano-enabled approaches in microbiological food safety and quality. Curr Opin Biotechnol 44:87–93
Epp A (2015) Nanoview—perception of nanotechnology by the German population and media coverage. In: 1st joint symposium on nanotechnology. BfR. http://www.bfr.bund.de/cm/349/nanoviewperception-of-nanotechnology-by-the-german-population-andmedia-coverage.pdf
Gallagher MJ, Allen A, Buchman JT, Qiu TA, Clement PL, Krause MOP, Gilbertson LM (2017) Research highlights: applications of life-cycle assessment as a tool for characterizing environmental impacts of engineered nanomaterials. Environ Sci-Nano 4:276–281
Ghaani M, Cozzolino CA, Castelli G, Farris S (2016a) An overview of the intelligent packaging technologies in the food sector. Trends Food Sci Technol 51:1–11
Ghaani M, Nasirizadeh N, Ardakani SAY, Mehrjardi FZ, Scampicchio M, Farris S (2016b) Development of an electrochemical nanosensor for the determination of gallic acid in food. Anal Methods 8:1103–1110
Gonzalez L, Sanderson BJS, Kirsch-Volders M (2011) Adaptations of the in vitro micronucleus assay for the genotoxicity assessment of nanomaterials. Mutagenesis 26(1):185–191
Haas KH (2013) Industrial relevant production processes for nanomaterials and nanostructures. Safety aspects of engineered nanomaterials. Pan Standford Publishing Pte. pp 30–66
Hristozova D, Gottardo S, Semenzin E, Oomen A, Bos P, Peijnenburg W, Tongeren MV, Nowack B, Hunt N, Brunelli A, Scott FJJ, Tran L, Marcomini A (2016) Environ Int 95:36–53
Hund-Rinke K, Schlich K (2014) The potential benefits and limitations of different test procedures to determine the effects of Ag nanomaterials and $AgNO_3$ on microbial nitrogen transformation in soil. Environ Sci Eur 26(1):28
ISO (2015) ISO/TS 80004-1:2015 nanotechnologies—vocabulary—Part 1: core terms ISO/TS 80004-1:2015 nanotechnologies—vocabulary—Part 1: core terms. Geneva Switzerland
Jeon CS (2016) Surface functionalization of bioanalytical applications: virus decorated gold microshells and modified synaptic cell adhesion molecules. PhD dissertation, Seoul National University
Jincy M, Djanaguiraman M, Jeyakumar P, Subramanian KS, Jayasankar S, Paliyath G (2017) Inhibition of phospholipase D enzyme activity through hexanal leads to delayed mango (*Mangiferaindica* L.) fruit ripening through changes in oxidants andantioxidant enzymes activity. SciHort 218:316–332

Justo HR, Dayan T (2014) European risk governance of nanotechnology: explaining the emerging regulatory policy. Res Pol 44:1527–1530

Katouzian I, Jafari SM (2016) Nano-encapsulation as a promising approach for targeted delivery and controlled release of vitamins. Trends Food Sci Technol 53:34–48

Kleandrova VV, Luan F, González-Díaz H, Ruso JM, Speck-Planche A, Natália M, Cordeiro DS (2014) Computational tool for risk assessment of nanomaterials: novel QSTR-perturbation model for simultaneous prediction of ecotoxicity and cytotoxicity of uncoated and coated nanoparticles under multiple experimental conditions. Environ Sci Technol 48:86–94

Klöpffer W, Curran MA, Frankl P, Heijungs R, Köhler A, Olsen SI (2007) Nanotechnology and life cycle assessment. A systems approach to nanotechnology and the environment: synthesis of results obtained at a workshop Washington, DC 2–3 October 2006. European Commission, DG Research, jointly with the Woodrow Wilson International Center for Scholars

Kumar LY (2015) Role and adverse effects of nanomaterials in food technology. J Toxicol Health 2:2

Laux P, Tentschert J, Riebeling C, Braeuning A, Creutzenberg O, Epp A, Fessard V, Haas K, Haase A, Hund-Rinke K, Jakubowski N, Kearns P, Lampen A, Rauscher H, Schoonjans R, Stormer A, Theilmann A, Muhle U, Luch A (2018) Nanomaterials: certain aspects of application, risk assessment and risk communication. Arch Toxicol 92(1):121–141

Linkov I, TrumpBD WenderBA, SeagerTP Kennedy AJ, Keisler JM (2017) Integrate lifecycle assessment and risk analysis results, not methods. Nature Nanotechnol 12:740–743

Magdolenova Z, Collins A, Kumar A, Dhawan A, Stone V, Dusinska M (2014) Mechanisms of genotoxicity: a review of in vitroand in vivo studies with engineered nanoparticles. Nanotoxicology 8(3):233–278

Martirosyan A, Schneider YJ (2014) Engineered nanomaterials in food: implications for food safety and consumer health. Int J Environ Res Pub Health 11:5720–5750

Meyer DE, Curran MA, Gonzalez M (2009) An examination of existing data for the industrial manufacture and use of nanocomponents and their role in the life cycle impact of nanoproducts. Environ Sci Technol 43(5):1256–1263

Mishra S, Keswani C, Abhilash PC, Fraceto LF, Singh HB (2017) Integrated approach of agri-nanotechnology: challenges and future trends front. Plant Sci 8:471. https://doi.org/10.3389/fpls.2017.00471

Oberdörster G (2012) Nanotoxicology: in vitro-in vivo dosimetry. Environ Health Perspect 120(1):A13. https://doi.org/10.1289/ehp.1104320

OECD (2009) Preliminary review of OECD test guidelines for their applicability to manufactured nanomaterials, ENV/JM/MONO(2009)21

OECD (2014) Report of the OECD expert meeting on the physical chemical properties of manufactured nanomaterials and test guidelines. Series on safety of manufactured nano-materials, 41, ENV/JM/MONO(2014)15

OECD (2016) Physical-chemical parameters: measurements and methods relevant for the regulation of nano-materials.OECD workshop report. Series on safety of manufactured nano-materials, 63, ENV/JM/MONO(2016)2

Pfuhler S, Elespuru R, Aardema MJ, Doak SH, Donner EM, Honma M, Kirsch-Volders M, Landsiedel R, Manjanatha M, Singer T, Kim JH (2013) Genotoxicity of nanomaterials: refining strategies and tests for hazard identification. Environ Mol Mutagen 54:229–239

Rajkishore SK, Subramanian KS, Natarajan N, Gunasekaran K (2013) Nanotoxicity at various trophic levels: a review. The Bioscan 8(3):975–982

Rasmussen K, Alez MG, Kearns P, Sintes JR, Rossi F, Sayre P (2016) Review of achievements of the OECD working party on manufactured nanomaterials' testing and assessment programme. From exploratory testing to test guidelines. Regul Toxicol Pharmacol 74:147–160

Rossi M, Cubadda F, Dini L, Terranova ML, Aureli F, Sorbo A, Passeri D (2014) Scientific basis of nanotechnology, implications for the food sector and future trends. Trends Food Sci Technol 40:127–148. https://doi.org/10.1016/j.tifs.2014.09.004

Sabourin V, Ayande A (2015) Commercial opportunities and market demand for nanotechnologies in agribusiness sector. J Technol Manag Innov 10:40–51

SCENIHR (2009) Risk assessment of products of nanotechnologies, 19 January

Sengül H, Theis TL, Ghosh S (2008) Toward sustainable nanoproducts: an overview of nanomanufacturing methods. J Indus Ecol 12(3):329–359

Sheehan B, Murphy IDF, Mullins M, Furxhi I, Costa AL, Simeone FC, Mantecca P (2018) Hazard screening methods for nanomaterials: a comparative study. Int J Mol Sci 19:649. https://doi.org/10.3390/ijms19030649

Shvedova A, Pietroiusti A, Kagan V (2018) Nanotoxicology ten years later: lights and shadows. Toxicol Appl Pharmacol 15(299):1–2

Sodano V, Verneau F (2014) Competition policy and food sector in the European Union. J Int Food Agribus Mark 26:155–172

Steinfeldt M, von Gleich A, Petschow U, Haum R (2007) Nanotechnologies, hazards and resource efficiency. Springer, Heidelberg

Subramanian KS, Manikandan A, Thirunavukkarasu M, Rahale CS (2015) In: Rai M et al (eds) Nanotechnologies in food and agriculture. Springer International Publishing, Switzerland, pp 69–80

Subramanian KS, Raja K, Marimuthu S (2018) Multifarious applications of nanotechnology for enhanced productivity in agriculture. In: Singh HB, Mishra S, Fraceto LF, de Lima R (eds) Emerging trends in agri-nanotechnology: fundamentals and applied aspects. CAB International, pp 56–77

Subramanian KS, Tarafdar JC (2009) Nanotechnology in soil science. In: Proceedings of the Indian society of soil science-platinum jubilee celebration, December 22–25, IARI Campus, New Delhi, pp 199

Tantra R, Bouwmeester H, Bolea E, Rey-Castro C, David CA, Dogné, JM, Jarman J, Laborda F, Laloy J, Robinson KN, Undas AK, van der Zande M (2015) Suitability of analytical methods to measure solubility for the purpose of nanoregulation. Nanotoxicology 22:1–12

TERI (2010) Nanotechnology development in India: building capability and governing the technology. The Energy and Research Institute (TERI) Briefing Paper, supported by IDRC, Canada

Thakur S, Thakur S, Kumar R (2018) Bio-nanotechnology and its role in agriculture and food industry. J Mol Genet Med 12:324. https://doi.org/10.4172/1747-0862.1000324

Tsai JW, Chen WY, Ju YR, Liao CM (2009) Bioavailability links mode of action can improve the long-term field risk assessment for tilapia exposed to arsenic. Environ Int 35(4):727–736

Tsang MP, Hristozov D, Zabeo A, Koivisto AJ, Jensen ACO, Jensen KA, Pang C, Marcomini A, Sonnemann G (2017) Probabilistic risk assessment of emerging materials: case study of titanium dioxide nanoparticles. Nanotoxicology 11:558–568

USEPA (US Environmental Protection Agency) (2007) US Environmental Protection Agency Nanotechnology White Paper. USEPA, Washington, DC

Viji N, Chinnamuthu CR (2015) Iron oxide nanoparticles to break the of the world's worst weed the *Cyperusrotandus*. Int J Agric Sci Res 5(3):259–266

WHO (2017) WHO guidelines on protecting workers from potential risks of manufactured nanomaterials

Yang Y, Lin Y, Liao H (2017) Toxicity-based toxicokinetic/toxicodynamic assessment of bioaccumulation and nanotoxicity of zerovalent iron nanoparticles in Caenorhabditiselegans. Int J Nanomed 12:4607–4621

Zion Market Research (ZMR) (2016) Nanomaterials Market (Metal Oxide, Metals, Chemicals & Polymers and Others) for Construction, Chemical Products, Packaging, Consumer Goods, Electrical and Electronics, Energy, Health Care, Transportation and Other Applications: Global Market Perspective, Comprehensive Analysis and Forecast, 2016–2022. https://www.zionmarketresearch.com/report/nanomaterials-market. Accessed on 22 February 2018

Chapter 17
Nanotechnological Regulations in Brazil

Wilson Engelmann, Raquel Von Hohendorff and Daniele Weber S. Leal

Abstract A new technological movement is being followed in current society, in which nanotechnologies are associated with innovation in form of products, proving the so-called Fourth Industrial Revolution. They are innovative products, with potential for effects and bringing both beneficial effects as well as the risk attached to uncertainties. In this context, it is necessary to discuss nanotechnologies, especially in the Brazilian scenario, in order to encourage the creation of new legal alternatives for these new demands. There are new rights, new complexities measured from the *nanos*. The present chapter discusses the regulatory situation of nanotechnology in Brazil, and also throws light on Principles listed in Nanoaction.

Keywords Nanotechnologies · Risk · Brazil · Regulation · Innovative products Principles

17.1 General Vision of Nanotechnology

Nanotechnologies represent a set of actions of research, development and innovation, obtained due to the development of special equipment, which can access the special properties of matter organized from structures of nanometric dimensions (PNUMA 2017). A nanometer corresponds to the billionth part of a meter or 10^{-9} scientific notation. The term nanotechnology derives from the Greek prefix *nános*,

W. Engelmann (✉)
Universidade do Vale do Rio dos Sinos – UNISINOS, Dois Irmãos, Brazil
e-mail: wengelmann@unisinos.br

R. V. Hohendorff
Universidade do Vale do Rio dos Sinos – UNISINOS, São Leopoldo, Brazil
e-mail: vetraq@gmail.com

D. W. S. Leal
Faculdades Integradas de Taquara – FACCAT and JUSNANO Reseacher (UNISINOS), Porto Alegre, Brazil
e-mail: weber.daniele@yahoo.com.br

M. Rai and J. K. Biswas (eds.), *Nanomaterials: Ecotoxicity, Safety, and Public Perception*, https://doi.org/10.1007/978-3-030-05144-0_17

which means dwarf, technically that equates to craft, and *logos* that expresses knowledge. Currently, the various technologies that make nano-scale operation feasible bring with them many uncertainties, especially concerning the possibility of highly harmful risks to health and the environment (Durán et al. 2006). Looking at the characteristics of the Fourth Industrial Revolution, it is noticed that its impact will be reflected in the interrelationships between man and technology (impacts on the individual, power, politics and economy and the reflexes on culture, society and the environment) (Engelmann 2017a).

Nanotechnology portrays a high degree of interdisciplinarity (Murphy et al. 2017). Biologists, chemists, physicists, physicians and engineers contribute their experiences and ideas to generate innovative applications and products for society. It requires intensive collaboration based on division of labor and a high level of understanding of other disciplines (Germany 2016), in which Law and other Applied Social Sciences are located, as well as the so-called Human Sciences.

In order to understand the implications, effects and possibilities that nanotechnologies may bring, it will be necessary to develop a "systemic thinking" in order to integrate and promote communication among several areas of knowledge, many of them still disconnected today. The "systemic thinking" can be developed through interdisciplinarity, which reveals a way of working that searches for the bases of the integration of Law in the researches and advances of the nano scale that are being developed by the so called Exact Areas.

In terms of nanoscience, we have a movement that goes from "concepts" to "application". The legal knowledge seeking to integrate the advances of nanotechnology revolution, helping to build systems of thought. This form of thinking will require the transposition of disciplinary barriers in order to perceive the challenges that are generated from the nano scale, but with effects on a global scale, their impacts and consequences (Ulijn and Rievo 2016). Despite this, expectations of increasing investments in nanotechnologies are surging (BCC Research 2018), the amount of products that reach the consumer market is expanding (Peters et al. 2016) what may generate future problems, given the uncertainties that exist at the moment (Van Wezel et al. 2018). At the same time, there are programs from several countries that focus on nanotechnologies for economic growth.

An important point about the theme refers to the conceptual contours of "nanotechnology" and "nanomaterial". There is no uniformity in the definition of these two terms in literature (Stone et al. 2018). There are arguments, like Maynard (2011), claiming that it is not recommendable to work with definitions, once that "basing regulations on a term without scientific justification will do more harm than good."

This book chapter adopts the concept edited by The International Standardization Organization (ISO), through its Technical Committee 229: nanotechnologies are the "understanding and control of matter and processes at nanoscale typically, but not exclusively, below 100 nm in one or more dimensions, in which the appearance of size-dependent phenomena generally allows new applications." This feature is further added by "using the properties of nanoscale materials that differ from the properties of individual atoms, molecules and mass matter to create improved materials, devices and systems that exploit these new properties." (ISO/TC 229 2018).

The first characteristic refers to the particle size and this may generate the risks (ISO/TC 229 2018) considering the physical-chemical changes that are generated in this size scale. In this scenario, in a publication of Nature Nanotechnology, it is written: "But despite these fundamental advances, nanotechnology faces a critical context. We are still seeking to highlight the fundamental advances reported in the scientific literature in tangible technological applications that can be appreciated by all." (Ulijn and Riedo 2016).

Therefore, a duplication of the problem is verified: "First, the properties of matter change when in bulk, just as they change when reduced to nanoscale; in particular, the level of control that can be exerted at the nanoscale or single-object level tends to decrease at meso and at macro scales or when dealing with a large number of objects. And second, **the industry is reluctant to invest money in the development of new, large-scale processes for the fabrication of nanomaterials unless a considerable profitable return is assured** (A matter of scale 2016). This is also Warheit's concern: the assessment of environmental and health risks of nanomaterials is a difficult task involving multidisciplinary knowledge and requiring global visions. Therefore, arguments and misunderstandings regarding nanotoxicity are common. The future *design* of environmentally friendly and biocompatible nanomaterials should receive special attention (Warheit 2010), once there is space for consequences that are still unknown.

17.2 Understanding the Nanotechnological Risk

Although the benefits of nanotechnology dominate our thinking in the current days, the potential of this technology for undesirable results in human health and the environment should not be underestimated. Because of its size, the nano materials are governed by physical laws that are extremely different from those with which the science is accustomed to, opening possibilities for the nanoparticles to present a higher degree of toxicity than in larger sizes. For this reason there is a need to evaluate the risks that exist due to the manipulation, development and application of these new technologies, taking into account toxicity, appropriate methods for toxicity testing, as well as the impacts on human health and environmental (Hohendorff and Engelmann 2014).

An increase in publications and research on the risks of nanoparticles (Kulinowski 2015) is currently occurring and this set of publications should also be considered because it represents what is already known about the behavior of some particles when manipulated at nanometer scale. By 2008, the US federal government was investing nearly $60 million a year into health research and the environmental impacts of nanotechnology. In 2016 US federal agencies proposed the investment of $105.4 million in research to understand and address potential risks of nanotechnology for the environment and health. This represents a massive 80% increase over the previous eight years and reflects continuing concerns about the potential hazards of purposely designed and engineered nanoscale materials (Maynard 2016).

Nanomaterials are challenging once they combine a database of relatively limited toxicity, great variability, and uncertainty about toxicity. Materials belonging to the same compound can vary considerably when it comes to shape, size distribution, functionalization, surface modification, aggregation of states and dissolution ability. Due to this material-based variability, science today is confronted with gaps in knowledge about the destiny and behavior of nanoengineered materials in test systems and the environment (Gottschalk et al. 2013).

Evaluation of nanomaterials requires the development of new analytical methodologies or the adaptation of existing ones. Nanometrology is a challenge for the regulation of nanotechnology based products, since there is no universal technique for characterizing the size distribution of nanomaterials (Hassellöv and Kaegi 2009).

A series of gaps have been identified in the current knowledge about processes of transformation of engineered nanoparticles into environmental matrices. The identified gaps are divided into four categories: environmental targeting processes, characterization and measurement methods, availability of data and information, and issues related to the environmental exposure model (Kjølholt et al. 2015).

Nanoworld is huge but it is just at the beginning of its exploration, and some of the physical, chemical, and biological rules already known for materials on a macro scale may have no force there. Such exploitation requires a highly interdisciplinary and open-minded background, in which the only certainty is failure. Somewhat paradoxically, nano can be a problem and a solution at the same time. But nanosafety is the password for the answer creation (Gatti and Montanari 2015).

Regarding the processes of environmental destination, once in the environment, the fate and behavior of the nanoparticles will be determined by their own intrinsic properties, the processes of transformation and the specific environmental conditions. The characterization and measurement methods cover gaps related to different limitations of methods to characterize and measure nanoparticle concentrations. Presently, the advancement in the understanding of the destiny and behavior of these particles is hindered or limited by the lack of appropriate characterization, and measurement methods and also methods for measuring dissolution, redox potential, adsorption as well as aggregation in environmentally realistic contexts. As for the availability of data and information, the common point for all mass-based models and frameworks is that they need a basic set of data and experimental/empirical information to be useful for estimating environmental exposure. This includes data on quantities of nanoparticles produced, specific uses and concentrations of use, as well as the release of quantification data during production, transport, usage and disposal. Currently, access to such information and data is very limited and little known about the routes of entry of these particles into the environment. In the case of environmental exposure modeling, it can be mentioned that the reliability and relevance of all mass-based models and structures depend on the availability and quality of the input data and that the necessity to know about the actual quantities of nanoparticle used, as well as a better understanding of the destiny and behavior of the engineered nanomaterials in all environmental compartments (Kjølholt et al. 2015).

In a paper published in August 2017 by the European Commission entitled *Assessing the Environmental Safety of Manufactured Nanomaterials* (European Commission 2017), it is noticeable that the current test guidelines for assessing the risk of nanoparticles are not always optimal and that there is a lack of data to validate existing risk assessment methods; therefore, some exposure limits or hazard data already produced may be questionable. First, this gap exists because the necessary scientific research to develop nanotechnology does not provide adequate data to assess the effects (hazards) of these products. Second, some of the available scientific data from industry is not adequate for regulatory risk assessment, which requires more robust datasets (Klaine et al. 2012).

Due to the large number of factors that can alter nanoparticles, people and the environment may receive different exposures of the same nanomaterials. Therefore, the knowledge in transformation and current exposure models for conventional substances is not likely to be appropriate for predicting exposure throughout different stages of their life cycle (Savolainen 2013).

When it comes to nanosafety, Fernando Galembeck, a professor at IQ-Unicamp and former director of the National Nanotechnology Laboratory (LNNano), warned that "In laboratories, during basic or applied research, it is easy to control and destroy these materials because the quantities are very small, but when the scale changes to an industrial system the situation is different (Galembeck 2017). "Alongside with LNNano, Galembeck (2017) mentions having worked with the idea that "on nanoproduct should be technologically developed without its toxicological and environmental risks being assessed, starting as soon as possible. We need to know what the risks are in the early stages of the research, because technological development involves large costs".

The field of nanomaterials is moving rapidly, as new materials, new applications for existing materials and new methods for the production of nanomaterials. However, the risk assessment associated with exposure to nanomaterials and risk characterization unaccompanied advances in nanotechnology. Soon, there are more questions than answers (Buzby 2010). Harmful impacts and potential risks to human and animal health, the environment and even human behavior are still poorly understood (ABDI 2010). For the evaluation of these aspects, it should be developed and improved tests in order to identify: (a) their physicochemical properties; (b) its potential for degradation and accumulation in the environment; (c) its environmental toxicity; and (d) their toxicity to mammals (ABDI 2010). As there is an increasing interest in the production of nanomaterials, the potential of these materials as environmental contaminants should be considered. Thus, there are still many unanswered questions that need to be discussed and reflected, such as:

> What are the main routes of contact/interaction between nanomaterials and biosystems? By what means (aquatic, aerial or terrestrial) nanomaterials penetrate the environment? What are the possible ways of dispersing and accumulating nanomaterials in the environment? Can these materials be transformed by biotic and/or abiotic factors? Do they interact with other contaminants already present in the environment? What is the chemical stability of the nanostructure? Are they soluble in water? Do they form stable suspensions or agglomerate in biological media? What are the by-products generated during the synthesis of nanomaterials? What are the byproducts generated in the degradation? What is the toxicity of these materials for bio systems? What interaction mechanisms are involved? Is it possible to remove or remedy nanomaterials from the environment in the event of an accident? What is the

> life cycle of products containing these nanometric materials? Within this context, two new areas of knowledge are emerging, called bionanotechnology and nanotoxicology, in which, through concrete multidisciplinary actions, the bionano interface can be explored, allowing to establish proactive responses to the potential/eventual harmful effects of nanomaterials on biosystems (Martinez and Alves 2013).

The number of scientific studies in nanoscience has grown almost exponentially, tripling over the last decade, confirming that nanoscience as a field of knowledge is developing much faster than scientific knowledge in other areas. As an emerging scientific area, nanomaterials have several inherent characteristics (uncertainty, lack of knowledge and potentially irreversible long-term adverse health effects) that are likely to generate concern, distrust or fear (EU-OSHA 2012). In the face of risks, precautionary measures must be taken, Drexler, in 1986, in his book Engines of Creation, mentions the need to limit the creation of nanotechnologies, once it was possible to generate consequences. In this way, Drexler (1986) wisely explains that-

> the laws of nature and the conditions of the world will limit what we do. Without limits, the future will be totally unknown, something misshapen making a mockery of our efforts in thinking and planning. With limits, the future is still turbulent uncertainty, but it is forced to fly within certain limits.

Aware of the risks and these unique characteristics of the nanos, an appropriate cycle of evaluation of the LCA nanomaterials must be defined and conducted before its commercialization, in order to be considered safe. In addition, appropriate resources should be devoted to research the processes, products and usage of materials that entail less risk, once the inappropriate disposal of nanowaste may generate to the ecosystem unprecedented risks and impacts (International Center for Technology Assessment 2007):

> Nanotoxicity arises as diverse nanomaterials are synthesized, handled and disposed in different environments, whether natural, urban or industrial, without proper control and regulation. Some reasons for attention and caution with nanomaterials are: a) increasing industrial production (increased risk of exposure); b) high surface area due to nanometric size (high chemical reactivity); c) enormous compositional and structural diversity (syntheses, preparations, modifications, functionalization, heterogeneity and impurities); d) traditional toxicological tests are not adapted and standardized for nanomaterials (International Center for Technology Assessment 2007).

In Brazil, it is important to highlight the work of Jorge Duprat Figueiredo Foundation for Safety and Occupational Medicine (FUNDACENTRO) with the participation of several institutions, to evaluate the impacts of nanotechnology on the health of workers and the environment to propose control measures, noting that there is still no national recommendation beyond these actions. Through this project are conducted actions of training workers and professionals of Safety and Health at Work and have already produced six comic magazines: Nanotechnology—The Transport for a New Universe, Nanotechnologies—Wonders and Uncertainties in the Universe of Chemistry, Nanotechnology—A Universe in Construction and Nanotechnology in the Field, Nanotechnologies in Foods, which has participation of UNISINOS as well and Nanotechnology in the Automotive Metallurgical Sector to address the issue with workers (Fundacentro 2017).

Exposure assessment is a crucial step in the assessment of risks in the work environment. Compared with the study on animal toxicology, for example, the nanoparticle effect of titanium dioxide on human health is delayed, partly due to the lack of exposure assessment. It is important to further estimate the exposure to particles at different stages of their life cycle, especially at the production stage, in order to have a real picture of the exposure of workers to nanoparticles (Xu et al. 2016).

A notable challenge in achieving an improved research situation and responsible development is how to better utilize the different skills needed to solve current and emerging research challenges, i.e. governance and the convergence of multiple scientific disciplines need to address this new challenge. The multidisciplinarity of nanotoxicology covers the fields of chemistry, biology, physics and engineering.

There are some recommendations in areas such as health, safety and environmental impacts; regulation, social and ethical aspects; dialogue between stakeholders and the public, aiming to ensure the responsible development of nanotechnologies. Among these recommendations, the following may be listed: (a) further research on possible adverse health, safety and environmental impacts of nanomaterials, which should be an integral part of the product innovation and design process, including nanomaterials; (b) avoid the release of nanoparticles and nanotubes manufactured in the environment, as far as possible; (c) regulatory authorities should consider whether existing regulations are adequate to protect humans and the environment and include future applications of nanotechnologies in their long-term programs to ensure timely identification of regulatory gaps; (d) consider that the ethical and social implications of advanced technologies (such as nanotechnologies) should be part of the formal training of all research students and employees working in these areas; and (e) develop comprehensive qualitative work involving members of the general public as well as members of interested sections of society and government funding of public dialogue on the development of nanotechnologies and establish a group bringing together representatives of a wide range of stakeholders to analyze technologies emerging markets and to identify, as early as possible, areas in which potential health, safety, environmental, social, ethical, and regulatory issues may arise and advise on how these can be addressed (Hansen et al. 2013).

17.3 The Nano Regulatory Initiatives in the Brazilian Scenario

Brazil has initiated very early actions to promote the development of nanotechnologies almost at the same time as many developed countries. The Ministry of Science and Technology was able to articulate a national policy for the sector, investing considerable in resources, which had efficient results in terms of organizing a large number of researchers, rapid improvement of the laboratory infrastructure and consequently boosting scientific production in the field. But this policy has been more closely articulated with industrial policy, consistently with its objectives of promoting national competitiveness and involving a growing group of government institutions. As a result, the country has ranked itself in a leading position in the region, and is often referred to as a country with potential in the global map of nanotechnology. Despite these advances, which can be read as a successful story, the Brazilian Nanotechnology Policy faces two types of challenges: the first in relation to achieving its own objectives, and secondly, in relation to the incorporation of aspects that were not only marginally (Invernizzi et al. 2012).

Thus, the Ministry of Science, Technology and Innovation instituted in 2012, the Interministerial Committee of Nanotechnology that should act as a focal point in the management of nanotechnology in the country. This Committee involves the following Ministries: Science, Technology and Innovation, Agriculture, Livestock and Supply, Defense, Development, Industry and Foreign Trade, Education, Environment, Mines and Energy and Health, thus demonstrating the need to deal with this issue with a transdisciplinary approach and searching for answers in the most different areas of knowledge (SBPC 2013). However, the question remains: Where is the Ministry of Justice? Moreover, for what reason the Ministry of Labor did not integrate the original formation of the committee, being inserted only at the end of 2012 after intense discussions?

In December 2012, a public hearing was held in the Chamber of Deputies in Brasilia, sponsored by the Commission for the Environment and Sustainable Development to discuss the current situation of nanotechnology in Brazil, whose objective was to know the researches in the area, the products being produced and the possible environmental and health damages. This event also mentioned the fact that the chemical sector of the State of São Paulo has a recommendation for the topic nanotechnology to be addressed in the SIPAT—Internal Week of Prevention of Work Accidents—of companies. Also, regarding the pharmaceutical industry of São Paulo, after three years of negotiation, workers were able to insert a clause on nanotechnology in the collective convention of the category, in April 2012. The expectation for 2013 was to extend this achievement to the entire chemical sector of the State (this achievement is paradigmatic for the national and international trade union movement, including the labor unions in the pharmaceutical industries of Switzerland and the United States that have been fighting for it for a certain time).

However, even an important victory is followed by defeat. The claim made by workers in 2008 about the inclusion of a clause in the collective agreement of São Paulo chemical industry, related to the access to information on the use of nanotechnologies in production processes in the chemical industries, was denied in 2013 for the sixth year consecutive. According to the representatives of the companies, the issue is still restricted to a few companies in Brazil, which can be understandable, given that the introduction of nanotechnologies into production processes has been occurring mainly in the central countries of capitalism, in which are the headquarters of transnational corporations (Jensen 2018). Still, companies mention that they have little information on the subject, but it is the companies themselves that deny the right to information that they seek to promote with governments, the National Bank for Economic and Social Development BNDES and the clients as innovative industries mentioning nanotechnologies (some of which even make use of the Sis-

NANO Laboratories). Nanotechnology brings many promises, but also many risks and uncertainties, considering that many entrepreneurs are fearful to link their products or production methods to this new technology.

At the public hearing on nanotechnologies in the country, Paulo Martins, president of Renanosoma (Research Network on Nanotechnology, Society and Environment) mentioned that since 2001 there have been registration of government agencies promoting research in the country, aimed at nanotechnology, but practically a totality of these resources were applied only in production (knowledge for new products) and almost nothing in science that researches the impacts of these products nor in the dissemination and discussion about the results with the society. He also mentioned the fact that no one, neither state, nor state agencies or development agencies, neither Brazilian corporate collegial bodies, effectively knows which and how many companies use nanotechnological inputs in their production processes, as well as the number of products that contain nanotechnology today in the Brazilian market. He also questioned the delay in the governmental actions on nanotechnology, because only after more than 10 years the Interministerial Committee was created and this first public hearing on the subject was held (information collected at the Public Hearing on the current situation of nanotechnologies in the country, held in the Federal Senate in Brasilia, on December 13, 2012). Pedro Binsfield, coordinator of regulatory affairs at the Ministry of Health, claimed that it is only possible to create a good regulatory framework by corroborating scientific knowledge and questioning how it will be possible to regulate if there are no elements, there are only indicators that may represent some risk. However, he affirmed the need to deal with regulatory frameworks, with some urgency, emphasizing the necessity for convergence of the scientific community to create knowledge (information collected at the Public Hearing on the current situation of nanotechnologies in the country, held in the Federal Senate in Brasilia, on December 13, 2012).

In August 2013, the Brazilian Ministry of Science, Technology and Innovation (MCTI) launched the Brazilian Nanotechnology Initiative (IBN), a set of actions aimed at creating, integrating and strengthening government activities and agents anchored in nanoscience and nanotechnology, to promote the scientific and technological development of the industry, focusing on innovation. It is a public policy program comprised of several strategies to make nanotechnology more innovative. Investments of approximately R$440 million were expected in 2013–2014. Chemist Oswaldo Luiz Alves, a professor at the State University of Campinas and one of the country's leading nanotechnology experts, mentions that the government's proposal should make Brazil competitive in the area and impel innovation in the country, emphasizing that it is essential that a regulatory framework for the sector to be elaborated as soon as possible, taking into consideration the international context and both the benefits of nanotechnology and its potential risks to health and the environment.

Regulation was also a matter for the public hearing, and it has been mentioned that the lack of legislation on the subject is also a concern, and the elaboration is hampered by the uncertainties related to nanotechnology. It is possible to identify that contemporary legislations are not prepared to guide decisions about nanotechnological products and processes in development, such as active nanosystems and nanostructures. Government agencies have failed in their capacity as regulator of matter, so that products containing engineered nanoparticles are widely available in the market (International Center for Technology Assessment 2007). As Wilson Engelmann explained, it is necessary to ask: how to regulate? What to regulate? Why regulate? Is it the Law that has to regulation? Sometimes Law is the problem, it freezes reality ("Law is always a photograph of the past that must be applied to events of the future"). Yet, according to Engelmann, we need a precautionary approach and not paralyzing precaution, parallel to development, with responsible control of effects: "Instead of the Law, with serious consequences, including a thickening of knowledge development, perhaps we could think of a consolidation of national and international normative structures (information collected at the Public Hearing on the current situation of nanotechnologies in the country, held in the Federal Senate in Brasilia, on December 13, 2012).

Regarding regulatory agencies, even in the Brazilian context there are institutions that present an initial dynamics of studies on nanos, which appropriate international studies and protocols to search for regulatory frameworks (Leal 2017). One indicator of regulation is found—in a more significant matter—and preoccupation with nanotechnologies in at least two Brazilian agencies, ANVISA, since it has already included in its schedule the possibility of regulation and surveillance of nanoscale products, as well as the Brazilian Industrial Development Agency (ABDI), which has a series of reports that derive from research in this area, and therefore provide an overview of what measures to take in response to their concern about (possible) environmental impacts (Leal and Engelmann 2017).

ANVISA inserted in the *Regulatory Agenda*, Biennial 2015–2016, published in the Official Gazette on April 30, 2015 (corresponding also to the 2nd Quadrennial Cycle Agenda (2013–2016) and also to the 6th Agenda published by ANVISA), priority issues for the agency's regulatory performance for the period, being composed of 172 sub-themes, grouped into 76 themes and divided into 15 macro themes. The topic of nanotechnology, specifically related to products and processes that require health surveillance, was explicitly included in item 74 (ANVISA 2015). On the other hand, ABDI—operates on fronts that impact on the competitiveness of companies, as a regulatory framework and dissemination of the use of nanotechnology as a differential. The project is coordinated with the Ministry of Industry, Foreign Trade and Services (MDIC), coordinator of the Nanotechnology Competitiveness Council, Secretariat for Technological Development and Innovation of MCTI and INMETRO, and seeks to continue the actions of nanotechnology identified in the Sector Technological Agendas (ABDI 2017).

According to a study posted on Waissmann's (2012) ABDI website, there is a need to encourage greater intersectorial integration, with volunteer regulations (soft Law) and new forms of public participation in decision-making. It also proposes the anticipatory technological evaluation, which should be part of a larger movement toward sustainability, and finally, the prospective governance, using new methodologies such as OECD, ISO, NIOSH, European Union. For many it will be sufficient, but validation is necessary. In this scenario, the international instruments of these various organizations, which develop specific studies and protocols on nanotechnologies, could be used in domestic Law as a viable and sufficient alternative.

Likewise, in the sphere of the legislature, the same initial and incipient movement of discussion of the nanos, with propositions and bills, both at state and national level, that need to be presented has been observed. It is a small number if we consider the great flow of national legislative proposals, but the intuit of regulation can not be discarded. First, before enter in the current national proposals, it is important to mention two legal initiatives that have no legislative sequence (Leal 2017):

(a) Bill No. 5,076/2005, which envisaged provisions on the research and use of nanotechnology in Brazil, creating a National Technical Committee on Nanosecurity (CTNano), and establishing a Fund for the Development of Nanotechnology (NDF) presented on 18/04/2005, by Federal Deputy Edson 7, and filed on November 28, 2008 (Duarte 2005);
(b) Senate Bill No. 131/2010, which proposed amending Decree-Law No. 986 of October 21, 1969, which sought to establish basic food standards, and Law No. 6,360, of September 23, 1976, which provides for sanitary surveillance to which medicines, drugs, pharmaceutical and related materials, cosmetics, sanitizers and other products are subjected, and it provides other measures to determine which labels, packaging, package inserts and advertising products elaborated with nanotechnology resources contain information on this fact. The initiative was rejected on 08/08/2013 (Viana 2010)

Currently, in the Brazilian Legislative Branch, it is possible to envisage a movement—albeit timid—in initiatives of regulatory frameworks specific to nanotechnologies, which are in action. At the moment these four legal initiatives are observed:

(a) proposal of draft Law No. 19/2014 (Rio Grande do Sul 2018), in Rio Grande do Sul;
(b) two bills in the federal scope (Chamber of Deputies), of no. 5133/2013 (Sarney Filho 2013a) and no. 6741/2013 (Sarney Filho 2013b);
(c) Draft Law No. 1456/2014 (Nishimoto 2015), in the State of São Paulo.

In the Federal Legislature, there is Bill 5133/2013, authored by Deputy Federal Sarney Filho (2018) (PV-MA), filed on 03/13/2013, with the purpose of "Regulate the labeling of nanotechnology products and products that make use of nanotechnology." The last moves occurred in December 2016, when the application no. 3,478/2015, which postulated the joint processing of PL No. 5,133/2013 to the other bill referent to nanos of no. 6,741/2013 (Araújo 2015). Thus, the last movement of this initiative occurred on 05/04/2017, determining the joining of PL 5,133/2013 to PL 6,741/2013, as well as the analysis by the Special Commission to the Draft Law no. 5,133/2013 (Sarney Filho 2013a), currently awaiting for continuance. Still in the Chamber of Deputies, there is another bill by the same deputy, Sarney Filho, presented at the end of 2013, precisely on 11/11/2013, to which was added the PL 5,133/2013. In a more comprehensive and meaningful way, it finally proposes the provision of a "National Nanotechnology Policy, research, production, destination of tailings and the use of nanotechnology in the country, and other measures," listed by the number 6,741/2013.

The structure of the bill is linked to several other laws, such as Law no. 6,938/1981 (National Environmental Policy), Law no. 8,080/1990 (Human Health Services), Law no. 9,605/1998 (Environmental Crimes), Law no. 10,973/2004 (Law of Innovation), among others. It is necessary to emphasize on this project that in a debate held on November 20, 2013 in the Committee on Economic Development, Industry and Commerce of the Chamber of Deputies, Representative Sarney Filho (PV-MA) defended the mandatory labeling of these materials and the project manager of ABDI, Cleila Guimarães Pimenta, claimed that the measure may confuse consumers (for her a symbol on a label will only confuse the consumer even more, who will not know if it represents a risk or a benefit). It was defended that more studies should be conducted about the technology before establishing any legislation on the subject (Pompeu 2013). The PL was submitted to the Committee of Environment and Sustainable Development (CMADS), and a favorable opinion was given (on 04/07/2015), by the rapporteur, Mr. Bruno Covas, who understood that it was important to follow step by step the results of the research, anticipating events, rather than being reactive to problems, as the legislative process tends to be (CMADS 2013). Therefore, since October 13, 2015, the Plenary was awaiting consideration, and on April 4, 2017 it was conferred the request for analysis of the PL 5,133/2015 and they await, the two bills together, for analysis in the Plenary of the Chamber of Deputies.

Regarding the initiatives of the State Legislative Branch, there are two movements: (a) proposal of draft Law no. 19/2014 (Rio Grande do Sul 2014), in Rio Grande do Sul; and (b) Bill n. 1456/2014 (Nishimoto 2015), in the State of São Paulo.

In the State of Rio Grande do Sul, Proposition no. 19/2014, filed on February 25, 2014, by Alexandre Postal, postulates "Making mandatory the labeling of products of nanotechnologies and products that make use of nanotechnologies". After more than 2 years of processing, there are 3 favorable opinions to the proposal, the last being given on June 14, 2016 (Rio Grande do Sul 2014), including the Constitution and Justice Commission voting favorably. However, in spite of all manifestations and appearances of that legislative house, the proposal was filed on August 2, 2016, without presenting any justification on the website of the Legislative Assembly of Rio Grande do Sul.

Following, in the State of São Paulo, Bill no. 1,456/2015, which is being discussed in the Legislative Assembly, by Hélio Nishimoto, with the same proposal to regulate and make mandatory the labeling of nanotechnology products and the ones, which make use of it. Its last steps had been analysis and approval, on September 3, 2016, by the *Commission for the Defense of Human Rights, Citizenship, Participation and Social Questions*, with the favorable opinion of all members, with subsequent publication of the decision, on March 15, 2016, since then, is eligible for continuation and voting (Nishimoto 2015). Therefore, this bill is in the final phase of approval, which demonstrates an evolution at least in the discussion on the regulation of nanotechnologies and their products, at least with regard to the right to information and the duty to report on such components on a nano scale.

Thus, the need to clarify the fundamental policies of the Brazilian government and the legal and regulatory boundaries are essential for Brazil to advance and affirm its innovative role in the field of nanotechnologies in the scientific world community. In this process, international organizations can play a facilitating role, encouraging the adoption by all countries of new tools, new nomenclatures and new measurement systems, making it important that countries such as Brazil, China, India, Russia (which have researches in nanotechnology) participate in this global regulatory effort on nanotechnology, which has been stimulated by ONU (ABDI 2011) itself.

Through all the exposed conjuncture, there is an initial incipient movement in the Brazilian scenario, so that both regulatory agencies and legislative houses develop, even if they are moving in very slow steps, regulatory milestones for nanotechnologies. Aiming for greater regulation thrive and the promotion of more safety in the main principles of NANOACTION (International Center for Technology Assessment 2007), and how they are inserted in this context and in the national scenario, will be seen in the next topic.

17.4 The Usage of the Principles of NanoAction

The Principles for the Supervision of Nanotechnologies and Nanomaterials was a work done by NanoAction, a group of the "International Center for Technology Assessment". In January 2007, the International Center for Technology Assessment and Friends of the Earth organization jointly organized the first strategic NGO summit on nanotechnology in Washington, DC, which attracted the public interest and organizations of workers, women's health, civil society, environmentalists, and grassroots organizations formed by citizens of North America to discuss and reach agreements on the fundamental principles for the evaluation and supervision of nanotechnology. Over the next six months, participants developed principles led by the NanoAction project of the International Center for Technology Assessment. So far, around 70 groups from six continents have already approved it (International Center for Technology Assessment 2007).

The respective document indicates that the current situation would give no hope that it will "get right" with nanotechnology because the manufacturing and laboratory conditions are developed without protective measures and adequate safety standards. Consumers are unintentionally exposed to nanomaterial ingredients without warning on product labels and consumers are not informed of possible risks. Nanomaterials are discarded or released into the environment, although the impact of the nanomaterials is unknown, and the lack of adequate means to detect, accompany or eliminate them. Governments and creators of nanotechnology offer scarce real opportunities for an informed public participation in debates and decisions about how or if it is convenient to continue with the nano transformation of the world. The NanoAction project of the International Center for Technology Assessment presents eight fundamental principles which are believed to proportionate a base to adequate and efficient evaluation and supervision of the emerging field of nanotechnology, including those materials whose utilization is commercial and generalized.

Those principles are: (I) Precaution Principle; (II) Principle about the Nano specific Mandatory Regulation; (III) Principle of protection to the health and safety of public and workers; (IV) Principle of environmental sustainability; (V) Principle of transparency; (VI) Principle of public participation; VII. Principle of inclusion of broad impacts; (VIII) Principle of responsibility of the producer (International Center for Technology Assessment 2007).

From these principles of NanoAction, the authors of this chapter of the book, drew up the following table, highlighting where these principles are already provided in the Brazilian Law System (Engelmann 2017b):

In this way, it is feasible, through the analysis of the current Brazilian legislation, to envisage the applicability of the principles listed in the study of NanoAction, which would reinforce the movement to promote safety in the development of nanotechnologies.

Some of the principles of the document are very relevant in the nanotechnology context, and they are already used by the doctrine and organizations in fostering nano regulation. It is possible to highlight the well-known Precautionary Principle. Given the particularities of nanotechnologies, their potential future damage to present and future generations, it is necessary to adopt precautionary measures, respecting the Precautionary Principle, together with the idea of responsibility and ethics inherent in the context of new technologies and risk (Leal 2017).

> The precautionary principle should be applied to nanotechnologies because current scientific research suggests that exposure to some nanomaterials, nano-devices or nanobiotechnology-derived products can cause serious damage to health or the environment. The small size of engineered nanomaterials can provide them with innovative, useful physical, chemical and biological properties; however, the high reactivity, mobility and other properties that accompany matter at the molecular level may generate unknown levels of toxicity in parallel. Recent research and studies on the impacts of nanomaterials on human health and the environment have raised the alarm that legitimates precautionary actions and more in-depth studies. The regulations must be rigorous, precise, comprehensible and implemented before marketing; as well as the unique properties of nanomaterials in their risk assessment. Thus, the potential toxicity of materials at the nanoscale can not be studied and predicted in relation to the bulk profile of the compounds, i.e. without being in 'nano-form' (International Center for Technology Assessment 2007).

The precautionary principle must be present in decision-making processes that involve risk situations—which are part of the nanotechnology era—always aiming at sustainability and the duty of solidarity that the public and private sectors have in relation to present and future generations (Hupffer 2013).

Sunstein (2005), an American author who has shown himself to be largely averse to such a principle, emphasizes that such a principle is often invoked in cases involving risks of irreversible damage—which falls within the context of the nanos—and the term irreversible is described in several ways in which the principle tends to avoid:

Principle	Normative equivalent in Brazil
Principle of Precaution	Federal Constitution, art. 200, I to VII; art. 225, *caput*, and IV, V; Art. 1°, Law n. 11,105/2005 (Biosafety)
Principle of Precaution to Health and Safety of the Public and Workers	Federal Constitution, art. 1°, III and IV (dignity of the human person; social value of work); defense of the consumer (CF, art. 5° XXXII, and 170, V; Consumer Protection Code), and art. 201, §10 (coverage of the risk of work accident); Federal Constitution, art. 225, §1°, V (risk); Federal Constitution, art. 196 (right to health)
Principle of Environmental Sustainability	Prevalence of the Human Rights (art. 4°, II; art. 170, VI; art. 200, VIII; art. 225, all articles of the Federal Constitution). National Policy of Solid Residues (destination and final disposal, which are environmentally adequate) (art. 3°, VII and VIII); sustainable patterns of consumption (art. 3°, XIII)
Transparency Principle	Solidarity (Federal Constitution, art. 3°, I; life cycle of the product (National Policy of Solid Residues, art. 3°, IV) and social control (National Policy of Solid Residues, art. 3°, VI)
Principle of Public Participation	Right/duty to information/inform (Federal Constitution, art. 5°, XIV) principle of publicity (Federal Constitution, art. 37, *caput*); Federal Constitution, art. 225, VI; Federal Constitution, art. 23: common competences among the Union, States and Municipality to: VI—protect the environment and combat pollution in any of its forms; and legislate (Federal Constitution, art. 24, VI and VIII)
Principle About the Consideration of Other Impacts	Objective of Brazil Federal Republic: promote well being to all citizens (Federal Constitution, art. 3°, IV); safety (Federal Constitution, art. 5°, *caput*); Federal Constitution, art. 218, §2°; National Policy of the Environment (Law n. 6,938/1981); National Policy of Solid Residues, art. 3°, X and XI, integrated management of solid residues
Principle of Responsibility of the Producer	Art. 931, Civil Code (development risk); principle of the polluter payer; Law of Environmental Crimes (Law n. 9,605/1998, §3°; Federal Constitution, art. 225; National Policy of Solid Residues, art. 3°, XVII—responsibility shared between manufacturers, importers, distributers, merchants and citizens

> The Precautionary Principle is often invoked in cases involving risks of irreversible harms. As we have seen, the term 'irreversible' appears in numerous descriptions on what the principle is designed to avoid. The intuition here is both straightforward and appealing: More steps should be taken to prevent harms that are effectively final than to prevent those that can be reversed at some cost.

Considering the context of nanotechnologies, the precautionary principle is one that should guide organizational behavior, especially as it requires decision-making at key moments throughout the research and development process in order to promote the: "Construction of a formula that integrates an ethical premise, based on the fundamental right to respect for precaution, understood as the calculation, evaluation and projection of consequences" (Engelmann 2011).

Specific regulations on such new technology are inexistent, consequently people and the environment remain at the mercy of unbridled development, which can lead to irreversible damage. Therefore, precaution should be taken to bioethical and ethical limits to protect future generations as well as preserve the present one. It is necessary for this scientific progress to be observed (read, braked) through the premises of bioethics, making an ethical reflection on its principles and premises. Facing the understanding of the innovations from it, the advance of nanotechnologies will be allowed to occur with caution and respecting human (Leal and Hohendorff 2016) dignity in the first place.

Since prudence is a mechanism of caution, protecting future (potential) damages, it is important to associate it with sustainability, another principle also listed in NanoAction. Relevant then is to present at least one international organization that has been studying the development of nanotechnologies associated with sustainability, in order to investigate the interaction between nanomaterials and ecosystem, promoting the evolution of nanotechnology with an environmental (The Center for Sustainable Nanotechnology 2015). This are the Center for Sustainable Nanotechnology (CSN) goals:

> The aim of the Center for Sustainable Nanotechnology is to develop and utilize an understanding of the molecular level of nanomaterial-biological interactions to enable the development of sustainable and socially beneficial nanotechnologies. Indeed, we intend to understand the molecular-level, chemical and physical principles governing how nanoparticles interact with living systems in order to provide the scientific fundamentals needed to ensure that the continuous evolution of nanotechnology can occur with the minimum environmental footprint and maximum benefit to society. CSN is not a physical center, but it is a central point for collaboration that links the complementary experience of researchers in 12 different institutions to achieve what none of us could do individually. We co-advise graduate students and often meet in cyberspace, taking advantage of the latest in web-enabled communications to achieve a high level of interaction. The funding for the CSN comes from the Division of Chemistry of the National Science Foundation through the Program of Chemical Innovation Centers (The Center for Sustainable Nanotechnology 2015).

Where there is no consistent scientific basis, legally binding decisions are required in scenarios of great uncertainty, especially in relation to risks of future damage, i.e., yes or no decisions on activities, products, substances or techniques, so that lawyers and researches must act with prudence and special common sense in the application of avoidance measures (Aragão 2008).

One way that Law can choose to anticipate the harmful potentialities of nano products has already been widely discussed for some years at the international level. It is about investing in the Right to Information: those who consume the products developed with nanotechnology have the right to know about the changes and their positive and negative effects, allowing a conscious choice of the products that carry this news. A number of countries, including France, Belgium, Sweden, Norway, Denmark, China and Canada, have made registration compulsory for companies selling products containing nanoparticles, whether produced or imported by them (Engelmann 2015). Also, countries such as Australia, the United States, Switzerland and Korea have elaborated operational definitions of nanomaterials, even if not always binding (Engelmann 2015).

Therefore, this principle takes great importance in the nanotechnological scenario, the Principle of the information duty. Not only should the precautionary principle be invoked and serve as a basis for decision-making, but also the principle of information needs to be mentioned and effectively applied. It is necessary that the information, arising from the study of risks, is widely disseminated and available to society so that the actors directly involved in the decisions as well as civil society are better qualified to face the challenges that arise with this new technology (Hohendorff 2018).

For nanotechnology, and more specifically for engineered nanomaterials, there is awareness that the information requirements established in regulations and related guidance may not fully cover the information needed to characterize the risks to human and environmental health. This discrepancy between information requirements and information needs may lead to uncertain estimates of risk (Hohendorff 2018).

It is urgent for the right to information to be strengthened in this new nanotechnological era, so that the consumer consciously decides what he or she acquires, and that they have the minimum knowledge to know how this product can impact their lives, especially when talking about nanocosmetics, because they are products that possibly will be used in a continuous way.

17.5 Conclusion

In Brazil, the Right to Information is made effective in the Federal Constitution of Brazil of 1988 in article 5, item XIV, which reads as follows: "Everyone is guaranteed access to information and safeguarding the secrecy of the source, when necessary for professional practice", it is constitutionally guaranteed the inalienable requirement of communication regarding the use of nanotechnologies in cosmetic products and their potential for risk and should be "recognized and established in the sphere of positive constitutional Law" (Sarlet 2018). In the words of Maria Tereza Marques, information is a fundamental right/duty "without which there is no participation, there is no freedom, equality collapses, it prevents the existence of democracy" (Marques 2000).

From the Right to Information, which is present in the Federal Constitution and in the Consumer Protection Code—in particular Article 31—, those who consume such products have the right to know the changes, the positive and negative effects, allowing a conscious choice of the products which contain such nanotechnological novelties, at the same time, it can be called a right/duty, since it is also the duty of the producer, to give access to information regarding the use of nanotechnologies applied to the products that they make available to the consumer market.

Thus, in the Brazilian scenario, the principles listed in the NanoAction are present in several normative mechanisms, and the Right to Information and Caution is more important in this context. Taking into account the urgency for greater security in the development and consumption of nano products, the use of this principle-logic base linked to other foreign self-regulating mechanisms, makes it feasible to promote regulations in Brazil and at the global level.

Therefore, it will be very important that the diffusion of the greater knowledge of nanotechnologies occurs, respecting the Principle of Information, which will enable a conscious choice in the use and consumption of nano products.

This is a perspective of the Brazilian scenario, an initial and incipient movement of nano regulation, with the existence of some Law Projects. There are not many advances in terms of effectiveness, but the very initiative of the debate in the political arena, as well as the Brazilian regulatory agencies, already indicates the encouragement of research and regulation of this new technology.

In addition, an important distinction was made between nanotechnological risk and how it will influence human life and the environment considering its particularities of interaction, effects and (possible) future damages.

In Brazil, the principles listed in the NanoAction are present in several normative mechanisms, and the Right to Information and the Precautionary Principle are of greater importance in this context. All the principles listed in the study of NanoAction lead to a more responsible development and consumption of nanotechnologies; with the adoption of these auxiliary mechanisms risk management is possible. It will consequently minimize (possible) future damages, which conducts to greater safety in the nanotechnological context. Therefore, taking into account such urgency in this adoption, the use of this principle base linked to other foreign self-regulating mechanisms, it becomes feasible to promote regulations in Brazil and at the global level.

References

ABDI. Agência Brasileira de Desenvolvimento Industrial (2010) Estudo prospectivo nanotecnologia. Brasília, DF (Série Cadernos da Indústria ABDI, v. 20). http://www.abdi.com.br/Estudo/Estudo%20Prospectivo%20de%20Nanotecnologia.pdf. Accessed 15 Feb 2018

ABDI. Agência Brasileira de Desenvolvimento Industrial (2011) Nanotecnologias: subsídios para a problemática dos riscos e regulação. Brasília, DF. Disponível em:. Accessed 18 Feb 2018

ABDI. Agência Brasileira de Desenvolvimento Industrial (2017) Ações ABDI: frentes de trabalho. Brasília, DF. http://www.abdi.com.br/Paginas/detalhamento_acao.aspx?f=Nanotecnologia. Accessed 25 Apr 2018
ANVISA. Agência Nacional de Vigilância Sanitária (2015) Agenda regulatória: ciclo quadrienal 2013–2016: biênio 2015–2016. Brasília, DF. http://portal.anvisa.gov.br/documents/33880/460876/Agenda+regulat%C3%B3ria+2015-2016/6b9f3a07-1be3-47df-bca2-9fc969849a23. Accessed 25 Apr 2018
A matter of scale (2016) Nat Nanotechnol 11(9). http://www.nature.com/nnano/journal/v11/n9/full/nnano.2016.180.html. Accessed 15 Feb 2018
Aragão A (2008) Princípio da precaução: manual de instruções. Magazine Centro de Estudos Direito do Ordenamento, do Urbanismo e do Ambiente - RevCEDOUA, Coimbra 11(2). https://digitalis.uc.pt/pt-pt/artigo/princípio_da_precaução_manual_de_instruções. Accessed 25 Apr 2018
Araújo J (2015) REQ 3478/2015 => PL 5133/2013b. Requires the joint processing of PL No. 5,133/2013 to PL No. 6,741/2013. http://www.camara.gov.br/proposicoesWeb/fichadetramitacao?idProposicao=2053251. Accessed 25 Apr 2018
BCC Research: Market Research Reports & Industry Analysis (2018) The global nanotechnology market should reach $90.5 billion by 2021 from $39.2 billion in 2016 at a compound annual growth rate (CAGR) of 18.2%, from 2016 to 2021. https://www.bccresearch.com/market-research/nanotechnology/nanotechnology-market-products-applications-report-nan031g.html. Accessed 22 Jan 2018
Buzby JC (2010) Nanotechnology for food applications: more questions than answers. J Consum Aff Malden 44(3). http://onlinelibrary.wiley.com/doi/10.1111/j.1745-6606.2010.01182.x/epdf. Accessed 15 Feb 2018
CMADS (2013) Comissão de Meio Ambiente e Desenvolvimento Sustentável. Bill no 6741, de 2013. Discusses the National Policies on Nanotechnology, research, production, destiny of waste and the usage of nanotechnology in the country, as well as giving other providences. http://www.camara.gov.br/proposicoesWeb/prop_mostrarintegra?codteor=1317650&filename=PRL+1+CMADS+%3D%3E+PL+6741/2013. Accessed 25 Apr 2018
Drexler E (1986) Engines of creation: the coming era of nanotechnology. Anchor Books, New York
Duarte E (2005) PL 5076/2005. Discusses the research and the usage of nanotechnology, creates the National Technical Commission of Nanosafety—CTNano, establishes The Nanotechnology Development Fund and offers other providences. http://www.camara.gov.br/proposicoesWeb/fichadetramitacao?idProposicao=282392. Accessed 25 Apr 2018
Durán N, Mattoso LHC, Morais PC (2006) Nanotecnologia: introdução, preparação e caracterização de nanomateriais e exemplos de aplicação. Artliber, São Paulo
Engelmann W (2015) Nanocosméticos e o Direito à Informação: Construindo os Elementos e as Condições para Aproximar o Desenvolvimento Tecnocientífico na Escala Nano da Necessidade de Informar o Público Consumidor, 1st ed. Erechim - Rs, Deviant
Engelmann W (Org) (2011) O princípio da precaução como um direito fundamental: os desafios humanos das pesquisas com o emprego da nanotecnologia. In: Souza IF, Vieira RdS (Org) Direitos fundamentais e estado: políticas públicas e práticas democráticas. Criciúma: Ed. UNESC
Engelmann W (2017a) O pluralismo das fontes do Direito como uma alternativa para a estruturação jurídica dos avanços gerados a partir da escala nanométrica. In: Streck LL, Rocha LS, Engelmann W (Org) Constituição, Sistemas Sociais e Hermenêutica – Yearbook of the Graduation Program in Law of Unisinos. Livraria do Advogado Editor, Porto Alegre, vol 13, pp 247–259
Engelmann W (2017b) Presentation: Os desafios da regulação das nanotecnologias. Seminanosoma. UFRGS, Porto Alegre, Rio Grande do Sul
European Commission (2017) Science for Environment Policy. Assessing the environmental safety of manufactured nanomaterials: in-depth report 14. Bristol, Aug. http://ec.europa.eu/environment/integration/research/newsalert/pdf/assessing_environmental_safety_nanomaterials_IR14_en.pdf. Accessed 16 Feb 2018

EU OSHA (2012) European agency for safety and health at work. Risk perception and risk communication with regard to nanomaterials in the workplace. European Risk Observatory: literature review, Bilbao. https://osha.europa.eu/en/tools-and-publications/publications/literature_reviews/risk-perception-and-risk-communication-with-regard-to-nanomaterials-in-the-workplace. Accessed 15 Feb 2018

FUNDACENTRO. Fundação Jorge Duprat Figueiredo de Segurança e Medicina do Trabalho (2017). Histórias em quadrinhos. São Paulo. http://www.fundacentro.gov.br/nanotecnologia/publicacoes. Accessed 15 Feb 2018

Galembeck F (2017) Medidas preventivas: estudos apresentam propostas para possíveis impactos de nanoprodutos na saúde humana e no meio ambiente. Magazine Pesquisa FAPESP, São Paulo, ed. 251, p 72, Jan. http://revistapesquisa.fapesp.br/wp-content/uploads/2017/01/070-073_Nano_251.pdf. Accessed 19 Feb 2018

Gatti AM, Montanari S (2015) Case studies in nanotoxicology and particle toxicology. Academic Press, Cambridge

Germany (2016) Federal Ministry of Education and Research. Action plan nanotechnology 2020: an inter-departmental strategy of the Federal Government. Rostock, Oct. https://www.bmbf.de/pub/Action_Plan_Nanotechnology.pdf. Accessed 15 Feb 2018

Gottschalk F, Kost E, Nowack B (2013) Engineered nanomaterials in water and soils: a risk quantification based on probabilistic exposure and effect modeling. Environ Toxicol Chem N Y 32(6). https://www.ncbi.nlm.nih.gov/pubmed/23418073. Accessed 16 Feb 2018

Hansen SF, Maynard A, Baun A, Tickner, JA, Bowman DM (2013) Nanotechnology—early lessons from early warnings. Late lessons from early warnings: science, precaution, innovation: EEA report, Luxembourg, no 1. https://www.eea.europa.eu/publications/late-lessons2. Accessed 15 Feb 2018

Hassellöv M, Kaegi R (2009) Analysis and characterization of manufactured nanoparticles in aquatic environments. In: Lead JR, Smith E (eds) Environmental and human health impacts of nanotechnology. Wiley, Chichester

Hohendorff RV, Engelmann W (2014) Nanotecnologias aplicadas aos agroquímicos no Brasil: a gestão do risco a partir do diálogo entre as fontes do direito. Juruá, Curitiba

Hohendorff, RV (2018) A Contribuição do *Safe by design* na estruturação autorregulatória da gestão dos riscos nanotecnológicos: Lidando com a Improbabilidade da Comunicação Inter-Sistêmica entre o Direito e a Ciência em Busca de Mecanismos para Concretar os Objetivos de Sustentabilidade do Milênio. 480 f. Thesis (Doctorate in Law)—Graduation in Law Program, Universidade do Vale do Rio dos Sinos (UNISINOS), São Leopoldo

Hupffer, HM, Martins JBB, Artmann M (2013). A experiência do Rio de Janeiro na implementação do ICMS verde. In: Hupffer HM, Weyermüller AR (Org) ICMS ecológico: instrumento de estímulo à conservação e à proteção ambiental. Entremeios, Porto Alegre

International Center for Technology Assessment (2007) Princípios para a supervisão de nanotecnologias e nanomateriais. Translated by Secretaria Regional Latino-Americana da União Internacional dos Trabalhadores na Alimentação, Agricultura, Hotelaria, Restaurantes, Tabaco e Afins (Rel-UITA). Washington: NanoAction: International Center for Technology Assessment, Jan. (NanoAction Project). http://www6.rel-uita.org/nanotecnologia/Principios_Supervision_NANOTECNOLOGIAS-por.pdf. Accessed 25 Apr 2018

International Organization for Standardization. ISO/TC 229: nanotechnologies. Geneva, 2005. http://www.iso.org/iso/standards_development/technical_committees/list_of_iso_technical_committees/iso_technical_committee.htm?commid=381983. Accessed 15 Feb 2018

Invernizzi N, Körbes C, Fuck MP (2012) Política de Nanotecnología em Brasil: a 10 años de las primeras redes. In: Foladori G, Invernizzi, N, Lau EZ (cords) Perspectivas sobre el desarrollo de las nanotecnologias em América Latina. Universidad Autónoma de Zacatecas, México, DF

Jensen TF (2018) Mais um ano sem informação sobre nanotecnologias na indústria química paulista. Porquê? http://www.relans.org/Nano_trabajadores.html. Accessed 22 Apr 2018

Kjølholt J, Gottschalk F, Brinch A, Lützhøft HCH, Hartmann NB, Nowack B, Baun A (2015) Environmental assessment of nanomaterial use in Denmark. The Danish Environmental Protection Agency, Copenhagen. (Environmental Project, no 1788). https://www.2.mst.dk/Udgiv/publications/2015/10/978-87-93352-71-1.pdf. Accessed 15 Feb 2018

Klaine SJ et al (2012) Paradigms to assess the environmental impact of manufactured nanomaterials. Environ Toxicol Chem N Y 31(1) http://onlinelibrary.wiley.com/doi/10.1002/etc.733/abstract. Accessed 17 Feb 2018

Kulinowski KM (2015). Tentación, tentación, tentación: ¿por qué es probable que respuestas simples sobre los riesgos de los nanomateriales sean erróneas? In: Foladori G, Hasmy A, Invernizzi N, Lau EZ (Coord) Nanotecnologías en América Latina: trabajo y regulación, vol 1. Miguel Ángel Porrúa, Universidad Autónoma de Zacatecas, México, DF, pp 149–154

Leal DWS (2017) A autorregulação da destinação final dos resíduos nanotecnológicos: a possibilidade de utilização do protocolo da OECD e do pluralismo jurídico (Teubner) para Gestão do Risco do Nanowaste, p 279. Dissertation (Master's in Law) – Graduation Program in Law, Universidade do Vale do Rio dos Sinos (UNISINOS), São Leopoldo. http://www.repositorio.jesuita.org.br/handle/UNISINOS/6755. Accessed 25 Apr 2018

Leal DWS, Engelmann W (2017) Nanotecnologias e meio ambiente: o movimento inicial de marcos regulatórios nacionais em face do contexto de (possibilidade) riscos. In: Bizawu SK, de Oliveira ML (Coord) IV Congresso Internacional de Direito Ambiental: constitucionalismo, economia e desenvolvimento sustentável. Organização Escola Superior Dom Helder. Belo Horizonte: ESDH. https://www.conpedi.org.br/publicacoes/t5ssa9m9/1r807z77/3cSqY9RcMsXdlKI3.pdf. Accessed 25 Apr 2018

Leal DWS, Hohendorff RV (2016) Evolução das nanotecnologias e a necessária aplicação da bioética como garantia de respeito à Dignidade humana. In: Alkimin MA, Gordilho HJdeS (Coord) XXV Congress of CONPEDI - Curitiba: biodireito e direitos dos animais II. Organization CONPEDI/UNICURITIBA. Florianópolis: National Council of Research and Graduation in Law (CONPEDI). https://www.conpedi.org.br/publicacoes/02q8agmu/z1gb219p/37TLs7OZn4G9xuCg.pdf. Accessed 25 Apr 2018

Marques MT (2000) Direito à informação: direito fundamental: base para democracia. Revista do Instituto de Pesquisas e Estudos, Bauru, no 28, pp 29–42, Apr/Jul. http://bdjur.stj.jus.br/jspui/handle/2011/19916. Accessed 25 Apr 2018

Martinez DST, Alves OL (2013) Interação de nanomateriais com biossistemas e a nanotoxicologia: na direção de uma regulamentação. Ciência e Cultura, São Paulo 65(3):32. http://cienciaecultura.bvs.br/pdf/cic/v65n3/a12v65n3.pdf. Accessed 15 Feb 2018

Maynard AD (2011) Don't define nanomaterials. Nature 475. http://www.nature.com/nature/journal/v475/n7354/full/475031a.html. Accessed 14 Feb 2018

Maynard AD (2016) Não falamos mais sobre riscos da nanotecnologia, mas isso não significa que eles desapareceram. Tecnologias Emergentes, Sociedade e Desenvolvimento, Curitiba, July 4. http://nanotecnologiasociedade.weebly.com/blog/nao-falamos-mais-sobre-riscos-da-nanotecnologia-mas-isso-nao-significa-que-desapareceram-por-andrew-maynard. Accessed 18 Feb 2018

Murphy P, Munshi D, Kurian P, Chakraborty M (2017) Nanotechnology, society, and environment. Mater Sci Mater Eng (Elsevier)

Nishimoto H (2015) Bill no 1456/2015. Regulates and makes mandatory the labeling of nanotechnology products and products that make use of nanotechnology. http://www.al.sp.gov.br/propositura/?id=1286924. Accessed 25 Apr 2018

Peters RJB et al (2016) Nanomaterials for products and application in agriculture, feed and food. Trends Food Sci Technol 54:155–164

PNUMA. Programa de las Naciones Unidas para el Medio Ambiente (2017) Fronteras 2017. Nuevos temas de interés ambiental. Programa de las Naciones Unidas para el Medio Ambiente, Nairobi

Pompeu C (2013) Deputado e gerente da ABDI discordam sobre alerta de uso de nanotecnologia. Câmara Notícias, Brasília, DF, Dec 5. http://www2.camara.gov.br/camaranoticias/noticias/

industria-e-comercio/457489-deputado-e-gerente-da-abdi-discordam-sobre-alerta-de-uso-de-nanotecnologia.html. Accessed 25 Apr 2018

Rio Grande do Sul (2014) Proposition: PL 19 2014. Regulates and makes mandatory the labeling of nanotechnology products and products that make use of nanotechnology. http://www.al.rs.gov.br/legislativo/ExibeProposicao/tabid/325/SiglaTipo/PL/NroProposicao/19/AnoProposicao/2014/Origem/Px/Default.aspx. Accessed 25 Apr 2018

Sarlet IW (2018) A Eficácia dos Direitos Fundamentais: uma teoria geral dos direitos fundamentais na perspectiva constitucional. 13th ed. rev. e atual. Livraria do Advogado, Porto Alegre

Sarney Filho J (2013a) PL 5133/2013. Regulates and makes mandatory the labeling of nanotechnology products and products that make use of nanotechnology. http://www.camara.gov.br/proposicoesWeb/fichadetramitacao?idProposicao=567257. Accessed 25 Apr 2018

Sarney Filho J (2013b) PL 6741/2013. Discusses the national policies on nanotechnology, research, production, destiny of waste and the usage of nanotechnology in the country, as well as giving other providences. http://www.camara.gov.br/proposicoesWeb/fichadetramitacao?idProposicao=600333. Accessed 25 Apr 2018

Savolainen K (Coord) (2013) Nanosafety in Europe 2015–2025: towards safe and sustainable nanomaterials and nanotechnology innovations. Helsinki: Edita; Finland: Finnish Institute of Occupational Health. http://www.nanowerk.com/nanotechnology/reports/reportpdf/report159.pdf. Accessed 15 Feb 2018

SBPC (2013) Jornal da Ciência of Dec 17. http://www.jornaldaciencia.org.br/Detalhe.php?id=91278. Accessed 21 Apr 2018

Stone V, Önlü S, Bergamaschi E, Carlander D, Costa A, Engelmann W, Ghanem A, Hartl S, Hristozov D, Scott-Fordsmand JJ, Jensen A, der Kammer FV, Sergent JA, Sharma M, Dusinska M, Nowack B, Sayre P, Vogel U, Tongeren MV, Campos SV, Wohlleben W (2018) Research priorities relevant to development or updating of nano-relevant regulations and guidelines. http://www.safenano.org/news/news-articles/eu-nanosafety-cluster-publishes-research-regulatory-roadmap/. Accessed 14 Feb 2018

Sunstein CR (2005) Laws of fear: beyond the precautionary principle. Cambrigde University Press, Cambrigde

The Center for Sustainable Nanotechnology (2015) About us. Madison. https://susnano.wisc.edu/about/. Accessed 25 Apr 2018

Ulijn RV, Riedo E (2016) Learning to 'think systems'. Nat Nanotechnol 11:824. http://www.nature.com/nnano/journal/v11/n9/full/nnano.2016.161.html. Accessed 17 Jan 2018

Viana T (2010) Senate Bill no 131, of 2010. Alters Decree-Law No. 986, of October 21, 1969, which establishes basic norms on food, and Law No. 6.360, of September 23, 1976, which provides for the sanitary surveillance to which medicines, drugs, pharmaceutical and related materials, cosmetics, sanitizers and other products are subject, and portrays other providences to determine that labels, packaging, package inserts and advertising materials for products made with the usage of nanotechnology contain information about this fact. https://www25.senado.leg.br/web/atividade/materias/-/materia/96840. Accessed 25 Apr 2018

Waissmann W (2012) Nanotecnologias e nanomateriais: aspectos sobre toxicidade, riscos, governança e regulação. Brasília, DF, Nov. http://www.abdi.com.br/Estudo/William%20Waissmann.pdf. Accessed 28 Apr 2018

Warheit DB (2010) Debunking some misconceptions about nanotoxicology. Nano Lett 10(12):4777–4782

Van Wezel AP, van Lente H, van de Sandt JJ, Bouwmeester H, Vandeberg, RL, Sips AJ (2018) Risk analysis and technology assessment in support of technology development: putting responsible innovation in practice in a case study for nanotechnology. Integr Environ Assess Manag 14(1):9–16

Xu H, Zhao L, Chen Z, Zhou J, Tang S, Kong F, Li X, Yan L, Zhang J, Jia G (2016) Exposure assessment of workplace manufacturing titanium dioxide particles. J Nanopart Res Neth 18. https://link.springer.com/article/ https://doi.org/10.1007/s11051-016-3508-9. Accessed 19 Feb 2018

Index

M. Rai and J.K. Biswas (eds.), *Nanomaterials Ecotoxicity, Safety, and Public Perception*, https://doi.org/10.1007/978-3-030-05144-0